International Comparisons of Information Communication Technologies:

Advancing Applications

Felix B. Tan
Auckland University of Technology, New Zealand

Managing Director:	Lindsay Johnston
Senior Editorial Director:	Heather Probst
Book Production Manager:	Sean Woznicki
Development Manager:	Joel Gamon
Development Editor:	Hannah Abelbeck
Acquisitions Editor:	Erika Gallagher
Typesetters:	Jennifer Romanchak
Print Coordinator:	Jamie Snavely
Cover Design:	Nick Newcomer, Greg Snader

Published in the United States of America by
Information Science Reference (an imprint of IGI Global)
701 E. Chocolate Avenue
Hershey PA 17033
Tel: 717-533-8845
Fax: 717-533-8661
E-mail: cust@igi-global.com
Web site: http://www.igi-global.com

Library of Congress Cataloging-in-Publication Data

International comparisons of information communication technologies: advancing applications / Felix B. Tan, editor.
    p. cm.
  Includes bibliographical references and index.
  Summary: "This book offers perspectives on international information management with particular emphasis on the strategies for the implementation and application of information technology in developed and developing countries"--Provided by publisher.
  ISBN 978-1-61350-480-2 (hbk.) -- ISBN 978-1-61350-481-9 (ebook) -- ISBN 978-1-61350-482-6 (print & perpetual access) 1. Information technology--Management. 2. Information technology--Cross-cultural studies. I. Tan, Felix B., 1959-
  HD30.2.I55653 2012
  303.48'33--dc23
                    2011044616

British Cataloguing in Publication Data
A Cataloguing in Publication record for this book is available from the British Library.

All work contributed to this book is new, previously-unpublished material. The views expressed in this book are those of the authors, but not necessarily of the publisher.

Table of Contents

Section 2
International Perspectives

Section 3
Government and Development Issues

Detailed Table of Contents

Section 1
The Role of Culture

Chapter 1

Xunhua Guo, Tsinghua University, China
Nan Zhang, Tsinghua University, China

Accumulated literature on technology adoption research has suggested that cultural factors have important impacts on the cognition and behavior of information systems users. In this paper, the authors argue that cultural factors should be treated as aggregate characteristics at the population level instead of personal attributes at the individual level. The authors also propose that theoretical models could be developed for specific cultural contexts when examining IT/IS user behavior. In this regard, a model for analyzing user attitude toward mandatory use of information systems is proposed. Drawing on generally recognized cultural characteristics of China, three factors are introduced as determinants of user attitude—leader support, technology experience, and perceived fit. An empirical study is conducted with regard to the acceptance of a mobile municipal administration system in Beijing, China, for validating the proposed model with survey data and analyzing the adoption mechanism of the target system. The moderating roles of gender, age, and education level on the model are explored by interaction effect analyses and the findings provide helpful insights for related studies in other cultural contexts.

Chapter 2

Dinesh A. Mirchandani, University of Missouri - St. Louis, USA
Albert L. Lederer, University of Kentucky, USA

Hofstede's national culture model has been applied in prior research to better understand the management of multinational firms. That research suggests that national culture may influence the information systems planning autonomy of the subsidiaries of multinational firms, but such an impact has not yet been tested empirically. A postal survey of 131 chief information officers and 103 senior non-IS managers of U.S. subsidiaries of such firms collected data to test hypotheses based on the model. Structural equation modeling using PLS-Graph 3.0 revealed that Individualism-Collectivism, Masculinity-Femininity, and Uncertainty Avoidance predicted autonomy for particular IS planning phases (as rated by the CIOs). On the basis of the supported hypotheses, the study provides evidence of the relevance of the national culture model to IS planning effectiveness and IS contribution. The study also suggests to subsidiary managers that an understanding of the national culture of their parent firm can help them gain an insight into the parent's management perspective.

Elizabeth White Baker, Virginia Military Institute, USA
Said S. Al-Gahtani, King Khalid University, Saudi Arabia
Geoffrey S. Hubona, Georgia State University, USA

This study investigates technology adoption behavior of Saudi Arabian knowledge workers using desktop computers within the context of TAM2, and the unique effects of Saudi culture on IT adoption within the developing, non-Western, country. Following the guidelines of the etic-emic research tradition, which encourages cross-cultural theory and framework testing, the study findings reveal that the TAM2 model accounts for 40.3% of the variance in behavioral intention among Saudi users, which contrasts Venkatesh and Davis' (2000) explained 34-52% of the variance in usage intentions among U.S. users. The model's explanatory power differs due to specific Saudi Arabian emic constructs, including its collectivist culture and the worker's focus on the managerial father figure's influence on individual performance, a stark difference from TAM findings in more individualistic societies. The authors' findings contribute to understanding the effects of cultural contexts in influencing technology acceptance behaviors, and demonstrate the need for research into additional cultural factors that account for technology acceptance.

Chapter 4

Hak-Jin Kim, Yonsei University, Korea
Hun Choi, Catholic University of Pusan, Korea
Jinwoo Kim, Yonsei University, Korea

This study examines the effects of uncertainty avoidance (UA) at the individual level on continuance behavior in the domain of mobile data services (MDS). It proposes a research model for post-expectation factors and continuance behavior that considers the moderating effect of UA, and verifies the model with online survey data gathered in Korea and Hong Kong. Post-expectation factors are classified as either intrinsic or extrinsic motivational factors, while respondents are classified according to their propensities into low-UA and high-UA groups. The results indicate that UA has substantial effects not only on the mean values of the post-expectation factors studied but also on the strength of those factors' impact on satisfaction and continuance intention. The effects of intrinsic motivational factors on satisfaction and continuance intention are stronger for the high-UA group than for the low-UA group. In contrast, the effects of extrinsic motivational factors are generally stronger for the low-UA group.

Trevor T. Moores, University of Nevada Las Vegas, USA

This article examines the relationship between Hofstede's national culture indices (IDV, PSI, MAS, and UAI), economic wealth (GNI), and national software piracy rates (SPR). Although a number of studies have already examined this relationship, the contribution of this article is two-fold. First, we develop a path model that highlights not only the key factors that promote software piracy, but also the inter-relationships between these factors. Second, most studies have used the dataset from the pre-2003 methodology which only accounted for business software and did not take into account local market conditions. Using the latest dataset and a large sample of countries (n=61) we find there is an important triadic relationship between PDI, IDV, and GNI that explains over 80% of the variance in software piracy rates. Implications for combating software piracy are discussed.

Section 2
International Perspectives

Chapter 6

L.G. Pee, Tokyo Institute of Technology, Japan
A. Kankanhalli, National University of Singapore, Singapore
V.C.Y. On Show, National University of Singapore, Singapore

To bridge the digital divide, subsidized access to information and communications technology (ICT) is often provided in less-developed countries. While such efforts can be helpful, their effectiveness depends on targeted users' willingness to utilize the ICT provided. This study examines the factors influencing individuals' use of one such ICT, public internet kiosks, in Mauritius. Findings from a survey indicate that individuals' self-efficacy, perceived ease of use, perceived usefulness, subjective norm, and perceived behavioral control have significant effects. This study contributes to research by highlighting key factors influencing the use of public internet kiosks and discussing how the factors' perception and assessment differ from those in the developed world. Mauritius also provides an interesting context for this study, as her government has been actively promoting the diffusion of ICT in the country yet there have been limited empirical studies on Mauritian and sub-Saharan African users in the digital divide research. Suggestions for promoting the use public internet kiosks in less-developed countries are also provided.

Chapter 7

Dahui Li, University of Minnesota Duluth, USA
Fujun Lai, University of Southern Mississippi, USA
Jian Wang, University of International Business and Economics, China

Despite the importance of international trade firms in China's economic development, there is only limited empirical evidence about how these firms assimilate Internet-based e-business in global supply chain operations. Using the Technology-Organization-Environment framework, this study investigates technological, organizational, and environmental factors which determine e-business assimilation in these firms. Based on survey data collected from 307 international trade firms in the Beijing area, we found that environmental uncertainty was negatively associated with e-business assimilation, while a firm's internal IT capability, relative advantage of e-business, learning orientation, and inter-organizational dependence were positive determinants of e-business assimilation. The effect of a firm's ownership type was also significant. Environmental uncertainty was the most important inhibitor, and IT capability and inter-organizational dependence were the most salient enablers of e-business assimilation.

    Mark F. Peterson, Florida Atlantic University, USA
    Stephanie J. Thomason, University of Tampa, USA
    Norm Althouse, University of Calgary, Canada
    Nicholas Athanassiou, Northeastern University, USA
    Gudrun Curri, Dalhousie University, Canada
    Robert Konopaske, Texas State University, USA
    Tomasz Lenartowicz, Florida Atlantic University, USA
    Mark Meckler, University of Portland, USA
    Mark E. Mendenhall, University of Tennessee-Chattanooga, USA
    Andrew A. Mogaji, Benue State University, Nigeria
    Julie I. A. Rowney, University of Calgary, Canada

This chapter extends communication and technology use theories about factors that predict e-mail use by explaining the reasons for cultural contingencies in the effects of managers' personal values and the social structures (roles, rules and norms) that are most used in their work context. Results from a survey of 576 managers from Canada, the English-speaking Caribbean, Nigeria, and the United States indicate that e-mail use may support participative and lateral decision making, as it is positively associated with work contexts that show high reliance on staff specialists especially in the U.S., subordinates, and unwritten rules especially in Nigeria and Canada. The personal value of self-direction is positively related to e-mail use in Canada, while security is negatively related to e-mail use in the United States. The results have implications for further development of TAM and media characteristic theories as well as for training about media use in different cultural contexts.

    She-I Chang, National Chung Cheng University, Taiwan
    Shin-Yuan Hung, National Chung Cheng University, Taiwan
    David C. Yen, Miami University, USA
    Pei-Ju Lee, National Chung Cheng University, Taiwan

Small and Medium-sized Enterprises (SMEs) play a vital and pervasive role in the current development of Taiwan's economy. Recently, the application of Enterprise Resource Planning (ERP) systems have enabled large enterprises to have direct contact with their clients via e-commerce technology, which has led to even fiercer competition among the SMEs. This study develops and tests a theoretical model including critical factors which influence ERP adoption in Taiwan's SMEs. Specifically, four dimensions, including CEO characteristics, innovative technology characteristics, organizational characteristics, and environmental characteristics, are empirically examined. The results of a mail survey indicate that the CEO's attitude towards information technology (IT) adoption, the CEO's IT knowledge, the employees' IT skills, business size, competitive pressure, cost, complexity, and compatibility are all important determinants in ERP adoption for SMEs. The authors' results are compared with research on IT adoption in SMEs based in Singapore and the United States, while implications of the results are also discussed.

Chapter 10

D. Li, Peking University, China
W.W. Huang, Ohio University, USA & Harvard University, USA
J. Luftman, Stevens Institute of Technology, USA
W. Sha, Pittsburg State University, USA

There have been periodical studies on key IS management issues facing the IT industry in North
America; however, an empirical investigation on key IS management issues in developing countries
has been largely ad hoc and inadequate. This paper identifies and analyzes important issues faced by
CIOs in the developing country of China. The results of this study are based on two national wide CIO
surveys in China, where the first was conducted in 2004 and followed by a more recent survey in 2008.
The authors provide insight for both IS practitioners and researchers who have interests in developing
countries. Data analysis indentified key IS management issues and demonstrated similarities as well
as differences between the two rounds of surveys. Although some strategic IS issues were still within
the top 10 on both the 2004 and 2008 lists, their importance ratings were different. Implications of the
findings are also discussed.

Chapter 11

Manlu Liu, Zhejiang University, China & Rochester Institute of Technology, USA
Xiaobo Wu, ZheJiang University, China
J. Leon Zhao, City University of Hong Kong, China
Ling Zhu, Long Island University, USA

Community source (a community-based open source) has emerged as an innovative approach to developing enterprise application software. Different from the conventional model of in-house development,
community source creates a virtual community that pools human, financial, and technological resources
from member organizations to develop custom software. Products of community source are available as
open source software to all members. To better understand community source, the authors studied the
Kuali project through interviewing its participants. The interview analysis revealed that community source
faced a number of challenges in project management, particularly in the areas of staffing management
and project sustainability. A viable solution to these issues, as supported by the findings in the interview
and the literature review on the drivers and expected benefits of outsourcing, is outsourcing software
development in community source projects. The authors accordingly proposed a research framework
and seven propositions that warrant future investigation into the relationship between community source
and software outsourcing.

Chapter 12

Wen Tian, USTC-CityU Joint Advanced Research Center, China
Douglas Vogel, City University of Hong Kong, China
Jian Ma, City University of Hong Kong, China
Jibao Gu, University of Science and Technology of China, China

In the first decade of the 21st century, China's Research Community (CRC) is struggling to achieve better performance by increasing growth in knowledge quantity (e.g., publications), but has failed to generate sound growth in knowledge quality (e.g., citations). An innovative E-government project, Internet-based Science Information System (ISIS), was applied nationwide in 2003 with a variety of embedded incentives. The system has been well received and supports the National Natural Science Foundation of China (NSFC) to implement managerial control to cope with pressing demands relating to China's research productivity. This paper explores the impact of Information Systems (IS) from the perspective of agency theory based on CRC empirical results. Since the nationwide application of ISIS in 2003, CRC outcomes have markedly improved. The discussion and directions for future research examine implications of IS for E-government implementation and business environment building in developing countries.

Section 3
Government and Development Issues

Chapter 13

Farid Shirazi, Ryerson University, Canada
Dolores Añón Higón, University of Valencia, Spain
Roya Gholami, Aston Business School, UK

This chapter investigates the impact of inward and outward FDI on ICT diffusion in the Asia-Pacific and Middle East regions for the period 1996-2008. The results indicate that while inward FDI has generally had a positive and significant impact on ICT diffusion in Asia-Pacific economies, its impact on the Middle Eastern countries has been detrimental. In contrast, the results of this study also show that outward FDI has had, in general, the inverse effect, it has been in general positive and significant for the Middle East countries but insignificant for Asia-Pacific economies.

Chapter 14

Ahmed Imran, The Australian National University, Australia
Shirley Gregor, The Australian National University, Australia

Least developed countries (LDCs), have been struggling to find a workable strategy to adopt information and communication technology (ICT) and e-government in their public sector organizations. Despite a number of high-level initiatives at national and international level, the progress is still unsatisfactory in this area. Consequently, the countries are failing to keep pace in the global e-government race, further increasing the digital divide. This chapter reports on an exploratory study in a least developed country, Bangladesh, involving a series of focus groups and interviews with key stakeholders. A lack of knowledge and entrenched attitudes and mindsets are seen as the key underlying contributors to the lack of progress. The analysis of the relationships among the major barriers to progress led to a process model, which suggests a pathway for e-government adoption in an LDC such as Bangladesh. The chapter introduces important directions for the formulation of long-term strategies for the successful adoption of ICT in the public sector of LDCs and provides a basis for further theoretical development.

Chapter 15

    Bijan Azad, American University of Beirut, Lebanon
    Samer Faraj, McGill University, Canada
    Jie Mein Goh, University of Maryland, USA
    Tony Feghali, American University of Beirut, Lebanon

Prior research has established the existence of a differential between industrialized and other countries
for e-Government diffusion. It attempts to explain this divide by identifying economic and technical
variables. At the same time, the role of national governance institutions in e-Government diffusion has
been relatively under-theorized and under-studied. The authors posit that, the existing national governance
institutions shape the diffusion and assimilation of e-Government in any country via associated institutions in three key sectors: government, private sector and non-governmental organizations. This paper
develops and tests a preliminary model of e-Government diffusion using the governance institutional
climate as represented via democratic practices, transparency of private sector corporate governance,
corruption perception, and the free press. The results indicate that the level of development of national
governance institutions can explain the level of e-Government diffusion over and above economic and
technical variables. The authors' research contributes to the literature by providing initial evidence
that the existing national governance institutions influence and shape e-Gov diffusion and assimilation
beyond the adoption stage.

Preface

To keep pace with the rapidly expanding world of global business, it is vital to stay abreast of the latest advances in management and technological strategies for a globalized market. This book of advances in management practices offers just those set of tools to a manager, IT professional, business student, or any member of the global business chain.

Narrowing the field of global business to give the most relevant and helpful practices exposure, "International Comparisons of Information Communication Technologies: Advancing Applications" offers fifteen chapters in three sections. The authors of these chapters hail from a dozen countries around the world, offering insight in case studies and comparisons of international enterprises.

The first section, "The Role of Culture" introduces some of the latest in the field of global information technology, its governance, and the trends in terminology and semantics surrounding these advances. Section one includes five chapters that introduce the book and lay a framework for implementing practices discussed in further chapters.

In chapter one, "User Attitude Towards Mandatory Use of Information Systems: A Chinese Cultural Perspective," by Xunhua Guo and Nan Zhang, the authors argue that cultural factors should be treated as aggregate characteristics at the population level instead of personal attributes at the individual level. The authors also propose that theoretical models could be developed for specific cultural contexts when examining IT/IS user behavior. In this regard, a model for analyzing user attitude toward mandatory use of information systems is proposed. Drawing on generally recognized cultural characteristics of China, three factors are introduced as determinants of user attitude—leader support, technology experience, and perceived fit. An empirical study is conducted with regard to the acceptance of a mobile municipal administration system in Beijing, China, for validating the proposed model with survey data and analyzing the adoption mechanism of the target system. The moderating roles of gender, age, and education level on the model are explored by interaction effect analyses and the findings provide helpful insights for related studies in other cultural contexts.

Chapter two, "The Impact of National Culture on Information Systems Planning Autonomy," by Dinesh A. Mirchandani and Albert L. Lederer, suggests that national culture may influence the information systems planning autonomy of the subsidiaries of multinational firms, but such an impact has not yet been tested empirically. A postal survey of 131 chief information officers and 103 senior non-IS managers of U.S. subsidiaries of such firms collected data to test hypotheses based on the model. Structural equation modeling using PLS-Graph 3.0 revealed that Individualism-Collectivism, Masculinity-Femininity, and Uncertainty Avoidance predicted autonomy for particular IS planning phases (as rated by the CIOs). On the basis of the supported hypotheses, the study provides evidence of the relevance of the national culture model to IS planning effectiveness and IS contribution. The study also suggests to subsidiary managers that an understanding of the national culture of their parent firm can help them gain an insight into the parent's management perspective.

Chapter three, "Cultural Impacts on Acceptance and Adoption of Information Technology in a Developing Country," by Elizabeth White Baker, Said S. Al-Gahtani, and Geoffrey S. Hubona investigates technology adoption behavior of Saudi Arabian knowledge workers using desktop computers within the context of TAM2, and the unique effects of Saudi culture on IT adoption within the developing, non-Western, country. Following the guidelines of the etic-emic research tradition, which encourages cross-cultural theory and framework testing, the study findings reveal that the TAM2 model accounts for 40.3% of the variance in behavioral intention among Saudi users, which contrasts Venkatesh and Davis' (2000) explained 34-52% of the variance in usage intentions among U.S. users. The model's explanatory power differs due to specific Saudi Arabian emic constructs, including its collectivist culture and the worker's focus on the managerial father figure's influence on individual performance, a stark difference from TAM findings in more individualistic societies. The authors' findings contribute to understanding the effects of cultural contexts in influencing technology acceptance behaviors, and demonstrate the need for research into additional cultural factors that account for technology acceptance.

Chapter four, "A Comparative Study of the Effects of Low and High Uncertainty Avoidance on Continuance Behavior," by Hak-Jin Kim, Hun Choi, and Jinwoo Kim, examines the effects of uncertainty avoidance (UA) at the individual level on continuance behavior in the domain of mobile data services (MDS). It proposes a research model for post-expectation factors and continuance behavior that considers the moderating effect of UA, and verifies the model with online survey data gathered in Korea and Hong Kong. Post-expectation factors are classified as either intrinsic or extrinsic motivational factors, while respondents are classified according to their propensities into low-UA and high-UA groups. The results indicate that UA has substantial effects not only on the mean values of the post-expectation factors studied but also on the strength of those factors' impact on satisfaction and continuance intention. The effects of intrinsic motivational factors on satisfaction and continuance intention are stronger for the high-UA group than for the low-UA group. In contrast, the effects of extrinsic motivational factors are generally stronger for the low-UA group.

Chapter five, "Untangling the Web of Relationships between Wealth, Culture, and Global Software Piracy Rates: A Path Model," by Trevor T. Moores, examines the relationship between Hofstede's national culture indices (IDV, PSI, MAS, and UAI), economic wealth (GNI), and national software piracy rates (SPR). Although a number of studies have already examined this relationship, the contribution of this article is two-fold. First, we develop a path model that highlights not only the key factors that promote software piracy, but also the inter-relationships between these factors. Second, most studies have used the dataset from the pre-2003 methodology which only accounted for business software and did not take into account local market conditions. Using the latest dataset and a large sample of countries (n=61) we find there is an important triadic relationship between PDI, IDV, and GNI that explains over 80% of the variance in software piracy rates. Implications for combating software piracy are discussed.

The second section, "International Perspectives," comprises the bulk of the book, containing seven chapters that draw experiences from different sectors of global industry and enterprise.

Chapter six, "ICT for Digital Inclusion: A Study of Public Internet Kiosks in Mauritius," by L.G. Pee, A. Kankanhalli, and V.C.Y. On Show examines the factors influencing individuals' use of one such ICT, public internet kiosks, in Mauritius. Findings from a survey indicate that individuals' self-efficacy, perceived ease of use, perceived usefulness, subjective norm, and perceived behavioral control have significant effects. This study contributes to research by highlighting key factors influencing the use of public internet kiosks and discussing how the factors' perception and assessment differ from those in the developed world. Mauritius also provides an interesting context for this study, as her government has been actively promoting the diffusion of ICT in the country yet there have been limited empirical studies on Mauritian and sub-Saharan African users in the digital divide research. Suggestions for promoting the use public internet kiosks in less-developed countries are also provided.

Chapter seven, "E-Business Assimilation in China's International Trade Firms: The Technology-Organization-Environment Framework," by Dahui Li, Fujun Lai, and Jian Wang, investigates technological, organizational, and environmental factors which determine e-business assimilation in these firms. Based on survey data collected from 307 international trade firms in the Beijing area, we found that environmental uncertainty was negatively associated with e-business assimilation, while a firm's internal IT capability, relative advantage of e-business, learning orientation, and inter-organizational dependence were positive determinants of e-business assimilation. The effect of a firm's ownership type was also significant. Environmental uncertainty was the most important inhibitor, and IT capability and inter-organizational dependence were the most salient enablers of e-business assimilation.

Chapter eight, "An International Comparative Study of the Roles, Rules, Norms, and Values that Predict Email Use," by Mark F. Peterson, Stephanie J. Thomason, Norm Althouse, Nicholas Athanassiou, Gudrun Curri, Robert Konopaske, Tomasz Lenartowicz, Mark Meckler, Mark E. Mendenhall, Andrew A. Mogaji, and Julie I. A. Rowney extends communication and technology use theories about factors that predict e-mail use by explaining the reasons for cultural contingencies in the effects of managers' personal values and the social structures (roles, rules and norms) that are most used in their work context. Results from a survey of 576 managers from Canada, the English-speaking Caribbean, Nigeria, and the United States indicate that e-mail use may support participative and lateral decision making, as it is positively associated with work contexts that show high reliance on staff specialists especially in the U.S., subordinates, and unwritten rules especially in Nigeria and Canada. The personal value of self-direction is positively related to e-mail use in Canada, while security is negatively related to e-mail use in the United States. The results have implications for further development of TAM and media characteristic theories as well as for training about media use in different cultural contexts.

Chapter nine, "Critical Factors of ERP Adoption for Small- and Medium- Sized Enterprises: An Empirical Study," by She-I Chang, Shin-Yuan Hung, David C. Yen, and Pei-Ju Lee, develops and tests a theoretical model including critical factors which influence ERP adoption in Taiwan's SMEs. Specifically, four dimensions, including CEO characteristics, innovative technology characteristics, organizational characteristics, and environmental characteristics, are empirically examined. The results of a mail survey indicate that the CEO's attitude towards information technology (IT) adoption, the CEO's IT knowledge, the employees' IT skills, business size, competitive pressure, cost, complexity, and compatibility are all important determinants in ERP adoption for SMEs. The authors' results are compared with research on IT adoption in SMEs based in Singapore and the United States, while implications of the results are also discussed.

Chapter ten, "Key Issues in Information Systems Management: An Empirical Investigation from a Developing Country's Perspective," by D. Li, W.W. Huang, J. Luftman, and W. Sha identifies and analyzes important issues faced by CIOs in the developing country of China. The results of this study are based on two national wide CIO surveys in China, where the first was conducted in 2004 and followed by a more recent survey in 2008. The authors provide insight for both IS practitioners and researchers who have interests in developing countries. Data analysis indentified key IS management issues and demonstrated similarities as well as differences between the two rounds of surveys. Although some strategic IS issues were still within the top 10 on both the 2004 and 2008 lists, their importance ratings were different. Implications of the findings are also discussed.

Chapter eleven, "Outsourcing of Community Source: The Case of Kuali," by Manlu Liu, Xiaobo Wu, J. Zhao, and Ling Zhu studied the Kuali project through interviewing its participants. The interview analysis revealed that community source faced a number of challenges in project management, particularly in the areas of staffing management and project sustainability. A viable solution to these issues, as supported by the findings in the interview and the literature review on the drivers and expected benefits of outsourcing, is outsourcing software development in community source projects. The authors accordingly proposed a research framework and seven propositions that warrant future investigation into the relationship between community source and software outsourcing.

Chapter twelve, "IS-Supported Managerial Control for China's Research Community: An Agency Theory Perspective," by Wen Tian, Douglas Vogel, Jian Ma, and Jibao Gu begins with an introduction to the subject matter, discussing the following: in the first decade of the 21st century, China's Research Community (CRC) is struggling to achieve better performance by increasing growth in knowledge quantity (e.g., publications), but has failed to generate sound growth in knowledge quality (e.g., citations). An innovative E-government project, Internet-based Science Information System (ISIS), was applied nationwide in 2003 with a variety of embedded incentives. The system has been well received and supports the National Natural Science Foundation of China (NSFC) to implement managerial control to cope with pressing demands relating to China's research productivity. This paper explores the impact of Information Systems (IS) from the perspective of agency theory based on CRC empirical results. Since the nationwide application of ISIS in 2003, CRC outcomes have markedly improved. The discussion and directions for future research examine implications of IS for E-government implementation and business environment building in developing countries.

The third section, "Government and Development Issues," concludes the book with three final chapters on topics such as ICT diffusion, successful e-government adoption, IT maturity, and more from research and case studies across the globe.

Chapter thirteen, "ICT Diffusion and Foreign Direct Investment: A Comparative study between Asia-Pacific and the Middle Eastern Economies," by Farid Shirazi, Dolores Higón, and Roya Gholami investigates the impact of inward and outward FDI on ICT diffusion in the Asia-Pacific and Middle East regions for the period 1996-2008. The results indicate that while inward FDI has generally had a positive and significant impact on ICT diffusion in Asia-Pacific economies, its impact on the Middle Eastern economies has been detrimental. In contrast, the results of this study also show that outward FDI has had, in general, the inverse effect, it has been in general positive and significant for the Middle East economies but insignificant for Asia-Pacific countries.

Chapter fourteen, "A Process Model for Successful E-Government Adoption in the Least Developed Countries: A Case of Bangladesh," by Ahmed Imran and Shirley Gregor, reports on an exploratory study in a least developed country, Bangladesh, involving a series of focus groups and interviews with key stakeholders. A lack of knowledge and entrenched attitudes and mindsets are seen as the key underlying contributors to the lack of progress. The analysis of the relationships among the major barriers to progress led to a process model, which suggests a pathway for e-government adoption in an LDC such as Bangladesh. The chapter introduces important directions for the formulation of long-term strategies for the successful adoption of ICT in the public sector of LDCs and provides a basis for further theoretical development.

Chapter fifteen, "What Shapes Global Diffusion of E-Government: Comparing the Influence of National Governance Institutions," by Bijan Azad, Samer Faraj, Jie Mein Goh, and Tony Feghali concludes the book. Prior research has established the existence of a differential between industrialized and other countries for e-Government diffusion. It attempts to explain this divide by identifying economic and technical variables. At the same time, the role of national governance institutions in e-Government diffusion has been relatively under-theorized and under-studied. The authors posit that, the existing national governance institutions shape the diffusion and assimilation of e-Government in any country via associated institutions in three key sectors: government, private sector and non-governmental organizations. This chapter develops and tests a preliminary model of e-Government diffusion using the governance institutional climate as represented via democratic practices, transparency of private sector corporate governance, corruption perception, and the free press. The results indicate that the level of development of national governance institutions can explain the level of e-Government diffusion over and above economic and technical variables. The authors' research contributes to the literature by providing initial evidence that the existing national governance institutions influence and shape e-Gov diffusion and assimilation beyond the adoption stage.

It is the sincere hope of the editor of this collection of case studies, research, and best practices that it will serve as an essential reference tool for practitioners of global business. The international perspectives offered here give pointed insight into the latest word on Information Systems and Technology within burgeoning global business.

Felix B. Tan
Auckland University of Technology, New Zealand

Section 1
The Role of Culture

Chapter 1
User Attitude towards Mandatory Use of Information Systems:
A Chinese Cultural Perspective

Xunhua Guo
Tsinghua University, China

Nan Zhang
Tsinghua University, China

ABSTRACT

Accumulated literature on technology adoption research has suggested that cultural factors have important impacts on the cognition and behavior of information systems users. In this paper, the authors argue that cultural factors should be treated as aggregate characteristics at the population level instead of personal attributes at the individual level. The authors also propose that theoretical models could be developed for specific cultural contexts when examining IT/IS user behavior. In this regard, a model for analyzing user attitude toward mandatory use of information systems is proposed. Drawing on generally recognized cultural characteristics of China, three factors are introduced as determinants of user attitude—leader support, technology experience, and perceived fit. An empirical study is conducted with regard to the acceptance of a mobile municipal administration system in Beijing, China, for validating the proposed model with survey data and analyzing the adoption mechanism of the target system. The moderating roles of gender, age, and education level on the model are explored by interaction effect analyses and the findings provide helpful insights for related studies in other cultural contexts.

INTRODUCTION

During the past two decades, the trends toward a globalized business environment have been accelerating and continuously affected companies by increasing the competitiveness of the marketplace, restructuring business relationships, re-defining organizational boundaries, and creating new challenges for managers who deal with multinational companies or international alliances (Friedman, 2005), especially when emerging technologies and applications featuring Web 2.0 started to bring

DOI: 10.4018/978-1-61350-480-2.ch001

the Internet and the world to a new era of social networks, ubiquitous computing, mobility, personalization, and virtualization in the past five years.

Correspondingly, cultural issues have become an important topic in the study of the information systems area (Chau, Cole, Massey, Montoya-Weiss, & O'Keefe, 2002; Davison & Martinsons, 2003; Evaristo, 2003; Gallivan & Srite, 2005; Myers & Tan, 2003), particularly when user behavior in adoption, diffusion, and infusion of new technologies and systems is concerned (Chau, 2008). Accumulated literature on technology adoption research has suggested that cultural factors potentially have significant impacts on the perception and behavior of information systems users (Chau et al., 2002; Lippert & Volkmar, 2007; Mao & Palvia, 2006; Zhang, Guo, Chen, & Song, 2009).

The cultural concepts incorporated in information systems research include not only organizational culture and societal culture, but also national culture and ethnic culture (Karahanna, Evaristo, & Srite, 2005). When national cultural issues are taken into account, Hofstede's well-known five dimensions model (Hofstede, 2001) provides the most influential theoretical framework, which has been widely cited for cross-cultural studies in various areas (Davison & Martinsons, 2003; Ford, Connelly, & Meister, 2003; Gallivan & Srite, 2005; Myers & Tan, 2003). Although frequently criticized for the lack of practical feasibility and incompleteness of the dimensions (Baskerville, 2003; Myers & Tan, 2003; Straub, Loch, Evaristo, Karahanna, & Srite, 2002), Hofstede's model has, to some extent, successfully consolidated a conceptual foundation for investigating cross-national differences in human and organizational behavior.

In the research area of IT/IS adoption, when considering cultural factors, existing efforts either incorporated cultural interpretation in cross-national comparison (Lippert & Volkmar, 2007; e.g., Straub, Keil, & Brenner, 1997), or borrowed variables from classical culture theories for constructing user behavioral models(Huang,

Lu, & Wong, 2003; Min, Li, & Ji, 2009; e.g., Srite, Thatcher, & Galy, 2008). Both of these two approaches have produced valuable theoretical and practical insights of significant relevance. Contrastively, however, few have tried to postulate theoretical models for specific cultural contexts when examining IT/IS user behavior, especially when Eastern nations, such as China, are concerned. By far, IT/IS adoption studies regarding Chinese users (e.g., Huang et al., 2003; Mao & Palvia, 2006) were mostly designed to analyze the special phenomena in China with Western theoretical frameworks, or to explain some unexpected empirical results with some Chinese cultural considerations. We believe that such a gap should not be ignored and it would be promising to construct or extend IT/IS user behavior models by introducing new factors originating in Eastern cultural values, so as to better reflect the IT/IS practice in these particular contexts.

Based on such understanding, this paper is focused on establishing a conceptual model for examining user attitude towards the use of new technologies from a Chinese cultural perspective. As a typical representative of the Eastern culture (Hofstede, 2001), China provides an ideal stage for culture-specific behavioral model development. Meanwhile, because the impact of IT/IS application in China might be even more profound than that in any other country, as China is changing from an isolated centrally-controlled economy to a market that opens to the global economy (Guo & Chen, 2005; Martinsons, 2005), it would be even more worthwhile to probe into the interactions between new technology adoption and cultural factors.

When a cultural lens is taken, it is also worth noting that cultural characteristics would have different effects in mandatory and voluntary adoption contexts (Zhang, Guo, Chen, & Song, 2008; Zhang et al., 2009). A large portion of the early researches in the information technology adoption area were conducted in the voluntary environment, especially those based on TAM and its derived models. An implicit assumption of such models is that users of IT have a choice about the extent to which they use the technology. However, when technology use is mandated, as it is in many organizations, it is expected that the underlying relationships of traditional technology acceptance models will be different (Brown, Massey, Montoya-Weiss, & Burkman, 2002; Melone, 1990; Nah, Tan, & Teh, 2004).

It is worth noting that mandatory IT adoption in China is prevalent, especially in governmental organizations. In government agencies at various levels, information systems are often implemented by obeying the orders from superior authorities (Zhang, Guo, Chen, & Chau, 2009). In such mandatory adoption contexts, a potential problem is that although users have apparently adopted the new system and use it frequently, they often do not actively seek to efficiently utilize it. Consequently, the effects of the new system on user productivity and organizational performance are limited. For instance, in the IT Application Office of a district government in Beijing, where one of the authors worked as a part-time consultant, more than 30 e-Government application systems have been implemented among the government's subsidiaries and citizens in the district since 2001. Each time when a new system was introduced, the government required the employees to use it. In most of the cases, although the users obeyed the orders and used the new system, no significant improvement in performance was achieved. About several months later, when the mandatory requirements were relaxed, users would gradually give up using the system. By 2007, none of the systems implemented before 2004 had been effectively utilized.

Therefore, we believe that it is worthwhile to investigate the mandatory adoption of information systems in Chinese organizations from a cultural perspective. Meanwhile, as culture is generally defined as the collective programming of a human group, we expect that the impacts of cultural factors be more consistent and distinct when the use

behavior is less discursive. In other words, when the choices of actions are somehow restricted, the effects of diversified personal attributes may be restrained and collective characteristics may play more important roles.

Consequently, this paper is aimed at developing a novel model for analyzing user attitude towards mandatory use of information systems from a Chinese cultural perspective. Drawing upon the generally recognized cultural characteristics of China, which include long power distance, low uncertainty avoidance, and long-term orientation, three factors are introduced as determinants of user attitude, namely leader support, technology experience, and perceived fit. An empirical study is conducted with regard to the acceptance of a mobile municipal administration system in Beijing, China, for validating the proposed model with survey data and analyzing the adoption mechanism of the target system. In the light of a sample consisting of 134 valid responses, results from structured equation model (SEM) analysis illustrate the ability of the model to interpret the user attitude of Chinese users in mandatory contexts. Moreover, the moderating roles of gender, age, and education level on the model are also investigated by interaction effect analyses. We believe that this model is sufficiently valid for explaining individual's mandatory IT/IS adoption and use behavior from a novel perspective in the Chinese cultural context, and may provide helpful insights for related studies in other cultural contexts.

LITERATURE REVIEW

As mentioned above, our research was aimed at developing a novel model for analyzing user attitude towards mandatory use of information systems. In this section, we will briefly review the related IT/IS adoption literature and discuss the theoretical foundation that we draw upon.

Cultural Issues in IT/IS Adoption Research

Studies about user adoption behavior towards new technologies and systems from the social psychological perspective have been prevailing in the information systems research area around the world for nearly thirty years (Benbasat & Barki, 2007; Legris, Ingham, & Collerette, 2003; Venkatesh, Davis, & Morris, 2007; Venkatesh, Morris, Davis, & Davis, 2003). Related efforts have produced a number of theoretical models, such as technology acceptance model (TAM) (Davis, 1989; Davis, Bagozzi, & Warshaw, 1989) and task-technology fit model (TTF) (Goodhue & Thompson, 1995), to interpret the psychological and social mechanisms that potentially drive the behavior of technology adoption. Among them, TAM has been the most influential model, based on which a large amount of derived models were development. Some of the related efforts were later integrated in the unified theory of acceptance and use of technology (UTAUT) (Venkatesh et al., 2003). As well, influential theories in other areas, including theory of planned behavior (TPB) (Mathieson, 1991; Taylor & Todd, 1995), social cognitive theory (SCT) (Compeau, Higgins, & Huff, 1999; Compeau & Higgins, 1995), and innovation diffusion theory (IDT) (Moore & Benbasat, 1991; Plouffe, Hulland, & Vandenbosch, 2001; Rogers, 1995), were borrowed into the studies of technology adoption. Meanwhile, large amounts of empirical studies were conducted in various regions of the world to validate the theories. It is believed that the achievements in technology adoption research could essentially help IT manufacturers and managers to better handle the development, application, and management of IT (Legris et al., 2003).

As empirical evidence accumulated, the results from different counties are often found conflicting with each other (Legris et al., 2003). Consequently, researchers also tried to explain such inconsistency by discussing the differences of national cultures (Straub, 1994; Straub et al., 1997). With very few exceptions, most cross-cultural IT/IS studies are based on Hofstede's culture dimensions (Ford et al., 2003; Leidner & Kayworth, 2006; Myers & Tan, 2003).

Hofstede's defined national culture as "the collective programming of the mind which distinguishes the members of one human group from another" and argued that national differences can be interpreted in terms of national culture (Hofstede, 2001). The original version of Hofstede's model includes four dimensions, namely power distance, uncertainty avoidance, individualism-collectivism, and masculinity-femininity. It was later found that these four dimensions provide limited ability to explain the characteristics of Eastern cultures, so the fifth dimension, long-term orientation, or Confucian dynamism, was added, referring to the extent to which a culture programs its members to accept delayed gratification of their material, social, and emotional need (Hofstede, 2001; Hofstede & Bond, 1988). The five dimensions in Hofstede's framework, especially power distance and individualism-collectivism, have been largely adopted in IT/IS adoption research (Huang et al., 2003; Lippert & Volkmar, 2007; Mao & Palvia, 2006; e.g., Mejias, Shepherd, Vogel, & Lazaneo, 1996). In existing efforts, the framework has basically been applied in two approaches. One is to use it as a theoretical foundation for interpreting different observations in different nations, and the other is to borrow variables from the dimensions for constructing user behavioral models.

The first approach has been proved to be effective. For instance, Chau et al. (2002) investigated the online behavior of consumers and explained the differences between US and Hong Kong users with cultural dimensions; Straub et al. (1997) tested the TAM model in Japan, Switzerland, and US, and attributed the different results to cultural factors; Lippert and Volkmar (2007) compared technology performance and utilization between US and Canada with the masculinity-femininity work value dimension. In such studies, Hofstede's cultural dimensions have demonstrated strong capability for explaining different findings across nations. In many cases, however, such explanation seems to be contingent and lacks necessary consistency at the theoretical level. Moreover, with this approach, instead of being treated as a research lens, cultural factors usually serve as supporting materials for discussion, so the insights that they can provide are restricted.

The second approach to cultural analysis is to take one or more variables from Hofstede's dimensions and integrate them into user behavioral models. To name a few examples among others, Srite et al. (2008) extended the TAM model and took masculinity-femininity and individualism-collectivism as determinant of user technology perceptions, such as perceived usefulness and perceived ease-of-use; in a mobile commerce adoption model based on the TPB theory, Min et al. (2009) suggested that uncertainty avoidance and individualism-collectivism have direct impacts on self-efficacy, while at the same time moderating the impact of subjective norm on behavioral intention. This type of efforts in model development also provided valuable contribution to related literature. A major problem with regard to this approach, however, is that the combined models thusly derived often lead to ambiguity between the population level and the individual level. In the light of its definition as a kind of "collective programming" (Hofstede, 2001), national cultural factors should be treated as aggregate characteristics at the population level, instead of personal attributes at the individual level. Individual behavior is influenced by cultural factor at more than one aggregation levels (Karahanna et al., 2005), but fundamentally there is no "collective programming" at the individual level. Therefore, it

would be farfetched to directly include a national cultural dimension variable in an individual level user perception model. In contrast, we believe that a behavioral model would be more solid if it is constructed based on the cultural values of the specific context.

Another limitation of existing cultural studies in IT/IS adoption research is most of the target systems that have been examined are simple technologies such as E-mail (Huang et al., 2003; Mao & Palvia, 2006; Straub, 1994; Straub et al., 1997), while little attention has ever been paid to new emerging technologies such as mobile computing.

In recent years, IT/IS researches that considered Chinese culture are also increasing (Chen, Wu, & Guo, 2007; Hempel & Kwong, 2001; Huang et al., 2003; Mao & Palvia, 2006; Zhang et al., 2009). Although these studies were conducted from different perspectives, they were commonly based on the fundamental judgment that at least some of the special characteristics in IT/IS application and management in China can be attributed to China's Eastern cultural nature and, consequently, the Chinese culture and Confucians values may essentially affect IT/IS development strategy and management practice in China. However, existing studies, especially those on IT/IS adoption by Chinese users (Huang et al., 2003; Mao & Palvia, 2006), were mostly designed to analyze the special phenomena in China with Western theoretical frameworks, or to explain some unexpected empirical results with some Chinese cultural considerations. In comparison with this, a previous study of us has attempted to extend the classical IT/IS adoption/evaluation model by introducing new factors originated from Chinese cultural values, so as to better reflect the IT/IS practice in China (Zhang et al., 2009). In this paper, we will continue this effort, aimed building a more comprehensive model in the light of the Chinese cultural context.

Mandatory Information Systems Adoption

As mentioned above, it is worth noting that cultural characteristics would have different effects in mandatory and voluntary adoption contexts. It is particularly meaningful to distinguish the adoption scenarios between voluntary and mandatory and, in turn, to recognize the different user behavioral structures between these two situations (Melone, 1990).

Most of the early researches in the information technology adoption area were conducted in the voluntary environment (Davis et al., 1989), following the conceptual path of "Determinant → Attitude → Intention → Actual Use". Later researches often omitted the factor of "Attitude", following the path as "Determinant → Intention → Actual Use" (Davis, 1989; Venkatesh et al., 2003; Venkatesh & Davis, 2000). However, in the mandatory adoption environment, it is frequently assumed that there is little variance in usage, making it difficult to empirically test the full impact path (Brown et al., 2002; Hartwick & Barki, 1994; Melone, 1990; Venkatesh et al., 2003).

To deal with this problem, some scholars introduced the concept of "Symbolic Adoption" in the mandatory environment to replace the actual use for exploring users' mental acceptance (Nah et al., 2004). This reveals that although users in the mandatory environment are usually required to use the system by their superior authorities, the extent of use may vary, which would further lead to the variance in user performance. Meanwhile, it has been empirical validated that attitude has a positive impact on symbolic adoption (Nah et al., 2004).

Conceptually, however, we feel that the two constructs of attitude and symbolic adoption are somehow overlapping, since both of them represent a user's mental recognition of the target system. Therefore, we do not feel that it is important to distinguish symbolic adoption from attitude. Instead, we suggest that the effect of attitude on user performance be examined. When the system is well accepted and efficiently used, the job performance of the user would increase. Thereby, voluntary and mandatory adoption scenarios are characterized with different conceptual influence path, as shown in Figure 1.

RESEARCH MODEL

In view of the above analysis, this paper is aimed at postulating a novel conceptual model from the Chinese cultural perspective for understanding the determinants of information systems acceptance. In this section, the proposed model will be discussed and research hypotheses will be presented.

As illustrated above, our research model is focused on mandatory adoption scenarios. Therefore, in our study, we propose that the influence path in mandatory environment is "Determinant → Attitude → User Performance". When a user has a positive attitude towards the target system, it is likely that he/she would actively seek to use the system more efficiently so as to better utilize it in his/her work. As a result, the job performance of the user would increase. Consequently, we have the following hypothesis:

H1: A user's attitude toward the target system has significantly positive influence on his/her work performance.

Aimed at developing a model situated in the Chinese cultural context, we strived to draw upon the generally recognized cultural characteristics of China. A focus group discussion (Morgan, 1997) was organized to verify the principal Chinese cultural values concluded from literature, as well as to explore the substantial factors reflecting Chinese cultural characteristics. It was revealed through the focus group discussion that the Chinese cultural context is characterized with long power distance, low uncertainty avoidance, and a great emphasis on harmony or *Hexie,* a Chinese word

Figure 1. Voluntary vs. mandatory adoption

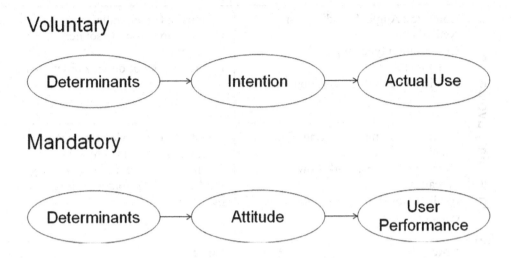

that can be approximately interpreted as interactive cooperation and synergy in a complex system that includes a large amount of subsystems or elements (Xi & Wang, 2006). This is consistent with the results from most of the national culture surveys (Hofstede, 2001; Hofstede & Bond, 1988; Tsui, 2006; Xi & Wang, 2006). Correspondingly, three factors are introduced as determinants of user attitude, namely leader support, technology experience, and perceived fit.

With regard to the other two dimensions in Hofstede's framework, namely individualism-collectivism and masculinity-femininity, no general agreement was achieved in the focus group discussion. In other words, we found that it is difficult to posit China along these two dimensions. On the aspect of individualism-collectivism, Chinese people are now highly diversified and the related societal norms are still evolving dynamically. Although it was once accepted that China is characterized with a high degree of collectivism (Hofstede & Bond, 1988; Jiacheng, Lu, & Francescob, 2009), the situation has changed significantly along with the progress of economic transformation. As Hofstede has pointed out, the degree of individualism may increase when the national wealth grows (Hofstede, 2001). With the

rapid development of Chinese economy, it can be expected that China is dynamically changing on the dimension of individualism-collectivism. Therefore, this dimension is not considered in our research model. Similar observations were found for masculinity-femininity. Furthermore, existing literature shows that this dimension has little effect in IT/IS adoption (Ford et al., 2003), so it is neither taken into account in our model.

Power distance (PD) and uncertainty avoidance (UA) are the two most important dimensions of the Hofstede's theoretical framework. PD is defined as "the extent to which the less powerful members in an organization accept and expect that power is distributed unequally", while UA is defined as "the extent to which a culture programs its members to feel either uncomfortable or comfortable in unstructured situations" (Hofstede, 2001). In Hofstede's famous UA-PD matrix, China is located in the up-right quadrant and described as a culture with long power distance and weak uncertainty avoidance (Figure 2).

In the situation of a high PD culture, users usually attach more importance to leaders' attitude toward the technology. The findings about key issues of information management in China have proved that "top management support" is gener-

Figure 2. Hofstede's UA-PD Matrix (Hofstede & Bond, 1988)

	Low	
Uncertainty avoidance	Countries: Anglo, Scandinavian, Netherlands Organization type: implicitly structured Implicit model of organization: *market*	Countries: China, India Organization type: personnel bureaucracy Implicit model of organization: *family*
	Countries: German-speaking, Finland, Israel Organization type: work-flow bureaucracy Implicit model of organization: *well-oiled machine*	Countries: Latin, Mediterranean, Islamic, Japan, some other Asian Organization type: full bureaucracy Implicit model of organization: *pyramid*
	High	
	Low	**Power distance** **High**

ally regarded as an important factor (Chen et al., 2007). Besides, in some former technology adoption research, the cultural effects related to PD have been validated in China and some other countries with similar cultural characteristics (Huang et al., 2003; Lim, 2004). Hence, there is a hypothesis as follow:

H2: A user's perception on leader support has significant positive influence on his/her attitude toward the target system.

In some early studies, Straub et al. asserted that the high degree of Uncertainty Avoidance (UA) in Japanese culture has decreased the speed of the diffusion of new technologies such as E-mail (Straub, 1994; Straub et al., 1997). In such environments, users often tend to avoid or resist the unfamiliar technologies. In China, the situation is exactly the opposite. Therefore, a new factor, technology experience, is introduced into our model. Technology experience represents a user's familiarity to the target system as well as other technologies related to the system. This factor may be deemed as closely related to self-efficacy, which is an individual's self-confidence in his/her ability to perform a behavior (Taylor

& Todd, 1995), since when a user is more familiar to related technologies, he/she would be more self-confident to use the system. However, technology experience is not fully equivalent to self-efficacy, because this factor is more subject to the influence of the social norms with regard to uncertainty avoidance. When a culture makes its members feel uncomfortable in unstructured situations (namely, when the level of UA is high), an individual is more likely to resist a new technology that he/she is not familiar with. On the contrary, when a culture makes its members feel easy in unstructured situations (namely, when the level of UA is low), an individual would tend to try a new technology even if he/she is not familiar with it. Living in the context of a low UA culture, Chinese people have very high passion for new technologies and the familiarity to the new technology is not as important as in other country. This can be illustrated by the fact that Chinese users are usually characterized with a high level of technological innovativeness, as reported in some recent surveys (Zhang, Guo, & Chen, 2006). Hence, we expect that technology experience would not significantly determine Chinese users' systems adoption, so we have the hypothesis as follows:

H3: A user's perception on his/her technology experience has no significant influence on his/her attitude toward the target systems.

To test the non-significant relationship, we perform a power analysis in the results section, following the procedure proposed by Venkatesh et al. (2003).

When Hofstede's original framework was found to be limited in the ability to explain the characteristics of Eastern cultures, the fifth dimension, long-term orientation (LTO), or Confucian dynamism, was added, referring to "the extent to which a culture programs its members to accept delayed gratification of their material, social, and emotional need" (Hofstede, 2001; Hofstede & Bond, 1988). However, this new dimension has not been really accepted as the other dimensions were, as it not only seems unfamiliar to Western people, but also tends to get lost in confusion for Chinese minds (Fang, 2003). Essentially, the dimension divides the Confucian values into the "positive" pole and the "negative" pole, in a similar way as the other four dimensions. Contrarily, in the traditional philosophy of China, the positive and negative aspects are often considered to be indivisible and depend on the specific contexts in which they are discussed. In other words, the ideal status of everything is to achieve a balance between positive and negative, and moreover, to achieve *Hexie*, the interactive cooperation and synergy in the whole system (Xi & Wang, 2006). We suggest that, compared with LTO, *Hexie* is a

better concept to reflect the traditional Chinese values.

Based on this consideration, a previous study of us introduced the new concept of perceived fit, which is regarded as equivalent of *Hexie* in the Chinese work values, referring to "the degree in user perception to which the target innovation fits to all aspects of the user's work style and environment" (Zhang et al., 2009). When a Chinese user perceives that a new system fits well with his/her work style and environment, he/she would be particularly active in attempting to adopt the system, since a *Hexie*-oriented culture makes its members feel better when technology can co-exist with other aspects of work/life without conflicts. In this paper, we continue to use this factor of "perceived fit" and propose the following hypothesis:

H4: User's perceived fit has significantly positive influence on his/her attitude toward the target system.

In summary, our research model about user attitude towards mandatory use of information systems from the Chinese cultural perspective is illustrated in Figure 3.

Empirical Test

In order to validate the proposed Chinese cultural model for systems acceptance, we conducted a survey among municipal patrol in Chaoyang District, Beijing China, and investigate their acceptance of a

Figure 3. Research model

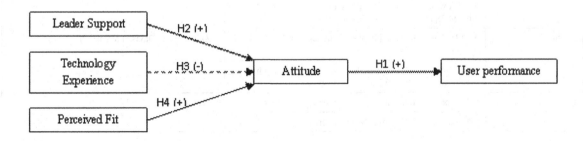

newly implemented mobile administration system, particularly with regard to their attitude towards the handset terminal for the system. Chaoyang District is a large district in Beijing, accounting for a big portion of the capital's economy. It is also where the Central Business District (CBD) of the city and the main venue of 2008 Olympic Games are located.

In order to utilize innovative technologies to improve the effectiveness and efficiency of municipal administration, the municipal administration commission (MAC) of Beijing has been actively promoting a so-called "Grid Administration Model" project in all administrative districts of the city since 2006. One of the critical supporting issues for this infrastructure system is the acquisition of high quality data. To achieve this gold, the Chaoyang district government established a new department, the Supervision and Commanding Center (SCC), to take the responsibility of data collection. SCC recruited 1000 patrols to help monitoring the status of municipal facilities

(such as drain covers). Each patrol is responsible for about 12 residential blocks, or about 180,000 square meters area, and up to 1400 public municipal facilities. The patrols are required to spot, check, and report the municipal administration related problems, as well as to take necessary steps to ensure the problems are properly solved. Each patrol is equipped with a bicycle and a smart mobile terminal, which is called "Chengguantong" in Chinese, with the meaning of "all-in-one city management".

Measurement Items

The measurement items used in the survey are listed in Table 1. All the items were translated into Chinese and adjusted in wording in the light of the characteristics of the technology for better understanding. Two researchers translated all the items independently and the two translation versions were then compared and discussed among a group of four researchers, so as to ensure the

Table 1. Measurement items

Construct	Items	Origins
Leader Support	**LS1:** My superior attaches great importance to the project of Chengguantong.	*Adapted from* (Lewis, Agarwal, & Sambamurthy, 2003)
	LS2: The top directors of the department (MAC) consider Chenguantong as an important project.	
Technology Experience	**TE1:** Before I became a patroller, I often used short message service (SMS) with my mobile phone.	Developed in our study
	TE2: Before I became a patroller, I often used personal computer.	
Perceived Fit	**FIT1:** Using the Chenguantong is compatible with my day-to-day work.	*Adapted from* (Zhang et al., 2009)
	FIT2: Using the Chenguantong fits with my work style.	
	FIT3: I would find the Chengguantong fits well with the other IT/IS applications used in my work.	
Attitude	**ATT1:** Generally speaking, the Chenguantong is a good system.	*Adapted from* (Davis et al., 1989)
	ATT2: Using the Chengguantong in my municipal administration work is a good idea.	
User performance	**UP1:** I am satisfied with my performance in my work.	Developed in our study
	UP2: The amount of reported cases per week.	

quality of the final Chinese version. Most of the items were measured using a five-point Likert-type scale, ranging from "strongly disagree" (1) to "strongly agree" (5).

As shown in Table 1, two sets of scales were developed in the current study. Among them, technology experience is a new construct introduced in our study and user performance is highly context-specific, so we found that the measurement items for these two construct cannot be easily adapted from existing literature. We developed the scale items for these two constructs through a pilot study among ten employees of SCC, who are the first group of people that experience the target system and were very familiar with it. Candidate items were proposed from discussions. Through a categorization-prioritization method (Davis, 1989), the items shown in Table 1 were finally selected for our survey.

Data Collection

In December 2007, SCC upgraded the software and hardware of the "Chengguantong" system to eliminate some technical obstacles for the users. The survey on the patrols' perceptions towards the mobile terminals was conducted one month later.

In total, 350 questionnaires were distributed and 194 responses were collected after two weeks. Eliminating the incomplete ones, 134 valid responses were used in data analysis, with a responding rate of 38.3 percent. The demographic information of valid respondents is shown in Table 2.

Reliability and Validity

The internal consistency reliability (ICR) was assessed by computing composite reliability coefficients (Fornell coefficients), of which values higher than 0.8 generally indicate acceptable reliability (Gefen & Straub, 2005). In our study, the ICR values range from 0.80 (for attitude) to 0.93 (for leader support). With none of the values

Table 2. Sample demographic information

	Amount	Percent
Gender		
Male	78	58.2%
Female	56	41.8%
Age		
20~29	12	9.0%
30~39	59	44.0%
40~49	54	40.3%
Elder than 50	9	6.7%
Education level		
Junior high school	41	30.6%
Senior high school	77	57.5%
College	16	11.9%

Table 3. Internal Consistency reliability (ICR)

	Att.	LS	TE	PF
ICR	0.80	0.93	0.88	0.88
Mean	3.84	4.15	3.54	3.95
St. Dev	0.81	0.80	0.99	0.84

for all 4 constructs less than 0.8 (see Table 3), the reliability of the scales can be accepted.

To assess discriminant validities, we examined item loadings to construct correlations (Gefen & Straub, 2005), as shown in Table 4. Convergent validity was evaluated by the average variance extracted (AVE). According to related studies, AVE values higher than 0.5 are acceptable (i.e., the square root of AVE is higher than 0.707). For a satisfactory degree of discriminant validity, the square root of AVE of a construct should exceed inter-construct correlations that reflect the variance shared between the construct and the other ones in the model (Gefen & Straub, 2005). In our research, as shown in Table 5, although some of the variables' inter-correlations are relatively high, convergent and discriminant validities of the model both attain a satisfying level, with all the AVE square root values above 0.8.

Common Method Bias Analysis

Because data on all independent and dependent variables were collected from the same source, there is a potential for common method biases. Following the statistical approach suggested by Podsakoff et al. (2003) and Liang et al. (2007), we used PLS to address this concern. Specifically, we included in the PLS model a common method factor whose indicators included all the principal constructs' indicators and calculated each indicator's variances substantively explained by the principal construct and by the method. As shown in Table 6, the results demonstrate that the average substantively explained variance of the indicators is 0.862, while the average method-based variance is -0.007. In addition, most method factor loadings are not significant. Given the small magnitude and insignificance of method variance, we contend that the method is unlikely to be a serious concern for this study.

RESULT AND DISCUSSION

We used the PLS-Graph software (version 3.0) for the analysis, utilizing the bootstrap re-sampling method (200 re-samples) to determine the significance of the paths within the structural model (Chin, Marcolin, & Newsted, 2003; Gefen & Straub, 2005). Figure 4 show the testing results of the proposed model and related hypotheses.

Although widely used in the IS field and some other research areas, Hofstede's theory has been criticized for its absoluteness (Ford et al., 2003; Gallivan & Srite, 2005). Different people often have different perceptions on some cultural factors because of their personal attributes, even when they are in the same cultural context (Karahanna et al., 2005). Following the method provided by Chin and colleagues (Chin et al., 2003), we try to explore this diversity by adding some moderators to the cultural model for acceptance, so as to examine their interaction effects on the relevant influence paths. The moderators considered in this study include gender, age, and education level.

For moderator analysis, we compared the values of the squared multiple correlation (R^2) for the interaction model and the "main effects" model, which exclude the moderators. The difference between the squared multiple correlations is used to access the overall effect size (f^2) for the interaction of the moderators, namely:

$$f^2=[R^2(\text{interaction model})- R^2 (\text{main effects model})/[1- R^2 (\text{main effects model})]$$

Interaction effect sizes are considered as small at the level of 0.02, medium at 0.15, and large at 0.35 (Chin et al., 2003). The testing results including the moderator are shown in Table 7.

The statistical results strongly support the impact of perceived fit on attitude, which is consistent the conclusion of previous research (Zhang et al., 2009). Moreover, none of the three mod-

Table 4. Items Loadings and Cross Loadings (CFA)

	PF	LS	TE	A
F1	**0.803**	0.576	0.308	0.493
F2	**0.873**	0.393	0.296	0.501
F3	**0.838**	0.419	0.421	0.511
LS1	0.434	**0.893**	0.292	0.435
LS2	0.453	**0.963**	0.282	0.510
TE1	0.377	0.309	**0.821**	0.371
TE2	0.333	0.276	**0.937**	0.463
A1	0.325	0.270	0.520	**0.728**
A2	0.511	0.522	0.358	**0.902**

Table 5. Average Variance Extracted (AVE)

	Att.	LS	TE	PF
Attitude	0.80			
Leader supp.	0.49	0.93		
Tech. exp.	0.48	0.36	0.89	
Perc. fit	0.61	0.54	0.39	0.84

Table 6. Common method bias analysis

Construct	Indicator	Substantive Factor Loading (R1)	R1²	Method Factor Loading (R2)	R2²
Perceived Fit	PF1	0.841**	0.707	0.013	0.000
	PF2	0.813**	0.661	-0.126	0.004
	PF3	0.857**	0.735	0.107	0.003
Leader Support	LS1	0.926**	0.858	-0.025	0.000
	LS2	0.929**	0.864	0.025	0.000
Technology Experience	TE1	0.883**	0.780	-0.005	0.000
	TE2	0.896**	0.803	0.004	0.000
Attitude	A1	0.761**	0.580	-0.275*	0.030
	A2	0.852**	0.726	0.222**	0.019
Average		0.862	0.746	-0.007	0.006

*p<.05; **p<0.01

erators show significant effect on this influence path. This result fully demonstrates that the factor of perceived fit, based on Confucian values and special element of *Hexie* in the Chinese culture, is a substantial determinant of Chinese users' attitude towards information systems.

The significant impact of leader support was also validated. An interesting phenomenon is that the degree of recognition on leader support of respondents increased with age. As a typical high power distance country, China has a long existing principle which is called "government by man". With the awakening consciousness about democracy in younger generations, however, even in the governmental organizations that

preserved more traditional culture characteristics than commercial companies, young and middle-aged employees have started to depend more on their own judgments rather than blindly follow the superior authorities. Such a change might be beneficial for overcoming the limitations of traditional highly centralized decision making mechanism in the process of diffusion of new technologies in organizations.

The testing result with regard to the impact of technology experience is opposite to our initial hypothesis in the model. The negative interacting effect of education level, however, shows that the path is more significant in the lower educated group. A critical challenge to the Hofstede's

Figure 4. Model testing results

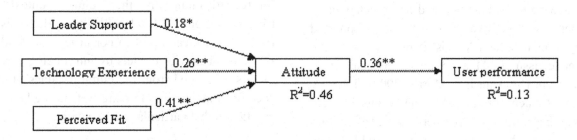

For determinants: *p<0.05, **p<0.01, ***p<0.001.

Table 7. PLS validating results

	Attitude		User Perf.	
	D only	D+1	D only	D+1
R²	0.46	0.49	0.13	0.15
Leader Supp.	0.18*	0.16		
Tech. Exp.	0.26**	0.28***		
Perceived Fit	0.41***	0.42***		
Attitude			0.36**	0.35**
LS*G		0.02		
LS*A		**0.17***		
LS*E		-0.06		
TE*G		-0.03		
TE*A		0.03		
TE*E		**-0.10***		
PF*G		0.04		
PF*A		0.09		
PF*E		0.01		
Att.*G				-0.15*
Att.*A				0.01
Att.*E				0.02

theory is that "whether the IBM employees could represent the culture of the whole country or not" (Gallivan & Srite, 2005). In the current study, most patrols were recruited from laid-off workers, and only 11 percent of them have experience of studying in colleges. Such a situation is obvious different from that of IBM employees that Hofstede has examined. Lack of knowledge often leads to estrangement from innovations.

We also found from the results that the path from attitude to user performance is significant, especially for men. The proposition that the relation between intention and usage behavior of men is stronger than woman has been supported by Venkatesh and Morris' famous work on the gender issue of IT adoption in voluntary environment (Venkatesh & Morris, 2000). To some extent, such similar validating results indicate that the "attitude → user performance" structure in the mandatory scenario is comparable to the "behavior intention → usage behavior" structure in the voluntary scenario.

CONCLUSION

In this paper, based on reviewing the existing literature on cultural issues in IT/IS adoption research, we argue that cultural factors should be treated as aggregate characteristics at the population level, instead of personal attributes at the individual level, and propose that it would be promising to postulate theoretical models for specific cultural contexts when examining IT/IS user behavior. Furthermore, we develop a novel model for analyzing user attitude towards mandatory use of information systems from the Chinese cultural perspective. Drawing upon the generally recognized cultural characteristics of China, which include long power distance, low uncertainty avoidance, and long-term orientation, three factors are introduced as determinants of user attitude, namely leader support, technology experience, and perceived fit. An empirical study is conducted with regard to the acceptance of a mobile municipal administration system in Beijing, China, for validating the proposed model with survey data and analyzing the adoption mechanism of the target system. In the light of a sample consisting of 134 valid responses, results from structured equation model (SEM) analysis illustrate the ability of the model to interpret the user attitude of Chinese users in mandatory contexts.

Our findings demonstrate that the proposed model is sufficiently valid for explaining individual's mandatory systems adoption and use behavior from a novel perspective in the Chinese cultural context, and may provide helpful insights for related studies in other cultural contexts. Meanwhile, our study also illustrates that it is feasible and meaningful to construct or extend IT/IS user behavior models by introducing new factors originated from the cultural values of a specific context, so as to better reflect the IT/IS practice in that particular environment.

ACKNOWLEDGMENT

The work was partly supported by the National Natural Science Foundation of China (70972029/70890081/70831003), China Postdoctoral Science Foundation (20080440030), and the Research Center for Contemporary Management of Tsinghua University. The authors contributed equally to this manuscript. The authors would like to thank Professors Guoqing Chen from Tsinghua University, Kai H. Lim from Hong Kong City University, and Patrick Y. K. Chau from Hong Kong University, for their valuable comments on the draft of this paper.

REFERENCES

Baskerville, R. F. (2003). Hofstede Never Studied Culture. *Accounting, Organizations and Society*, *28*(1), 1–14. doi:10.1016/S0361-3682(01)00048-4

Benbasat, I., & Barki, H. (2007). Quo Vadis, TAM? *Journal of AIS*, *8*(4), 211–218.

Brown, S. A., Massey, A. P., Montoya-Weiss, M. M., & Burkman, J. R. (2002). Do I really have to? User acceptance of mandated technology. *European Journal of Information Systems*, *11*(4), 283–295. doi:10.1057/palgrave.ejis.3000438

Chau, P. Y. K. (2008). Cultural Differences in Diffusion, Adoption, and Infusion of Web 2.0. *Journal of Global Information Management*, *16*(1), I–III.

Chau, P. Y. K., Cole, M., Massey, A. P., Montoya-Weiss, M., & O'Keefe, R. M. (2002). Cultural differences in the online behavior of consumers. *Communications of the ACM*, *45*(10), 138–143. doi:10.1145/570907.570911

Chen, G., Wu, R., & Guo, X. (2007). Key Issues in Information Systems Management in China. *Journal of Enterprise Information Management*, *20*(2), 198–208. doi:10.1108/17410390710725779

Chin, W. W., Marcolin, B. L., & Newsted, P. R. (2003). Partial Least Squares Latent Variables Modeling Approach for Measuring Interaction Effects: Results from a Monte Carlo Simulation Study and an Electronic-Mail Emotion/Adoption Study. *Information Systems Research*, *14*(2), 189–217. doi:10.1287/isre.14.2.189.16018

Compeau, D. R., & Higgins, C. A. (1995). Application of Social Cognitive Theory to Training for Computer Skills. *Information Systems Research*, *6*(2), 118–143. doi:10.1287/isre.6.2.118

Compeau, D. R., Higgins, C. A., & Huff, S. L. (1999). Social Cognitive Theory and Individual Reactions to Computing Technology: A Longitudinal Study. *Management Information Systems Quarterly*, *23*(2), 145–158. doi:10.2307/249749

Davis, F. D. (1989). Perceived Usefulness, Perceived Ease of Use, and User Acceptance of Information Technology. *Management Information Systems Quarterly*, *13*(3), 319–340. doi:10.2307/249008

Davis, F. D., Bagozzi, R. P., & Warshaw, P. R. (1989). User Acceptance of Computer Technology: a Comparison of Two Theoretical Models. *Management Science*, *35*(8), 982–1003. doi:10.1287/mnsc.35.8.982

Davison, R. M., & Martinsons, M. G. (2003). Cultural Issues and IT Management: Past and Present. *IEEE Transactions on Engineering Management*, *50*(1), 3–7. doi:10.1109/TEM.2003.808249

Evaristo, R. (2003). Cross-cultural research in IS. *Journal of Global Information Management*, *11*(4), i–iii.

Fang, T. (2003). A Critique of Hofstede's Fifth National Culture Dimension. *International Journal of Cross Cultural Management*, *3*(3), 347–368. doi:10.1177/1470595803003003006

Ford, D. P., Connelly, C. E., & Meister, D. B. (2003). Information Systems Research and Hofstede's Culture's Consequences: An Uneasy and Incomplete Partnership. *IEEE Transactions on Engineering Management, 50*(1), 8–25. doi:10.1109/TEM.2002.808265

Friedman, T. L. (2005). *The World Is Flat: A Brief History of the Twenty-first Century*. New York: Farrar, Straus and Giroux.

Gallivan, M., & Srite, M. (2005). Information technology and culture: Identifying fragmentary and holistic perspectives of culture. *Information and Organization, 15*(4), 295–338. doi:10.1016/j.infoandorg.2005.02.005

Gefen, D., & Straub, D. (2005). A Practical Guide to Factorial Validity Using PLS-Graph: Tutorial and Annotated Example. *Communications of AIS, 16*(1), 91–109.

Goodhue, D. L., & Thompson, R. L. (1995). Task-Technology Fit and Individual Performance. *Management Information Systems Quarterly, 19*(2), 213–226. doi:10.2307/249689

Guo, X., & Chen, G. (2005). Internet Diffusion in Chinese Companies. *Communications of the ACM, 48*(4), 54–58. doi:10.1145/1053291.1053318

Hartwick, J., & Barki, H. (1994). Explaining the Role of User Participation in Information System Use. *Management Science, 40*(4), 425–440. doi:10.1287/mnsc.40.4.440

Hempel, P. S., & Kwong, Y. K. (2001). B2B e-Commerce in emerging economies: i-metal.com's non-ferrous metals exchange in China. *The Journal of Strategic Information Systems, 10*(4), 335–355. doi:10.1016/S0963-8687(01)00058-0

Hofstede, G. (2001). *Culture's Consequences: Comparing Values, Behaviors, Institutions, and Organizations across Nations* (2nd ed.). Thousand Oaks, CA: Sage Publications.

Hofstede, G., & Bond, M. H. (1988). The Confucius Connection: From Cultural Roots to Economic Growth. *Organizational Dynamics, 16*(4), 5–21. doi:10.1016/0090-2616(88)90009-5

Huang, L., Lu, M., & Wong, B. K. (2003). The Impact of Power Distance on Email Acceptance: Evidence from the PRC. *Journal of Computer Information Systems, 44*(1), 93–101.

Jiacheng, W., Lu, L., & Francescob, C. A. (2009). A cognitive model of intra-organizational knowledge-sharing motivations in the view of cross-culture. *International Journal of Information Management*, 10–1016.

Karahanna, E., Evaristo, J. R., & Srite, M. (2005). Levels of Culture and Individual Behavior: An Integrative Perspective. *Journal of Global Information Management, 13*(2), 1–20.

Legris, P., Ingham, J., & Collerette, P. (2003). Why do people use information technology? A critical review of the technology acceptance model. *Information & Management, 40*(3), 191–204. doi:10.1016/S0378-7206(01)00143-4

Leidner, D. E., & Kayworth, T. (2006). A Review of Culture in Information Systems Research: Toward a Theory of Information Technology Culture Conflict. *Management Information Systems Quarterly, 30*(2), 357–399.

Lewis, W., Agarwal, R., & Sambamurthy, V. (2003). Sources of Influence on Beliefs about Information Technology Use: An Empirical Study of Knowledge Workers. *Management Information Systems Quarterly, 27*(4), 657–678.

Liang, H., Saraf, N., Hu, Q., & Xue, Y. (2007). Assimilation of enterprise systems: The effect of institutional pressures and the mediating role of top management. *Management Information Systems Quarterly, 31*(1), 59–87.

Lim, J. (2004). The Role of Power Distance and Explanation Facility in Online Bargaining Utilizing Software Agents. *Journal of Global Information Management, 12*(2), 27–43.

Lippert, S. K., & Volkmar, J. A. (2007). Cultural Effects on Technology Performance and Utilization: A Comparison of U.S. and Canadian Users. *Journal of Global Information Management, 15*(2), 56–90.

Mao, E., & Palvia, P. C. (2006). Testing an Extended Model of IT Acceptance in the Chinese Cultural Context. *The Data Base for Advances in Information Systems, 37*(2/3), 20–32.

Martinsons, M. G. (2005). Transforming China. *Communications of the ACM, 48*(4), 44–48. doi:10.1145/1053291.1053316

Mathieson, K. (1991). Predicting User Intentions: Comparing the Technology Acceptance Model with the Theory of Planned Behavior. *Information Systems Research, 2*(3), 173–191. doi:10.1287/isre.2.3.173

Mejias, R. J., Shepherd, M. M., Vogel, D. R., & Lazaneo, L. (1996). Consensus and perceived satisfaction levels: a cross-cultural comparison of GSS and non-GSS outcomes within and between the United States and Mexico. *Journal of Management Information Systems, 13*(3), 137–161.

Melone, N. P. (1990). A Theoretical Assessment of the User-Satisfaction Construct in Information Systems Research. *Management Science, 36*(1), 76–91. doi:10.1287/mnsc.36.1.76

Min, Q., Li, Y., & Ji, S. (2009, June 27-28). *The effects of individual-level culture on mobile commerce adoption: An empirical study.* Paper presented at the 2009 Eighth International Conference on Mobile Business, Liaoning, China.

Moore, G. C., & Benbasat, I. (1991). Development of an instrument to measure the perceptions of adopting an information technology innovation. *Information Systems Research, 2*(3), 192–222. doi:10.1287/isre.2.3.192

Morgan, D. L. (1997). *Focus Group as Qualitative Research.* Thousand Oaks, CA: Sage Publications.

Myers, M. D., & Tan, F. B. (2003). Beyond Models of National Culture in Information Systems Research. *Journal of Global Information Management, 10*(2), 14–29.

Nah, F. F., Tan, X., & Teh, S. H. (2004). An Empirical Investigation on End-Users' Acceptance of Enterprise Systems. *Information Resources Management Journal, 17*(3), 32–53.

Plouffe, C. R., Hulland, J. S., & Vandenbosch, M. (2001). Richness versus Parsimony in Modeling Technology Adoption Decisions--Understanding Merchant Adoption of a Smart Card-Based Payment System. *Information Systems Research, 12*(2), 208–222. doi:10.1287/isre.12.2.208.9697

Podsakoff, P. M., MacKenzie, S. B., Jeong-Yeon, L., & Podsakoff, N. P. (2003). Common Method Biases in Behavioral Research: A Critical Review of the Literature and Recommended Remedies. *The Journal of Applied Psychology, 88*(5), 879. doi:10.1037/0021-9010.88.5.879

Rogers, E. M. (1995). *Diffusion of Innovations* (4th ed.). New York: The Free Press.

Srite, M., Thatcher, J. B., & Galy, E. (2008). Does Within-Culture Variation Matter? An Empirical Study of Computer Usage. *Journal of Global Information Management, 16*(1), 1–25.

Straub, D. (1994). The effect of culture on IT diffusion: E-mail and FAX in Japan and the U.S. *Information Systems Research, 5*(1), 23–47. doi:10.1287/isre.5.1.23

Straub, D., Keil, M., & Brenner, W. (1997). Testing the Technology Acceptance Model Across Cultures: A Three Country Study. *Information & Management, 33*(1), 1–11. doi:10.1016/S0378-7206(97)00026-8

Straub, D., Loch, K., Evaristo, R., Karahanna, E., & Srite, M. (2002). Toward a Theory-Based Measurement of Culture. *Journal of Global Information Management, 10*(1), 13–23.

Taylor, S., & Todd, P. A. (1995). Understanding Information Technology Usage: A Test of Competing Models. *Information Systems Research, 6*(2), 144–176. doi:10.1287/isre.6.2.144

Tsui, A. S. (2006). Contextualization in Chinese Management Research. *Management and Organization Review, 2*(1), 1–13. doi:10.1111/j.1740-8784.2006.00033.x

Venkatesh, V., & Davis, F. D. (2000). A Theoretical Extension of the Technology Acceptance Model: Four Longitudinal Field Studies. *Management Science, 46*(2), 186–204. doi:10.1287/mnsc.46.2.186.11926

Venkatesh, V., Davis, F. D., & Morris, M. G. (2007). Dead or Alive? The Development, Trajectory and Future of Technology Adoption Research. *Journal of AIS, 8*(4), 267–286.

Venkatesh, V., & Morris, M. G. (2000). Why Don't Men Ever Stop To Ask For Directions? Gender, Social Influence, and Their Role in Technology Acceptance and Usage Behavior. *Management Information Systems Quarterly, 24*(1), 115–239. doi:10.2307/3250981

Venkatesh, V., Morris, M. G., Davis, G. B., & Davis, F. D. (2003). User Acceptance of Information Technology: Toward a Unified View. *Management Information Systems Quarterly, 27*(3), 425–478.

Xi, Y., & Wang, D. (2006, 0507-10-01). *Harmony configuration and Hexie management: An empirical examination of Chinese.* Paper presented at the 13th International Conference on Management Science and Engineering.

Zhang, N., Guo, X., & Chen, G. (2006). *Extended Initial Technology Acceptance Model and Its Empirical Test.* Paper presented at the Fifth Wuhan International Conference on E-Business (WHICEB2006), Wuhan, China.

Zhang, N., Guo, X., Chen, G., & Chau, P. Y. K. (2009). Impact of Perceived Fit on e-Government User Evaluation: A Study with a Chinese Cultural Context. *Journal of Global Information Management, 17*(1), 49–69.

Zhang, N., Guo, X., Chen, G., & Song, G. (2008). *The Cultural Perspective of Mobile Government Terminal Acceptance – An Exploratory Study in China.* Paper presented at the 12th Pacific Asia Conference on Information Systems (PACIS 2008), Suzhou, China.

Zhang, N., Guo, X., Chen, G., & Song, G. (2009). *A MCT Acceptance Model from the Cultural Perspective and Its Empirical Test in the Mobile Municipal Administrative System Application.* Paper presented at the Eighth International Conference on Mobile Business.

This work was previously published in International Journal of Global Information Management, Volume 18, Issue 4, edited by Felix B. Tan, pp. 1-18, copyright 2010 by IGI Publishing (an imprint of IGI Global).

Chapter 2
The Impact of National Culture on Information Systems Planning Autonomy

Dinesh A. Mirchandani
University of Missouri - St. Louis, USA

Albert L. Lederer
University of Kentucky, USA

ABSTRACT

Hofstede's national culture model has been applied in prior research to better understand the management of multinational firms. That research suggests that national culture may influence the information systems planning autonomy of the subsidiaries of multinational firms, but such an impact has not yet been tested empirically. A postal survey of 131 chief information officers and 103 senior non-IS managers of U.S. subsidiaries of such firms collected data to test hypotheses based on the model. Structural equation modeling using PLS-Graph 3.0 revealed that Individualism-Collectivism, Masculinity-Femininity, and Uncertainty Avoidance predicted autonomy for particular IS planning phases (as rated by the CIOs). On the basis of the supported hypotheses, the study provides evidence of the relevance of the national culture model to IS planning effectiveness and IS contribution. The study also suggests to subsidiary managers that an understanding of the national culture of their parent firm can help them gain an insight into the parent's management perspective.

INTRODUCTION

Fifty-three of the 100 largest "economies" in the world today are not nations, but instead are multinational corporations. Of the more than 60,000 such corporations, one thousand account for 80% of the world's industrial production. These firms employ 90 million people, pay more than $1.5 trillion in wages, and contribute 25% of the gross world product (Gabel & Bruner, 2003).

However, each country in which a multinational firm operates has a distinct economic, political, legal, industrial, competitive market, and cultural context. Businesses must be responsive to this diversity, especially when integrating their different operations (Akmanligil & Palvia, 2004; Makino, Isobe, & Chan, 2004; Martinsons & Davison, 2007). Today's growing globalization, with the corporate parent in one culture and the subsidiaries in another, is increasing the need to

DOI: 10.4018/978-1-61350-480-2.ch002

manage the impact of culture on organizational choices (Kankanhalli et al., 2004). Likewise, today's growing dependence on information systems is increasing the need to make the choices that provide the systems that contribute the most to the organization (Grover & Segars, 2005). The consequences of choosing information systems that fail to meet needs are too great. Not surprisingly therefore, both culture and information systems planning have been a subject of interest to researchers (Mohdzain & Ward, 2007).

One of the most widely used models of culture was developed by Hofstede (1980, 1984). According to him, all people carry "patterns of thinking, feeling and potential acting which were learned throughout their lifetime" (Hofstede, 1991, p. 4). Hofstede analyzed data on the individual personal values of 116,000 IBM employees in 72 countries spanning more than 20 languages, and he identified four major dimensions of culture. He called the dimensions Power Distance, Individualism-Collectivism, Uncertainty Avoidance, and Masculinity-Femininity.[1]

Several researchers have studied the impact of national culture in information systems. They have done so in terms of the impact on users, information systems developers, and information systems management. *User research* has included user attitudes (Igbaria & Zviran, 1991; Igbaria, 1992; Harris & Davison, 1999); group support systems usage (Watson et al., 1994); electronic communication media selection (Straub, 1994); adaptation and usage characteristics (Igbaria & Zviran, 1996); self-efficacy (Igbaria & Iivari, 1995); human behavior in information systems usage (Walczuch et al., 1995; Burn et al., 1993; Burn et al., 1997; Gefen & Straub, 1997; Robichaux & Cooper, 1998; Davison & Jordan, 1998; Hanke & Teo, 2003; Myers & Tan, 2002; Straub et al., 2002: Tan et al., 1998; Smith, 2001); e-commerce buying behavior (Chai & Pavlou, 2004; Pavlou & Chai, 2002; Pereira, 1998); online auction (Ho et al., 2007); performance (Lippert & Volkmar, 2007); information overload (Kock et al., 2008); and

satisfaction (Reinig et al., 2009). *Information systems developer* research has included perceptions of system usability (Del Galdo & Nielson, 1996); icon and interface effectiveness (Ito & Nakakoji, 1996; Nakakoji, 1996; Sturrock & Kirwan, 1996; McDougall et al., 1998); technical, political, and social values of IS developers (Kankanhalli et al., 2004); cross-border ecommerce (Sinkovics et al., 2007); and knowledge communication (Siau et al., 2007). *Information systems management* research has included information privacy (Milberg et al., 2000); software project management (Tan et al., 2003); and outsourcing (Ramingwong & Sajeev, 2007). Thus, the IS literature has seen an abundance of cross-cultural research (Ford et al., 2003).

The current study sought to extend this prior research to information systems planning in multinational firms. An understanding of the impact of national culture on such planning has emerged as a pressing need of IS management in such firms (Ford, Connelly, & Meister, 2003; Palvia, 1998; Palvia, Palvia, & Whitworth, 2002). An important dimension of IS planning is the autonomy that parent firms grant their subsidiaries.

Such autonomy remains critical to general management, that is, to planning and control, with a wide range of it afforded to the subsidiary in practice and sophistication (Hoffman, 1988; Young & Tavares, 2004). The decision to centralize or decentralize (i.e., withhold or grant autonomy to) the management and processing of information is a key strategic issue in multinational firms (Deans et al., 1991; Anthes, 2001), but the study of autonomy in IS planning has been neglected (Peppard, 1999).

Autonomy in IS planning at the subsidiary level enables the subsidiary to choose the information systems that help it meet its business objectives and thereby contribute to overall corporate needs whereas the lack of autonomy for the subsidiary may force it to implement systems that fail to meet those needs, waste resources, and miss opportunities (Mohdzain & Ward, 2007). National culture might influence the willingness of the parent to grant autonomy to the subsidiary and thereby might influence the use of information systems. Thus, for example, if a culture high on a particular dimension grants less autonomy and if autonomy engenders IS planning effectiveness, then an organization in such a culture might benefit from understanding how that culture is impeding its use of information systems.

Because, moreover, the national culture of subsidiary managers typically differs from that of their parent, the parent might grant a different degree of IS planning autonomy to the subsidiary than expected or desired in the subsidiary culture. The consequences of the lack of autonomy at the subsidiary are that local business requirements are not adequately addressed (Mohdzain & Ward, 2007). According to case studies of the subsidiaries of multinational organizations with varying levels of IS planning autonomy, "both IT and business managers believe their IS do not meet their business needs and thus are not satisfactory" in those with less autonomy (Mohdzain & Ward,

2007, p. 324). The production planning manager in one of the organizations noted, "The way that we are handling or using the software planned by our corporate headquarters … we need to suit our requirements with the available systems, not the other way around" (p. 339). However, research has yet to investigate the impact of national culture on the granting of subsidiary IS planning autonomy, and the granting of subsidiary IS planning autonomy on IS planning effectiveness.

The purpose of this research was to assess the link between national culture (in terms of Hofstede's (1980, 1984, 1991) dimensions) and the autonomy of subsidiary managers for IS planning, the link between such autonomy and IS planning effectiveness, and the link between IS planning effectiveness and IS contribution. Little understanding exists to date about the impact of national culture on subsidiary information systems planning autonomy, and the impact of such autonomy on planning effectiveness, impacts investigated herein. Because IS planning may be

Figure 1. The research model

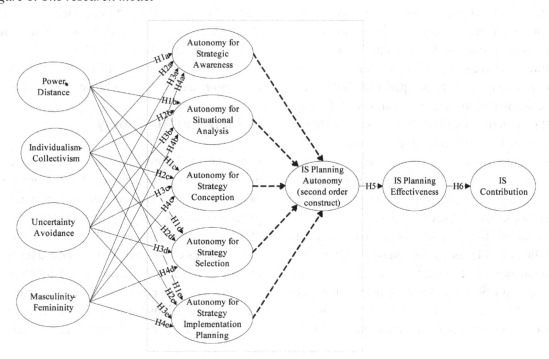

conceptualized as comprised of five phases, i.e., *strategic awareness*, *situational analysis*, *strategy conception*, *strategy selection*, and *strategy implementation planning*, the influence of the culture dimensions on each phase can provide a richer understanding of the role of culture. Figure 1 shows the research model.

The next section describes the research constructs. Subsequent sections identify the hypotheses, research methodology, and findings. The paper concludes with a discussion of the findings and implications for research and practice.

RESEARCH CONSTRUCTS

Hofstede's Indices for National Culture of the Parent Firm[2]

Hofstede (1991) identified four major dimensions of national culture and developed indices to measure each. The four were Power Distance, Individualism-Collectivism, Uncertainty Avoidance, and Masculinity-Femininity. Although the four may be correlated (Hofstede, 1980), management researchers have extensively employed them to better understand differences in the impact of national culture on work-related behavior.

Power Distance: This dimension describes how a society accepts unequal distribution of power in institutions and organizations (Hofstede, 1991). Societies with high Power Distance have pronounced inequalities of power and wealth within them. Citizens of such societies are subservient to those who have more power, and are also comfortable with the power differential. These societies are thus more likely to follow a caste system that disallows significant upward mobility of citizens and emphasizes hierarchy in organizations (Teboul, Chen, & Fritz, 1994). Hofstede's (1991) score for this dimension ranges from 11 to 104, with a higher score representing a more developed Power Distance in the culture.

Individualism-Collectivism: This dimension describes how people in a culture relate to each other. People from Individualistic cultures live in loosely knit societies. They care primarily for themselves and their immediate families while individuals from Collectivist cultures look after the interest of their group before that of themselves. The Hofstede score for this dimension ranges from 8 to 112, with a higher score representing a more individualistic culture.

Uncertainty Avoidance: This dimension describes how a society accommodates uncertainty and ambiguity in the environment. Individuals from cultures with high Uncertainty Avoidance are more risk averse, prefer greater structure in every situation, and are more conscious of societal norms, while individuals from cultures with low Uncertainty Avoidance are more risk taking and open to deviations from such norms. The Hofstede score for this dimension ranges from 6 to 91, with a higher score representing lower inclination to take risk and more regulation in the culture.

Masculinity-Femininity: This dimension describes how gender roles are allocated in the culture. Individuals from Masculine cultures value material success and assertiveness while individuals from Feminine cultures stress relationships and concern for others. The Hofstede score for this dimension ranges from 5 to 95, with a higher score representing a more Masculine culture.

IS Planning Autonomy

IS planning is the process of assessing the external and internal environments to identify new computer applications that support the business objectives of the organization (Premkumar & King, 1994; Segars & Grover, 1998). Such planning also includes the specification of new databases, telecommunications facilities, and information systems skills and positions to create and support the new applications. It further entails formulating IS objectives, defining strategies and policies to achieve them, and developing detailed plans to achieve the objectives (Teo et al., 1997). Mentzas (1997) described it in terms of the five aforementioned phases of activities (*strategic awareness, situational analysis, strategy conception, strategy selection*, and *strategy implementation planning*). These phases have been widely accepted in strategic management (Thompson, 1967) and marketing (Cravens, 1988), and have also been applied in IS research (Newkirk, Lederer, & Srinivasan, 2003).

The objective of the *strategic awareness* phase is to increase management commitment to the planning process and to provide initial answers to major questions about the objectives of the organization and the issues facing it. The phase thus provides direction for the overall planning process.

The major purposes of the *situational analysis* phase are to arrive at a clear and documented diagnosis of the existing business and IT conditions in the organization, to identify problems and inefficiencies, and to understand the internal and external environments. The phase synthesizes a range of views on the strengths and weaknesses of the organization in the context of its environment, as well as on the management and planning of its information resources.

The goal of *strategy conception*, the third phase of the process, is to envision alternative future uses of information systems. It thus includes the identification of opportunities for competitive and performance advantages, and of scenarios for future growth.

The purpose of *strategy selection* is to choose new systems and processes. The phase includes specifications for the development and maintenance of data and applications, as well as for the architecture to be implemented. *Strategy selection* thus elucidates the functional, organizational, and technical models.

The objective of *strategy implementation planning* is to increase the likelihood of carrying out the plan. The phase includes the definition of concrete actions, the evaluation of budgetary requirements, and the study of time and organizational constraints. It also concerns change management, plan coordination, and migration.

Autonomy in IS planning refers to the extent of independence in choosing or adjusting new information systems to fit the needs of the subsidiary. Although autonomy across strategic business units influences the effectiveness of business strategies by motivating managers to be more committed and perform more diligently (Gupta, 1987), freedom can never be absolute. In the case of IS planning autonomy, constraints include the need to maintain security; to control costs via economies of scale while avoiding duplication of effort, equipment, and training; and to provide sufficient compatibility for consolidated financial reporting by the parent. Thus IT applications cannot be designed completely independent of corporate oversight. However, similarly to end user development planning, within the constraints, it would lead to more subsidiary effectiveness. Autonomy for the subsidiary acknowledges that what may be great for the parent may be suboptimal for organization overall performance, while what may be great for the subsidiary may likewise be suboptimal for that performance. After all, parent policies, which make local financial reporting more difficult, can impede local efficiency and effectiveness, and can reduce the effectiveness of information systems planning and thus the IS contribution.

Doz and Prahalad (1984) illustrated the problem of excessive parent control of information systems planning and the need for greater subsidiary autonomy when they wrote of Xerox, saying "… while the focus of strategy making was Xerox's world headquarters in the U.S., the design and blending of management tools [e.g., data management tools] was partly delegated to the European headquarters. Concerned with the excessive autonomy of subsidiary managers, corporate executives may have overlooked the needs for responsiveness and let more nimble competitors, with competitive strategies more differentiated from country to country, capture larger shares of the reproduction equipment market than they otherwise could have" (Doz & Prahalad, 1984, p. 66).

The five phases and their major activities appear in Table 1. They were chosen to measure IS planning autonomy in the current study because they represent a contemporary, succinct and comprehensive description of the IS planning process; because they had been used in prior research (Newkirk, Lederer, & Srinivasan, 2003); and because the authors found no other such comparable detailed measure of the IS planning process for which autonomy could be assessed.

IS Planning Effectiveness

The effectiveness of IS planning can be assessed in terms the extent that the goals of information systems in the organization are met (Doherty, Marples, & Suhaimi, 1999). That effectiveness has typically been operationalized as either a single- or multi-factor construct in terms of the achievement of planning objectives (Raghunathan & Raghunathan, 1991, 1994; Segars & Grover, 1998, 1999). One study meticulously validated a single factor of nine planning objectives (Raghunathan & Raghunathan, 1991, 1994). These objectives and explanations of them appear in Table 2. Due to their rigorous validation, brevity, usage in an empirical study by Raghunathan and Raghunathan (1991), and adaptability to fit into the lengthy survey in the current study, they were used in this study to measure IS planning effectiveness.

IS Contribution

Commonly cited indicators of IS contribution are the alignment of IS plans with business plans (King, 1988) and the competitive advantage provided by IS to the organization (Ives & Learmonth, 1984; McFarlan, 1984; Porter & Millar, 1985; Wiseman, 1985). This study therefore used two items to measure IS contribution, i.e., the extent

Table 1. IS planning phases and their activities

Phases	Activities
Strategic Awareness	Determining key planning issues Defining planning objectives Organizing the planning team(s) Obtaining top management commitment
Situational Analysis	Analyzing current business systems Analyzing current organizational systems Analyzing current information systems Analyzing the current external business environment Analyzing the current external IT environment
Strategy Conception	Identifying major IT objectives Identifying opportunities for improvement Evaluating opportunities for improvement Identifying high level IT strategies
Strategy Selection	Identifying new business processes Identifying new IT architectures Identifying specific new projects Identifying priorities for new projects
Strategy Implementation Planning	Defining change management approaches Defining action plans Evaluating action plans Defining follow-up and control procedures

to which the organization has been able to "Align information technology with business needs" and "Gain a competitive advantage from information technology" (Lederer & Sethi, 1996).

HYPOTHESES

In high Power Distance cultures, managers expect to use the privileges inherent in their position to give orders. Such managers work actively to maintain their status and control over others because the use of a participative leadership style might be viewed as a sign of incompetence or irresponsibility. For example, in terms of writing memos, running meetings, interviewing candidates and

Table 2. Objectives of IS planning

Objectives of IS Planning	Rationale
Prediction of future trends	While operating in a rapidly changing technological environment, predicting future trends can be crucial in minimizing costs such as those caused by obsolescence. Planning has been justified as a formal mechanism to prepare an organization to cope with changes.
Improved short-term IS performance	One of the main reasons for undertaking IS planning is to improve IS performance. The IS plan can also serve as a control mechanism in that it furnishes goals and objectives for later evaluation of performance.
Improved long-term IS performance	The objective for undertaking IS planning is to improve IS performance. In addition to the goal of improving performance, the IS plan can also serve as a control mechanism in that it furnishes goals and objectives for later evaluation of performance.
Improved decision making	Planning for information systems includes the objective of higher levels of use by managers. The resulting improvement in managerial decisions can be regarded as a measure of the effectiveness of IS.
Avoidance of problem areas	Effective planning systems are adaptive learning systems. Thus an effective planning system should enable an organization to avoid problem areas and minimize the recurrence of errors.
Increased user satisfaction	User satisfaction has been used as a measure of IS success and hence can be a measure of IS effectiveness.
Improved systems integration	The systems planning phase of IS planning is directed towards integrating various individual systems and ensuring the compatibility of hardware and software. Therefore the level of achievement of systems integration can be considered a measure of effectiveness.
Improved resource allocation	A well conceived IS plan should facilitate optimal allocation of resources among competing IS projects.
Enhanced management development	The IS plan should facilitate management development.

preparing presentations, these managers control the content of their communication without the involvement of subordinates (Munter, 1993). In such cultures, subordinates too tend to be uncomfortable with a management-by-objectives system or other forms of participative goal setting (Kreitner & Kinicki, 2001; Luthans, 2002).

On the other hand, in low Power Distance cultures, managers only rarely give signals of their status and employees normally do not feel inferior to them. These managers try to minimize the differences between themselves and their subordinates by delegating and sharing power to the greatest extent possible. For example, they reward subordinates for taking personal initiative. In such cultures, organizations tend to be non-hierarchic, subordinates refer to managers by their first name at work, and more social interactions exist between them (Kreitner & Kinicki, 2001; Luthans, 2002).

As examples of national cultures, managers in Russia and China (with high Power Distance scores) expect their subordinates to accept having very little power, and do not grant much power to them. In contrast, managers in Austria and Denmark (with lower Power Distance scores) expect their subordinates to want more power, and they grant it to them.

Thus, in multinational firms from higher Power Distance cultures, parent managers would be expected to grant less autonomy to their subsidiary managers than would those in firms from lower Power Distance cultures (Sherer, 2007). Such granting and withholding of autonomy would pervade the parent firm, and include autonomy in all information systems planning phases.

For example, in the strategic awareness phase of planning, parent managers in multinationals from higher power distance cultures would expect to use the privileges inherent in their position to give orders. The would view a participative leadership style as a sign of incompetence or irresponsibility, expect their subordinates to accept having very little power, and not grant much autonomy to those subordinates in determining key planning issues, defining planning objectives, organizing the planning team(s), and obtaining top management commitment. They would expect subordinates to accept parent-defined planning objectives, would communicate their instructions unambiguously, and demand subordinates follow those instructions. Hence, we hypothesize:

H1a: The higher the Power Distance score of the national culture of the parent company, the lower the subsidiary manager's autonomy for information systems planning for strategic awareness.

For the situational analysis phase, parent managers from multinationals from higher power distance cultures would likewise grant less autonomy to their subsidiaries in analyzing their current business systems, organizational systems, information systems, external business environment, and external IT environment. They would expect subordinates to accept their authority, and for example, accept greater direction in their analysis of current information systems. Hence:

H1b: The higher the Power Distance score of the national culture of the parent company, the lower the subsidiary manager's autonomy for information systems planning for situational analysis.

For the strategy conception phase, parent managers in multinationals from higher power distance cultures would grant less autonomy to their subsidiaries in identifying major IT objectives, identifying opportunities for improvement, evaluating opportunities for improvement, and identifying high level IT strategies. They would expect the subsidiaries to, for example, be more accepting of major IT objectives proposed by the parent. Hence:

H1c: The higher the Power Distance score of the national culture of the parent company, the lower the subsidiary manager's autonomy for information systems planning for strategy conception.

For the strategy selection phase, parent managers of multinationals from higher power distance cultures would grant less autonomy to their subsidiaries in identifying new business processes, IT architectures, projects, and project priorities. They would thus, for example, give close guidance and supervision to the subsidiary in its identification of its specific new projects. Hence:

H1d: The higher the Power Distance score of the national culture of the parent company, the lower the subsidiary manager's autonomy for information systems planning for strategy selection.

For the strategy implementation phase, parent managers in multinationals from higher power distance cultures would grant less autonomy to their subsidiaries in defining change management approaches, defining action plans, evaluating action plans, and defining follow-up and control procedures. They would thus closely supervise subsidiary strategy implementation planning. Hence:

H1e: The higher the Power Distance score of the national culture of the parent company, the lower the subsidiary manager's autonomy for information systems planning for strategy implementation planning.

In a society high on Individualism-Collectivism, ties between individuals are loose. People emphasize looking after themselves and their own immediate families. In the workplace, they act in their own economic best interest and therefore employer-employee relationships are merely viewed as transactions in a labor market. Individualists perceive themselves as relatively free to follow their own wishes regardless of the views of others (Pavlou & Chai, 2002). Hence, managers from such cultures are self-reliant and independent, and thus fairly demanding that subordinates adhere to their directives.

In a society low on Individualism-Collectivism, people from birth onwards are integrated into strong, cohesive groups, often extended families, who continue protecting them in exchange for unquestioning loyalty. In the workplace, relationships prevail over tasks and therefore employers and employees view their relationship as a moral contract. Managers from such cultures are more likely to follow group consensus, have favorable attitudes toward teamwork, and prefer reward systems that provide incentives for group achievement. Hence, these managers are participatory and egalitarian, and thus less demanding that subordinates adhere to their directives.

Managers in Canada and the United Kingdom, with high Individualism-Collectivism scores, focus on achieving their own objectives. In contrast, managers in Taiwan and Malaysia, with lower Individualism-Collectivism scores, focus on the interests of their subordinates.

Thus, in multinational firms from Individualistic cultures, parent managers would be expected to grant less autonomy to their foreign subsidiary managers than would those in firms from Collectivist cultures. Such granting of autonomy would pervade the parent firm, and include autonomy for IS planning. For example, in the strategic awareness phase, parent managers in multinationals from high Individualism-Collectivism cultures would disrespect the independence of the subsidiary, and grant less autonomy to the subsidiary management in determining key planning issues, defining planning objectives, organizing the planning team(s), and obtaining top management commitment. Hence, we hypothesize:

H2a: The higher the Individualism-Collectivism score of the national culture of the parent company, the lower the subsidiary manager's autonomy for information systems planning for strategic awareness.

For the situational analysis phase, parent managers in multinationals from higher Individualism-Collectivism cultures, in disrespecting the independence of the subsidiary, would grant less autonomy in analyzing their current business systems, organizational systems, information systems, external business environment, and external IT environment. Hence:

H2b: The higher the Individualism-Collectivism score of the national culture of the parent company, the lower the subsidiary manager's autonomy for information systems planning for situational analysis.

For the strategy conception phase, parent managers in multinationals from higher Individualism-Collectivism cultures would disrespect the subsidiary's independence, and grant less autonomy in identifying major IT objectives, identifying opportunities for improvement, evaluating opportunities for improvement, and identifying high level IT strategies. Hence:

H2c: The higher the Individualism-Collectivism score of the national culture of the parent company, the lower the subsidiary manager's autonomy for information systems planning for strategy conception.

For the strategy selection phase, parent managers in multinationals from higher Individualism-Collectivism cultures would grant less autonomy to their subsidiaries in identifying new business processes, IT architectures, projects, and project priorities. Hence:

H2d: The higher the Individualism-Collectivism score of the national culture of the parent company, the lower the subsidiary manager's autonomy for information systems planning for strategy selection.

Finally, for the strategy implementation phase, parent managers in multinationals from higher Individualism-Collectivism cultures would grant less autonomy to their subsidiaries in defining change management approaches, defining action plans, evaluating action plans, and defining follow-up and control procedures. Hence:

H2e: The higher the Individualism-Collectivism score of the national culture of the parent company, the lower the subsidiary manager's autonomy for information systems planning for strategy implementation planning.

Uncertainty Avoidance refers to the extent to which people within a culture are uncomfortable with situations they perceive as unstructured, unclear, or unpredictable. It indicates the extent to which members of a culture feel threatened by uncertain and ambiguous situations, and attempt to avoid those situations by establishing more formal rules, forbidding deviant behaviors, and espousing absolute truths. The predisposition of such members to avoid uncertainty profoundly affects organizational management (Straub, 1994). Managers from high Uncertainty Avoidance cultures thus instill strict laws, formal reporting rules, safety, and security measures in the organization, and they expect subordinates to follow their plans.

Uncertainty accepting cultures try to have as few rules as possible, and are relativist on philosophical and religious issues. People within these cultures are more phlegmatic and contemplative, and not expected to express emotions. Managers from such cultures are more likely to allow subordinates to make independent decisions.

Managers in Japan and Greece, with high Uncertainty Avoidance scores, are more reluctant to tolerate any deviations from their rules. In contrast, managers in Denmark and Sweden (with lower Uncertainty Avoidance scores) are less reluctant to tolerate such deviations by their subordinates.

Thus, in multinational firms from cultures high in Uncertainty Avoidance, parent managers would be expected to grant less autonomy to their subsidiary managers than would those in firms from uncertainty accepting cultures not only in general management, but also for the phases of information systems planning.

For example, in the strategic awareness phase, parent managers in multinationals from high Uncertainty Avoidance cultures would feel uncomfortable with foreign cultures which they would perceive as unstructured, unclear, or unpredictable. They would feel threatened by uncertain and ambiguous situations in those cultures, and try to reduce the uncertainty and ambiguity by establishing more rules and other controls. They would thus grant less autonomy to subsidiary management in determining key planning issues, defining planning objectives, organizing the planning team(s), and obtaining top management commitment. Hence, we hypothesize:

H3a: The higher the Uncertainty Avoidance score of the national culture of the parent company, the lower the subsidiary manager's autonomy for information systems planning for strategic awareness.

For the situational analysis phase, parent managers in multinationals from more Uncertainty Avoidance cultures, in reducing uncertainty and ambiguity, would grant less autonomy in analyzing their current business systems, organizational systems, information systems, external business environment, and external IT environment. Hence:

H3b: The higher the Uncertainty Avoidance score of the national culture of the parent company, the lower the subsidiary manager's autonomy for information systems planning for situational analysis.

For the strategy conception phase, parent managers in multinationals from more Uncertainty Avoidance cultures would reduce uncertainty, and grant less autonomy in identifying major IT objectives, identifying opportunities for improvement, evaluating opportunities for improvement, and identifying high level IT strategies. Hence:

H3c: The higher the Uncertainty Avoidance score of the national culture of the parent company, the lower the subsidiary manager's autonomy for information systems planning for strategy conception.

For the strategy selection phase, parent managers in multinationals from more Uncertainty Avoidance cultures would grant less autonomy to their subsidiaries in identifying new business processes, new IT architectures, specific new projects, and priorities for new projects. Hence:

H3d: The higher the Uncertainty Avoidance score of the national culture of the parent company, the lower the subsidiary manager's autonomy for information systems planning for strategy selection.

Finally, for the strategy implementation phase, parent managers in multinationals from more Uncertainty Avoidance cultures would grant less autonomy to their subsidiaries in defining change management approaches, defining action plans, evaluating action plans, defining follow-up and control procedures. Hence:

H3e: The higher the Uncertainty Avoidance score of the national culture of the parent company, the lower the subsidiary manager's autonomy for information systems planning for strategy implementation planning.

In cultures high in Masculinity-Femininity, men are expected to be assertive and focused on material needs while women are expected to be tender, modest, and concerned about the quality of life. These cultures value achievement and the aggressive pursuit of goals over good relationships, harmony, and care for people. In the workplace, managers expect their subordinates to be decisive, assertive, and independent, and to resolve conflicts by fighting them out.

In cultures low in Masculinity-Femininity, both men and women are expected to be tender, modest, and concerned about quality of life. These cultures are nurturing and value relationships. In the workplace, managers depend on their intuition and strive for consensus within the organization. They believe in using compromise and negotiation to resolve conflicts.

Managers in Japan and Switzerland (with high Masculinity-Femininity scores) focus on organizational performance and results, and expect their subordinates to use any means to achieve them. In contrast, managers in Denmark and Sweden (with lower Masculinity-Femininity scores) focus on quality of life in the workplace.

Thus, in multinational firms from Masculine cultures, parent managers would be expected to grant more autonomy to their subsidiary managers than would those in firms from Feminine cultures. For example, in the strategic awareness phase of planning, multinationals from higher Masculinity-Femininity cultures would value achievement and the aggressive pursuit of goals, and expect their subordinates to be decisive, assertive, and independent. They would grant more autonomy to their subsidiaries in determining key planning issues, defining planning objectives, organizing the planning team(s), and obtaining top management commitment. They would expect subordinates to act decisively, assertively, and independently in determining planning objectives, would communicate their instructions unambiguously, and demand subordinates follow that instruction. Hence, we hypothesize:

H4a: The higher the Masculinity-Femininity score of the national culture of the parent company, the higher the subsidiary manager's autonomy for information systems planning for strategic awareness.

For the situational analysis phase, parent managers in multinationals from more Masculine cultures would grant more autonomy to their subsidiaries in analyzing their current business systems, organizational systems, information systems, external business environment, and external IT environment. They would expect subordinates act more decisively, and expect them to do so when they, for example investigate and evaluate their current information systems. Hence:

H4b: The higher the Masculinity-Femininity score of the national culture of the parent company, the higher the subsidiary manager's autonomy for information systems planning for situational analysis.

For the strategy conception phase, parent managers in multinationals from more Masculine cultures would grant more autonomy to their subsidiaries in identifying major IT objectives, identifying opportunities for improvement, evaluating opportunities for improvement, and identifying high level IT strategies. They would expect the subsidiaries to, for example, identify major IT objectives consistent with initiative, hence would give grant more autonomy to the subsidiary in its identification of those objectives. Hence:

H4c: The higher the Masculinity-Femininity score of the national culture of the parent company, the higher the subsidiary manager's autonomy for information systems planning for strategy conception.

For the strategy selection phase, parent managers in multinationals from more Masculine cultures would grant more autonomy to their subsidiaries in identifying new business processes, IT architectures, projects, and project priorities. They would thus, for example, give greater guidance and supervision to the subsidiary in its identification of its specific new projects. Hence:

H4d: The higher the Masculinity-Femininity score of the national culture of the parent company, the higher the subsidiary manager's autonomy for information systems planning for strategy selection.

For the strategy implementation phase, parent managers in multinationals from more Masculine cultures would grant more autonomy to their subsidiaries in defining change management approaches, defining action plans, evaluating action plans, and defining follow-up and control procedures. They would thus less closely supervise subsidiary strategy implementation planning. Hence:

H4e: The higher the Masculinity-Femininity score of the national culture of the parent company, the higher the subsidiary manager's autonomy for information systems planning for strategy implementation planning.

Each of a multinational firm's subsidiaries in a different country operates in its own distinct economic, political, legal, cultural, industrial, and competitive context. The multinational must be responsive to these differences (Makino et al., 2004). Prahalad (1975), Doz (1976), Doz, Bartlett, and Prahalad (1981), and Roth and Morrison (1990) suggest that pressures for local responsiveness necessitate local, context-sensitive, strategic decision-making and quick responses to each local market or industrial setting. Thus, greater autonomy granted by the parent to subsidiary top, IS, and business managers for IS planning in a multinational firm would be expected to better facilitate IS planning to meet those local needs, and therefore result in improved IS planning effectiveness at the subsidiary level. Hence, we hypothesize,

H5: The greater the autonomy for IS planning, the greater the effectiveness of IS planning for the subsidiary.

Effective IS planning can enhance the IS contribution via IS support for business strategy and via competitive advantage. Effective IS planning can improve alignment (Baets, 1992; Bowman et al., 1983; Das et al., 1991; Henderson & Venkatraman, 1993; King, 1978). It can do so by helping facilitate the acquisition and deployment of information technology that is congruent with the organization's competitive needs rather than with existing patterns of usage within the organization (Segars & Grover, 1998). Such planning heightens the visibility and stature of IS within the organization, and through the resulting alignment, facilitates the financial and managerial support for effectively implementing new, innovative systems (Chan & Huff, 1992; Das et al., 1991; Henderson et al., 1987).

Effective IS planning can facilitate the competitive advantage potentially provided by the organization's IT (Mata, Fuerst, & Barney, 1995). Such planning can enable IT to reduce a firm's costs and increase its revenues, to increase customer switching costs (Clemons & Kimbrough, 1986; Clemons & Row, 1987, 1991; Feeny & Ives, 1990), and to help implement a wide range of competitive strategies, including cost leader-

ship, product differentiation, strategic alliance, diversification, and vertical integration to provide sustainable competitive advantage (Barney, 1996).

We thus hypothesize that effective IS planning will lead to greater IS contribution in terms of greater alignment and competitive advantage. Hence:

H6: The greater the effectiveness of IS planning for the subsidiary, the greater the contribution of IS in the subsidiary.

METHODOLOGY

Instrument

A field survey of CIOs and other senior non-IS managers of U.S. subsidiaries of multinational manufacturing firms provided the data for this research. The U.S. was chosen because it is among the world's largest recipients of foreign direct investment inflow (UNCTAD, 2004) and would thus have more potential respondents, and because the researchers had easier access to subsidiaries of multinationals within it. Moreover, the use of a single country for subsidiaries facilitated comparison by keeping the subsidiary national culture constant and varying only that of its parent. The relatively sparser research literature available on multinational firms based outside the U.S. as well a growing interest in such firms also motivated their choice for this study.

The survey asked the senior non-IS managers to identify the nationality of the parent firm. Indices for the national culture (in terms of Power Distance, Individualism-Collectivism, Uncertainty Avoidance, and Masculinity-Femininity) of the parent firms were obtained from Hofstede (2001) for the independent variables. (Thus the survey did not ask individual managers about their own cultural values or about the values of the culture of the parent.)

The survey contained five-point Likert scale items to measure the extent of subsidiary autonomy for the items of each planning phase as shown in Table 1 and the extent of IS effectiveness in terms of the achievement of nine planning objectives as shown in Table 2. It also included the IS contribution items that were to be rated by a senior non-IS manager who was either the CEO or was identified by the CIO as being knowledgeable about the IS function. Finally, the survey contained demographic questions about the participant, the subsidiary company, and its parent company. (Appendix A shows the wording of the autonomy items in the survey, and Appendix B shows that of the IS planning effectiveness and IS contribution items.)

The unit of analysis was the company (i.e., the survey instructions said "please circle the number to indicate the extent of autonomy your local company has for each phase and task"). Subsidiary managers thus described their local organization's autonomy, effectiveness, and contribution rather than their own autonomy as individuals. Hofstede's measures likewise represented their parent organization's national culture rather than the national culture of the individual manager. The study thus had "ecological validity" via its consistent use of the organization rather than the individual as the unit of analysis in all constructs (Hofstede, 1991; Srite & Karahanna, 2006).

Pilot Study

Before mailing, the instrument was pilot tested with top executives of five U.S. manufacturing subsidiaries of multinational firms based outside the U.S. The senior author met with each participant independently at his or her workplace and reviewed the purpose of the survey. The top IS executive was asked to read the sample cover letter, complete the survey, make any comments while doing so, and identify a top non-IS manager knowledgeable about the IS function. The non-IS manager was then asked to read a second cover letter and complete a very short survey with the questions about IS contribution and demographics.

After the completion of each survey, the senior author discussed the survey with the participant and asked for comments regarding the presentation of the questions and the utility of the cover letter in gaining responses. The pilot testing suggested minor revisions to the questions and the cover letter. They were duly incorporated in the final version.

Data Collection

A mailing list was obtained from Applied Computer Research (ACR) in Phoenix, AZ for a one-time fee based use. It included the name, telephone number and mailing address of the CIOs of about 11,000 manufacturing companies based in the U.S. This list was compared to a list of foreign owned subsidiaries in the U.S. from a directory of foreign firms operating in the U.S. The directory listed about 4,000 foreign owned subsidiaries. Comparison of the ACR list with the directory enabled the identification of 416 CIOs of foreign owned subsidiaries in the U.S to whom the pair of surveys was mailed with a cover letter asking the CIO to forward the second survey to the CEO or other top executive of the subsidiary.

To increase the sample, 490 manufacturing subsidiaries with more than 50 employees but not shown in the ACR list were identified in the directory. This group of firms had contact information specified for the CEO rather than the CIO. The surveys were mailed to the CEO with a cover letter requesting the CIO's questionnaire be forwarded to that individual; both the CEO and CIO survey had an identification number for matching responses to ensure both provided data for the same organization.

Within six weeks of the initial mailing, 21 of the 906 (416 + 490) firms had returned completed surveys. To determine why firms had not returned more and to encourage greater participation, the senior author and an assistant phoned non-respondents. Each non-respondent received an average of three calls. When the individual was unreachable, the caller left no message but phoned again after a few days. On the fourth try, the caller left a message requesting participation and mentioning that another survey would be mailed within a few days. In all, of the 906 possible subjects, 137 were unreachable because the subject had moved, or the company had closed, or was no longer a subsidiary. Of the remaining, 98 declined to participate because of corporate policy. A total of 131 CIO surveys were received from the mailings. After dropping one unusable survey and two others due to an outlier analysis, the effective response rate was about 19.1% (128/671). The rate for the non-IS managers was 15.4% (103/671) (Ignoring those who decline to participate and who are unreachable in such

response rate calculations is illustrated by Ba & Pavlou, 2002; Teo, Wei, & Benbasat, 2003; Karahanna, Straub, & Chervany, 1999).

Demographics

Table 3 displays the national groups represented in the sample with multiple cases, and the number of cases for each. Their scores and descriptive statistics in the table were obtained from Hofstede (2001). Although the number of cases per country in the study varied widely, a correlation between those numbers and the corresponding country's direct U.S. investment (r=.80, p<.001) suggests that the sample is representative of the population.

The industries of the respondent companies included fabricated metal products (16.8%), industrial and commercial machinery and computer equipment (11.5%), electronic and electrical

Table 3. Countries represented in the sample

Country	Number of cases	Power Distance Score	Individualism-Collectivism Score	Uncertainty Avoidance Score	Masculinity-Femininity Score
Australia	1	36	90	51	61
Belgium	4	65	75	94	54
Canada	15	39	80	48	52
Finland	3	33	63	59	26
France	4	68	71	86	43
Germany	26	35	67	65	66
Italy	2	50	76	75	70
Japan	29	54	46	92	95
Mexico	1	81	30	82	69
Netherlands	7	38	80	53	14
Sweden	6	31	71	29	5
Switzerland	8	34	68	58	70
UK	20	35	89	35	66
Thailand	1	64	20	64	34
South Korea	1	60	18	85	39
	Range	11-104	6-91	8-112	5-95
	Mean	56.8	43.1	65.5	48.7
	Median	60	38	68	49

equipment and components (10.7%), transportation equipment (8.4%), chemical and allied products (7.6%), and others.

Of the respondent subsidiary firms, 44% had local sales revenues of less than $100 million. About 33% had such revenues between $100 million and $250 million, while the remainder had more than $250 million. Fifty percent had less than 500 employees locally, 19% had between 500 and 1,000 employees, and the rest had more than 1,000 employees.

Of the parent firms, about 35% had worldwide sales revenues of less than $1 billion. Another 33% had sales between $1 billion and $5 billion, while the remainder had sales of more than $5 billion. Fifty-seven percent had fewer than 5,000 employees worldwide, 20% had between 5,000 and 15,000 employees, and the rest had more than 15,000 employees.

The CIOs who responded to the survey had been with their companies for an average of 9 years. They had nearly 14 years of experience in IS and 15 years in manufacturing. Sixty-seven percent had at least a 4-year college degree. The senior non-IS managers had been with their companies for 11 years and had 16 years in manufacturing while 72% had at least a 4-year degree.

The means and standard deviations of the constructs appear in Table 4.

DATA ANALYSIS

Non-Response Bias

Non-response bias poses a serious concern in information systems research (Sivo et al., 2006). Two post hoc approaches to test the effects of such bias are now described. One compared late respondents to non-respondents using survey items and the other compared all respondents to non-respondents using organization characteristics from an independently published directory.

Table 4. Means and standard deviations of constructs

Factor	Mean	SD
Autonomy for ...		
.... Strategic Awareness	4.18	.99
.... Strategic Analysis	4.17	.93
.... Strategic Conception	4.16	.90
.... Strategic Selection	3.99	.99
.... Strategic Implementation Planning	4.23	.91
IS effectiveness	3.41	.71
IS contribution	2.84	.79

The returned surveys were thus first examined for non-response bias by treating late respondents as surrogates for non-respondents (Armstrong & Overton, 1977). Multivariate analysis of variance evaluated whether differences between early and late subjects were associated with different responses (Compeau & Higgins, 1995). The analysis indicated no significant differences in the key variables tested (subsidiary employees, subsidiary sales, parent employees, and parent sales). This is consistent with the absence of non-response bias (Wilks' Λ =.83, p =.28).

Respondents to the survey were also compared to actual non-respondents in terms of their number of PCs, IS budget, and IS employees (using data from the directory), as well as sales, gross profit, income, return on assets, and return on investment (using data from the *Compustat* database). P-values ranged from .38 to .81 except for income (p=.08). These findings are also consistent with the absence of the bias.

Content Validity

The IS planning autonomy, IS planning effectiveness, and IS contribution scales used in this study were derived from prior research (Mentzas, 1997; Raghunathan & Raghunathan, 1994; Lederer & Sethi, 1996). They were pilot tested with experts, i.e., CIOs and other non-IS managers of multinational subsidiaries in the United States. Such evidence of content validity made them suitable for the current study.

Convergent and Discriminant Validity

All the constructs in the current study were deemed formative because the items defined their constructs and were not interchangeable (Jarvis et al., 2003; Diamantopolous & Winkhofer, 2001; Petter et al., 2007).[3] Validation therefore followed Loch, Straub, and Kamel's (2003) procedure where the item scores for each subject were first multiplied by the PLS item weights (which were occasionally negative) to create a weighted measure for each item. The weighted measures were then summed to create a composite score for each construct. Convergent validity was demonstrated because the weighted measures in the same construct correlated significantly with each other and with their composite construct scores. Discriminant validity was demonstrated because weighted measures in the same construct correlated more highly with the weighted measures in their own construct than with such measures in other constructs with few exceptions, and because the weighted measures in the same construct generally correlated more highly with their own composite construct score than with other composite construct scores. A few such exceptions are acceptable in a large matrix (Loch et al., 2003; Campbell & Fiske, 1957). Weighted-measure-to-composite correlations appear in Appendix C. Weighted measure correlations appear in Appendix D.

Second Order IS Planning Autonomy

A second order construct representing overall IS planning autonomy was created by modeling the paths from the first order (i.e., autonomy for *strategic awareness, situational analysis, strategy conception, strategy selection* and *strategy implementation planning*) to the second order construct (Edwards, 2001). The relative weights

of the first order constructs were derived from a principal components analysis (Pavlou & Feygenson, 2006). The second order construct was thus a linear combination of the five first order constructs. The psychometric properties and validity of this second order construct were assessed through its first order constructs (Staples & Seddon, 2004).

Structural Model

The bootstrap procedure with 1,000 samples was used to calculate the significance of the path coefficients for the research model (Chin, 1998). The five main industry groups in the sample were included as dummy variables to control for industry, and the number of employees was included to control for subsidiary size; only one of the control variables was significant (fabricated metal industry). Table 5 shows the coefficients, t-values, and significance levels for each hypothesis.

The study found support for nine of the twenty component hypotheses and for H5 and H6. It found support for the hypotheses that Individualism-Collectivism predicts autonomy for *strategic awareness* (H2a, p<.05), *situational analysis* (H2b, p <.05), *strategy selection* (H2d, p<.05) and *strategy implementation planning* (H2e, p<.05) (or in other words, for all except H2c with a dependent variable of *strategy conception*), that Uncertainty Avoidance predicts autonomy for *strategic awareness* (H3a, p<.05) alone, and that Masculinity-Femininity predicts autonomy for *strategic awareness* (H4a, p<.05), *situational analysis* (H4b, p<.05), *strategy conception* (H4c, p<.05) and *strategy selection* (H4d, p<.05) (or in other words, for all except H4e with a dependent variable of *strategy implementation planning*). Furthermore, the proportion of variance explained in each of the planning phases was .14 for *strategic awareness*, .18 for *situational analysis*, .15 for *strategy conception*, .16 for *strategy selection*, and .11 for *strategy implementation planning*.

Table 5. Results of the structural model

Independent Construct		Dependent Construct Autonomy for…	Path	T-value
H1a	Power Distance	Strategic Awareness	.11	.72
H1b		Situational Analysis	.12	.80
H1c		Strategy Conception	.15	1.11
H1d		Strategy Selection	-.01	.03
H1e		Strategy Implementation Planning	.03	.21
H2a	Individualism-Collectivism	Strategic Awareness	-.39	2.44*
H2b		Situational Analysis	-.32	2.09*
H2c		Strategy Conception	-.23	1.57
H2d		Strategy Selection	-.35	2.37*
H2e		Strategy Implementation Planning	-.29	1.97*
H3a	Uncertainty Avoidance	Strategic Awareness	-.49	2.38*
H3b		Situational Analysis	-.26	1.43
H3c		Strategy Conception	-.23	1.13
H3d		Strategy Selection	-.19	.88
H3e		Strategy Implementation Planning	-.28	.92
H4a	Masculinity-Femininity	Strategic Awareness	.30	2.23*
H4b		Situational Analysis	.29	2.25*
H4c		Strategy Conception	.26	2.08*
H4d		Strategy Selection	.21	1.96*
H4e		Strategy Implementation Planning	.18	1.08
H5	IS Planning Autonomy	IS Planning Effectiveness	.51	2.66**
H6	IS Planning Effectiveness	IS Planning Contribution	.44	2.27*

*p <.05, **p <.01

The study also found support for the impact of IS Planning Autonomy on IS Planning Effectiveness (H5, p<.01), and for the impact of IS Planning Effectiveness on IS Planning Contribution (H6, p<.05). The proportion of variance explained was .46 and .38 for H5 and H6 respectively. Finally, a power analysis for each supported hypothesis using regression (as suggested by Cohen, 1988) found an average power of .88, thus exceeding the frequently suggested criterion of .80.

DISCUSSION

The analysis found some support for the Hofstede's model of national culture. In particular, it confirmed nine of the 20 hypotheses about the impact of culture on autonomy. It also confirmed the hypotheses about the impact of autonomy on effectiveness and of effectiveness on contribution. The support for these hypotheses coincides, of course, with the theory, but the lack of support for the others may provide more insight into the impact of national culture on IS planning.

The analysis showed no support for the expectation that greater Power Distance would predict less autonomy (H1a-H1e). In high Power Distance cultures, managers expect to exercise control because failure to do so might make them appear incompetent and irresponsible. However, perhaps they do not delegate less autonomy to subsidiary IT managers because they believe that the subsidiary IT managers have greater experience in their own particular IT needs, and that taking on the responsibility themselves might make them appear even more incompetent and irresponsible if the planning does not succeed as intended. Regardless of the reason for the lack of support for these hypotheses, other research similarly found Power Distance to be unassociated with corporate adoption of IT infrastructure (Png et al., 2001).

The study found support for hypotheses 2a, 2b, 2d, and 2e which had predicted that parent firms from high Individualism-Collectivism cultures grant their subsidiaries lower autonomy for *strategic awareness* (p <.05), *situational analysis* (p <.05), *strategy selection* (p <.05), and *strategy implementation planning* (p <.05). Such support is consistent with the expectation that in an individualistic society, ties between individuals are loose, and employer-employee relationships are viewed as transactions in a labor market wherein managers are fairly demanding that subordinates adhere to their directives. However, hypothesis 2c (*strategy conception*) was not supported. *Strategy conception* (identifying opportunities and strategies for improvement) is probably the most creative phase, and is thus not only necessary but also has unique management challenges (Cooper, 2000). Perhaps the parent realizes that it is best left to the subsidiary because control over the creative aspect of planning might in fact stifle creativity whereas autonomy would nurture the creativity and resourcefulness of the planners (Nidumolu & Subramani, 2003). In other words, the employer-employee relationship in *strategy conception* is viewed as a special labor market transaction requiring more autonomy for the subsidiary. (Perhaps interestingly, hypotheses concerning *strategy conception* were not supported for three of the four national culture indices (H1c, H2c, and H3c), giving further credence to the unusual nature of the most creative planning activity.)

The study found little support for the expectation that parent firms from high Uncertainty Avoidance cultures grant lower autonomy to their subsidiaries. Hypotheses 3b, 3c, 3d, and 3e were not supported. Only hypothesis 3a, which had predicted that parent firms from high Uncertainty Avoidance cultures would grant lower autonomy for *strategic awareness* (p <.05) was supported. That support is consistent with the expectation that in Uncertainty Avoiding cultures, people are uncomfortable with unstructured, unclear, or unpredictable situations. Thus managers instill strict laws, formal reporting rules, safety, and security measures in the organization, and expect subordinates to follow their plans. Because *strategic awareness* (determining planning objectives, organizing the planning team(s), and obtaining top management commitment) initiates new planning activities in the subsidiary, parent managers might especially want to control it.

However, the other Uncertainty Avoidance hypotheses predicting that high uncertainty avoidance leads to low autonomy were not supported though all the coefficients were negative as expected. Perhaps the parent recognizes that *situational analysis, strategy conception, strategy selection*, and *strategy implementation planning* are best handled by subsidiary managers given their familiarity with the local environment, needs of the subsidiary, and difficulties implementing plans. Perhaps lack of knowledge of local conditions creates the unstructured, unclear, or unpredictable environments that parent managers seek to avoid. Prior researchers examining U.S. subsidiaries of multinationals headquartered in Germany and Japan (cultures high in Uncertainty Avoidance) have in fact speculated that the control engendered by Uncertainty Avoidance may stifle innovation, creativity, and the search for new knowledge in the subsidiaries (Ambos & Reitsperger, 2002). Perhaps parent managers from such cultures recognize this danger from prior experiences, and loosen control over their subsidiaries in the subsequent, more detailed phases of planning. They do, however, want to control the direction of the planning process and the choice of subsidiary managers (via *strategic awareness*) who will deal with the local uncertainty.

The study found support for hypotheses 4a-4d which had predicted that parent firms from high Masculinity-Femininity cultures grant their subsidiaries higher autonomy for *strategic awareness* (p <.05), *situational analysis* (p <.05), *strategy conception* (p <.05) and *strategy selection* (p <.05). Such support is consistent with the expectation that in Masculine cultures, achievement and the aggressive pursuit of goals are valued over good relationships, harmony, and care for people. Managers expect their subordinates to be decisive, assertive, and independent, and to resolve conflicts by fighting them out.

Hypothesis 4e was not supported. *Strategy implementation planning* (defining change management and action plans, evaluating action plans, and defining follow-up and control procedures), the final planning phase, delivers the results for all of the previous phases. It is, however, considered the phase most likely to fail (Gottschalk, 1999a, 1999b, 1999c). Because cultures high in Masculinity-Femininity not only value independence, but also focus on organizational performance and results, parent managers may want to control this phase rather than delegate autonomy to the subsidiary for it. Hence, perhaps the lack of support is not so surprising.

Support for H5 (the impact of IS planning autonomy on effectiveness) is consistent with the expectation that local responsiveness will permit local, context-sensitive, strategic decision-making and quick responses to each local market or industrial setting (Bartlett & Ghoshal, 1989; Doz, 1976; Prahalad & Doz, 1987) so that greater autonomy granted by the parent to subsidiary local managers for IS planning would better facilitate IS planning to meet those local needs, and therefore result in improved IS planning effectiveness. Support for H6 (the impact of IS planning effectiveness on IS contribution) is consistent with expectations that IS planning improves the alignment of the IS strategy and business strategy (Baets, 1992; Bowman et al.,1983; Das et al., 1991; Henderson & Venkatraman, 1993; Henderson et al., 1987; King, 1978). and increases the competitive advantage provided by the organization's IT (Mata, Fuerst, & Barney, 1995). Moreover, the support for H5 and H6 in the context of the model gives greater credence to the importance of the impact of national culture on IS planning autonomy.

IMPLICATIONS FOR RESEARCH

The current study extends Hofstede's model of national culture to a new domain, namely IS planning in the subsidiaries of multinational firms, thereby supporting the assertion that a cultural perspective be added to existing IS theories and practices (Tan, Wei, Watson, & Walczuch, 1998; Straub et al., 2002). The study also validates instruments for measuring IS planning autonomy, IS planning effectiveness, and IS contribution which future researchers can use more confidently in their own work. More importantly, it facilitates a better understanding of the management of IS planning.

Although many hypotheses were supported, some were not. We speculated about reasons for the exceptions. That is, we speculated that high Power Distance parent managers might refrain from withholding autonomy to avoid appearing incompetent if planning fails; that high Individualism-Collectivism parent managers might refrain from withholding autonomy for s*trategy conception* to avoid quashing local manager creativity; that high Uncertainty Avoidance parent managers might refrain from withholding autonomy after planning project initiation to let local managers apply their local knowledge; and, finally, that high Masculinity-Femininity parent managers might refrain from granting autonomy in *strategy implementation planning* because that phase delivers the planning results and has more direct impact on organizational performance. Future researchers might investigate those proposed reasons and other possible explanations.

Further data collection could help in this effort. In particular, interviews with managers might provide alternate explanations.

The finding for H5, the autonomy-on-effectiveness hypothesis, represents the views of the subsidiary IS managers (i.e., they supplied both independent and dependent variables). Perhaps subsidiary IS managers perceive that seemingly arbitrary demands from the parent prevent them from being as effective as they might otherwise be. In other words, the reader may interpret the H5 support to reflect subsidiary manager parochial perception. That interpretation in itself may be seen as a contribution of this research. Regardless, future research might consider the perspective of parent managers about autonomy.

The support for much of the Hofstede's model in this study demonstrates its meaningfulness in IS research where autonomy may be a key element. Future researchers might thus apply the model in such other areas of IS as help desks, systems analysis and design, programming, and other more technical endeavors.

The current study characterized planning in terms of specific actions. While such an approach has been popular, in future cross-cultural research the characterization of it in terms of such process dimensions as comprehensiveness, formalization, focus, flow, participation, and consistency might provide an interesting alternative perspective (Segars & Grover, 1999).

The current study focused its attention exclusively on the U.S. manufacturing subsidiaries of multinational firms. Future research should vary the cultures of subsidiaries to confirm generalizability of the current findings. It should also test the hypotheses within service-sector subsidiaries and within individual industries.

The current study followed typical national culture research. For example, it assumed that the U.S. lacks regional differences. Conventional wisdom suggests, however, that the U.S. does not have a single prevailing culture – the common set of values on the West Coast may differ significantly from those in the Midwest or the South. Future research might use regional measures of the constructs in this study, and perhaps find different results.

The current study did not consider varying degrees of subsidiary need for autonomy. Future research might thus investigate the proposition that the greater the need for information to support management decision making in the subsidiary (as in the subsidiaries of the multi-national enterprise pursuing a multi-domestic strategy), the greater the need for IS planning autonomy at the subsidiary.

This study had its limitations. For example, similarly to most other planning research, it relied heavily on managerial perceptions rather than objective measures. Moreover, although Hofstede's dimensions came from his research, the IS contribution measure came from non-IS managers, and testing for common source variance did not detect a problem with the single subject assessment of both IS planning autonomy and effectiveness, multiple subject assessment of those constructs would still have been preferable.

IMPLICATIONS FOR PRACTICE

This study underscores the importance of the national culture of the multinational parent in the management of information systems. It confirms the impact of Hofstede's dimensions – especially of Individualism-Collectivism and Masculinity-Femininity – on autonomy in various phases of IS planning. Subsidiary managers can use the results of this study to better understand the autonomy granted by their parent organization by helping them realize that the national culture of the parent may be at least in part the reason the parent either grants more or less autonomy than expected or desired. Such a realization might help the subsidiary manager accept the extent of autonomy or it might encourage the manager to take action to attempt to change it. The results further suggest to local managers that they consider the national culture of the parent if they choose to argue for more or less autonomy.

CONCLUSION

This research has contributed towards a better understanding of the management of information systems planning in the foreign subsidiaries of multinational firms. It has extended an emerging cultural model by demonstrating that its dimensions are related to IS planning autonomy. It has done so while confirming the impact of such autonomy on IS planning effectiveness and of such effectiveness on IS contribution. It further suggests the possibility of employing the dimensions in other areas of IS research.

More specifically, the study found strong support for the expected impact on IS planning autonomy of Individualism-Collectivism (with the exception of *strategy conception*, suggesting perhaps that the unusual demands of the most creative planning phase inspire the parent to grant more autonomy) and of Masculinity-Femininity (with the exception of *strategy implementation planning*, suggesting perhaps the great challenges and risks of this planning phase force the parent not to grant more autonomy). The study did not find support for the expected impact of Power-Distance (perhaps due to parent managers' concern that lack of knowledge of the subsidiary's own particular IT needs might make them appear less competent) and Uncertainty Avoidance (perhaps due to parent managers' concern that lack of knowledge of the subsidiary's local culture might make them be less competent) with the exception of *strategic awareness* for the latter (where perhaps that parent managers might feel they must maintain control at least for planning initiation).

These findings do not suggest that IT applications be designed independently of corporate oversight. But they do suggest that national culture matters. The empirical confirmation of its impact reminds both investigators and managers alike that it must not be ignored in their information systems research and practice.

REFERENCES

Akmanligil, M., & Palvia, P. (2004). Strategies for Global Information Systems Development. *Information & Management*, *42*(1), 45–59.

Ambos, B., & Reitsperger, W. (2002). Governing Knowledge Processes in MNCs: The Case of German R&D Units Abroad. In *Proceedings of the 28th European International Business Academy Meeting*, Athens, Greece.

Anthes, G. (2001). Think globally, act locally. *Computerworld*, *35*(22), 36–37.

Armstrong, J., & Overton, T. (1977). Estimating Non-Response Bias in Mail Surveys. *JMR, Journal of Marketing Research*, *14*(8), 396–402. doi:10.2307/3150783

Ba, S., & Pavlou, P. A. (2002). Evidence of the Effect of Trust Building Technology in Electronic Markets: Price Premiums and Buyer Behavior. *Management Information Systems Quarterly*, *26*(3), 243–268. doi:10.2307/4132332

Baets, W. (1992). Aligning information systems with business strategy. *The Journal of Strategic Information Systems*, *1*(4), 205–213. doi:10.1016/0963-8687(92)90036-V

Barney, J. (1996). *Gaining and Sustaining Competitive Advantage*. Reading, MA: Addison Wesley.

Bartlett, C., & Ghoshal, S. (1989). *Managing Across Borders: The Transnational Solution*. Boston: Harvard Business School Press.

Bowman, B., Davis, G., & Wetherbe, J. (1983). Three Stage Model of MIS Planning. *Information & Management*, *6*(1), 11–25. doi:10.1016/0378-7206(83)90016-2

Burn, J., Davison, R., & Jordan, E. (1997). The Information Society - A Cultural Fallacy. *The Journal of Failures and Lessons Learned in IT Management*, *1*(4), 219–232.

Burn, J., Saxena, K., Ma, L., & Cheung, H. (1993). Critical Issues of IS Management in Hong Kong: A Cultural Comparison. *Journal of Global Information Management*, *1*(4), 28–37.

Campbell, D. T., & Fiske, D. W. (1959). Convergent and Discriminant Validation by the Multitrait-Multimethod Matrix. *Psychological Bulletin*, *56*(2), 81–105. doi:10.1037/h0046016

Chai, L., & Pavlou, P. (2004). From 'Ancient' to 'Modern': A Cross-Cultural Investigation of Electronic Commerce Adoption in Greece and the United States. *Journal of Enterprise Information Management*, *17*(6), 416–423. doi:10.1108/17410390410566706

Chan, Y., & Huff, S. (1992). Strategy: An Information Systems Research Perspective. *The Journal of Strategic Information Systems*, *1*(4), 191–204. doi:10.1016/0963-8687(92)90035-U

Chin, W. (1998). The partial least squares approach for structural equation modeling. In Marcoulides, G. (Ed.), *Modern Methods for Business Research*. Hillsdale, NJ: Lawrence Erlbaum Associates.

Clemons, E., & Row, M. (1991). Information Technology at Rosenbluth Travel: Competitive Advantage in a Rapidly Growing Global Service Company. *Journal of Management Information Systems*, *8*(2), 53–79.

Clemons, E. K., & Kimbrough, S. O. (1986). Information Systems, Telecommunications, and Their Effects on Industrial Organization. In *Proceedings of the 7th International Conference on Information Systems* (pp. 99-108).

Clemons, E. K., & Row, M. (1987, December 1-9). Structural Differences among Firms: A Potential Source of Competitive Advantage in the Application of Information Technology. In *Proceedings of the 8th International Conference on Information Systems*.

Compeau, D., & Higgins, C. (1995). Computer Self-efficacy: Development of a Measure and Initial Test. *Management Information Systems Quarterly*, *19*(2), 189–212. doi:10.2307/249688

Cooper, R. B. (2000). Information technology development creativity: A case study of attempted radical change. *Management Information Systems Quarterly*, *24*(2), 245–276. doi:10.2307/3250938

Cravens, D. (1988). Gaining strategic marketing advantage. *Business Horizons*, *31*(5), 44–54. doi:10.1016/0007-6813(88)90054-7

Das, S., Zahra, S., & Warkentin, M. (1991). Integrating the Content and Process of Strategic MIS Planning with Competitive Strategy. *Decision Sciences*, *22*(1), 953–984.

Davison, R., & Jordan, E. (1998). Group Support Systems: Barriers to Adoption in a Cross-Cultural Setting. *Journal of Global Information Technology Management*, *1*(2), 37–50.

Deans, P. C., Karwan, K., Goslar, M., Ricks, D., & Toyne, B. (1991). Identification of Key International Information Systems Issues in U.S-based Multinational Corporations. *Journal of Management Information Systems*, *7*(4), 27–50.

Del Galdo, E., & Nielsen, J. (Eds.). (1996). *International User Interfaces*. New York: John Wiley and Sons.

Diamantopoulos, A., & Winklhofer, H. (2001). Index Construction with Formative Indicators: An Alternative to Scale Development. *JMR, Journal of Marketing Research*, *38*(2), 269–277. doi:10.1509/jmkr.38.2.269.18845

Doherty, N. F., Marples, C. G., & Suhaimi, A. (1999). The Relative Success of Alternative Approaches to Strategic Information Systems Planning: An Empirical Analysis. *The Journal of Strategic Information Systems*, *8*(3), 263–283. doi:10.1016/S0963-8687(99)00024-4

Doz, Y. (1976). *National Policies and Multi-national Management*. Unpublished doctoral dissertation, Harvard Business School, Boston.

Doz, Y., Bartlett, C., & Prahalad, C. (1981). Global Competitive Pressures and Host Country Demands: Managing Tensions in MNCs. *California Management Review*, *23*(3), 63–74.

Doz, Y., & Prahalad, C. K. (1984). Patterns of Strategic Control within MNCs. *Journal of International Business Studies*, *15*(2), 55–72. doi:10.1057/palgrave.jibs.8490482

Edwards, J. (2001). Multidimensional constructs in organizational behavioral research: An integrative analytical framework. *Organizational Research Methods*, *4*(2), 144–192. doi:10.1177/109442810142004

Fang, T. (2003). A critique of Hofstede's fifth national culture dimension. *International Journal of Cross Cultural Management*, *3*(3), 347–368. doi:10.1177/1470595803003003006

Feeny, D., & Ives, B. (1990). In Search of Sustainability: Reaping Long-Term Advantage from Investments in Information Technology. *Journal of Management Information Systems*, *7*(1), 27–46.

Ford, D. P., Connelly, C. E., & Meister, D. B. (2003). Information Systems Research and Hofstede's Culture's Consequences: An Uneasy and Incomplete Partnership. *IEEE Transactions on Engineering Management*, *50*(1), 8–25. doi:10.1109/TEM.2002.808265

Gabel, M., & Bruner, H. (2003). *Global Inc.: An Atlas of the Global Corporation: A Visual Exploration of The History, Scale, Scope, and Impacts of Multinational Corporations*. New York: New Press.

Gefen, D., & Straub, D. W. (1997). Gender Differences in the Perception and Use of E-mail: An Extension to the Technology Acceptance Model. *Management Information Systems Quarterly*, *21*(4), 389–400. doi:10.2307/249720

Gottschalk, P. (1999a). Implementation predictors of strategic information systems plans. *Information & Management*, *36*(2), 77–91. doi:10.1016/S0378-7206(99)00008-7

Gottschalk, P. (1999b). Strategic information systems planning: The IT strategy implementation matrix. *European Journal of Information Systems*, *8*(2), 107–118. doi:10.1057/palgrave.ejis.3000324

Gottschalk, P. (1999c). Implementation of formal plans: The case of information technology strategy. *Long Range Planning*, *32*(3), 362–372. doi:10.1016/S0024-6301(99)00040-0

Graen, G. B. (2006). In the eye of the beholder: Cross-cultural lesson in leadership from project GLOBE: A response viewed from the third culture bonding (TCB) model of cross-cultural leadership. *The Academy of Management Perspectives*, *20*(4), 95–101.

Grover, V., & Segars, A. H. (2005). An Empirical Evaluation of Stages of Strategic Information Systems Planning: Patterns of Process Design and Effectiveness. *Information & Management*, *42*(5), 761–779. doi:10.1016/j.im.2004.08.002

Gupta, A. (1987). SBU strategies, corporate-SBU relations, and SBU effectiveness in strategy implementation. *Academy of Management Journal*, *30*(3), 477–500. doi:10.2307/256010

Hanke, M., & Teo, T. (2003). Meeting the Challenges in Globalizing Electronic Commerce at United Airlines. *Journal of Information Technology Cases and Applications*, *5*(4), 21.

Harris, R., & Davison, R. (1999). Anxiety and Involvement: Cultural Dimensions of Attitudes Toward Computers in Developing Societies. *Journal of Global Information Management*, *7*(1), 26–38.

Henderson, J. C., Rockart, J. F., & Sifonis, J. G. (1987). Integrating Management Support Systems into Strategic Information Systems Planning. *Journal of Management Information Systems*, *4*(1), 5–24.

Henderson, J. C., & Venkatraman, N. (1993). Strategic Alignment: Leveraging Information Technology for Transforming Organizations. *IBM Systems Journal*, *32*(1), 4–16. doi:10.1147/sj.382.0472

Ho, K. K. W., Yoo, B., Yu, S., & Tam, K. Y. (2007). The Effect of Culture and Product Categories on the Level of Use of Buy-It-Now (BIN) Auctions by Sellers. *Journal of Global Information Management*, *15*(4), 1–19.

Hoffman, R. (1988). The general management of foreign subsidiaries in the USA: An exploratory study. *Management Information Review*, *28*(2), 41–55.

Hofstede, G. (1980). *Culture's Consequences: International Differences in Work-Related Values*. Beverly Hills, CA: Sage Publications.

Hofstede, G. (1984). *Culture's Consequences: International Differences in Work-Related Values*. Beverly Hills, CA: Sage Publications.

Hofstede, G. (1991). *Cultures and Organizations: Software of the Mind*. London: McGraw-Hill.

Hofstede, G. (2001). *Culture's Consequences* (2nd ed.). Thousand Oaks, CA: Sage Publications.

Hofstede, G. (2006). What did GLOBE really measure? Researchers' minds versus respondents' minds. *Journal of International Business Studies*, *37*(6), 882–896. doi:10.1057/palgrave.jibs.8400233

Hofstede, G., & Bond, M. (1988). The Confucian Connection: From Cultural Roots to Economic Growth. *Organizational Dynamics*, *16*(4), 4–21. doi:10.1016/0090-2616(88)90009-5

House, R. J., Hanges, P. J., Javidan, M., Dorfman, P. W., & Gupta, V. (2004). *Culture, Leadership, and Organizations: The GLOBE Study of 62 Societies*. Thousand Oaks, CA: Sage Publications.

Igbaria, M. (1992). An examination of microcomputer usage in Taiwan. *Information & Management*, *22*(10), 19–28. doi:10.1016/0378-7206(92)90003-X

Igbaria, M., & Iivari, J. (1995). The Effects of Self-efficacy on Computer Usage. *OMEGA International Journal of Management Science*, *23*(6), 587–605. doi:10.1016/0305-0483(95)00035-6

Igbaria, M., & Zviran, M. (1991). End-User Effectiveness: A Cross-Cultural Examination. *Omega*, *19*(5), 369–379. doi:10.1016/0305-0483(91)90055-X

Igbaria, M., & Zviran, M. (1996). Comparison of End-User Computing Characteristics in the US, Israel, and Taiwan. *Information & Management*, *30*(1), 1–13. doi:10.1016/0378-7206(95)00044-5

Ito, M., & Nakakoji, K. (1996). Impact of Culture on User Interface Design. In Galdo, E. M., & Nielsen, J. (Eds.), *International User Interfaces* (pp. 41–73). New York: Wiley.

Ives, B., & Learmonth, G. (1984). The Information System as a Competitive Weapon. *Communications of the ACM*, *27*(12), 1193–1201. doi:10.1145/2135.2137

Jarvis, C., MacKenzie, S., & Podsakoff, P. (2003). A Critical Review of Construct Indicators and Measurement Model Misspecification in Marketing and Consumer Research. *The Journal of Consumer Research*, *30*(2), 199–218. doi:10.1086/376806

Kankanhalli, A., Tan, B. C. Y., Wei, K. K., & Holmes, M. (2004). Cross-Cultural Differences and Information Systems Developer Values. *Decision Support Systems*, *38*(2), 183–195. doi:10.1016/S0167-9236(03)00101-5

Karahanna, E., Straub, D. W., & Chervany, N. (1999). Information Technology Adoption across Time: A Cross-Sectional Comparison of Pre-Adoption and Post-Adoption Beliefs. *Management Information Systems Quarterly*, *23*(2), 183–213. doi:10.2307/249751

King, W. (1978). Strategic planning for MIS. *Management Information Systems Quarterly*, *2*(1), 27–37. doi:10.2307/249104

King, W. (1988). How effective is your IS planning? *Long Range Planning*, *21*(5), 103–112. doi:10.1016/0024-6301(88)90111-2

Kock, N., Parente, R., & Verville, J. (2008). Can Hofstede's Model Explain National Differences in Perceived Information Overload? A Look at Data from the US and New Zealand. *IEEE Transactions on Professional Communication*, *51*(1), 33–49. doi:10.1109/TPC.2007.2000047

Kreitner, R., & Kinicki, A. (2001). *Organizational Behavior* (5th ed.). New York: McGraw-Hill.

Lederer, A., & Sethi, V. (1996). Key Prescriptions for Strategic Information Systems Planning. *Journal of Management Information Systems*, *13*(1), 35–62.

Lippert, S. K., & Volkmar, J. A. (2007). Cultural Effects on Technology Performance and Utilization: A Comparison of U.S. and Canadian Users. *Journal of Global Information Management*, *15*(2), 56–80.

Loch, K., Straub, D., & Kamel, S. (2003). Diffusing the Internet in the Arab World: The Role of Social Norms and Technological Culturation. *IEEE Transactions on Engineering Management*, *50*(1), 45–63. doi:10.1109/TEM.2002.808257

Luthans, F. (2002). *Organizational Behavior* (9th ed.). New York: McGraw-Hill.

Makino, S., Isobe, T., & Chan, C. (2004). Does Country Matter? *Strategic Management Journal*, *25*(10), 1027–1043. doi:10.1002/smj.412

Martinsons, M. G., & Davison, R. M. (2007). Culture's Consequences for IT Application and Business Process Change: A Research Agenda. *International Journal of Internet and Enterprise Management*, *5*(20), 158–177. doi:10.1504/IJIEM.2007.014087

Mata, F. J., Fuerst, W. L., & Barney, J. B. (1995). Information Technology and Sustained Competitive Advantage: A Resource-based Analysis. *Management Information Systems Quarterly*, *19*(4), 487–505. doi:10.2307/249630

McDougall, S., Curry, M., & Bruijn, O. (1998). Understanding What Makes Icons Effective: How Subjective Rating Can Inform Design. In Hanson, J. (Ed.), *Contemporary Ergonomics* (pp. 285–289). London: Taylor & Francis.

McFarlan, F. W. (1984). Information technology changes the way you compete. *Harvard Business Review*, *62*(3), 98–103.

Mentzas, G. (1997). Implementing an IS strategy - A team approach. *Long Range Planning*, *30*(1), 84–95. doi:10.1016/S0024-6301(96)00099-4

Milberg, S., Smith, H. J., & Burke, S. (2000). Information Privacy: Corporate Management and National Regulation. *Organization Science*, *11*(1), 35–57. doi:10.1287/orsc.11.1.35.12567

Mohdzain, M. B., & Ward, J. M. (2007). A Study of Subsidiaries' Views of Information Systems Strategic Planning in Multinational Organizations. *The Journal of Strategic Information Systems*, *16*(4), 324–352. doi:10.1016/j.jsis.2007.02.003

Munter, M. (1993). Cross-cultural communication for managers. *Business Horizons*, *36*(3), 69–78. doi:10.1016/S0007-6813(05)80152-1

Myers, M., & Tan, F. (2002). Beyond Models of National Culture in Information Systems Research. *Journal of Global Information Management*, *10*(1), 24–32.

Nakakoji, K. (1996). Beyond language translation: Crossing the cultural divide. *IEEE Software*, *13*(6), 42–46. doi:10.1109/52.542293

Newkirk, H., Lederer, A., & Srinivasan, C. (2003). Strategic Information Systems Planning: Too Little or Too Much. *The Journal of Strategic Information Systems, 12*(3), 201–228. doi:10.1016/j.jsis.2003.09.001

Nidumolu, S., & Subramani, M. (2003). The Matrix of Control: Combining Process and Structure Approaches to Managing Software Development. *Journal of Management Information Systems, 20*(3), 159–196.

Palvia, P. (1998). Research issues in global information technology management. *Information Resources Management Journal, 11*(2), 27–36.

Palvia, P., Palvia, S., & Whitworth, J. (2002). Global Information Technology: A Meta Analysis of Key Issues. *Information & Management, 39*(5), 403–414. doi:10.1016/S0378-7206(01)00106-9

Pavlou, P., & Chai, L. (2002). What Drives Electronic Culture Across Cultures – A Cross-Cultural Empirical Investigation. *Journal of Electronic Commerce Research, 3*(4), 240–253.

Pavlou, P., & Fygenson, M. (2006). Understanding and Predicting Electronic Commerce Adoption: An Extension of the Theory of Planned Behavior. *Management Information Systems Quarterly, 30*(1), 115–143.

Peppard, J. (1999). Information management in the global enterprise: An organizing framework. *European Journal of Information Systems, 8*(2), 77–94. doi:10.1057/palgrave.ejis.3000321

Pereira, R. (1998). Cross-cultural influences on global electronic commerce. In *Proceedings of the AIS*, Baltimore, MD (pp. 318-320).

Peterson, M. F., & Castro, S. L. (2006). Measurement Metrics at Aggregate Levels of Analysis: Implications for Organization Culture Research and the GLOBE Project. *The Leadership Quarterly, 17*(5), 506–521. doi:10.1016/j.leaqua.2006.07.001

Petter, S., Straub, S., & Rai, A. (2007). Specification and Validation of Formative Constructs in IS Research. *Management Information Systems Quarterly, 31*(4), 623–656.

Png, I. P. L., Tan, B. C. Y., & Wee, K.-L. (2001). Dimensions of National Culture and Corporate Adoption of IT Infrastructure. *IEEE Transactions on Engineering Management, 48*(1), 36–45. doi:10.1109/17.913164

Porter, M. E., & Millar, V. E. (1985). How Information Gives You Competitive Advantage. *Harvard Business Review, 63*(4), 149–160.

Prahalad, C. (1975). *The Strategic Process in a Multinational Corporation.* Unpublished doctoral dissertation, Harvard Business School, Boston.

Prahalad, C., & Doz, Y. (1987). *The Multinational Mission: Balancing Local Demands and Global Vision.* New York: The Free Press.

Premkumar, G., & King, W. (1994). Organizational Characteristics and Information Systems Planning: An Empirical Study. *Information Systems Research, 5*(2), 75–109. doi:10.1287/isre.5.2.75

Raghunathan, B., & Raghunathan, T. (1991). The Relationship between Information Systems Planning and Planning Effectiveness: An Empirical Analysis. *OMEGA: The International Journal of Management Science, 19*(1), 125–135. doi:10.1016/0305-0483(91)90022-L

Raghunathan, B., & Raghunathan, T. (1994). Adaptation of a Planning System Success Model to Information Systems Planning. *Information Systems Research, 5*(3), 326–340. doi:10.1287/isre.5.3.326

Ramingwong, S., & Sajeev, A. S. M. (2007). Offshore Outsourcing: The Risk of Keeping Mum. *Communications of the ACM, 50*(8), 101–103. doi:10.1145/1278201.1278230

Reinig, B. A., Briggs, R. O., & de Vreede, G.-J. (2009). Satisfaction as a Function of Perceived Change in Likelihood of Goal Attainment: A Cross-Cultural Study. *International Journal of e-Collaboration, 5*(2), 61–74.

Robichaux, B. P., & Cooper, R. B. (1998). GSS Participation: A Cultural Examination. *Information & Management, 33*(6), 287–290. doi:10.1016/S0378-7206(98)00033-0

Roth, K., & Morrison, A. J. (1990). An Empirical Analysis of the Integration-Responsiveness Framework in Global Industries. *Journal of International Business Studies, 21*(4), 541–605. doi:10.1057/palgrave.jibs.8490341

Segars, A., & Grover, V. (1998). Strategic Information Systems Planning Success: An Investigation of the Construct and Its Measurement. *Management Information Systems Quarterly, 22*(2), 139–163. doi:10.2307/249393

Segars, A., & Grover, V. (1999). Profiles of Strategic Information Systems Planning. *Information Systems Research, 10*(3), 199–232. doi:10.1287/isre.10.3.199

Sherer, S. A. (2007). Comparative study of IT investment management processes in U.S. and Portugal. *Journal of Global Information Management, 15*(3), 43–68.

Siau, K., Nah, F. F.-H., & Ling, M. (2007). National Culture and Its Effects on Knowledge Communication in Online Virtual Communities. *International Journal of Electronic Business, 5*(5), 518–532. doi:10.1504/IJEB.2007.015450

Sinkovics, R. R., Yamin, M., & Hossinger, M. (2007). Cultural Adaptation in Cross Border E-Commerce: A study of German Companies. *Journal of Electronic Commerce Research, 8*(4), 221–236.

Sivo, S. A., Saunders, C., Chang, Q., & Jiang, J. J. (2006). How Low Should You Go? Low Response Rates and the Validity of Inference in IS Questionnaire Research. *Journal of the AIS, 7*(6), 361–414.

Smith, H. J. (2001). Information privacy and marketing: What the U.S. should (and shouldn't) learn from Europe. *California Management Review, 43*(2), 8–33.

Srite, M., & Karahanna, E. (2006). The Role of Espoused National Cultural Values in Technology Acceptance. *MIS Quarterly, 30*(3), 2006, 679-704.

Staples, D., Sandy, P., & Seddon, P. (2004). Testing the Technology-to-Performance Chain Model. *Journal of Organizational and End User Computing, 16*(4), 17–36.

Straub, D. (1994). The effect of culture on IT diffusion: E-mail and FAX in Japan and the US. *Information Systems Research, 5*(1), 23–47. doi:10.1287/isre.5.1.23

Straub, D., Loch, K., Evaristo, J., Karahanna, E., & Srite, M. (2002). Toward a Theory Based Measurement of Culture. *Journal of Global Information Management, 10*(1), 13–23.

Sturrock, F., & Kirwan, B. (1996). Interface Display Designs Based on Operator Knowledge Requirements. In Hanson, J. (Ed.), *Contemporary Ergonomics* (pp. 280–284). London: Taylor & Francis.

Tan, B. C. Y., Smith, H. J., Keil, M., & Montealegre, R. (2003). Reporting Bad News about Software Projects: Impact of Organizational Climate and Information Asymmetry in an Individualistic and a Collectivistic Culture. *IEEE Transactions on Engineering Management, 50*(1), 64–77. doi:10.1109/TEM.2002.808292

Tan, B. C. Y., Wei, K. K., Watson, R., & Walczuch, R. (1998). Reducing Status Effects with Computer-Mediated Communication: Evidence from Two Distinct National Cultures. *Journal of Management Information Systems, 15*(1), 119–141.

Teboul, J., Chen, L., & Fritz, L. (1994). Intercultural Organizational Communication Research in Multinational Organizations. In Wiseman, R. L., & Shuter, R. (Eds.), *Communicating in Multinational Organizations* (pp. 12–29). Thousand Oaks, CA: Sage.

Teo, H. H., Wei, K. K., & Benbasat, I. (2003). Predicting Intention to Adopt Interorganizational Linkages: An Institutional Perspective. *Management Information Systems Quarterly, 27*(1), 19–49.

Teo, T. S. H., Ang, J. S. K., & Pavri, F. N. (1997). The State of Strategic IS Planning Practices in Singapore. *Information & Management, 33*(1), 13–23. doi:10.1016/S0378-7206(97)00033-5

Thompson, J. (1967). *Organizations in Action: Social Science Bases of Administrative Theory.* New York: McGraw-Hill.

UNCTAD. (2004). *World Investment Report 2004: The Shift towards Services.* Geneva, Switzerland: United Nations.

Walczuch, R. M., Singh, S. K., & Palmer, T. S. (1995). An Analysis of the Cultural Motivations for Transborder Data Flow Legislation. *Information Technology & People, 8*(2), 37–57. doi:10.1108/09593849510087994

Watson, R., Ho, T., & Raman, K. (1994). Culture: A Fourth Dimension of Group Support Systems. *Communications of the ACM, 37*(10), 44–55. doi:10.1145/194313.194320

Wiseman, C. (1985). *Strategy and Computers: Information Systems as Competitive Weapons.* New York: Jones-Irwin.

Young, S., & Tavares, A. (2004). Centralization and Autonomy: Back to the Future. *International Business Review, 13*(2), 215–237. doi:10.1016/j.ibusrev.2003.06.002

ENDNOTES

[1] Hofstede and Bond (1988) identified a fifth dimension, Long Term Orientation, but others have suggested that it may be flawed (Fang 2003), and hence the current research does not consider it.

[2] House et al. (2004) have created an interesting, new alternative set of national culture indices. However, critics have challenged their validity and generalizability (Graen 2006; Peterson and Castro, 2006), and the new indices have not yet been widely applied. Moreover, evidence suggests considerable similarity between the new indices and the earlier ones (Hofstede, 2006). Hence the current study used the much more established and recognized Hofstede approach.

[3] An analysis of the same data using reflective constructs (i.e., with the validation of composite reliabilities, loadings, average variance extracted, and the square root of average variance extracted) yielded very similar results; however, the authors felt that the use of the formative ones was theoretically more appropriate.

APPENDIX A: IS PLANNING AUTONOMY

For the following five strategic IS planning phases and their tasks, please circle the number to indicate the extent of autonomy your local company has for each phase and task. (Note: The SAn, SITn, SCn, SSn, and SIPn designators below refer to the phases and tasks in Table 1, and are used in Appendix C and D, but did not appear on the instrument itself.)

	Extent of autonomy in your local company		
	No Autonomy		Full Autonomy
1. Planning the IS planning process (SA*)		1 2 3 4 5	
Determining key planning issues (SA1)		1 2 3 4 5	
Defining planning objectives (SA2)		1 2 3 4 5	
Organizing the planning team(s) (SA3)		1 2 3 4 5	
Obtaining top management commitment (SA4)		1 2 3 4 5	
2. Analyzing the current environment (SIT*)		1 2 3 4 5	
Analyzing current business systems (SIT1)		1 2 3 4 5	
Analyzing current organizational systems (SIT2)		1 2 3 4 5	
Analyzing current information systems (SIT3)		1 2 3 4 5	
Analyzing the current external business environment (SIT4)		1 2 3 4 5	
Analyzing the current external IT environment (SIT5)		1 2 3 4 5	
3. Conceiving strategy alternatives (SC*)		1 2 3 4 5	
Identifying major IT objectives (SC1)		1 2 3 4 5	
Identifying opportunities for improvement (SC2)		1 2 3 4 5	
Evaluating opportunities for improvement (SC3)		1 2 3 4 5	
Identifying high level IT strategies (SC4)		1 2 3 4 5	
4. Selecting strategy (SS*)		1 2 3 4 5	
Identifying new business processes (SS1)		1 2 3 4 5	
Identifying new IT architectures (SS2)		1 2 3 4 5	
Identifying specific new projects (SS3)		1 2 3 4 5	
Identifying priorities for new projects (SS4)		1 2 3 4 5	
5. Planning strategy implementation (SIP*)		1 2 3 4 5	
Defining change management approaches (SIP1)		1 2 3 4 5	
Defining action plans (SIP2)		1 2 3 4 5	
Evaluating action plans (SIP3)		1 2 3 4 5	
Defining follow-up and control procedures (SIP4)		1 2 3 4 5	

APPENDIX B: IS PLANNING EFFECTIVENESS AND CONTRIBUTION

For the following objectives of IS planning, please circle the number to indicate the extent to which your local company has achieved each objective from IS planning. (Note: The ISPEn designators below refer to the objectives in Table 2, and the ISCn designators refer to the IS Contribution measures discussed in the text. Both sets of designators are used in Appendix C and D, but did not appear on the instrument itself.)

	Extent that your local company has achieved this objective
	No Extent
For the CIO:	
Predict future trends (ISPE1)	1 2 3 4 5
Improve short-term information systems performance (ISPE2)	1 2 3 4 5
Improve long-term information systems performance (ISPE3)	1 2 3 4 5
Improve decision making (ISPE4)	1 2 3 4 5
Avoid problem areas (ISPE5)	1 2 3 4 5
Increase user satisfaction (ISPE6)	1 2 3 4 5
Improve systems integration (ISPE7)	1 2 3 4 5
Enhance management development (ISPE8)	1 2 3 4 5
Improve resource allocation (ISPE9)	1 2 3 4 5
For the CEO:	
Align information technology with business needs (ISC1)	1 2 3 4 5
Gain a competitive advantage from information technology (ISC2)	1 2 3 4 5

APPENDIX C: WEIGHTED-MEASURE-TO-COMPOSITE CORRELATIONS

	SA	SIT	SC	SS	SIP	ISPE	ISC
SA*	**.94**	.70	.70	.71	.75	.30	.37
SA3	**.96**	.74	.66	.62	.78	.22	.24
SA4	**.83**	.60	.60	.60	.66	.13	.15
SIT*	-.69	**-.86**	-.70	-.70	-.69	-.29	-.27
SIT2	.73	**.93**	.70	.69	.71	.36	.31
SIT3	.62	**.87**	.70	.72	.68	.34	.32
SIT5	.67	**.92**	.77	.75	.70	.43	.35
SC*	.67	.78	**.94**	.80	.77	.46	.32
SC1	.63	.71	**.93**	.77	.69	.32	.27
SC2	.63	.71	**.93**	.77	.69	.32	.27
SC3	.71	.78	**.90**	.73	.75	.41	.37
SC4	.64	.73	**.93**	.82	.68	.37	.35
SS*	.69	.74	.77	**.91**	.77	.33	.30
SS1	.63	.72	.74	**.92**	.78	.42	.30
SS2	.56	.71	.81	**.91**	.70	.46	.22
SS3	.72	.77	.78	**.91**	.75	.33	.10
SS4	.58	.64	.65	**.84**	.69	.29	.16
SIP*	.75	.69	.80	.79	**.96**	.32	.18
SIP1	.79	.75	.74	.77	**.94**	.40	.29
SIP2	-.75	-.67	-.64	-.66	**-.84**	-.18	-.04
SIP3	.78	.70	.66	.72	**.93**	.29	.27
SIP4	.71	.66	.63	.71	**.88**	.25	.23
ISPE1	.16	.23	.29	.28	.21	**.44**	.22
ISPE2	-.23	-.20	-.26	-.28	-.27	**-.35**	-.08
ISPE3	.25	.35	.38	.41	.35	**.86**	.26
ISPE4	.34	.36	.33	.32	.29	**.62**	.19
ISPE5	-.17	-.16	-.10	-.09	-.13	**-.29**	-.02
ISPE6	.20	.17	.20	.22	.25	**.52**	.11
ISPE8	-.13	-.16	-.22	-.28	-.21	**-.47**	-.16
ISPE9	.20	.33	.37	.37	.31	**.76**	.31
ISC1	.34	.33	.44	.30	.33	.33	**1.00**
ISC2	.06	.27	.15	.09	.06	.23	**.63**

SA*, SIT*, SC*, SS*, and SIP* were the global items for the phases in the left column of Table 1; SAn, SITn, SSn, and SIPn refer to items in the right column in Table 1. Both sets of abbreviations also appear with their items in Appendix A.)

APPENDIX D: WEIGHTED-MEASURE CORRELATIONS

	SA*	SA3	SA4	SIT1	SIT2	SIT3	SIT5	SC	SC1	SC2	SC3	SC4
SA*	1	0.82	0.71	-0.70	0.66	0.67	0.66	0.68	0.60	0.60	0.69	0.64
SA3	0.82	1	0.81	-0.68	0.74	0.59	0.66	0.63	0.59	0.59	0.68	0.58
SA4	0.71	0.81	1	-0.55	0.61	0.55	0.49	0.55	0.60	0.60	0.56	0.55
SIT1	-0.70	-0.68	-0.55	1	-0.85	-0.78	-0.79	-0.69	-0.65	-0.65	-0.73	-0.62
SIT2	0.66	0.74	0.61	-0.85	1	0.75	0.76	0.69	0.60	0.60	0.74	0.62
SIT3	0.67	0.59	0.55	-0.78	0.75	1	0.72	0.70	0.63	0.63	0.71	0.61
SIT5	0.66	0.66	0.49	-0.79	0.76	0.72	1	0.76	0.71	0.71	0.72	0.75
SC*	0.68	0.63	0.55	-0.69	0.69	0.70	0.76	1	0.87	0.87	0.78	0.84
SC1	0.60	0.59	0.60	-0.65	0.60	0.63	0.71	0.87	1	1	0.73	0.86
SC2	0.60	0.59	0.60	-0.65	0.60	0.63	0.71	0.87	1	1	0.73	0.86
SC3	0.69	0.68	0.56	-0.73	0.74	0.71	0.72	0.78	0.73	0.73	1	0.73
SC4	0.64	0.58	0.55	-0.62	0.62	0.61	0.75	0.84	0.86	0.86	0.73	1
SS*	0.73	0.61	0.60	-0.77	0.68	0.73	0.70	0.77	0.70	0.70	0.70	0.73
SS1	0.66	0.57	0.52	-0.60	0.62	0.63	0.70	0.68	0.66	0.66	0.67	0.70
SS2	0.57	0.51	0.50	-0.61	0.60	0.64	0.70	0.78	0.75	0.75	0.66	0.81
SS3	0.73	0.66	0.66	-0.76	0.70	0.73	0.74	0.71	0.75	0.75	0.69	0.74
SS4	0.59	0.56	0.61	-0.64	0.60	0.65	0.59	0.62	0.65	0.65	0.60	0.62
SIP*	0.71	0.73	0.66	-0.61	0.63	0.62	0.64	0.78	0.74	0.74	0.73	0.71
SIP1	0.75	0.75	0.65	-0.68	0.72	0.62	0.69	0.75	0.62	0.62	0.70	0.63
SIP2	-0.64	-0.77	-0.74	0.65	-0.67	-0.56	-0.60	-0.64	-0.61	-0.61	-0.57	-0.56
SIP3	0.69	0.78	0.62	-0.68	0.68	0.59	0.63	0.64	0.57	0.57	0.65	0.53
SIP4	0.63	0.72	0.64	-0.67	0.64	0.56	0.60	0.61	0.55	0.55	0.58	0.54
ISPE1	0.17	0.13	0.09	-0.23	0.21	0.19	0.23	0.27	0.24	0.24	0.27	0.25
ISPE2	-0.17	-0.22	-0.30	0.18	-0.17	-0.17	-0.18	-0.22	-0.24	-0.24	-0.22	-0.24
ISPE3	0.24	0.22	0.17	-0.20	0.30	0.27	0.31	0.33	0.27	0.27	0.36	0.36
ISPE4	0.36	0.25	0.24	-0.31	0.34	0.34	0.33	0.30	0.22	0.22	0.31	0.29
ISPE5	-0.16	-0.16	-0.12	0.12	-0.17	-0.19	-0.10	-0.08	-0.06	-0.06	-0.11	-0.12
ISPE6	0.15	0.20	0.16	-0.09	0.14	0.09	0.13	0.13	0.16	0.16	0.19	0.17
ISPE8	-0.18	-0.07	-0.13	0.10	-0.15	-0.17	-0.11	-0.21	-0.18	-0.18	-0.18	-0.23
ISPE9	0.25	0.17	0.12	-0.27	0.28	0.30	0.34	0.39	0.29	0.29	0.34	0.32
ISC1	0.39	0.24	0.15	-0.26	0.31	0.31	0.34	0.32	0.26	0.26	0.37	0.34
ISC2	0.08	0.14	0.12	-0.27	0.25	0.28	0.27	0.15	0.24	0.24	0.17	0.23

APPENDIX D: WEIGHTED-MEASURE CORRELATIONS (CONTINUED)

	SS*	SS1	SS2	SS3	SS4	SIP*	SIP1	SIP2	SIP3	SIP4
SA*	0.73	0.66	0.57	0.73	0.59	0.71	0.75	-0.64	0.69	0.63
SA3	0.61	0.57	0.51	0.66	0.56	0.73	0.75	-0.77	0.78	0.72
SA4	0.60	0.52	0.50	0.66	0.61	0.66	0.65	-0.74	0.62	0.64
SIT1	-0.77	-0.60	-0.61	-0.76	-0.64	-0.61	-0.68	0.65	-0.68	-0.67
SIT2	0.68	0.62	0.60	0.70	0.60	0.63	0.72	-0.67	0.68	0.64
SIT3	0.73	0.63	0.64	0.73	0.65	0.62	0.62	-0.56	0.59	0.56
SIT5	0.70	0.70	0.70	0.74	0.59	0.64	0.69	-0.60	0.63	0.60
SC*	0.77	0.68	0.78	0.71	0.62	0.78	0.75	-0.64	0.64	0.61
SC1	0.70	0.66	0.75	0.75	0.65	0.74	0.62	-0.61	0.57	0.55
SC2	0.70	0.66	0.75	0.75	0.65	0.74	0.62	-0.61	0.57	0.55
SC3	0.70	0.67	0.66	0.69	0.60	0.73	0.70	-0.57	0.65	0.58
SC4	0.73	0.70	0.81	0.74	0.62	0.71	0.63	-0.56	0.53	0.54
SS*	1	0.79	0.77	0.85	0.79	0.76	0.73	-0.65	0.69	0.68
SS1	0.79	1	0.75	0.82	0.76	0.74	0.74	-0.54	0.61	0.58
SS2	0.77	0.75	1	0.77	0.72	0.67	0.64	-0.57	0.62	0.63
SS3	0.85	0.82	0.77	1	0.84	0.75	0.68	-0.67	0.65	0.65
SS4	0.79	0.76	0.72	0.84	1	0.65	0.64	-0.61	0.64	0.64
SIP*	0.76	0.74	0.67	0.75	0.65	1	0.85	-0.84	0.85	0.81
SIP1	0.73	0.74	0.64	0.68	0.64	0.85	1	-0.79	0.83	0.79
SIP2	-0.65	-0.54	-0.57	-0.67	-0.61	-0.84	-0.79	1	-0.87	-0.89
SIP3	0.69	0.61	0.62	0.65	0.64	0.85	0.83	-0.87	1	0.91
SIP4	0.68	0.58	0.63	0.65	0.64	0.81	0.79	-0.89	0.91	1
ISPE1	0.30	0.23	0.27	0.22	0.12	0.19	0.21	-0.11	0.14	0.11
ISPE2	-0.30	-0.24	-0.21	-0.28	-0.25	-0.28	-0.21	0.29	-0.24	-0.24
ISPE3	0.30	0.40	0.42	0.28	0.20	0.30	0.34	-0.21	0.27	0.26
ISPE4	0.33	0.25	0.25	0.32	0.25	0.23	0.24	-0.25	0.25	0.22
ISPE5	-0.14	-0.04	-0.07	-0.14	-0.09	-0.12	-0.08	0.19	-0.16	-0.20
ISPE6	0.18	0.20	0.19	0.08	0.09	0.20	0.19	-0.16	0.29	0.20
ISPE8	-0.22	-0.27	-0.27	-0.19	-0.18	-0.16	-0.19	0.12	-0.21	-0.15
ISPE9	0.29	0.34	0.36	0.36	0.34	0.24	0.31	-0.12	0.24	0.20
ISC1	0.30	0.30	0.22	0.09	0.16	0.20	0.29	-0.04	0.27	0.22
ISC2	0.14	0.12	0.19	0.09	0.05	-0.01	0.13	-0.09	0.16	0.17
	ISPE1	ISPE2	ISPE3	ISPE4	ISPE5	ISPE6	ISPE8	ISPE9	ISC1	ISC2
SA*	0.17	-0.17	0.24	0.36	-0.16	0.15	-0.18	0.25	0.39	0.08
SA3	0.13	-0.22	0.22	0.25	-0.16	0.20	-0.07	0.17	0.24	0.14
SA4	0.09	-0.30	0.17	0.24	-0.12	0.16	-0.13	0.12	0.15	0.12
SIT1	-0.23	0.18	-0.20	-0.31	0.12	-0.09	0.10	-0.27	-0.26	-0.27
SIT2	0.21	-0.17	0.30	0.34	-0.17	0.14	-0.15	0.28	0.31	0.25
SIT3	0.19	-0.17	0.27	0.34	-0.19	0.09	-0.17	0.30	0.31	0.28

SIT5	0.23	-0.18	0.31	0.33	-0.10	0.13	-0.11	0.34	0.34	0.27
SC*	0.27	-0.22	0.33	0.30	-0.08	0.13	-0.21	0.39	0.32	0.15
SC1	0.24	-0.24	0.27	0.22	-0.06	0.16	-0.18	0.29	0.26	0.24
SC2	0.24	-0.24	0.27	0.22	-0.06	0.16	-0.18	0.29	0.26	0.24
SC3	0.27	-0.22	0.36	0.31	-0.11	0.19	-0.18	0.34	0.37	0.17
SC4	0.25	-0.24	0.36	0.29	-0.12	0.17	-0.23	0.32	0.34	0.23
SS*	0.30	-0.30	0.30	0.33	-0.14	0.18	-0.22	0.29	0.30	0.14
SS1	0.23	-0.24	0.40	0.25	-0.04	0.20	-0.27	0.34	0.30	0.12
SS2	0.27	-0.21	0.42	0.25	-0.07	0.19	-0.27	0.36	0.22	0.19
SS3	0.22	-0.28	0.28	0.32	-0.14	0.08	-0.19	0.36	0.09	0.09
SS4	0.12	-0.25	0.20	0.25	-0.09	0.09	-0.18	0.34	0.16	0.05
SIP*	0.19	-0.28	0.30	0.23	-0.12	0.20	-0.16	0.24	0.20	-0.01
SIP1	0.21	-0.21	0.34	0.24	-0.08	0.19	-0.19	0.31	0.29	0.13
SIP2	-0.11	0.29	-0.21	-0.25	0.19	-0.16	0.12	-0.12	-0.04	-0.09
SIP3	0.14	-0.24	0.27	0.25	-0.16	0.29	-0.21	0.24	0.27	0.16
SIP4	0.11	-0.24	0.26	0.22	-0.20	0.20	-0.15	0.20	0.22	0.17
ISPE1	1	-0.37	0.47	0.35	-0.36	0.39	-0.39	0.43	0.20	0.37
ISPE2	-0.37	1	-0.53	-0.42	0.46	-0.62	0.39	-0.44	-0.07	-0.16
ISPE3	0.47	-0.53	1	0.49	-0.40	0.60	-0.49	0.52	0.25	0.24
ISPE4	0.35	-0.42	0.49	1	-0.57	0.43	-0.55	0.51	0.19	0.15
ISPE5	-0.36	0.46	-0.40	-0.57	1	-0.56	0.40	-0.45	0.00	-0.17
ISPE6	0.39	-0.62	0.60	0.43	-0.56	1	-0.47	0.48	0.10	0.15
ISPE8	-0.39	0.39	-0.49	-0.55	0.40	-0.47	1	-0.53	-0.15	-0.13
ISPE9	0.43	-0.44	0.52	0.51	-0.45	0.48	-0.53	1	0.30	0.29
ISC1	0.20	-0.07	0.25	0.19	0.00	0.10	-0.15	0.30	1	0.57
ISC2	0.37	-0.16	0.24	0.15	-0.17	0.15	-0.13	0.29	0.57	1

This work was previously published in International Journal of Global Information Management, Volume 18, Issue 3, edited by Felix B. Tan, pp. 1-34, copyright 2010 by IGI Publishing (an imprint of IGI Global).

Chapter 3
Cultural Impacts on Acceptance and Adoption of Information Technology in a Developing Country

Elizabeth White Baker
Virginia Military Institute, USA

Said S. Al-Gahtani
King Khalid University, Saudi Arabia

Geoffrey S. Hubona
Georgia State University, USA

ABSTRACT

This study investigates technology adoption behavior of Saudi Arabian knowledge workers using desktop computers within the context of TAM2, and the unique effects of Saudi culture on IT adoption within the developing, non-Western, country. Following the guidelines of the etic-emic research tradition, which encourages cross-cultural theory and framework testing, the study findings reveal that the TAM2 model accounts for 40.3% of the variance in behavioral intention among Saudi users, which contrasts Venkatesh and Davis' (2000) explained 34-52% of the variance in usage intentions among U.S. users. The model's explanatory power differs due to specific Saudi Arabian emic constructs, including its collectivist culture and the worker's focus on the managerial father figure's influence on individual performance, a stark difference from TAM findings in more individualistic societies. The authors' findings contribute to understanding the effects of cultural contexts in influencing technology acceptance behaviors, and demonstrate the need for research into additional cultural factors that account for technology acceptance.

INTRODUCTION

Advances in information systems technology (IT) are rapidly modernizing the way we live and work across the globe. Despite incredible advances in technology, organizations are still facing the problems of underutilization or rejection of implemented technologies. In developing countries, the adoption and use of technology in organizational settings is a topic of increasing interest. The vast number of findings on adoption and use of organizational information systems in

DOI: 10.4018/978-1-61350-480-2.ch003

developed nations are not necessarily applicable to less developed nations. With the accelerating trend of transnational globalization of companies, in particular the extensive economic developments in the countries within the Gulf Cooperation Council (GCC), it is necessary to understand how models of technology acceptance apply to organizations that are located in these regions. Such an understanding will assist the growing number of organizations in these regions to further their economic development in the globalized economy through the adoption of information technology.

Due to differences in cultural social norms, beliefs and behaviors, it is reasonable to expect that the impact of social influence processes, such as subjective norm, voluntariness and image, on an individual's acceptance of technologies in developing countries, such as the GCC, might differ substantively from industrialized Western nations, such as those in North America and western Europe (Hubona, Truex, Wang, & Straub, 2006). Using a specific example of a GCC nation, Saudi Arabia, technology acceptance success factors have been reported to differ from those in developed nations (i.e., individual, technology and organizational factors (Al-Gahtani, 2004)). These disparate findings encourage further research into technology acceptance factors in developing nations. Along these lines of inquiry, this individual research was part of a larger group of studies, including the Theory of Planned Behavior (TPB) (Baker, Al-Gahtani, & Hubona, 2007) and the Unified Theory of Acceptance and Use of Technology (UTAUT) (Al-Gahtani, Hubona, & Wang, 2007), to further investigate technology acceptance factors in a developing country with a non-Western culture, specifically Saudi Arabia. What makes Saudi Arabia unique is their societal structure; they are strongly attached among each other through family bonds and obligations, limiting their geographic

and occupational mobility (Palmer, Alghofaily, & Alminir, 1984). Hill et al. (1998) suggest that while selectively borrowing ideas from Western culture, Arab culture and history is a complex cultural system with contradictions and opposing forces. Arab culture seems to exert a stronger social influence than Western culture on its society through the development and enforcement of social norms and common beliefs (Al-Gahtani & Shih, 2009). In addressing the interaction between organizational culture and IT from the social identity perspective (Tajfel, 1978), Gallivan and Srite (2005) argue that in-group relationships among members would initially cause fragmented and differentiated interpretations of their identities, which would ultimately shape their beliefs and behaviors of IT in the cognitive process.

Using the extended Technology Acceptance Model (TAM2) (Venkatesh & Davis, 2000), this study investigates how social influence factors and cognitive instrumental factors affect technology acceptance in Saudi Arabia. The primary contribution of this research is to utilize TAM2 to predict technology acceptance of information systems in Saudi Arabia, a developing GCC country, while also examining any Saudi cultural influences that might affect technology acceptance. The next section relates the theoretical background of TAM2 for investigating technology acceptance in a developing country and discusses the primary TAM2 constructs and how they might be influenced by Saudi culture. The third section details the research methodology and explains the survey sample characteristics and measures. The fourth section describes the data analysis procedure and presents the results of the study. The fifth section considers the implications of the findings for both researchers and practitioners. Finally, the last section presents and discusses the conclusions of the study.

THEORY AND BACKGROUND

Technology Acceptance and its Importance in Developing Countries

Heavy investment in information technology can present a tremendous economic burden on developing countries, especially those without vast resources. However, significant investment in IT and IT's subsequent acceptance can be enormously beneficial to a developing country's economy, as potential gains in productivity can offset the high investment cost in IT. According to the latest Organization for Economic Cooperation and Development (OECD) Information Technology Outlook (2008), the world IT market (hardware, software and services) grew at an annual rate of approximately 5% a year over 2000-06 in current US dollars, and was slightly slower at 4% in 2008. The information and communication technologies (ICT) industry contributes over 8% of total business value added and employs more than 15 million people directly throughout OECD countries. IT spending continues to be strong, with the revenues of the top ICT firms now over 42% above the 2000 figures. These statistics reported by OECD show that many developing countries are far behind developed countries in spending on IT acquisition, although as more and more developing countries continue to build their IT infrastructure, their IT spending continues to rise proportionally with that infrastructure building. As the developing country exemplar in this study, Saudi Arabia engaged in $8 billion in ICT goods trade in 2007, with 12.2% growth from 2000-2006, with a total growth of 100% year over year through 2009 (Organization for Economic Cooperation and Development (OECD), 2008). With a rapidly rising population, there is expected strong growth in the Saudi IT market for the foreseeable future.

To enhance technology acceptance and diffusion, the Saudi government has implemented a policy of Saudization to motivate native employees to work in technical fields, offering a bonus of 25% of base salary to Saudi nationals who are specialists in the information technology field. Within the past three years, Saudi Telecommunications has become one of the top 100 ICT firms globally, a reflection of the ongoing globalization of ICT production and the progressive technology policies of the Kingdom. In this same period the total growth in the value of Saudi electronics production has increased 3.93% (Organization for Economic Cooperation and Development (OECD), 2008). The advent of the Internet in global business and the impetus from government authorities toward e-commerce and e-government are essential catalysts for individuals and organizations toward IT adoption for their competitive advantage as themselves and for their country. Yet, social contexts and political environments still implicitly control Arab society and govern the development of organizational culture with respect to IT. This government-sanctioned technology adoption policy accentuates Saudi Arabia as a unique instance of developing country IT adoption patterns.

The IT literature on developing countries has progressed to include a range of forces driving technology acceptance. Initially, studies indicated significant resistance to adopting and using computer resources in developing countries based on social and cultural factors (Abdul-Gader, 1999). Rose and Straub (1998) conducted one of the first studies of technology acceptance and use in the Arab world. Using a cross-sectional survey of 274 knowledge workers in five Arab nations, including Egypt, Jordan, Saudi Arabia, Lebanon, and the Sudan, they applied a modified version of TAM to assess the diffusion of personal computing (PC). Their model explained 40% of the variance of PC use in these Arab nations. Subsequently, Straub et al. (2001) developed a *cultural influence model* of IT transfer and suggested that Arab cultural beliefs were a strong predictor of resistance to IT transfer. Loch et al. (2003) applied the cultural influence model to examine enablers and impediments to the adoption and use of the Internet in the Arab world. They showed that both *social norms* and the degree of *technological culturation* can impact the individual and organizational acceptance and use of the Internet. Other studies (Hill et al., 1998; Hill, Straub, Loch, Cotterman, & El-Sheshai, 1994) have also indicated that both cultural beliefs and the degree of technological culturation significantly impact the transfer of IT to the Arab world.

Approaching Cross-Cultural Research

There are many theoretical frameworks of the factors promoting/inhibiting the adoption and use of IT in organizational settings, but few have been validated in non-Western cultures. The lack of frameworks that have been demonstrated to be robust across cultures can limit the development of theoretical extensions in this area (Maheswaran & Shavitt, 2000). The choice of *emic* research, indigenous and conducted on the basis of culture-specific frameworks, versus *etic* research, which examines cultural differences using previously established universal frameworks as benchmarks, can confuse the issue about the most appropriate orientation of cross-cultural research (Maheswaran & Shavitt, 2000; Morris, Leung, Ames, & Lickel, 1999; Peng, Peterson, & Shyi, 1991).

In conducting this research, the investigation follows Berry's (1989) five-step process as a basis for an integrated etic/emic approach to studying IT adoption differences among cultures. The first step is to examine the research problem in one's own culture, developing a conceptual framework and a relevant instrument. The study conducted by Venkatesh et al. (2000) provides the foundation for an initial emic study of the adoption of IT among professional workers. The second step Berry recommends is to transport this measurement model so as to examine the same issue in another culture, as an imposed etic study. Accordingly, an objective of our study is to report the findings of an imposed etic study of

predicted IT use in the Saudi workforce. According to Berry (1989), the third step is to enrich the imposed etic framework with unique aspects of the second culture, and to then examine the two, culturally-diverse sets of findings for comparability. Accordingly, the findings from this study can be leveraged in future technology acceptance studies to continue to investigate predicted IT use in culturally-diverse settings.

Saudi Arabia and the Importance of Technology Acceptance

Saudi Arabia is a developing country where IT adoption and individual technology acceptance is being influenced by explicit government policy in the attempt to enhance national organizational productivity. Saudi Arabia encompasses over 2 million square kilometers with a population of 24.2 million in 2007 (World Bank Group, 2009), including approximately 5.6 million foreign residents. The mobile cellular subscriptions per 100 people had risen from 13 in 2005 to 117 in 2007. Furthermore, Saudi Internet users were estimated to number 6.3 million in 2007. Unlike many developing countries, the Saudis do not suffer from financial resource limitations, although despite these abundant fiscal resources, Saudi Arabia has historically been characterized by the underutilization of available computing capacity (Atiyyah, 1989; Yavas, Luqmani, & Quraeshi, 1992).

With systemic policy support from the Saudi government, IT adoption should be promoted throughout Saudi Arabia. However, Arab workers are also heavily influenced by the existing social structure, and by the associated norms, values and expectations of the populace (Bjerke & Al-Meer, 1993). Atiyyah (1989) found that information technology transfer (ITT) is often hampered by technical, organizational, and human problems in Saudi Arabia. Cultural conflicts arise from the clash of management styles characteristic of Western and Arabic institutional leaders. These conflicts have affected workers and have impacted the system development process, fostering unsuccessful approaches to computer use and policy (Ali, 1990; Atiyyah, 1989; Goodman & Green, 1992). A deeper understanding of how the Saudi worker interacts with computing environments could facilitate the adoption of IT throughout the Kingdom.

Social and cultural characteristics of Arab and Muslim societies differ from those of the West. Saudi Arabia in particular is a conservative country where Islamic teachings and Arabian cultural values are dominant. The country falls along a spectrum of cultural characteristics of GCC countries, distinctly tribal, conservative in its adherence to Islam and influenced by significant exposure to the West (Dadfar, Norberg, Helander, Schuster, & Zufferey, 2003; Dadfar, 1990). Hofstede (2001) found that the large power distance value (80) combined with the large uncertainty avoidance value (68) reflects the Saudi society dominated by the patriarchs of society (the elites) and as a whole not readily accepting of change and very risk averse. Additionally, Saudi Arabia exhibits a highly rule-oriented culture with laws, rules, controls and expectations to reduce the amount of uncertainty experienced. In this way, the acceptance of Islam religion has influenced social values and practices in Saudi Arabia, as Sunni Muslim practice stresses collectivity (Kabasakal & Bodur, 2002). With the individualism ranking at 38, compared to a world average of 64, this overarching collectivity of Saudi society manifests in a close, long-term commitment to the member-group of a family or extended family, led by a patriarchal figure entrusted with the well-being and future care of his group. Loyalty to this group in a collectivist culture like Saudi Arabia is paramount and overrides most other societal rules (Hofstede, 2001). This leads to a loyalty to the father figure relationship that pervades Saudi organizational culture (Pillai, Scandura, & Williams, 1999), as opposed to individual initiatives and achievement (Yavas, 1997).

Each country also has different policies with respect to IT, ranging from positive and supportive of IT to not so distinctly so (Abdul-Gader, 1999). Most IT policies are part of larger economic policies aimed at spurring development. Specifically, there are social and cultural characteristics that impact individual technology acceptance (Abdul-Gader, 1999; Loch et al., 2003; Straub et al., 2001), and particularly, the study is concerned with the following question: How do social and cultural characteristics, particularly those related to beliefs, attitudes, and social norms, affect the technology acceptance among Saudi Arabian professional workers? Our methodology to test this question probes whether specific characteristics of the Saudi workforce influence technology acceptance and whether these characteristics have differential ramifications for existing IT acceptance models tested in developed countries (Ein-Dor, Segev, & Orgad, 1992).

It is not immediately evident that these tested technology acceptance models would be adequate to describe technology acceptance in developing countries, particularly Saudi Arabia. Al-Gahtani (2004), who researched technology acceptance success factors in Saudi Arabia, reported that anxiety played a significant role in determining individuals' intention to use technology. However, Venkatesh et al. (2003) found that anxiety did not have a significant effect on intention to use technology (in a North American context), and they accordingly eliminated the anxiety construct from their overall model. These disparate findings with respect to the role of anxiety on intention to use can be explained in terms of differences in the cultural environment. These disparate findings also suggest the possibility that applications of technology acceptance models in non-North American contexts may reveal significant differences in the underlying causal factors, thereby promoting our understanding of the mechanisms whereby individuals accept or reject technologies globally.

In our study, TAM2 (Venkatesh & Davis, 2000) is tested in the context of Saudi Arabia, a technologically-developing Arab nation. Since the technology acceptance model is built on the foundation of an individual's beliefs, specifically perceived usefulness and perceived ease of use, it is rational to expect that the application of any given technology acceptance model in the two different cultural contexts could produce different results. It is hoped that the findings of this study shall shed light upon the antecedents of technology acceptance and use in Saudi Arabia, thereby providing us an opportunity to refine the model to suit that country's unique cultural context and offering a path to investigate technology adoption characteristics in other developing countries.

Theories of Technology Acceptance: Technology Acceptance Model 2

The original TAM (Davis, 1989; Davis, Bagozzi, & Warshaw, 1989) proposed that two belief constructs, perceived usefulness and perceived ease of use, are primary determinants of an individual's behavioral intention to use an information technology. Behavioral intention was postulated as the predictor of actual usage behavior. TAM2, proposed by Venkatesh and Davis (2000), was the first extension of the original TAM model that postulated particular antecedent constructs as predictors of perceived usefulness, the primary determinant of intention to use a system. These TAM2 antecedent constructs for perceived usefulness included social influence process variables (e.g., subjective norm, voluntariness, and image), as well as cognitive instrumental process variables (e.g., job relevance, output quality, and result demonstrability). Subjective norm, a social influence process, was also postulated to be a direct antecedent variable of intention to use a system. TAM2 is shown in Figure 1.

Social Influence Processes

TAM2 describes the impact of three social influence processes, subjective norm, voluntariness and image, as affecting an individual's decision to adopt or reject a new system. *Subjective norm* is defined as "the perceived social pressure to perform or not to perform the behavior" by the individual (Fishbein & Ajzen, 1975). Studies prior to TAM2 that examined the effect of subjective norm on intention yielded mixed results. For instance, Mathieson (1991) found no significant effect of subjective norm on intention, whereas Taylor and Todd (1995) reported a significant effect. TAM2 captured this compliance-based direct effect of subjective norm on intention over and above perceived usefulness and perceived ease of use, but found that it occurred only in mandatory use situations.

In contrast to TRA (Ajzen & Fishbein, 1980) and the theory of planned behavior (TPB) (Ajzen, 1991), which explained only the compliance-based effects of subjective norm on intention to use a system, TAM2 incorporated two additional theoretical mechanisms, in the form of 'internalization' and 'identification' to explain the same effect. 'Internalization' refers to the process by which, when one perceives that an important referent thinks he should use a system, one incorporates the referent's belief into one's own belief structure (Kelman, 1958; Warshaw, 1980). In the case of internalization, subjective norm has an indirect effect on intention through perceived usefulness, as opposed to a direct effect on intention. TAM2 theorized that internalization, unlike compliance, will occur whether the context of system use is mandatory or voluntary. Additionally, TAM2 also theorized that subjective norm would positively influence image. This source of social influence is referred to as 'identification'. TAM2 theorized that 'identification', like 'internalization', but unlike compliance, will occur whether the context of system use is mandatory or voluntary.

Figure 1. TAM2 model

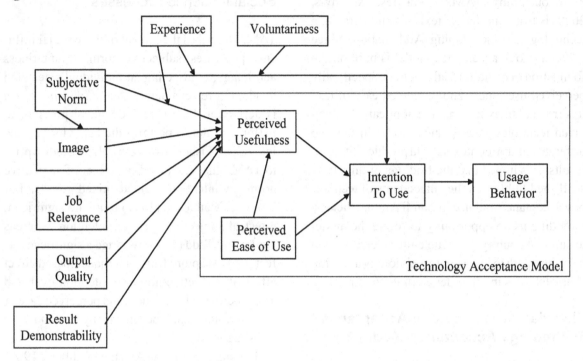

According to TAM2, the direct effect of subjective norm on both intention and perceived usefulness would attenuate over time. In the first case, the direct effect of subjective norm on intention would weaken with increasing personal experience using a system, because that experience provides a growing basis for intention towards ongoing use, regardless of the initial influence of normative social cues. In the second case, the direct effect of subjective norm on perceived usefulness would weaken with increasing experience using a system, because this experience will furnish concrete sensory information that will decrease the individual's reliance on social cues to form perceptions of the usefulness of a given system.

Cognitive Instrumental Processes

TAM2 also reflects the effects of four cognitive instrumental processes (job relevance, output quality, result demonstrability and perceived ease of use) that people use for assessing the match be-

tween important work goals and their perceptions of the usefulness of a given system. Job relevance is an "individual's perception regarding the degree to which the target system is applicable to his or her job," and output quality is an "individual's perception regarding how well the system performs the required tasks." (Venkatesh & Davis, 2000, p.191) Result demonstrability is the "tangibility of the results of using the innovation." (Venkatesh & Davis, 2000, p.191) Perceived ease of use is an individual's perception about how easy it is to use a given system. TAM2 theorizes that the above defined cognitive instrumental processes do not change (i.e., become stronger or weaker) with increasing experience.

In the published study by Venkatesh and Davis (2000), the application of TAM2 across four longitudinal field studies revealed that both the social influence processes (subjective norm, voluntariness and image) and cognitive instrumental processes (job relevance, output quality, result demonstrability, and perceived ease of use) significantly influenced user acceptance. The results showed that the model accounted for 40-60% of the variance in usefulness perceptions and 34-52% of the variance in usage intentions. While the most recent reformulation of TAM is reflected in UTAUT (Venkatesh et al., 2003), this particular emic study using TAM2 focuses on the social influence processes and the cognitive instrumental processes within a developing country and their impact through a cultural lens on technology adoption. (As noted earlier, additional research has been done on UTAUT in developing countries (Al-Gahtani et al., 2007)).

RESEARCH MODEL AND HYPOTHESES

Our research model differs from TAM2 in two key respects. In our study, a large multi-organizational survey, data on the model constructs were collected at a single point in time (see section IV, Method), as opposed to collecting several measurements over time. Our focus is on how well the social influence processes and cognitive instrumental processes predict technology adoption, in particular cultural effects on these processes which affect technology adoption. Also, the usage construct of the TAM2 model was not included, as the focus was targeted on cultural effects on behavioral intention to use technology (Figure 2).

In a rigidly hierarchical society, such as Saudi Arabia, where social expectations for behavior are rigid and generally accepted (Dadfar et al., 2003; Dadfar, 1990), workers would be subject to social pressure to conform to appropriate behavior in the workplace, including adopting new technologies, if those above them in the organizational hierarchy were also adopting the new technologies (Bjerke & Al-Meer, 1993). Indeed, responding to hierarchical influences within the culture that transfer to the workplace, workers in

Figure 2. Research model

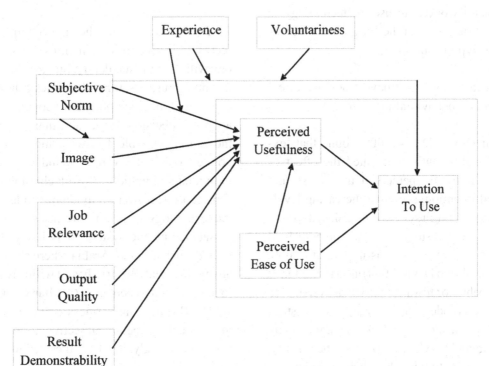

Saudi culture would perceive a system as useful if others who influence the worker's behavior thought they should use the system (Ali, 1990; Atiyyah, 1989). Therefore, it is hypothesized that:

H₁: Subjective norm will have a positive direct effect on perceived usefulness.

Additionally, TAM2 shows that subjective norm exerts a significant direct effect on usage intentions over and above perceived usefulness and perceived ease of use for mandatory systems, but not voluntary systems (Venkatesh & Davis, 2000). In the Saudi culture where there is a high power distance with subordinates likely to acknowledge the power of others simply based on where they are situated in certain formal, hierarchical positions (Hofstede, 1984, 2001), there is distinct social and cultural pressure from their superiors in the organization to conform. Therefore, it is anticipated that Saudi workers would behave as if technology acceptance were for a mandatory system, more so than employees in workplaces where power distance is lower, as it is mandatory for the Saudi workers to use the technology to maintain close ties with the 'in-group'. Accordingly, it is hypothesized that:

H₂: Subjective norm will have a positive direct effect on behavioral intention.

Venkatesh and Davis (2000) found that 'experience' significantly moderated the effect of subjective norm on perceived usefulness and behavioral intention, such that increasing levels of experience using a system diminished the positive direct effect of subjective norm on perceived usefulness and intention. It is important to note that Venkatesh and Davis (2000) used individual's adoption behavior toward 'particular systems' to test the TAM2 model. Our research, however, studies the adoption and use of desktop computers by Saudi Arabian knowledge workers. Specifically, this study is not studying the adoption and use of

a particular application or system, but the general use of desktop computers in the context of multiple software applications. In this non-system-specific context, and with the pervasiveness of the Saudi hierarchical conforming society in the workplace (Bjerke & Al-Meer, 1993), it is unlikely that the effect of subjective norm on intention to use a system will diminish with increasing experience. However, as with any workers regardless of cultural heritage, experience with a new technology does moderate the effect of subjective norm on perceived usefulness (Compeau & Higgins, 1995; Moore & Benbasat, 1993). A worker will not look to those around him to demonstrate the usefulness of a new technology if he has used the system himself and knows it to be useful. Accordingly, it is hypothesized that:

H₃: Experience will not significantly moderate the positive effect of subjective norm on behavioral intention.

H₄: The positive direct effect of subjective norm on perceived usefulness will attenuate with increased experience.

TAM2 theorizes that the direct compliance effect of subjective norm on intention to use technology will occur in mandatory, but not in voluntary, technology usage settings. It defines voluntariness (the extent to which potential adopters view the adoption decision to be non-mandatory) as a moderating variable. Hartwick and Barki (1994) pointed out that, even in mandatory settings, individuals' intention to use technology varies. This can result from individuals' underutilizing the technology, or in extreme cases, by rejecting or sabotaging the system. It is likely that in a society such as Saudi Arabia where social norms are hierarchical and strictly and willfully adhered to for societal acceptance (Dadfar et al., 2003; Dadfar, 1990), voluntariness will not, however, moderate the effect of subjective norm on intention. Accordingly, it is hypothesized that:

H₅: Voluntariness will not significantly moderate the effect of subjective norm on behavioral intention to use computers.

Image is defined as "the degree to which use of an innovation is perceived to enhance one's status in one's social system" (Moore & Benbasat, 1991). TAM2 theorizes that subjective norm will positively affect image because, if influential members of a person's social group at work believe that the person should perform a behavior, then performing it will tend to elevate the person's standing within the group (Pfeffer, 1982; Venkatesh & Davis, 2000). Within Saudi Arabia's high power distance cultural setting, this identification effect is enhanced, as a worker might perceive that using a particular system will lead to improvements in job performance (the definition of perceived usefulness) indirectly due to his enhanced image in the eyes of his organizational superiors. These benefits from image are over and above any performance benefits directly attributable to actual system usage. This identification effect is presented in TAM2 by the effect of subjective norm on image, and image on perceived usefulness. Accordingly, it is hypothesized that:

H₆: Subjective norm will have a positive direct effect on image.

H₇: Image will have a positive direct effect on perceived usefulness.

In TAM2 where there are no cultural factors to attenuate the effects of the proposed model, it is expected that the model will perform as it has in previous studies of technology adoption. In particular, the hypotheses that address the cognitive instrumental process variables (H₈₋₁₀) will not be affected by cultural or social indicators and are considered to be independent of cultural environment. Accordingly, it is hypothesized that:

H₈: Output quality will have a positive direct effect on perceived usefulness.

H₉: Job relevance will have a positive direct effect on perceived usefulness.

H₁₀: Results demonstrability will have a positive direct effect on perceived usefulness.

METHOD

Population Sample

The population sample for this study consisted of white collar knowledge workers reporting on "any hands on uses of computers for the purpose of their work." Participants in the study were knowledge workers within 56 private and public sector organizations in Saudi Arabia, including banking, merchandising, manufacturing, and petroleum industries, engaged in the use of desktop "computers for the purpose of their work." Originally, 136 public and private organizations were contacted. Eventually, 56 organizations participated in the study, of which 66.4% were from the public sector, and 33.6% were from the private sector. The returned usable responses from participants numbered 1190, with a response rate of slightly over 62%. Of the 1190 participants, 339 respondents were from the private sector (28.5%) and 851 (71.5%) were from the public sector, with 8.5% non-Saudi respondents. 79.3% of the respondents were men, while 20.7% were women. We recognize the imbalance of male gender representation in our sample. However, due to the cultural preponderance of working males in Saudi Arabia, there was nothing that could be done to correct this imbalance. Respondents ranged in age from 18 to 58, with the mean age of the respondents being 38.2 years old. The survey instrument is presented in Appendix A.

Measurement Items

All survey items, originally published in English, were adapted for this study by translating them into Arabic using Brislin's (1986) back translation method. The items were translated back and forth between English and Arabic by several bilingual professors. The process was repeated until both versions converged.

Table 1 indicates the survey items used to measure the Saudi predictor latent constructs. Similar to Venkatesh and Davis (2000), seven-point Likert scales with anchors ranging from 'strongly disagree' to 'strongly agree' were used in measuring the items.

Table 2 indicates the survey items used to measure the Saudi moderator variables. Experience was measured using five ordinal categories. Voluntariness was measured using the response of Saudi knowledge workers to three statements:

(1) My use of computers is voluntary (as opposed to being required by my superiors or job description); (2) My boss does not require me to use computers; and (3) Using computers or other alternatives is completely up to me. These were likert scale items with anchors ranging from 'strongly disagree' to 'strongly agree'.

Table 3 indicates the items used to measure the Saudi latent predicted variables. Measures of reliability and validity of these constructs are reported in the 'Results' section.

RESULTS

The research model shown in Figure 2 was analyzed using SmartPLS 2.0, a Partial Least Squares (PLS), structural equation modeling (SEM) software tool (Ringle, Wende, & Will, 2005). SmartPLS simultaneously assesses the

Table 1. Saudi predictor latent construct items

Subjective Norm (SN)
SN1: Most people who are important to me think I should use computers.
SN2: Most people who are important to me would want me to use computers
SN3: People whose opinions I value would prefer me to use computers.
Image (IMG)
IMG1: People in my organization who use computers have a high profile.
IMG2: Using computers is a status symbol in my organization.
IMG3: People in my organization who use computers have more prestige than those who do not.
Job Relevance (JR)
JR1: Using computers is compatible with aspects of tasks of my job.
Output Quality (OQ)
OQ1: Overall, the quality of computer systems / package(s) at my work is: (seven point scale anchored from 1 = poor to 7 = excellent).
OQ2: According to your job requirements, please indicate each task you use computers to perform (count of all that apply)?: (1) Letters and memos; (2) Producing reports; (3) Data storage and retrieval; (4) Making decisions; (5) Analyzing trends; (6) Planning and forecasting; (7) Analyzing problems and alternatives; (8) Budgeting; (9) Controlling and guiding activities; (10) Electronic communications with others; (11) Others (please indicate...).
Result Demonstrability (RD)
RD1: I would have no difficulty telling others about the results of using a computer.
RD2: I believe I could communicate to others the consequences of using a computer.
RD3: The results of using a computer are apparent to me.

Table 2. Saudi moderating (interacting) variables

Experience: (EXP) For how many years have you been using computers?: (1) Less than a year; (2) 1-3 years; (3) 4-7 years; (4) 8-10 years; (5) More than 10 years.
Voluntariness: (VOL)
VOL1: My use of computers is voluntary (as opposed to being required by my superiors or job description).
VOL2: My boss does not require me to use computers.
VOL3: Using computers or other alternatives is completely up to me.

Table 3. Saudi predicted latent construct items

Perceived Usefulness (PU)
PU1: I find computers useful in my job.
PU2: Using computers in my job enables me to accomplish tasks more quickly.
PU3: Using computers in my job increases my productivity.
PU4: Using computers enhances my effectiveness on the job.
Perceived Ease of Use (PEOU)
PEU1: My interactions with computers are clear and understandable.
PEU2: It is easy for me to become skillful using computers.
PEU3: I find computers easy to use.
PEU4: Learning to use computers is easy for me.
Behavioral Intention (BI)
BI1: To do my work, I would use computers rather than any other means available.
BI2: My intention would be to use computers rather than any other means available.
BI3: What are the chances in 100 that you will continue as a computer user?: (1) Zero; (2) 1-10%; (3) 11-30%; (4) 31-50%; (5) 51-70%; (6) 71-90%; or (7) More than 90%.

psychometric properties of the *measurement model* (i.e., the reliability and validity of the scales used to measure each variable), as well as estimates the parameters of the *structural model* (i.e., the strength of the path relationships among the model variables). The measurement model and the structural model results are discussed below.

The Measurement Model

The results obtained from testing the measurement model provide evidence of the robustness of the measures as indicated by their internal consistency reliabilities (indexed by the composite reliabilities). These are shown in Table 4. The composite reliabilities of the measures range from 0.89 to 0.95, with the exception of job relevance, a one-indicator construct which, accordingly, has a composite reliability of 1.00. All of these reliabilities far exceed the recommended threshold of 0.70 suggested by Nunnally (1978). Also, the average variances extracted (AVEs) for the measurement constructs range from 0.68 to 0.87 (again excluding the one-indicator job relevance construct which has an AVE of 1.0). Consistent with the recommendation of Fornell and Larcker (1981), the AVE for each measure well exceeds the lower bound threshold value of 0.50.

Table 5 provides evidence of the discriminant validity of the measures used in the study. The bolded elements in the matrix diagonals, representing the square roots of the AVEs, are greater in all cases than the off-diagonal elements in their corresponding row and column, supporting the discriminant validity of our scales.

We tested the convergent validity of the scales by extracting the factor loadings (and cross loadings) of all indicators to their respective constructs. These results shown in Table 6 indicate that all items loaded: (1) on their respective constructs from a lower bound of 0.77 to an upper bound of 0.95 (excluding job relevance); and (2) more highly on their respective construct than on any other construct (the non-bolded factor loadings). A common rule of thumb to indicate convergent validity is that all items should load greater than 0.7 on their own construct, and should load more highly on their respective construct than on the other constructs (Yoo & Alavi, 2001). Further-

Table 4. Assessment of the measurement model

Variable Constructs	The Composite Reliability (Internal Consistency Reliability)	Average Variance Extracted/Explained
Subjective Norm	0.95	0.87
Image	0.93	0.81
Job Relevance	1.00	1.00
Result Demonstrability	0.91	0.76
Perceived Usefulness	0.90	0.68
Perceived Ease of Use	0.89	0.68
Behavioral Intention	0.90	0.74

Table 5. Discriminant validity (inter-correlations) of variable constructs

Latent Variables	1	2	3	4	5	6	7
1. Subjective Norm	**.93**						
2. Image	.48	**.90**					
3. Job Relevance	.21	.19	**1.00**				
4. Result Demonstrability	.30	.23	.37	**.87**			
5. Perceived Usefulness	.32	.27	.61	.44	**.82**		
6. Perceived Ease of Use	.27	.20	.44	.57	.48	**.82**	
7. Behavioral Intention	.35	.36	.43	.45	.48	.55	**.86**

more, each item's factor loading on its respective construct was highly significant ($p < 0.001$). The loadings presented in Table 4 confirm the convergent validity of the measures for these latent constructs.

The Structural Model

Figure 3 visually presents the results of the structural model. The beta values of the path coefficients, indicating the direct influences of the exogenous constructs on the endogenous constructs, are presented. Each path's significance was estimated by a bootstrapping procedure (Ravichandran & Rai, 2000) using 500 resamples, which tends to provide reasonable standard error estimates (Chin, 2001). *Subjective norm* (SN) exhibited a strong positive direct influence (beta = 0.10, $p < 0.01$) on *perceived usefulness* (PU). *Image* (IMG) exhibited a positive direct influence on PU (beta = 0.07, $p <$ 0.05). SN exhibited a strong positive direct effect on IMG (beta = 0.47, $p < 0.001$). The direct effect of *output quality* (OQ) on PU was not significant and is therefore not represented in the structural model results. *Job relevance* (JR) (beta = 0.45, $p < 0.001$) and *result demonstrability* (RD) (beta = 0.12, $p < 0.01$) also had positive direct influences on PU (beta = 0.45, $p < 0.001$). *Perceived ease of use* (PEOU) exhibited a strong positive direct influence on PU (beta = 0.17, $p < 0.001$). PEOU also exhibited a strong direct effect on *behavioral intention* (BI) (beta = 0.38, $p < 0.001$). PU exhibited a strong positive direct influence on BI (beta = 0.24, $p < 0.001$). The influence of PU on BI also represents the indirect influences of SN, IMG, JR, and RD on BI, mediated through PU.

For the moderating (interacting) variables, statistically significant beta coefficients are indicated. For sake of clarity, statistically non-significant ($p > 0.05$) moderating paths from the research

Table 6. Factor loadings (bolded) and cross loadings

	Subj. Norm	Image	Job Relevance	Result Demon.	Percvd. USE	Percvd. EOU	Beh. Int.
SN1	**0.94**	0.45	0.19	0.28	0.29	0.28	0.34
SN2	**0.95**	0.45	0.20	0.30	0.29	0.24	0.31
SN3	**0.91**	0.46	0.21	0.27	0.30	0.24	0.33
IMG1	0.43	**0.88**	0.19	0.23	0.26	0.20	0.36
IMG2	**0.41**	**0.90**	**0.18**	**0.20**	**0.23**	**0.18**	**0.31**
IMG3	0.42	**0.91**	0.15	0.20	0.23	0.17	0.30
JR1	0.13	0.19	**1.00**	0.37	0.41	0.44	0.43
RD1	0.27	0.18	0.31	**0.85**	0.34	0.48	0.38
RD2	0.27	0.21	0.30	**0.90**	0.39	0.49	0.39
RD3	0.26	0.20	0.36	**0.87**	0.42	0.52	0.40
PU1	0.25	0.19	0.39	0.36	**0.78**	0.41	0.38
PU2	0.26	0.22	0.48	0.35	**0.87**	0.38	0.41
PU3	0.27	0.23	0.51	0.38	**0.87**	0.39	0.40
PU4	0.22	0.25	0.61	0.37	**0.77**	0.40	0.40
PEU1	0.19	0.16	0.33	0.46	0.38	**0.82**	0.42
PEU2	0.23	0.19	0.43	0.53	0.42	**0.87**	0.48
PEU3	0.22	0.20	0.36	0.44	0.41	**0.84**	0.46
PEU4	0.19	0.21	0.32	0.46	0.37	**0.77**	0.45
BI1	0.19	0.25	0.39	0.43	0.47	0.51	**0.85**
BI2	0.31	0.35	0.30	0.34	0.31	0.44	**0.85**
BI3	0.28	0.34	0.41	0.38	0.44	0.47	**0.89**

model (Figure 2) are omitted from the results presented in Figure 3. *Voluntariness* (VOL) did not exhibit any significant interaction effect on the effect of SN upon BI. *Experience* (EXP) had a significant *negative* (beta = - 0.08, $p < 0.05$) interacting effect with SN upon PU.

The direct influences of SN, IMG, JR, RD, and PEOU, together with the moderating influence of EXP with SN, accounted for 30.1% of the variance in PU ($R^2 = 0.301$). The direct influence of PU on BI (which mediates the indirect influences of SN, IMG, JR and RD on BI through PU), combined with the direct influences of SN and PEOU on BI, accounted for 40.3% of the variance in BI ($R^2 = 0.403$).

Table 7 presents the conclusions with respect to the ten hypotheses. The last column indicates whether or not the hypotheses were supported. The beta coefficient and p values are also reported. These findings are discussed in the next section.

DISCUSSION AND IMPLICATIONS

We draw on the guidelines provided by Chin et al. (1996) in interpreting the results presented in Figure 3. The social influence processes and cognitive instrumental processes that predict technology adoption in developed countries were shown also to do so for developing countries

Figure 3. Structural model results

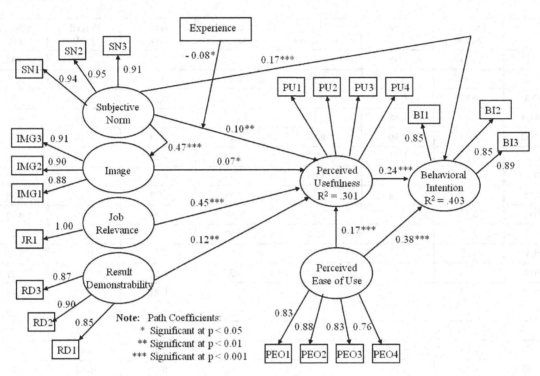

when using the TAM2 model. This study found, similar to Venkatesh and Davis (2000), that SN has a positive direct influence on PU and IMG. Thus, hypotheses H_1 and H_6 were supported. SN also had a positive direct influence on BI (beta = 0.17, p < 0.001). Thus, hypothesis H_2 was supported. With respect to the cognitive instrumental processes outlined in TAM2, JR and RD each exhibited a significant positive effect on PU, supporting hypotheses H_7, H_9 and H_{10}, again consistent with the findings of Venkatesh and Davis (2000). Finally, consistent with the majority of the prior technology acceptance research including TAM2, PEOU had a positive direct effect on PU, and PU had a positive direct effect on BI. Notably, OQ had no significant effect on PU, a finding that is not consistent with the original TAM2 findings. Hypothesis H_8 was not supported. The TAM2 model predicts comparable variance in BI between developed ($R^2 = 34$-52%) and developing countries ($R^2 = 40.3\%$ in this study).

What is most striking about our findings is that the moderating variables to social influence processes which were shown to have an effect in the TAM2 model when applied to Western cultural contexts do not apply in non-Western cultural contexts, in particular, Saudi Arabia. There was no interaction of voluntariness on the effect of SN on BI (hypothesis H_5 was supported). In contrast, Venkatesh and Davis (2000)*did* find an interaction of voluntariness on the influence of SN on BI. Specifically, they found that SN influenced BI only in mandatory usage contexts, and that in voluntary usage contexts, SN had no direct effect on BI, over and above what was explained by PU and PEOU. We contend that the significant effect of SN on BI (H_2), in both mandatory and voluntary usage contexts, can be explained within Saudi Arabia's cultural context. We argue that individuals belonging to Saudi culture would show a high regard for authority and would conform to the expectations of those who occupy elevated

Table 7. Conclusions with respect to the hypotheses

HYPOTHESIS	CONCLUSION
H_1: Subjective norm will have a positive direct effect on perceived usefulness.	Supported (beta = 0.10, p < 0.01) SN found to affect PU significantly
H_2: Subjective norm will have a positive direct effect on behavioral intention.	Supported (beta = 0.17, p < 0.001) SN found to affect BI significantly
H_3: Experience will not significantly moderate the positive effect of subjective norm on behavioral intention.	Supported EXP did not moderate the affect of SN on BI
H_4: The positive direct effect of subjective norm on perceived usefulness will attenuate with increased experience.	Supported (beta = - 0.08, p < 0.05) EXP *negatively* moderated the affect of SN on PU)
H_5: Voluntariness will not significantly moderate the effect of subjective norm on behavioral intention to use computers.	Supported VOL did not moderate the affect of SN on BI
H_6: Subjective norm will have a positive direct effect on image.	Supported (beta = 0.47, p < 0.001) SN found to affect IMG significantly
H_7: Image will have a positive direct effect on perceived usefulness.	Supported (beta = 0.07, p < 0.05) IMG found to affect PU significantly
H_8: Output quality will have a positive direct effect on perceived usefulness.	Not supported OQ had no significant effect on PU
H_9: Job relevance will have a positive direct effect on perceived usefulness.	Supported (beta = 0.45, p < 0.001) JR found to affect PU significantly
H_{10}: Results Demonstrability will have a positive direct effect on perceived usefulness.	Supported (beta = 0.12, p < 0.01) RD found to affect PU significantly

roles in the organizational hierarchy, especially in light of heavy promotion of technology adoption through governmental policy (Abdul-Gader & Al-Angari, 1995; Looney, 2004; Malek & Al-Shoaibi, 1998) and where there are typically more rigid cultural and organizational structures of authority between managers and subordinates (Bjerke & Al-Meer, 1993). Saudi Arabia has a culture that values collective achievements and interpersonal relationships (Dadfar et al., 2003; Dadfar, 1990), and in such a culture, the de-emphasis of individual achievement, and the greater importance attached to collective achievement and group success provides additional rationale to anticipate a strong relationship between subjective norm and behavioral intention. In light of these Saudi cultural impacts, even if the context of usage was voluntary, SN would continue to be a significant predictor of BI. Indeed our findings do confirm a pronounced effect of SN on BI in Saudi Arabia, a finding which is not usually observed in Western contexts.

As postulated in hypothesis H_3, but unlike the findings of Venkatesh and Davis (2000), EXP did not moderate the influence of SN on BI. This result could have stemmed from the context of our study, in that the adoption and use of a particular application or system was not studied, but, instead, the general use of desktop computers for 'work-related purposes.' In this context, it is less likely that the effect of subjective norm on intention to use a system will diminish with increasing experience, as reported in Venkatesh et al. (2000). However, this could have also resulted from the fact that the study did not capture the influence of increasing experience using a longitudinal study, as done by Venkatesh and Davis. Rather, the investigation measured experience as the respondents' self-reports of the number of years that they had been using computers. Nevertheless, this is a valid measure of experience in the context of our study. However, EXP did *negatively* moderate the positive influence of SN on PU. Hypothesis H_4 is supported. We interpret this negative coefficient as an indication that, in the Saudi data set, more experienced individuals are *less* inclined to be affected by significant others' opinions regarding the perceived usefulness of computers in the workplace.

In many such developing countries with distinct cultural traditions from the West, there is a fundamental challenge to synthesizing existing pre-technological, socio-cultural systems with ready-made imported technological products from other cultures (Tibi, 1990). In a high power distance society such as Saudi Arabia, it is less likely that 'rank and file' individuals adopting technology will have much influence on the society overall. Instead, it is more likely that the adoption beliefs and actions of the elites of this society will influence the attitudes and subjective norms with respect to using technology (Hashmi, 1988; Idris, 2007). If the elites of developing nations are able to plan a road map built from the perspective of "progress" for the country to cope culturally with the advent of widespread technological adoption, it will be impressed upon the general population that IT adoption is a positive outcome (Baker et al., 2007). This perspective is preferable to the perception that IT adoption is 'forced' from other cultures, like external reformation programs, that might be seen as a danger to spiritual authority and national and cultural independence (Tibi, 1990). By analogy, although less sensitive than reform, IT adoption and acceptance can be cultivated culturally with a higher probability of success using elites with an internal 'progress' perspective within a developing country's beliefs and traditions than using external entities to force IT adoption.

There is much of the variance of the TAM2 model that is unexplained in these results, and this warrants a closer investigation into the impact of Saudi management effectiveness in Saudi workplaces. Overall in the Saudi workplace, workers defer to the superior's orders and seek their guidance, and the superiors in return protect and give patronage and affection to their subordinates, in accordance with cultural values and customs (Abdalla & Al-Homoud, 2001). Therefore, managers in this collectivist society must be able to use their relationships and power over their group to spur workers to adopt technology in the workplace, as managers, playing the role of tribe leaders, have virtually ultimate power and authority in this societal domain (Abdalla & Al-Homoud, 2001; Hofstede, 2001). The TAM2 model overall has as its cultural basis that individuals as a part of their organizational 'in-group' will find value in increased technology adoption because the individual worker perceives the technology will be easy to use and useful to reflect improved individual performance (Venkatesh & Davis, 2000). Individuals in a collectivist workplace are not motivated toward organizational performance in this way (Alanazi & Rodrigues, 2003; Lundgren, 1998). Instead, these workers are spurred to technology adoption by the in-group behavior and their manager's vision that using this technology will be beneficial to the organization to decide for the worker that this technology is to be used (Bhuian, Abdulmuhmin, & Kim, 2001; Pillai et al., 1999). As a result, the specific variance not explained in these results could very plausibly be in factors that motivate employees in a collectivist society to perform overall, such as those identified by Idris (2007), with a subset of this being technology adoption in the workplace. Further investigations might focus on the success of the manager in influencing the workers to fully explain all of the factors that influence technology adoption in the Saudi (or any collectivist culture's) workplace.

Limitations

Our study has its limitations. This study is not an exact replication of the study conducted by Venkatesh and Davis (2000), which was longitudinal in design and included subjects from four different organizations as part of the sample. Venkatesh and Davis (2000) used regression as the statistical technique to analyze the data. In contrast, this study utilized PLS-based structural equation modeling (PLS-SEM), a more powerful technique to conduct the analysis. We hope that this provides additional support for the validity of the results of our model.

With this imposed etic study only have been tested in one developing nation, the generalizability of these findings are quite limited as well. Additional studies applying TAM2 to developing nations would have to be conducted to further support the findings of this study.

CONTRIBUTIONS AND RECOMMENDATIONS FOR FUTURE RESEARCH

As a model investigating the influences of social influence processes and the cognitive instrumental processes on technology adoption, the TAM2 model performs reasonably well in predicting intention to use computers among Saudi knowledge workers. This model accounted for approximately 40.3% of the variance in intention to use computers for Saudi workers using desktop computers in their work environment. Yet, the moderating variables to social influence processes which were shown to have an effect in the TAM2 model when applied to Western cultural contexts do not apply in non-Western cultural contexts, in particular, Saudi Arabia. These findings are an indication that cultural characteristics of a strongly hierarchical society tend to overshadow the individual attributes of experience and individual choice in the usage of a system (voluntariness) and that the impact of a Saudi manager's vision and outlook toward technology adoption could have much more influence on technology adoption than any individual performance motivation attributes. What makes these findings significant is their impact on the process of technology adoption in developing countries.

Within the cultural context of a rigidly hierarchical society, it is quite likely that the typical office knowledge worker will be heavily influenced by the adoption beliefs and actions of the elites/managers, or those at the apex of the society, who influence the social influence processes with respect to adopting technology. If the Saudi Arabian elites (or the elites of any developing country where there is a hierarchical and collectivist culture) are able to enact enlightened governmental policy from the perspective of "progress" for the country and lead technology adoption into the workplace, it will be impressed upon the Saudi Arabian people that technology adoption is a positive outcome. Technology adoption can be cultivated culturally with a higher probability of success using Saudi elites/managers with an internal "progress" perspective within the Saudi Arabian cultural beliefs and traditions than using external entities to force technology adoption. Of course, the underlying assumption here is that top Saudi executives overcome their concerns about their lack of control over technology's political and economic benefits and the threat to existing Arab society from Western values. Rather, they might appreciate how rapidly progress has been achieved in Saudi Arabia through the use of this technology without diminishing the distinctive Islamic identity of the country (Al-Farsy, 1996).

The number of TAM studies that have been conducted outside North American contexts constitutes but a small proportion of the total TAM studies published to date. The findings of each of these non-North American studies have underscored the need to take into account and address the cultural context of the country where the study is conducted. There is an opportunity for researchers to test and refine TAM and its extensions by studying factors governing technology acceptance outside of North America. There is a need, now more than ever, to understand the cultures of countries we partner with in the increasing trend towards the globalization of firms and industries and the continued economic development of many countries of the world. Such research could be of great practical value for managers, who, in this era of globalization and outsourcing, are required to interface with personnel in foreign countries and implement policies for increasing productivity among the workforce. An understanding of the interplay of technology acceptance and culture shall also equip them to deal with foreign employees serving in organizations here in North America.

REFERENCES

Abdalla, I., & Al-Homoud, M. (2001). Exploring the Implicit Leadership Theory in Arabian Gulf States. *Applied Psychology: An International Review, 50*(4), 506–531. doi:10.1111/1464-0597.00071

Abdul-Gader, A. H. (1999). *Managing Computer-Based Information Systems in Developing Countries: A Cultural Perspective*. Hershey, PA: IGI-Global.

Abdul-Gader, A. H., & Al-Angari, K. (1995). *Information Technology Assimilation in the Government Sector: An Empirical Study*. King Abdul-Aziz City for Science and Technology.

Ajzen, I. (1991). The Theory of Planned Behavior. *Organizational Behavior and Human Decision Processes, 50*(2), 179–211. doi:10.1016/0749-5978(91)90020-T

Ajzen, I., & Fishbein, M. (1980). *Understanding Attitudes and Predicting Social Behavior*. Englewood Cliffs, NJ: Prentice-Hall.

Al-Farsy, F. (1996). *Modernity and Tradition: The Saudi Equation*. London: Panarc International Ltd.

Al-Gahtani, S. S. (2004). Computer Technology Acceptance Success Factors in Saudi Arabia: An Exploratory Study. *Journal of Global Information Technology Management, 7*(1), 5–29.

Al-Gahtani, S. S., Hubona, G. S., & Wang, J. (2007). Information Technology (IT) in Saudi Arabia: Culture and the Acceptance and Use of IT. *Information & Management, 44*(8), 681–691. doi:10.1016/j.im.2007.09.002

Al-Gahtani, S. S., & Shih, H.-P. (2009). The Influence of Organizational Communication Openness on the Post-Adoption of Computers: An Empirical Study in Saudi Arabia. *Journal of Global Information Management, 17*(3), 20–41.

Alanazi, F., & Rodrigues, A. (2003). Power Bases and Attribution in Three Cultures. *The Journal of Social Psychology, 143*(3), 375–395. doi:10.1080/00224540309598451

Ali, A. J. (1990). Management Theory in a Transitional Society: The Arab's Experience. *International Studies of Management and Organization, 20*(3), 7–35.

Atiyyah, H. S. (1989). Determinants of Computer System Effectiveness in Saudi Arabian Public Organizations. *International Studies of Management and Organization, 19*(2), 85–103.

Baker, E. W., Al-Gahtani, S. S., & Hubona, G. S. (2007). The Effects of Gender and Age on New Technology Implementation in a Developing Country: Testing the Theory of Planned Behavior. *Information Technology & People, 20*(4), 352–375. doi:10.1108/09593840710839798

Berry, J. W. (1989). Imposed etics-emics-derived etics: The operationalization of a compelling idea. *International Journal of Psychology, 24*, 721–735.

Bhuian, S., Abdulmuhmin, A., & Kim, D. (2001). Business Education and its Influence on Attitudes to Business, Consumerism and Government Saudi Arabia. *Journal of Education for Business, 76*(4), 226–230. doi:10.1080/08832320109601315

Bjerke, B., & Al-Meer, A. (1993). Culture's Consequences: Management in Saudi Arabia. *Leadership and Organization Development Journal, 14*(2), 30–35. doi:10.1108/01437739310032700

Brislin, R. (1986). The Wording and Translation of Research Instruments. In Lonner, W., & Berry, J. (Eds.), *Field Methods in Cross-Cultural Research* (pp. 137–164). Beverly Hills, CA: Sage Publications.

Chin, W. W. (2001). *PLS-Graph User's Guide, Version 3.0*. Unpublished manuscript.

Chin, W. W., Marcolin, B. L., & Newsted, P. R. (1996). *A Partial Least Squares Latent Variables Modelling Approach for Measuring Interaction Effects: Results from a Monte-Carlo Simulation Study and Voice Mail Emotion/Adoption Study.* Paper presented at the International Conference on Information Systems, Cleveland, OH.

Compeau, D. R., & Higgins, C. A. (1995). Computer Self-Efficacy: Development of a Measure and Initial Test. *Management Information Systems Quarterly, 19*(2), 189–211. doi:10.2307/249688

Dadfar, A., Norberg, R., Helander, E., Schuster, S., & Zufferey, A. (2003). *Intercultural Aspects of Doing Business with Saudi Arabia.* Linkoping, Sweden: Linkoping University.

Dadfar, H. (1990). *Industrial Buying Behavior in the Middle East.* Linkoping, Sweden: Linkoping University.

Davis, F. D. (1989). Perceived Usefulness, Perceived Ease of Use, and User Acceptance of Information Technology. *Management Information Systems Quarterly, 13*(3), 319–340. doi:10.2307/249008

Davis, F. D., Bagozzi, R. P., & Warshaw, P. R. (1989). User Acceptance of Computer Technology: A Comparison of Two Theoretical Models. *Management Science, 35*(8), 982–1003. doi:10.1287/mnsc.35.8.982

Ein-Dor, P., Segev, E., & Orgad, M. (1992). The Effect of National Culture on IS: Implications for International Information Systems. *Journal of Global Information Management, 1*(1), 33–44.

Fishbein, M., & Ajzen, I. (1975). *Belief, Attitude, Intention and Behavior: An Introduction to Theory and Research.* Reading, MA: Addison-Wesley Publishing Company.

Fornell, C. R., & Larcker, D. F. (1981). Evaluating Structural Equation Models with Unobservable Variables and Measurement Error. *JMR, Journal of Marketing Research, 18*(1), 39–50. doi:10.2307/3151312

Gallivan, M. S. (2005). Information Technology and Culture: Identifying Fragmentary and Holistic Perspectives of Culture. *Information and Organization, 15*(4), 295–338. doi:10.1016/j. infoandorg.2005.02.005

Goodman, S. E., & Green, J. D. (1992). Computing in the Middle East. *Communications of the ACM, 35*(8), 21–25. doi:10.1145/135226.135236

Hartwick, J., & Barki, H. (1994). Explaining the Role of User Participation in Information System Use. *Management Science, 40*(4), 440–465. doi:10.1287/mnsc.40.4.440

Hashmi, M. (1988). *National Culture and Management Practices: United States and Saudi Arabia Contrasted.* Paper presented at the Proceedings of the Annual Eastern Michigan University Conference on Languages for Business and the Professions, Ann Arbor, MI.

Hill, C. E., Loch, K. D., Straub, D. W., & El-Sheshai, K. (1998). A Qualitative Assessment of Arab Culture and Information Technology Transfer. *Journal of Global Information Management, 6*(3), 29–38.

Hill, C. E., Straub, D. W., Loch, K. D., Cotterman, W. W., & El-Sheshai, K. (1994). *The Impact of Arab Culture on the Diffusion of Information Technology: A Culture-Centered Model.* Paper presented at the The Impact of Informatics on Society: Key Issues for Developing Countries, IFIP 9.4, Havana, Cuba.

Hofstede, G. (1984). *Culture's Consequences: International Differences in Work-Related Values*. Beverly Hills, CA: SAGE Publications.

Hofstede, G. (2001). *Culture's Consequences: Comparing Values, Behaviors, Institutions, and Organizations Across Nations*. Thousand Oaks, CA: Sage Publications.

Hubona, G. S., Truex, D. P., Wang, J., & Straub, D. W. (2006). Cultural and Globalization Issues Impacting the Organizational Use of Information Technology. In Galletta, D. F., & Zhang, P. (Eds.), *Human-Computer Interaction and Management Information Systems - Applications*. Armonk, NY: M. E. Sharpe.

Idris, A. M. (2007). Cultural Barriers to Improved Organizational Performance in Saudi Arabia. *SAM Advanced Management Journal, 72*(2), 36–53.

Kabasakal, H., & Bodur, M. (2002). Arabic Cluster: a Bridge between East and West. *Journal of World Business, 37*, 40–54. doi:10.1016/S1090-9516(01)00073-6

Kelman, H. C. (1958). Compliance, Identification and Internalization: Three Processes of Attitude Change. *The Journal of Conflict Resolution, 2*(1), 51–60. doi:10.1177/002200275800200106

Loch, K. D., Straub, D. W., & Kamel, S. (2003). Diffusing the Internet in the Arab World: The Role of Social Norms and Technological Culturation. *IEEE Transactions on Engineering Management, 50*(1), 45–63. doi:10.1109/TEM.2002.808257

Looney, R. (2004). Saudization and Sound Economic Reforms: Are the Two Compatible? *Strategic Insights, 3*(2).

Lundgren, L. (1998). The Technical Communicator's Role in Bridging the Gap between Arab and American Business Environments. *Journal of Technical Writing and Communication, 28*(4), 335–343. doi:10.2190/U8AH-MQWD-F9L7-QAFA

Maheswaran, D., & Shavitt, S. (2000). Issues and New Directions in Global Consumer Psychology. *Journal of Consumer Psychology, 9*(2), 59–66. doi:10.1207/S15327663JCP0902_1

Malek, M., & Al-Shoaibi, A. (1998). *Information Technology in the Developing Countries: Problems and Prospects*. Paper presented at the Conference on Administrative Sciences: New Horizons and Roles in Development, King Fahd University of Petroleum and Minerals, Dhahran, Saudi Arabia.

Mathieson, K. (1991). Predicting User Intentions: Comparing the Technology Acceptance Model with the Theory of Planned Behavior. *Information Systems Research, 2*(3), 173–191. doi:10.1287/isre.2.3.173

Moore, G., & Benbasat, I. (1993). *An empirical examination of a model of the factors affecting utilization of information technology by end-users (Tech. Rep.)*. Vancouver, Canada: University of British Columbia, Faculty of Commerce.

Moore, G. C., & Benbasat, I. (1991). Development of an Instrument to Measure the Perceptions of Adopting an Information Technology Innovation. *Information Systems Research, 2*(3), 173–191. doi:10.1287/isre.2.3.192

Morris, M. W., Leung, K., Ames, D., & Lickel, B. (1999). Views from Inside and Outside: Integrating Emic and Etic Insights about Culture and Justice Judgment. *Academy of Management Review, 24*(4), 781–796. doi:10.2307/259354

Nunnally, J. C. (1978). *Psychometric Theory* (2nd ed.). New York: McGraw-Hill.

Organization for Economic Cooperation and Development (OECD). (2008). *Information Technology Outlook*.

Palmer, M., Alghofaily, I., & Alminir, S. (1984). The Behavioral Correlates of Rentier Economies: A Case Study of Saudi Arabia. In Stookey, R. (Ed.), *The Arabian Peninsula: Zone of Ferment*. Stanford, CA: Hoover Institution Press.

Peng, T. K., Peterson, M. F., & Shyi, Y.-P. (1991). Quantitative Methods in Cross-National Management Research: Trends and Equivalence. *Journal of Organizational Behavior, 12*(2), 87–107. doi:10.1002/job.4030120203

Pfeffer, J. (1982). *Organizations and Organization Theory*. Marshfield, MA: Pitman.

Pillai, R., Scandura, T., & Williams, E. (1999). Leadership and Organizational Justice: Similarities and Differences. *Journal of International Business Studies, 30*(4), 763–779. doi:10.1057/palgrave.jibs.8490838

Ravichandran, T., & Rai, A. (2000). Quality Management in Systems Development: An Organizational System Perspective. *Management Information Systems Quarterly, 24*(3), 381–415. doi:10.2307/3250967

Ringle, C. M., Wende, S., & Will, A. (2005). *SmartPLS 2.0 (beta)*. Retrieved from http://www.smartpls.de

Rose, G., & Straub, D. W. (1998). Predicting General IT Use: Applying TAM to the Arabic World. *Journal of Global Information Management, 6*(3), 39–46.

Straub, D. W., Loch, K. D., & Hill, C. E. (2001). Transfer of Information Technology to the Arab World: A Test of Cultural Influence Modeling. *Journal of Global Information Management, 9*(4), 6–28.

Tajfel, H. (Ed.). (1978). *Differentiation between Social Groups: Studies in the Social Psychology of Intergroup Relations*. London: Academic Press.

Taylor, S., & Todd, P. A. (1995). Understanding Information Technology Usage: A Test of Competing Models. *Information Systems Research, 6*(2), 144–176. doi:10.1287/isre.6.2.144

Tibi, B. (1990). *Islam and the Cultural Accommodation of Social Change* (Krojzl, C., Trans.). Boulder, CO: Westview Press.

Venkatesh, V., & Davis, F. D. (2000). A Theoretical Extension of the Technology Acceptance Model: Four Longitudinal Field Studies. *Management Science, 46*(2), 186–204. doi:10.1287/mnsc.46.2.186.11926

Venkatesh, V., Morris, M. G., Davis, G. B., & Davis, F. D. (2003). User Acceptance of Information Technology: Toward a Unified View. *Management Information Systems Quarterly, 27*(3), 425–478.

Warshaw, P. R. (1980). A New Model for Predicting Behavioral Intentions: An Alternative to Fishbein. *JMR, Journal of Marketing Research, 17*(2), 153–172. doi:10.2307/3150927

World Bank Group. (2009). *Saudi Arabia Data Profile*. World Development Indicators Database.

Yavas, U. (1997). Management Know-How Transfer to Saudi Arabia: A Survey of Saudi Managers. *Industrial Management & Data Systems, 7*, 280–286. doi:10.1108/02635579710192959

Yavas, U., Luqmani, M., & Quraeshi, Z. A. (1992). Facilitating the Adoption of Information Technology in a Developing Country. *Information & Management, 23*(2), 75–82. doi:10.1016/0378-7206(92)90010-D

Yoo, Y., & Alavi, M. (2001). Media and Group Cohesion: Relative Influences on Social Presence, Task Participation, and Group Consensus. *Management Information Systems Quarterly, 25*(3), 371–390. doi:10.2307/3250922

APPENDIX A: SURVEY INSTRUMENT

Please indicate your level of agreement with the following statements, where 1 is extremely disagree and 7 is extremely agree.

Subjective Norm

(SN1) Most people who are important to me think I should use computers.
(SN2) Most people who are important to me would want me to use computers.
(SN3) People whose opinions I value would prefer me to use computers.

Image

(IMG1) People in my organization who use computers have a high profile.
(IMG2) Using computers is a status symbol in my organization.
(IMG3) People in my organization who use computers have more prestige than those who do not.

Job Relevance

(JR1) Using computers is compatible with aspects of tasks of my job.

Voluntariness

(VOL1) My use of computers is voluntary (as opposed to being required by my superiors or job description)
(VOL2) My boss does not require me to use computers.
(VOL3) Using computers or other alternatives is completely up to me.

Perceived Usefulness

(PU1) I find computers useful in my job.
(PU2) Using computers in my job enables me to accomplish tasks more quickly.
(PU3) Using computers in my job increases my productivity.
(PU4) Using computers enhances my effectiveness on the job.

Perceived Ease of Use

(PEOU1) My interactions with computers are clear and understandable.
(PEOU2) It is easy for me to become skillful using computers.
(PEOU3) I find computers easy to use.
(PEOU4) Learning to use computers is easy for me.

Behavioral Intention

(BI1) To do my work, I would use computers rather than any other means available.

(BI2) My intention would be to use computers rather than any other means available.

(BI3) What are the chances in 100 that you will continue as a computer user?: (1) Zero; (2) 1-10%; (3) 11-30%; (4) 31-50%; (5) 51-70%; (6) 71-90%; or (7) More than 90%.

Experience

Experience: (EXP) For how many years have you been using computers?: (1) Less than a year; (2) 1-3 years; (3) 4-7 years; (4) 8-10 years; (5) More than 10 years.

Output Quality

(OQ1) Overall, the quality of computer systems/package(s) at my work is: (seven-point scale anchored from 1 = poor to 7 = excellent).

(OQ2) According to your job requirements, please indicate each task you use computers to perform (count of all that apply): (1) Letters and memos; (2) Producing reports; (3) Data storage and retrieval; (4) Making decisions; (5) Analyzing trends; (6) Planning and forecasting; (7) Analyzing problems and alternatives; (8) Budgeting; (9) Controlling and guiding activities; (10) Electronic communications with others; (11) Others (please indicate…).

Demographic (Moderator) Variables

Country affiliation: (Choose one.)			
O	Saudi	O	Non-Saudi
Gender: (Choose one.)			
O	Male	O	Female
Sector affiliation: (Choose one.)			
O	Public Sector	O	Private Sector

This work was previously published in International Journal of Global Information Management, Volume 18, Issue 3, edited by Felix B. Tan, pp. 35-58, copyright 2010 by IGI Publishing (an imprint of IGI Global).

Chapter 4
A Comparative Study of the Effects of Low and High Uncertainty Avoidance on Continuance Behavior[1]

Hak-Jin Kim
Yonsei University, Korea

Hun Choi
Catholic University of Pusan, Korea

Jinwoo Kim
Yonsei University, Korea

ABSTRACT

This study examines the effects of uncertainty avoidance (UA) at the individual level on continuance behavior in the domain of mobile data services (MDS). It proposes a research model for post-expectation factors and continuance behavior that considers the moderating effect of UA, and verifies the model with online survey data gathered in Korea and Hong Kong. Post-expectation factors are classified as either intrinsic or extrinsic motivational factors, while respondents are classified according to their propensities into low-UA and high-UA groups. The results indicate that UA has substantial effects not only on the mean values of the post-expectation factors studied but also on the strength of those factors' impact on satisfaction and continuance intention. The effects of intrinsic motivational factors on satisfaction and continuance intention are stronger for the high-UA group than for the low-UA group. In contrast, the effects of extrinsic motivational factors are generally stronger for the low-UA group.

INTRODUCTION

The rapid development of information communication technology (ICT) has spawned a variety of new products and services for business and communication among which users may select according to their preferences. An ICT user is thus frequently in a position to decide whether to keep using a given current service or to switch to another (Parthasarathy & Bhattacherjee, 1998). His or her decision to continue or discontinue use of a service is referred to as post-adoption

DOI: 10.4018/978-1-61350-480-2.ch004

behavior (Parthasarathy & Bhattacherjee, 1998). This paper specifically focuses on post-adoption continuance behavior, that is, the choice whether to keep using products and services presently in use. Continuance behavior is an important factor for service providers to understand, simply because the cost of acquiring new customers—searching for them, setting up new accounts, and initiating them to the information systems—is five times that of retaining existing customers (Bhattacherjee, 2001; Parthasarathy & Bhattacherjee, 1998).

Studies of post-adoption behavior have found that cultural characteristics play an important role (De Mooij, 2004; Straub, 1994; Van Slyke, Lou, Belanger, & Sridhar, 2004; McCoy, Everard, & Jones, 2005), because culture has a strong effect on how a user interprets a system's content and functions (Hiller, 2003). A system feature appropriate for users in one culture may not be appropriate for users in others without significant adaptation.

Uncertainty avoidance (UA) is one cultural factor known to exert a strong influence on post-adoption behavior (Frank, Sundqvist, Puumalainen, & Taalikka, 2001). UA is a measure of how well individuals tolerate unpredictable and unstructured situations or contexts (Hofstede, 1997; Veiga, Floyd, & Dechant, 2001). In the context of current ICT, two factors make it especially important to examine the impact of UA on post-adoption behavior. First, UA has been found to influence substantially users' initial adoption behaviors with new services or products (Veiga et al., 2001). People with high UA tend to stick with traditional technologies and to be slow in accepting new ones. Conversely, people with low UA tend to adopt new technologies quickly and easily (Hofstede, 1997). This distinctive effect of UA on *initial adoption* behavior strongly suggests that UA may also influence *post-adoption* behavior significantly (Lippert & Volkmar, 2007). Second, newer ICT services, including mobile data services (MDS), are part of a ubiquitous computing environment, and are thus used in a far broader range of contexts than traditional ICT services, which

are restricted to relatively familiar environments like offices and houses (Evers & Day, 1997). This diversity of use and circumstance entails greater uncertainty in the use of information technologies; new and complicated connection methods, abstract or unfamiliar icons, and high usage fees further increase the uncertainty (Albers & Kim, 2000; Chae, Kim, Kim, & Ryu, 2002). The trend of growing uncertainty in the ICT environment suggests UA may have substantial effects on post-adoption behavior with current ICT services.

The goals of this study are, first, to construct a theoretical model to analyze the impact of UA on continuance behavior at the level of the individual user in the MDS domain, and, second, to identify specific effects of UA by conducting empirical tests through the proposed model. While many studies have examined UA at the *national* level, the present study focuses on UA at the level of the *individual* user. UA heterogeneity within countries differs significantly from that between countries, and understanding the former will be more helpful to service providers who want to identify homogeneous market segments in a given country.

We have selected mobile data services (MDS) among the many ICT-based services currently available as our research domain. MDS may be defined relatively narrowly as the range of digital data services accessible from a mobile device specifically through a mobile communications network (e.g., CDMA, TDMA, GPRS, or GSM) (Hong & Tam, 2006). The present study adopts this definition of MDS, and further confines its research to MDS accessed through cellular phones. Thus access to the Internet through the wireless LAN (Wi-Fi) capability of laptop computers is not considered in this study.

MDS are selected as the target ICT for two reasons. First, MDS in some countries, including those considered here, Korea and Hong Kong, are services in which users determine both the extent and duration of subscription. Conventional phone services require up-front acquisition of new devices and multi-year subscriptions to service plans. This significant initial investment may discourage users from discontinuing services even when they are dissatisfied with them. MDS, however, are treated in the Korean and Hong Kong markets as ancillary services, using mobile phones and primary subscriptions users already have. For instance, services such as ringtone or photo downloads, location of points of interest, and friend positioning services are used for enjoyment at the user's discretion, and users pay fees according to usage time and transferred data volume, rather than through a fixed subscription. Thus they are free to discontinue use of a service at any time (Chae et al., 2002). This may explain why only 32.6% of initial adopters continue using MDS, even though more than 30 million Koreans (63% of the population) have used them at least once (Sir, Cho, Yang, An, & Kim, 2004). The ease with which service is discontinued makes MDS an ideal domain in which to study continuance behavior.

Second, using MDS in daily life involves uncertainties that may affect continuance behavior. One source of uncertainty is the physical constraints imposed by MDS devices: small screens, low resolutions, and obscure abbreviations or icons make it more difficult to find target information effectively, increasing the risk one will incur high usage fees (Albers & Kim, 2000; Chae et al., 2002). Another source is the immatureness of MDS. Since MDS are newer and less matured than conventional phone services, their specifications are not completely settled. Thus service providers may change parts or functions of their MDS that users have used, and the change makes the functions not working anymore. Moreover, rapid changes in IT may produce uncertainty for MDS users at the post-adoption stage. There are now, in the data services market, many competitive alternative services based on various wireless technologies, such as Wi-Fi broadband and WiMAX, and more are sure to come. Service providers may provide new MDS, based on these new technologies, with more attractive merits, which are not available

to MDS users with older services. This would be another risk factor of post adopting MDS. All of these considerations at the post-adoption stage heighten the uncertainty involved in MDS use. Thus we consider MDS an appropriate domain in which to study the moderating effects of UA on continuance behavior.

Five sections follow this introduction. The next section presents the theoretical background of the study. The third lays out our research model and hypotheses, while the fourth explains our survey and data analysis method. The fifth presents the study results. The final section discusses the limitations and implications of the results.

THEORETICAL BACKGROUND

Determinants of Continuance Behavior

Rogers's innovation diffusion theory (1995) is widely used in explaining the adoption of new technologies. Rogers accounts for the entry of a new technology into the market in terms of an initial innovative user group and subsequent diffusion to a majority of users. In order to explain the initial adoption and post-adoption behaviors surrounding new technologies, he modeled the innovation decision process in terms of five consecutive stages: knowledge, persuasion, decision, implementation, and confirmation (Rogers, 1995). The decision stage is the point at which one chooses either to adopt or to reject an innovation, while the confirmation stage is the point where one either reinforces or reverses an innovation-decision already made by continuing or discontinuing use (Rogers, 1995). What we have called initial-adoption and post-adoption behaviors correspond respectively to Rogers's decision and confirmation stages.

Many researchers have focused on Rogers's decision stage and have studied the initial adoption of innovations (Davis, Bagozzi, & Warshaw,

1992; Hong, Thong, Wong, & Tam, 2001; Morris & Dillon, 1997; Segars & Grover, 1993; Straub, 1994). More specifically, several studies have examined initial adoption of MDS (Bruner & Kumar, 2005; Hung, Ku, & Chang, 2003; Pedersen & Ling, 2003; Teo & Pok, 2003; Xiaowen, Chan, Brzezinski, & Xu, 2005). More recently, studies have turned their attention to post-adoption behaviors at the confirmation stage (Bhattacherjee, 2001; Huh & Kim, 2008; Khalifa & Liu, 2003; Spiller, Vlasic, & Yetton, 2007; Zhu & Kraemer, 2005). However, studies of continuance behavior at the confirmation stage *in the MDS domain* are relatively few (Choi et al., 2003).

As Spiller et al. (2007) observe, there are two distinct streams in the literature on post-adoption behavior. The first is based on expectation-confirmation theory, where factors used in initial adoption research are generally extended to the circumstances of post-adoption, with consideration given to relevant differences between initial adoption and post-adoption factors. The second stream, proposed by Jasperson, Carter, and Zmud (2005), combines reasoned action theory with diffusion theory, and posits "prior use," "habit," and "feature-centric view of technology" as factors specifically relevant to continuance. The first stream is said to be more appropriate to consumer technology adoption, the second to organizational technology adoption (Spiller et al., 2007). Second stream studies have found that engineers in an organization scrutinize product features and adopt different features at different times. This reasoned action is required because handling such IT products in an organization is a critical job for engineers (Jasperson et al., 2005). This sort of deliberative decision-making is rarely seen in consumers' adoption behavior. Thus the approach offered by the second stream may be peripheral to the case of MDS, whose users behave as individual consumers rather than as organizational employees.

Expectation-confirmation theory explains continuance behavior well in information systems (IS) and in repurchase behavior in marketing (e.g., Oliver, 1980). For example, Bhattacherjee (2001) developed a continuance model for online banking systems. He found that consumer pre-use expectations differed from post-use expectations in that the former were typically based on the impressions of others disseminated through mass media, while the latter were created more realistically through the consumers' own experiences (Bhattacherjee, 2001). Although this research has developed a continuous behavior model, the model is limited to systems with functional purposes, such as online banking systems, and thus considers only one post-expectation factor, perceived usefulness. It is difficult, however, to explain the continuance behavior of MDS users with reference only to perceived usefulness, since MDS are used for a variety of purposes, many of them hedonistic rather than utilitarian. In fact, the effect of post-usefulness on continuance behavior has been found to be smaller with MDS than with traditional IS (Hong & Tam, 2006). Moreover, Bhattacherjee's research does not consider UA, which has been shown to influence continuance behavior substantially. Other continuance behavior studies pose similar problems, either considering too narrow a range of post-expectation factors, or disregarding the effect of cultural differences (Khalifa & Liu, 2001; Khalifa & Liu, 2003; Zhu & Kraemer, 2005). Our study addresses this gap in the literature by building and testing empirically a new research model specific to MDS, a model that adopts a range of post-use expectation factors (post-expectation factors, hereafter) found relevant in prior studies of continuance behavior and also takes UA into account.

The variety of post-expectation factors in the literature can be classified according to motivation. Many post-adoption studies treat expectation and motivation as closely linked (Kim & Han, 2009;

Thang, Hong, & Tam, 2006), and indeed the two appear to be mutually entailed, in that expectation incites a motivation to perform an action, while motivation implies the existence of an underlying expectation. Ryan and Deci (2000) distinguish different types of motivation based on different reasons or goals that give rise to actions. A basic function of their self-determination theory is classification of motivations as intrinsic or extrinsic (Ryan & Deci, 2000). Extrinsic motivations value outcomes achieved through an activity, rather than finding value in the activity itself, and focus on performance (Ryan & Deci, 2000; Shang, Chen, & Shen, 2005). Conversely, intrinsic motivations focus on the activity itself as a locus of inherent satisfaction, achievement, or fulfillment, rather than on consequences separable from the activity (Ryan & Deci, 2000; Shang et al., 2005).

Motivation involves an internal process that gives a certain *energy* and *direction* to behavior (Lee, Cheung, & Chen, 2005). Energy relates to the strength, intensity, and persistence of the behavior, direction to the behavior's specific purpose (Lee, Cheung, & Chen, 2005). For the purposes of this study, motivation involves the energy and direction of a user's behavior in using the technology (Lee, Cheung, & Chen, 2005). Davis et al. (1992) and Venkatesh and Brown (2001) have explored technology adoption from the perspective of motivation, finding that intrinsic and extrinsic motivation factors substantially influence adoption behavior with personal computers in the home. Van der Heijden (2004) finds that extrinsic motivational factors are more salient in utilitarian systems, intrinsic factors in hedonistic systems. As yet, however, few studies have investigated post-adoption behavior from the perspective of motivation factors, let alone the specific moderating role played by UA, even though both are likely to affect post-adoption behavior significantly.

Uncertainty Avoidance

Uncertainty avoidance is "the extent to which the members of a culture feel threatened by uncertain or unknown situations" (Hofstede, 1980, p. 161). People in high uncertainty-avoidance cultures are uncomfortable with uncertainty and show a low tolerance for risk (Moores & Gregory, 2000). They try to avoid ambiguous situations by believing in fixed truths and expertise, seeking stability, establishing formal rules, and rejecting deviant ideas and behaviors. In contrast, people in low uncertainty avoidance cultures deal well with uncertainty and can be characterized as risk-takers.

Cross-cultural studies have revealed that UA may influence IS usage behavior (McCoy, Galletta, & King, 2005; Straub, Loch, Evaristo, Karahanna, & Srite, 2002). Because people with high UA have a tendency to eliminate risk factors before use (Simon, 2001; Straub, 1994; Veiga, Floyd, & Dechant, 2001), they may adopt innovation later than people with low UA (Hasan & Ditsa, 1999). For example, Png, Tan, and Wee (2001) have demonstrated that businesses in high-UA countries are less likely to adopt new IT infrastructure. Conversely, people in low-UA cultures may feel more comfortable with the unknown and thus accept uncertainty and risk more easily (Hofstede, 1997; Marcus & Emilie, 2000). Acceptance of new technologies is quicker in a low-UA than in a high-UA culture because uncertainty and ambiguity are more likely to be tolerated and even to be embraced as a positive aspect of change (Hofstede, 1980; Liu, Olivier, & Subharshan, 2001; Veiga et al., 2001).

Although a number of studies have examined the effect of UA at the *initial-adoption* stage, relatively few have investigated the influence of UA at the *post-adoption* stage (Lee, Choi, Kim, & Hong, 2007). Since the behavioral effects of UA may differ between the two stages, this study focuses on the influence of UA on continuance (i.e., post-adoption) behavior.

In most IS studies of cultural issues, UA has been operationalized at the country level. According to Hofstede (1980), countries with high-UA cultures include Korea (Hofstede index = 85), Japan (92) and France (86). Singapore (Hofstede index = 8), Denmark (23) and Hong Kong (29) are examples of low-UA cultures. Using the same index, Lee, Kim and Kim's (2005) cultural study of MDS use identified Korea and Japan as high-UA countries, Hong Kong and Taiwan as low-UA countries.

However, treating country as co-extensive with culture is reductive. Within-country heterogeneity is in fact sometimes greater than between-country heterogeneity (Lee, Choi, Kim, & Hong, 2007). Even though definitions of national culture may adequately explain macro-level behavior, they lack the precision to explain behavior at the individual level (Straub et al., 2002), because the cultural characteristics an individual owes to his or her national culture may be influenced and modified by membership in other ethnic, religious, and social groups, each of which has its own distinctive culture. According to Straub et al. (2002), an individual's cultural profile is the product of several layers of culture—super-national, national, professional, organizational, and group—that interact with each other. An individual who belongs to different professional, organizational, and group cultures from his or her neighbors will likely have a different cultural profile, even though he or she shares their nationality (Erez & Gati, 2004). Thus a national culture affects but does not determine the cultural characteristics of individuals. Conceptualization of culture at the individual level better captures the multiplicity of interacting cultural values as they are compounded in an individual self—a compound that can be distinguished from values and traits that operate *only* on the individual level (Erez & Gati, 2004).

To account for cultural UA characteristics at the level of the individual user, our study needs not only to include multiple countries that vary in terms of UA, but also to measure UA propensities at the individual level. This study draws respondents from Korea, a high-UA country, and Hong Kong, a low-UA country, but also measures the UA propensities of study participants, in order to focus on cultural UA differences at the individual level.

Self-Perception Theory

Self-perception theory is often invoked as a conceptual background for post-adoption phenomena (Bajaj & Nidumolu, 1998; Kim & Malhotra, 2005; Ouellette & Wood, 1998). The theory assumes that people do not form specific evaluations of their routine behaviors until they are asked to do so (Kim & Malhotra, 2005). Once a routine behavior becomes a focus of attention, users may adjust their perceptions continually as they acquire new information about the behavior, and these updated perceptions may provide a basis for subsequent behaviors (Hsu, Yen, Chiu, & Chang, 2006). In such cases, a modified expectation replaces the initial expectation in their cognitive memory as the basis for subsequent behavior (Bem, 1972; Bhattacherjee, 2000). Moreover, when explicitly asked, people may arrive at their answers as if they were external observers. For example, people may respond to a question about their attitudes toward using a service by saying, "I use the service every day; therefore, I think I like it." As Bem (1972) has noted, this evaluation process resembles the way people infer others' internal states from their behavior: "I guess they like the service; they are always using it." According to Kim and Malhotra (2005), self-perception theory predicts that more frequent usage will make for more favorable user evaluations.

The evidence supporting self-perception theory is considerable. For example, teenagers' attitudes toward others have been shown to become more caring and considerate after participation in repeated and sustained volunteering services (Brunelle, 2001). Applying self-perception theory to post-adoption behavior, Kim and Malhotra (2005) have developed a longitudinal model of how users' evaluations and behavior evolve as they gain experience with Web-based IS use in non-experimental settings. They proposed and empirically verified a feedback mechanism in which behavior influences beliefs and beliefs in turn influence behavior. Because self-perception has been related to post-adoption behavior (Kim & Malhotra, 2005; Hong, Kim, & Lee, 2008; Hsu et al., 2006), it may play an important role in the present study, which tests its explanatory power in the context of continuance behavior and UA in the domain of MDS.

RESEARCH MODEL AND HYPOTHESES

This paper proposes a continuance model based on previous studies and seeks to investigate how the model is affected by UA in the domain of MDS. The following section explains the continuance model and the anticipated effects of UA on the model. Figure 1 presents the model and our hypotheses.

Continuance Model in the Domain of MDS

A *continuance intention*, adapted from Bhattacherjee's (2001) continuance model, is for this study the extent to which a user intends to continue using a given technology. Continuance intention in the domain of MDS is thus the extent to which a user intends to continue using MDS.

Satisfaction is the degree to which a user's feeling about use of a technology is positive (Bhattacherjee, 2001). Satisfaction with MDS can thus be defined as the extent to which a user's reaction to prior MDS usage is positive. Satisfaction and continuance intention derived from using MDS will be closely related to the four post-expectation factors described below.

Post-usefulness is the extent to which people feel benefits after using a particular technology (Bhattacherjee, 2001). Usefulness is the utility value found in continued use of a system, and may thus be thought of as an extrinsic motivation factor (Davis et al., 1992; Teo, Lim, & Lai, 1999). As

Figure 1. Research model and study hypotheses

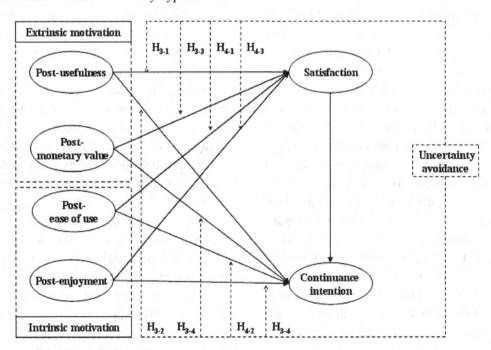

users identify MDS as a new technology resulting from the fusion of the existing Internet and mobile phones (Chae et al., 2002), post-usefulness appears to be a key factor determining MDS use, especially when its usefulness differentiates MDS from traditional Internet services (Pedersen, Methlie, & Thorbjonsen, 2002). For example, a recent study found that people use MDS only when the usefulness of its mobility really matters (Jarvenpaa, Lang, Takeda, & Tuunainen, 2003).

Post–monetary value is the user's evaluation of a service in terms of monetary value after using that service (Vlahos, Ferratt, & Knoepfle, 2000). It is a determination of whether use of a system is worth the monetary cost, and is thus, like post-usefulness, an extrinsic motivation factor. Existing research on information systems reveals that perceived monetary value is strongly related to user satisfaction (Mawhinney & Lederer, 1990; Vlahos & Ferratt, 1995). For example, Vlahos et al.'s (2000) study found that perceived monetary value is closely related to IS usage. Oh (1999) found that when a high value was attributed to a specific service, the behavioral intention to use that service in the future was greater. When MDS levy charges according to frequency of usage and data volume, as they do in this study's target countries, the monetary value of a service becomes more important to users (Lee, Kim, Lee, & Kim, 2002).

Post–ease-of-use is the degree to which users feel at ease (Preece et al., 1994) and comfortable (Davis, Bagozzi, & Warshaw, 1989; Segars & Grover, 1993) after learning and using (Davis et al., 1992; Karahanna, Straub, & Chervany, 1999; Morris & Dillon, 1997; Segars & Grover, 1993) a particular technology. Ease-of-use has been thought of as an intrinsic motivation factor (Atkinson & Kydd, 1997; Teo, 2002) because it focuses on effort expended during the use process (Davis et al., 1992; Teo et al., 1999), not on outcomes of use. Research following the technology acceptance model (TAM) has identified ease-of-use as one of the most important determinants of satisfaction (Lu, Yu, Liu, & Yao, 2003). Ease-of-use becomes

more critical in the case of MDS because of the small screens and inconvenient input methods of mobile devices (Chae et al., 2002). Recent studies have confirmed that ease-of-use is among the most important factors in determining MDS adoption (Hong & Tam, 2006; Lee, Choi, Kim, & Hong, 2007; Venkatesh, Ramesh, & Massey, 2003). Service providers are coming to believe that ease-of-use is in fact the key to retaining customers (Nachum, 2003).

Post-enjoyment is the degree to which someone finds use of a system enjoyable in its own right (Karahanna, Straub, & Chervany, 1999; Morris & Dillon, 1997; Segars & Grover, 1993; Yi & Hwang, 2003). It follows that enjoyment is a second intrinsic motivation factor, as pleasure is produced by use of the system itself, not by any outcomes of use (Yi & Hwang, 2003). Previous TAM research says that enjoyment is one of the most important determinants of satisfaction (Dabholkar, 1996; Yi & Hwang, 2003). For example, Nysveen et al.'s (2005) study found that enjoyment was an important determinant of MDS adoption. Many MDS users use entertainment-oriented MDS services frequently (HCI Lab, 2004), suggesting that the enjoyment factor is an essential ingredient in MDS continuance (Chae et al., 2002; Choi et al., 2003).

In summary, the continuance model consists of continuance intention, satisfaction, and four post-expectation factors, of which two are extrinsic—post-usefulness and post–monetary value—and two are intrinsic: post–ease-of-use and post-enjoyment.

Research Hypotheses

Two groups of research hypotheses are proposed. One concerns the effect of UA on the *mean values* of the post-expectation factors and the two dependent variables, satisfaction and continuance intention. The other concerns the effect of UA on the *path strengths* from the post-expectation factors to the dependent variables. Through the former, we can see and compare the absolute values of factors and dependent variables. Through the latter, we can see how the strength of each factor's influence on the dependent variables differs between the two UA groups. Though the hypotheses concerning path strength are more likely to offer meaningful implications, those concerning mean values may also provide insights into the effects of UA on continuance.

Effects of Uncertainty Avoidance on Mean Values of Post-Use Expectation Factors, Satisfaction, and Continuance Intention

During the initial-adoption stage, people with low UA tend to accept new products or services more easily than people with high UA, and thus show a stronger inclination to switch to a new product or service (Hofstede, 1980; Liu et al., 2001; Veiga et al., 2001). In the post-adoption stage, people with high UA are expected to have stronger post-expectations than people with low UA, and higher levels of satisfaction and continuance intention as well.

The relationship between UA propensity and post-expectations can be explained as follows. First of all, people with high UA may attain high levels of performance from services after adoption by removing, during the initial use period, uncertain factors that inhibit task performance. Although people with low UA adopt new technologies faster and enjoy using them, they make less of an effort to reduce uncertainties in their usage, because their interest is focused on the early experience itself. They pay less attention to improving task performance when using the technology. In contrast, people with high UA are task-oriented, and although they adopt technologies later, once they have done so, they are inclined to organize their usage experiences and to develop better use habits in order to reduce uncertainties and to improve task performance. In other words, they work to find stable ways to apply a given technology to

their tasks, once they have used it successfully beyond the initial adoption stage. Their efforts may lead to greater accuracy and speed in using services at the post-adoption stage and thus to higher usefulness.

Hypothesis 1-1: High-UA users will show a higher mean post-usefulness than low-UA users.

The accuracy and the speed with which one uses a service affect its monetary value. If people with high UA take less time to accomplish their tasks using MDS, and if service fees are governed by usage time or data transfer volume, they should get more value for their money. Studies in service marketing have made findings similar to these hypotheses. For example, studies by Donthu and Yoo (1998) and Tsikriktsis (2002) found that people with high UA had higher post-monetary value than people with low UA. Though research on issues of general service marketing cannot be directly applied to MDS, these findings do add circumstantial support for our hypotheses.

Hypothesis 1-2: High-UA users will show a higher mean post–monetary value than low-UA users.

People with high UA will also make fewer mistakes and have more fun because of their fluency with services they have adopted, resulting in higher values for the intrinsic motivation factors as well (Veiga et al., 2001).

Hypothesis 1-3: High-UA users will show higher mean post–ease-of-use than low-UA users.
Hypothesis 1-4: High-UA users will show higher mean post-enjoyment than low-UA users.

When people with high UA adopt a service, they take pains to resolve uncertainties involved in its use. Switching to another service would demand that they invest further time and effort to resolve new uncertainties. This demand is likely to strengthen the motivation to retain a service and to discourage switching to another. According to the self-perception theory, they think they would like to retain a service because they are more satisfied with the service. A stronger motivation to retain a service will thus lead to greater satisfaction with that service (De Mooij & Hofstede, 2002; Liu et al., 2001).

Hypothesis 2-1: High-UA users will show higher mean satisfaction than low-UA users.

People with high UA tend to be reluctant to try new services because of risks and uncertainties, and are inclined to keep using a service once they have adopted it (De Mooij & Hofstede, 2002; Liu et al., 2001), suggesting that they may have higher levels of continuance intention. In contrast, those with low UA might have lower levels of continuance intention because of a stronger predilection to switch to new services.

Hypothesis 2-2: High-UA users will show higher mean continuance intention than low-UA users.

With reference to continuance intention, the feedback mechanism identified by self-perception theory provides a supplementary account of post-expectations. Even as they allow their attitude toward a product or service to guide their actions regarding it, consumers may also *infer* their attitudes *from* actions they have already taken regarding that product or service (Bem, 1972). If the above hypotheses are correct, people with high UA have a higher continuance intention than people with low UA. They may then infer, from their own behavior, that the service currently in use is useful, easy to use, enjoyable, and valuable, and that these benefits explain why they have continued using it. Thus high-UA users may show higher post-expectation levels and higher satisfaction than low-UA users. Conversely, people with low UA have a tendency to try new services, from which they may infer that a service currently in use is inadequate—less useful, more difficult to use, less enjoyable, and/or less valuable than another might be. That is, they may attribute their intention to switch services not to their low UA propensity but to low post-expectation and satisfaction levels for the services they use.

Effects of Uncertainty Avoidance on Path Strengths between Post-Use Expectation Factors and Dependent Variables

Uncertainty avoidance correlates closely with adopter type. Early adopters show less propensity than late adopters to avoid uncertainty. Late adopters fear change more and tend to refrain from adopting new technologies (Rogers, 1995). Studying UA on the national level, Maitland (1999) argued that the diffusion of new technologies is slower in high-UA than in low-UA cultures, and Gales (2008) found that high-UA cultures are more resistant to introducing a new technology because the process increases anxiety. The correspondence between adopter type and UA propensity allows us to predict, on the basis of findings on the post-adoption behaviors of late adopters, the relative path strengths of extrinsic and intrinsic motivation factors. Specifically, Rogers (1995) has shown that late adopters focus on how using new technologies will help them in the end, and thus tend to put more emphasis on the outcome than on the process of using of a new technology (Rogers, 1995; Venkatesh & Brown, 2001). Late adopters are thus influenced more by extrinsic motivational factors such as usefulness and monetary value than by intrinsic factors like enjoyment or ease-of-use. Because high UA and late adoption correspond so closely, we predict that the same pattern will be found among high-UA users—that the path strengths between extrinsic motivational factors and the dependent variables will be stronger for them than for low-UA users:

Hypothesis 3-1: The influence of post-usefulness on satisfaction will be stronger for the high UA-group than for the low-UA group.

Hypothesis 3-2: The influence of post-usefulness on continuance intention will be stronger for the high-UA group than for the low-UA group.

Hypothesis 3-3: The influence of post–monetary value on satisfaction will be stronger for the high-UA group than for the low-UA group.

Hypothesis 3-4: The influence of post–monetary value on continuance intention will be stronger for the high-UA group than for the low-UA group.

Early adopters are, in contrast, more influenced by the process of using new technologies than by the outcome of use, because they are more curious about technological change and view it more favorably than late adopters do (Rogers, 1995). Early adopters actively explore and strive for proficiency with newly adopted technologies and emphasize the pleasure they derive from the process itself (Rogers, 1995; Venkatesh & Brown, 2001). They tend, in other words, to be motivated by intrinsic factors such as ease-of-use and enjoyment. Just as with late adopters and people with high UA, there is a strong correlation between early adoption and low UA. Early adopters tend to respond to uncertainties and risks more actively and to handle them better than late adopters (Rogers, 1995). Maitland (1999) saw more rapid diffusion of new technology in low-UA than in high-UA cultures, and Gales (2008) has argued that low-UA cultures do not consider new technology as a potential threat to customary procedures and easily embrace it. Because early adoption and low UA correspond so closely, we predict that the emphasis on intrinsic motivation seen in the one group will also be seen in the other—in other words, that the path strengths between post–ease-of-use and post-enjoyment and the dependent variables will be stronger for low-UA than for high-UA users:

Hypothesis 4-1: The influence of post–ease-of-use on satisfaction will be stronger for the low UA-group than for the high-UA group.

Hypothesis 4-2: The influence of post–ease-of-use on continuance intention will be stronger for the low-UA group than high for the UA-group.

Hypothesis 4-3: The influence of post-enjoyment on satisfaction will be stronger for the low UA-group than for the high-UA group.

Hypothesis 4-4: The influence of post-enjoyment on continuance intention will be stronger for the low-UA group than for the high UA-group.

RESEARCH METHODOLOGY

Two nationwide online surveys were conducted in Korea and Hong Kong to test the research hypotheses empirically.

Development Questionnaire

Questions for the seven constructs (post-usefulness, post–monetary value, post–ease-of-use, post-enjoyment, satisfaction, continuance intention, and uncertainty avoidance) were gathered through a review of the extant literature (Bhattacherjee, 2001; Davis et al., 1989; Lee, Choi, Kim, Hong, & Tam, 2004; Yi & Hwang, 2003). In-depth interviews with sixteen experts in MDS industries were then conducted to confirm the content validity of the selected questions. The interviews demonstrated that some questions did not fit well in the context of MDS, and these were deleted from the questionnaire. To ensure the validity of the remaining questions, a pre-test was conducted with 208 MDS users. The pre-test allowed us to delete several questions that did not contribute to the research constructs.[2] The final questionnaire, presented in Appendix 1, consisted of four questions for post-usefulness, three for post–monetary value, four for post–ease-of-use, four for post-enjoyment, four for satisfaction, three for continuance intention, and three for uncertainty avoidance.

Data Collection

This study used the online survey method through the Internet.[3] Large-scale online surveys were conducted in Korea and Hong Kong, using the same questionnaire at around the same time. The back-translation method was used to maintain the linguistic integrity of the survey items between the two countries. In a typical survey, researchers establish a sample group using email lists as invitation lists. However, multiple email addresses for individual users, multiple responses from individual users, and invalid email addresses all pose problems for this method (Wright, 2005). To avoid such problems, we solicited respondents via banner advertisements on websites from several popular portals. Survey data was collected as follows. Before presenting the questionnaire, we emphasized that only those who had used MDS more than four times in the past month were eligible to respond to the survey. This was the criterion used to identify MDS continuers in a prior study (Kim, Lee, & Kim, 2008). Respondents were asked to provide their mobile phone numbers, which the telecommunications companies would use to verify the identities of respondents and their prior experience with MDS. Only those who permitted the authors to check their usage log data with telecommunications companies were allowed to participate. We enforced accurate reporting of phone numbers by explaining that the lottery prizes for which all participants were eligible would be sent to the billing addresses of the reported phone numbers. This approach helps reduce the problem of multiple responses, as it is harder to have multiple phone numbers than multiple email addresses, and phone numbers can be identified easily through data possessed by the telecommunication companies. Another issue with online surveys is response rate. Although it is known that response rates with email invitations are at least as good as with traditional mailed surveys (Wright, 2005), our study used a common technique for improving response rate, namely, offering a financial incentive. One hundred respondents, selected by lottery, were awarded a gift for their participation. Because financial incentives may encourage multiple responses, we double-checked respondents' phone numbers later through the telecommunication companies. Respondents went on to answer each of the survey questions on a seven-point Likert scale ranging from strongly disagree (1) to strongly agree (7), as shown in Appendix 2. They concluded the survey by providing demographic information.

A total of 1,458 people participated in Korea and 1,605 in Hong Kong. Once subjects had finished the survey, their phone numbers and survey responses were sent to the telecommunication companies for data verification. The companies checked whether the phone numbers reported were legitimately registered and whether the owners of the phone numbers had used the MDS at least once in the past.[4] Those who failed either test were deleted from the data set. The final sets of effective respondents numbered 1,262 in Korea and 1,195 in Hong Kong. Demographic profiles for the respondents in the two countries are shown in Table 1.

Validity and Reliability of Questions

Here we present the results of a confirmatory factor analysis for our base model before conducting a multi-group invariance analysis, because reliable and valid base models are pre-requisite for the multi-group invariance analysis.

Table 1. Demographic information

Country	Average age	Gender	
		Male	Female
Korea	24.34	462	800
Hong Kong	28.98	514	681

Table 2. Results of reliability and convergent validity tests (*p < 0.01)

	Post-usefulness	Post–ease of use	Post-enjoyment	Post–monetary value	Satisfaction	Continuance intention	Uncertainty avoidance
Cronbach α	0.943	0.925	0.907	0.860	0.929	0.909	0.706
Construct reliability	0.945	0.926	0.907	0.862	0.934	0.912	0.683
Item1	0.71*	0.89*	0.87*	0.84*	0.83*	0.95*	0.67*
Item2	0.84*	0.90*	0.86*	0.92*	0.86*	0.96*	0.63*
Item3	0.83*	0.93*	0.87*	0.88*	0.86*	0.81*	0.64*
Item4	0.74*	0.88*	0.88*		0.82*		

A confirmatory factor analysis was conducted to test convergent and discriminant validity.[5] The fit indices of the measurement model, such as goodness of fit (GIF = 0.94), adjusted goodness of fit (AGFI = 0.93) and root mean square error approximation (RMSEA = 0.054), indicate that the measurement model provides an adequate fit with the data.

As shown in Table 2, the questions converged well for all seven factors: post-usefulness, post–ease-of-use, post-enjoyment, post–monetary value, satisfaction, continuance intention, and UA. Cronbach α (threshold value 0.7) is tested to determine the reliability of a measure based on internal consistency. The Cronbach α coefficients for all constructs but one are well above the threshold value of 0.70, indicating that each question faithfully represents the corresponding construct, and that the measurement model as a whole is reliable. Uncertainty avoidance of value 0.683 seems reliable since it is fairly close to 0.70 and above 0.6 is considered as an acceptable level of reliability by many researchers.[6] Convergent validity can be measured by factor loadings of observation variables. When t-values for each loading are significant ($t > 2.00$), the question ensures convergent validity (Anderson & Gerbing, 1988). The t-values of each corresponding construct were significant at the level of 0.01 with high factor loading, confirming the convergent validity of the survey instrument.

Discriminant validity was assessed using the average variance extracted (AVE) for each construct (Bhattacherjee, 2001). The AVE for each construct should exceed the squared correlation between that and any other construct. As can be seen in Table 3, test results confirmed the discriminant validity of all seven constructs.

Table 3. Correlations matrix and average variance extracted (AVE)

	AVE	PU	PEOU	PE	PMV	Satisfaction	Continuance Intention	UA
Post-Usefulness (PU)	**0.78**	1						
Post–Ease of Use (PEOU)	**0.90**	0.40	1					
Post-Enjoyment (PE)	**0.87**	0.55	0.51	1				
Post-monetary value (PMV)	**0.88**	0.42	0.19	0.29	1			
Satisfaction	**0.84**	0.41	0.38	0.41	0.39	1		
Continuance intention	**0.91**	0.36	0.46	0.36	0.24	0.56	1	
UA	**0.65**	0.08	0.12	0.10	-0.13	0.09	0.19	1

Common method bias, considered a latent problem in survey research (Flynn, Sakakibara, Schroeder, Bates, & Flynn, 1990), was checked for using Harmon's one-factor test (McFarlin & Sweeney, 1992; Podsakoff & Organ, 1986; Podsakoff, MacKenzie, Lee, & Podsakoff, 2003). Unrotated principal components analysis revealed seven factors with eigen-values greater than 1, accounting for 79.21% of the total variance. The first factor did not account for the most variance (13.91%). Therefore, no single dominant factor emerged or accounted for most of the variance, suggesting that common method bias is not a problem in this study.

Next, a multi-group invariance analysis was conducted to test whether the instruments provided equivalent measurement between the two countries in the base model (Doll, Deng, Raghunathan, Gholamreza, & Xia, 2004; Doll, Hendrickson, & Deng, 1998). We tested the equivalence of factor loadings because that is the basic condition for factorial invariance (Doll et al., 2004; Doll et al., 1998). In addition, equality of factor loadings is generally a higher priority than equality of other parameters (Bollen, 1989). To test for equal factor loadings, an equal factor loading constraint was added to the base model, creating a more restrictive model that is a subset of the base model. For testing equivalence, Little (1997) and Doll et al. (1998) suggest examining differences in subjective fit indices, such as CFI, NNFI, and RMSEA,

between the constrained and the unconstrained models. Although no absolute thresholds have been established, a difference of less than 0.04 between the values of the fit indices for the constrained and unconstrained models is generally taken to indicate invariance of the measurement models across the groups (Little, 1997; Rahim & Magner, 1995). As shown in Table 4, the CFI, NNFI, and RMSEA scores demonstrate changes of less than 0.01 between the constrained and the unconstrained models. These minor differences in subjective fit indices indicate invariance of the measurements across the two countries. Therefore, the invariance results allow us to merge data from the two countries and to analyze them as a single population.

Classification into High-UA and Low-UA Groups and Multi-Group Analysis

Since the invariance analysis showed insignificant differences between responses from Korea and Hong Kong, we merged the two data sets and re-classified them according to the grand means of UA measures. In other words, respondents were classified as high-UA or low-UA according to their individual UA propensities, using as criterion the grand mean value of UA including both Korean and Hong Kong respondents. This classification has been used widely in prior studies that employ

Table 4. Multi-group invariance analysis

Model-Test	Chi-square	df	Sig.Level	CFI	NNFI	RMSEA
All-λ's-unconstrained (Baseline Model)	2452.89	388		0.96	0.96	0.059
PU-constrained	2475.34	391	$p < 0.05$	0.96	0.96	0.059
PEOU-constrained	2473.03	391		0.96	0.96	0.059
PMV-constrained	2481.39	390		0.96	0.96	0.059
PE-constrained	2471.40	391		0.96	0.96	0.059
SAT-constrained	2510.60	391		0.96	0.96	0.060
INT-constrained	2455.46	390		0.96	0.96	0.059
All-λ's-constrained	2604.15	404		0.96	0.96	0.060

continuous variables as moderating variables (Aronson, Reilly, & Lynn, 2006; Carbonell & Rodriguez, 2006; Evanschitzky & Wunderlich, 2006; Laroche, Bergeron, & Goutaland, 2003). For example, Aronson et al. (2006) classified respondents into low- or high-uncertainty groups based on information provided by the respondents to two items. In our study, the high-UA group contained 1,278 subjects, with 852 Korean respondents and 426 from Hong Kong, while the low-UA group contained 1,179 subjects, with 410 Korean respondents and 769 from Hong Kong. These results confirm prior findings that Korea is, overall, a higher-UA culture than Hong Kong (Hofstede, 1997), but also that within each country, there are individual differences at variance with national traits (Hwang, 2004).

In summary, this study identifies cultural differences in UA at the individual level by merging respondents from the two countries and reclassifying them according to individual UA propensities, a procedure validated by the results of the multi-group invariance analysis. This reclassification allows the study to focus on the impact of UA without consideration for differences between Hong Kong and Korea.

A multi-group analysis with LISREL may be performed to investigate moderating effects in a structural model when independent data samples are available for each group of the moderating variables (Baron & Kenny, 1986). In this study, multi-group analysis was conducted to test whether the model, which showed a good fit for the entire sample, was different between the high-UA and low-UA groups. Byrne (1989) and Lai and Li (2005) suggest the following steps: first, the model that fits the data in the entire sample well is tested separately for high-UA and low-UA groups; second, the model is tested simultaneously for both UA propensities without constraints; third, equality constraints are imposed on the variance, covariance, and factor structure of the models, and the constrained model is test

simultaneously for both samples; finally, the chi-square difference between the unconstrained model and the constrained model is checked for statistical significance.

RESULTS

The results of the general continuance model are presented below. Two further sets of results follow, one for the effects of UA on the mean values of the model variables, and one for the effects of UA on the path coefficients.

A Structural Equation Model for the General Continuance Model

A structural equation model for the general continuance model including both the high-UA and low-UA groups is presented first, in order to reveal the relationships between post-expectation factors and satisfaction and continuance intention in the domain of MDS. As shown in Table 5, the fit indices indicate that the continuance model provides an adequate fit to the data for the entire survey population, except for the chi-square statistic and the normed chi-square (χ^2/df) (Hair, Anderson, Tatham, & Black, 1998). The p-value of chi-square and χ^2/df ratio statistic are absolute indices sensitive to sample size (Marsh, Balla, & McDonald, 1988). We believe that the high values for these items are a result of our sample size, and do not indicate that the models are inappropriate.

Figure 2 presents the final continuance model, which uses the maximum likelihood technique. Four post-expectation factors are found to influence satisfaction, while only post-usefulness and post–ease-of-use influence continuance intention. Also, post–monetary value is found to have more influence than any other post-expectation factor on satisfaction. Post–ease-of-use, however, is found to have the strongest effect on continuance intention.

Table 5. Goodness-of fit indices of the structural equation model

	χ^2	df	χ^2/df	GFI	AGFI	NFI	CFI	RMSEA
Recommendation Value	not significant		< 3.0	≥ 0.80	≥ 0.90	≥ 0.80	≥ 0.90	≥ 0.90
High-UA Group	874.45	194	4.507	0.94	0.92	0.98	0.98	0.052
Low-UA Group	917.85	194	4.731	0.93	0.91	0.98	0.98	0.056
Total	1602.92	194	8.262	0.94	0.93	0.98	0.98	0.054

Effects of UA on Mean Values of Post-Expectation Factors, Satisfaction, and Continuance Intention

The study conducted an independent samples t-test between the high-UA and low-UA groups on those factors affecting continuance behavior in the MDS domain. As shown in Table 6, the results show significant differences between the two groups in five out of six variables. The high-UA group was found to have statistically higher values for post-usefulness, post–ease-of-use, post-enjoyment, satisfaction, and continuance

intention. Therefore, hypotheses 1-1, 1-3, and 1-4 were supported, as well as hypotheses 2-1 and 2-2. Contrary to expectation, there was no difference between the two groups in terms of post–monetary value. Thus hypothesis 1-2 is not supported.

Effects of UA on Path Strengths from Post-Expectation Factors to Dependent Variables

The results of a multi-group analysis, using nested chi-square tests to compare the high- and low-UA groups, are shown in Table 7.

Figure 2. Path diagram for MDS

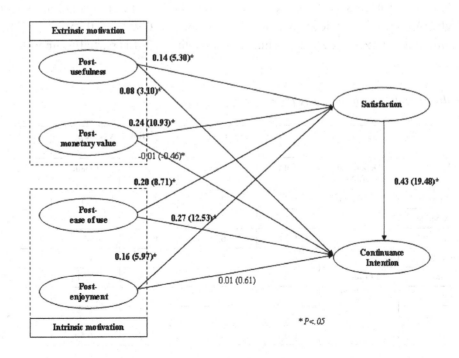

Table 6. Summary of independent samples t-test

	UA	Average Mean	Standard Deviation	Sig. (2-tailed)
Post-usefulness	Low group	4.3227	1.01080	0.000
	High group	4.5477	1.14523	
Post–Monetary Value	Low group	3.6319	1.00270	0.126
	High group	3.5642	1.18811	
Post–Ease-of-Use	Low group	4.9302	1.14701	0.000
	High group	5.3901	1.14380	
Post-Enjoyment	Low group	4.7691	1.06039	0.000
	High group	5.0409	1.07046	
Satisfaction	Low group	4.2487	0.90074	0.000
	High group	4.5661	0.94236	
Continuance Intention	Low group	4.8403	1.20068	0.000
	High group	5.3772	1.01103	

The results show that post-usefulness, an extrinsic factor, affects satisfaction more significantly in the high-UA group than in the low-UA group ($\Delta\chi^2 = 3.85$, $p < 0.05$). Therefore, hypothesis 3-1 is supported. Contrary to expectation, the results show no meaningful difference between the two groups in terms of the effect of post-usefulness on continuance intention. Although a difference appears, and moves in the direction predicted (high UA > low UA), it does not reach statistical significance ($\Delta\chi^2 = 0.75$, *n.s.*). Thus hypothesis 3-2 is not supported. Post–monetary value, another extrinsic factor, is found to affect satisfaction more strongly in the high-UA group than in the low-UA group ($\Delta\chi^2 = 3.67$, $p < 0.1$), supporting hypothesis 3-3. Although post–monetary value is found to affect continuance intention more strongly in the high-UA group than in the low-UA group ($\Delta\chi^2 = 6.63$, $p < 0.05$), path coefficients for both groups are not significant ($\beta = -0.03$, $t = -0.95$; $\beta = 0.05$, $t = 1.97$). Therefore, hypothesis 3-4 is partially supported.

Table 7. Results of multi-group analysis (**p < 0.05, *p < 0.1)

		Low-UA Group		High-UA Group		Δ chi-square (Δ *df*)
Independent Variable	Dependent Variable	Path	*t*-value	Path	*t*-value	Uncon (1792.31) *df* (388)
Post-Usefulness	Satisfaction	0.08	2.08	0.17	4.65	**3.85 (1)****
Post–Monetary Value		0.19	5.55	0.30	10.19	**3.67 (1)***
Post–Ease-of-Use		0.24	6.75	0.14	4.75	**5.76 (1)****
Post-Enjoyment		0.16	4.20	0.14	4.05	0.52 (1)
Post-Usefulness	Continuance Intention	0.06	1.69	0.09	2.81	0.75 (1)
Post–Monetary Value		-0.03	-0.95	0.05	1.97	**6.63 (1)****
Post–Ease-of-Use		0.28	8.19	0.21	7.96	**6.55 (1)****
Post-Enjoyment		0.03	0.87	-0.02	-0.59	1.39 (1)

In contrast, the paths from post–ease-of-use, an intrinsic factor, to satisfaction and continuance intention are stronger for the low-UA group than for the high-UA group ($\Delta\chi^2 = 5.76$; $p < 0.05$; $\Delta\chi^2 = 6.55$; $p < 0.05$). Therefore, hypotheses 4-1 and 4-2 are supported. The effects of post-enjoyment, another intrinsic factor, on satisfaction and continuance intention showed no significant difference between the two groups ($\Delta\chi^2 = 0.52$, *n.s.*; $\Delta\chi^2 = 1.39$, *n.s.*). Both effects are stronger for the low-UA group, as expected, but the differences between the two groups do not reach statistical significance. Consequently, hypotheses 4-3 and 4-4 are not supported. Table 8 summarizes the hypotheses on continuance behavior in the domain of MDS.

CONCLUSION AND DISCUSSION

This research investigates empirically the moderating effects of uncertainty avoidance on the continuance behavior of MDS users. The results indicate that in most cases people with high UA have higher post-expectations than people with low UA. The one exception is that no statistically significant difference was observed for the post-expectation factor of monetary value. One plausible explanation is that because the fees for MDS are higher than those for PC Internet access, both groups attribute a low post–monetary value to MDS. The fixed Internet is virtually free once a monthly access fee has been paid, whereas, to take one example, a user pays around $3 to download a single brief ringtone through a mobile data service. In other words, because of the high cost

Table 8. Hypothesis summary

	Hypothesis	Support
Hypothesis 1-1	High-UA users will show a higher mean post-usefulness than low-UA users	**Supported**
Hypothesis 1-2	High-UA users will show a higher mean post–monetary value than low-UA users	**Supported**
Hypothesis 1-3	High-UA users will show higher mean post–ease-of-use than low-UA users	**Supported**
Hypothesis 1-4	High-UA users will show higher mean post-enjoyment than low-UA users	Not supported
Hypothesis 2-1	High-UA users will show higher mean satisfaction than low-UA users	**Supported**
Hypothesis 2-2	High-UA users will show higher mean continuance intention than low-UA users	**Supported**
Hypothesis 3-1	The influence of post-usefulness on satisfaction will be stronger for the high UA-group than for the low-UA group	**Supported**
Hypothesis 3-2	The influence of post-usefulness on continuance intention will be stronger for the high-UA group than for the low-UA group	Not supported
Hypothesis 3-3	The influence of post–monetary value on satisfaction will be stronger for the high-UA group than for the low-UA group	**Supported**
Hypothesis 3-4	The influence of post–monetary value on continuance intention will be stronger for the high-UA group than for the low-UA group	**Partially supported**
Hypothesis 4-1	The influence of post–monetary value on continuance intention will be stronger for the high-UA group than for the low-UA group	**Supported**
Hypothesis 4-2	The influence of post–ease-of-use on continuance intention will be stronger for the low-UA group than high for the UA-group	**Supported**
Hypothesis 4-3	The influence of post-enjoyment on satisfaction will be stronger for the low UA-group than for the high-UA group	Not supported
Hypothesis 4-4	The influence of post-enjoyment on continuance intention will be stronger for the low-UA group than for the high UA-group	Not supported

of using MDS, post–monetary value may be too low to be differentiated between the two groups. The survey results partially support this explanation, in that for both groups, post–monetary value received the lowest score (3.632 for the low UA group, 3.564 for the high UA group) among the four post-expectation factors.

The paths from extrinsic expectation factors to satisfaction and continuance intention accorded for the most part with our expectations. They were found to be stronger in most cases for people with high UA than for people with low UA; the one exception was the path from post-usefulness to continuance intention, where the difference observed, though in the predicted direction, did not reach statistical significance. A more mixed result was obtained for the paths from the intrinsic expectation factors to the dependent variables. The paths from post–ease-of-use to satisfaction and to continuance intention were, as predicted, stronger for people with low UA than for those with high UA. The impact of post-enjoyment on satisfaction and continuance intention, however, revealed no significant differences between the high UA and low UA groups. Once more, some differences were observed in the predicted direction, but none that reached statistical significance. The effect of post-enjoyment on satisfaction and continuance intention was indistinguishable between the two groups. These findings may be explained by the fact that currently MDS provide far more pleasure-oriented services than productivity-oriented ones (Lee, Choi, Kim, & Hong, 2007). For both groups, consequently, pleasure may be highly important to satisfaction, with no significant difference between them (Van der Heijden, 2004). Moreover, although users may take considerable pleasure in using MDS, post-enjoyment may have a hard time affecting continuance intention directly in either group, because of the high cost of using MDS. That would explain why the path coefficients to continuance intention are insignificant for both groups.

This study has several limitations. First, the sampling method of the online survey is known to create uncertainty about validity (Wright, 2005). Problems may arise from technical difficulties in implementation or from a self-selection bias (Lefever, Dal, & Matthíasdóttir, 2006; Pitkow & Kehoe, 1996). Because our survey was conducted by soliciting participants over the Internet, rather than with randomly selected subjects, we cannot be certain that respondents represent the whole population of MDS users. Another limitation regarding to the sampling method is that we could not verify how many times our respondents actually used the MDS; we could verify that they had used at least once. In order to increase reliability of the study results, therefore, alternative data collection methods, such as stratified random sampling with more detailed verification methods, should be used in future research. Another limitation concerns the low construct reliability of the uncertainty avoidance scale which is 0.683 (fairly close to 0.70). However, many studies suggest 0.6 as a good reliability threshold. Therefore, we believed that low construct reliability of the uncertainty avoidance scale is acceptable.

The second limitation is that this research was conducted in only two countries, Hong Kong and Korea. In order to increase the external validity of the research results, future research might be extended to other countries.

The third is that users' responses in this study are possibly biased because all the survey questions (see Appendix 1) are in the affirmative mode. This limitation is an artifact of our method of gathering survey items, namely, adapting questions tested and validated in prior studies. Subsequent research should develop survey items in the negative mode, so that questionnaires can correct for potential bias by mixing positive and negative questions.

Fourth, because MDS have mostly been used for personal purposes (Kim, Choi, & Han, 2009), this study focuses on satisfaction and continuance intention. However, MDS have recently started to be used for business and organizational purposes, and future research on MDS may require dependent variables different from those used here. For instance, prior studies in work environments have focused on optimization of task performance and of organization processes (Ahuja & Thatcher, 2005). Ahuja and Thatcher (2005) have argued that researchers should understand the activities of using innovative IT, rather than personal use intention, if we want to describe continuance behavior in work environments accurately. Thus future studies should adopt additional dependent variables, such as actual usage, appropriate to continuance behavior in work environments.

Finally, this study used four post-expectation factors (post-usefulness, post–monetary value, post–ease-of-use and post-enjoyment) that have significant effects on satisfaction and continuance intention in the domain of MDS. The relationship of post-expectation and continuance behavior is modeled in many prior studies, and the four proposed expectation factors may explain fairly well continuance behavior in a domain that has both hedonistic and utilitarian qualities. However, other models have proposed post-expectation variables different in some respects from those used here. For example, Jasperson et al. (2005), in their conceptualization of post-adoption behavior in work systems, propose three factors influencing post-adoptive behavior: prior use, habit, and feature-centric view of technology. While our research framework does not deal with these variables explicitly, it may do so indirectly. For just as Bhattacherjee's (2001) model, in examining usefulness, implicitly considers prior use (as Jasperson et al. [2005] observe), our study, in measuring ease-of-use, implicitly considers two of Jasperson et al.'s (2005) three factors, namely, prior use and habit. People with high UA are reluctant to adopt new technology initially because of their wish to avoid uncertainty. When they do adopt a new technology, they select features and develop better use habits to improve their performance and to reduce uncertainty. After the

initial stage of use, these actions can be reflected in perceived ease of use. Our framework differs from that of Jasperson et al. (2005) in its inclusion of factors they ignore, namely, enjoyment and monetary value. These factors were taken into consideration because our study domain, MDS, can be used for hedonistic purposes by individuals, while Jasperson et al. (2005) consider only work systems used in companies for utilitarian purposes. Monetary value is not considered a critical issue in company work systems, as Jasperson et al. (2005) point out, but for individuals on limited budgets, as in the case of the MDS, it may indeed be critical. Nonetheless, future studies of MDS should include other post-expectation factors, such as connectivity, need for uniqueness, and service availability, some of which have been identified as important variables in the domain of MDS (Chae et al., 2002).

Despite these limitations, the study has a number of implications, among them four important theoretical implications. First, we propose a continuance model in the domain of MDS. While previous research on MDS has focused mostly on the initial acceptance stage, MDS markets in several countries, including Korea and Hong Kong, have already progressed to the confirmation stage, when users decide whether or not to continue current services. This study of continuance is thus appropriately timed and may help future ICT-related research determine whether a market is at the initial adoption or the post-adoption stage. Second, while previous research has mostly approached cultural difference at the national level, this study examines continuance behavior at the level of individual cultural traits, by measuring the UA propensities of each subject. It thus models a way for future studies to study other cultural traits at the individual level. Third, while many prior studies have suggested contextual, technological, cognitive, affective, and motivational factors to explain continuance behavior, among motivational factors they have measured only post-usefulness. Our study opens the field of possible motivation considerably, distinguishing between intrinsic and extrinsic motivations, proposing two new intrinsic post-expectation factors (post–ease-of-use and post-enjoyment), and adding to post-usefulness a second extrinsic factor (post–monetary value). The added factors are closely related to characteristics of MDS, which are multipurpose ICT services that provide a suite of utilitarian and hedonistic functions in a ubiquitous computing context. This enhanced conception of motivation should be useful to future studies of other multipurpose ICT services. Finally, although previous studies have verified empirically that motivational factors affect user behavior, few have focused on the relationship between post-expectation motivation factors and cultural differences of UA at the individual level. This research categorizes users' post-expectation factors according to motivation type, and reveals that the effects of motivational factors on satisfaction and continuance intention are different according to UA, one of the most important cultural characteristics of MDS users, at the individual level.

The study has two major practical implications. The first is that it sheds light on the continuance behavior of MDS users, enabling MDS providers to enhance customer satisfaction and continuance intention by developing more user-oriented services. For instance, in order to increase continuance intention, it would be most effective for providers to focus on post–ease-of-use, the factor found to have the strongest impact on continuance intention. Second, the study results may help MDS providers customize development strategies by taking individual differences of UA into account. The within-country individual differences of UA observed here mean that inter-market segmentation strategies may be possible. For instance, if an MDS company provides its service to a segment of low-UA users in one country, and wants to enter and target a segment in another country, it is more likely to find similar continuance behavior in the new segment if it targets one that also has low-UA characteristics. Moreover, when

upgrading services used primarily by people with low UA, it may be more effective to emphasize intrinsic factors like post–ease-of-use than either post-usefulness or post–monetary value.

In conclusion, we have found that uncertainty avoidance is an important cultural characteristic of MDS users, exerting a substantial effect on continuance behavior. MDS should thus be designed and adjusted to fit the individual UA traits of their target users. This study adds to our understanding of the intricate connections between an important cultural dimension, uncertainty avoidance, and continuance behavior with MDS.

REFERENCES

Ahuja, M. K., & Thatcher, J. B. (2005). Moving beyond intentions and toward the theory of trying: Effects of work environment and gender on post-adoption information technology use. *Management Information Systems Quarterly*, *29*(3), 427–459.

Albers, M. J., & Kim, L. (2000). *User Web browsing characteristics using palm handheld for Information Retrieval*. Paper presented at the Proceedings of IEEE Professional Communication Society International Professional Communication Conference and Proceedings of the 18th Annual ACM International Conference on Computer Documentation: Technology and Teamwork, Cambridge, UK.

Anderson, J., & Gerbing, D. (1988). Structural equation modeling in practice: A review and recommended two-step approach. *Psychological Bulletin*, *103*(3), 411–423. doi:10.1037/0033-2909.103.3.411

Aronson, Z. H., Reilly, R. R., & Lynn, G. S. (2006). The impact of leader personality on new product development teamwork and performance: The moderating role of uncertainty. *Journal of Engineering and Technology Management*, *23*(3), 221–247. doi:10.1016/j.jengtecman.2006.06.003

Atkinson, M. A., & Kydd, C. (1997). Individual characteristics associated with World Wide Web use: An empirical study of playfulness and motivation. *The Data Base for Advances in Information Systems*, 28(2), 53–62.

Ayoun, B. M., & Moreo, P. J. (2008). The influence of the cultural dimension of uncertainty avoidance on business strategy development: A cross-national study of hotel managers. *International Journal of Hospitality Management*, 27(1), 65–75. doi:10.1016/j.ijhm.2007.07.008

Bajaj, A., & Nidumolu, S. R. (1998). A feedback model to understand information system usage. *Information & Management*, 33(4), 213–224. doi:10.1016/S0378-7206(98)00026-3

Baron, R. M., & Kenny, D. A. (1986). The moderator-mediator variable distinction in social psychological research: conceptual, strategic, and statistical considerations. *Journal of Personality and Social Psychology*, 51(1), 173–182.

Bem, D. J. (1972). *Self-perception theory*. New York: Academic Press.

Bhattacherjee, A. (2001). Understanding information systems continuance: An expectation- confirmation model. *Management Information Systems Quarterly*, 25(3), 351–370. doi:10.2307/3250921

Bollen, K. A. (1989). *Structural equations with latent variables*. New York: John Wiley & Sons.

Brunelle, J. P. (2001). The impact of community service on adolescent volunteers' empathy, social responsibility, and concern for others. Unpublished doctoral dissertation, Virginia Common Wealth University, Virginia.

Bruner, G. C., & Kumar, A. (2005). Explaining consumer acceptance of handheld Internet devices. *Journal of Business Research*, 58(5), 553–558. doi:10.1016/j.jbusres.2003.08.002

Byrne, B. M., Shavelson, R. J., & Muthen, B. (1989). Testing for the equivalence of factor covariance and mean structures: the issue of partial measurement invariance. *Psychological Bulletin*, 105(3), 456–466. doi:10.1037/0033-2909.105.3.456

Carbonell, P., & Rodriguez, A. I. (2006). Designing teams for speedy product development: The moderating effect of technological complexity. *Journal of Business Research*, 59(2), 225–232. doi:10.1016/j.jbusres.2005.08.002

Chae, M. H., Kim, J. W., Kim, H. Y., & Ryu, H. S. (2002). Information quality for mobile data services: A theoretical model with empirical validation. *Electronic Markets*, 12(1), 38–46. doi:10.1080/101967802753433254

Chin, W. W. (1998). Issues and opinion on structural equation modeling. *Management Information Systems Quarterly*, 22(1), 7–16.

Choi, M., Lee, I. S., Choi, H., & Kim, J. W. (2003). *A cross-cultural study on the post-adoption behavior of mobile internet users*. Paper presented at the Proceedings of the Digit 2003 (Pre-ICIS), Seattle, WA.

Dabholkar, P. A. (1996). Consumer evaluations of new technology-based self-service options: An investigation of alternative models. *International Journal of Research in Marketing*, 13(1), 29–51. doi:10.1016/0167-8116(95)00027-5

Davis, F. D., Bagozzi, R. P., & Warshaw, P. R. (1989). User acceptance of computer technology a comparison of two theoretical models. *Management Science*, 35(8), 982–1003. doi:10.1287/mnsc.35.8.982

Davis, F. D., Bagozzi, R. P., & Warshaw, P. R. (1992). Extrinsic and intrinsic motivation to use computers in the workplace. *Journal of Applied Social Psychology*, 22(14), 1111–1132. doi:10.1111/j.1559-1816.1992.tb00945.x

De Mooij, M. (2004). *Consumer behavior and culture: Consequences for global marketing and advertising*. Thousand Oaks, CA: Sage Publications.

De Mooij, M., & Hofstede, G. (2002). Convergence and divergence in consumer behavior: implications for international retailing. *Journal of Retailing, 78*(1), 61–69. doi:10.1016/S0022-4359(01)00067-7

Dodds, W. B., & Monroe, K. B. (1985). The effect of brand and price information on subjective product evaluations. *Advances in Consumer Research. Association for Consumer Research (U. S.), 12*(1), 85–90.

Doll, W. J., Deng, Z., Raghunathan, T. S., Gholamreza, T., & Xia, W. (2004). The meaning and measurement of user satisfaction: A multi-group invariance analysis of the end-user computing satisfaction instrument. *Journal of Management Information Systems, 21*(1), 227–262.

Doll, W. J., Hendrickson, A., & Deng, X. (1998). Using Davis' perceived usefulness and ease of use instrument for decision making: A confirmatory and multigroup invariance analysis. *Decision Sciences, 29*(4), 839–870. doi:10.1111/j.1540-5915.1998.tb00879.x

Donthu, N., & Yoo, B. (1998). Cultural influences on service quality expectations. *Journal of Service Research, 1*(2), 178–186. doi:10.1177/109467059800100207

Erez, M., & Gati, E. (2004). A dynamic, multi-level model of culture: From the micro level of the individual to the macro level of a global culture. *Applied Psychology: An International Review, 53*(4), 583–598. doi:10.1111/j.1464-0597.2004.00190.x

Evanschitzky, H., & Wunderlich, M. (2006). An examination of moderator effects in the four-stage loyalty model. *Journal of Service Research, 8*(4), 330–345. doi:10.1177/1094670506286325

Evers, V., & Day, D. (1997). *The role of culture in interface acceptance*. Paper presented at the Proceedings of the IFIPTC13 International Conference on Human-Computer Interaction, London, UK.

Flynn, B. B., Sakakibara, S., Schroeder, R. G., Bates, K. A., & Flynn, E. J. (1990). Empirical research methods in operations management. *Journal of Operations Management, 11*(4), 339–366. doi:10.1016/S0272-6963(97)90004-8

Frank, L., Sundqvist, S., Puumalainen, K., & Taalikka, S. (2001). *Cross-cultural comparison of innovators: Empirical evidence from wireless services in Finland, Germany and Greece*. Paper presented at the ANZMAC Conference, Auckland, New Zealand.

Gales, L. (2008). The role of culture in technology management research: National character and cultural distance frameworks. *Journal of Engineering and Technology Management, 25*(1-2), 3–22. doi:10.1016/j.jengtecman.2008.01.001

Hair, J. F. Jr, Anderson, R. E., Tatham, R. L., & Black, W. C. (1998). *Multivariate data analysis* (5th ed.). Upper Saddle River, NJ: Prentice Hall.

Hasan, H., & Ditsa, G. (1999). The impact of culture on the adoption of IT: An interpretive study. *Journal of Global Information Management, 7*(1), 5–15.

HCI-Lab. (2004). *4th Worldwide mobile data study report*. South Korea: Yonsei University.

Hiller, M. (2003). The role of cultural context in multilingual website usability. *Electronic Commerce Research and Applications, 2*(1), 2–14. doi:10.1016/S1567-4223(03)00005-X

Hofstede, G. (1980). *Culture's consequences: International differences in work-related values.* Beverly Hills, CA: Sage Publications.

Hofstede, G. (1997). *Cultures and organizations: Software of the mind.* New York: McGraw-Hill.

Hong, S. E., Kim, J. K., & Lee, H. S. (2008). Antecedents of use-continuance in information systems: Toward an integrative view. *Journal of Computer Information Systems, 48*(3), 61–73.

Hong, S. J., & Tam, K. Y. (2006). Understanding the adoption of multipurpose information appliances: The case of mobile data services. *Information Systems Research, 17*(2), 162–179. doi:10.1287/isre.1060.0088

Hong, W., Thong, J. Y. L., Wong, W. M., & Tam, K. Y. (2001). Determinants of user acceptance of digital libraries: An empirical examination of individual differences and system characteristics. *Journal of Management Information Systems, 18*(3), 97–124.

Hsu, M. H., Yen, C. H., Chiu, C. M., & Chang, C. M. (2006). A longitudinal investigation of continued online shopping behavior: An extension of the theory of planned behavior. *International Journal of Human-Computer Studies, 64*(9), 889–904. doi:10.1016/j.ijhcs.2006.04.004

Huh, Y. E., & Kim, S. H. (2008). Do early adopters upgrade early? Role of post-adoption behavior in the purchase of next-generation products. *Journal of Business Research, 61*(1), 40–46. doi:10.1016/j.jbusres.2006.05.007

Hung, S. Y., Ku, C. Y., & Chang, C. M. (2003). Critical factors of WAP services adoption: An empirical study. *Electronic Commerce Research and Applications, 2*(1), 42–60. doi:10.1016/S1567-4223(03)00008-5

Hwang, Y. J. (2004). *An empirical examination of individual-level cultural orientation as an antecedent to ERP systems adoption.* Paper presented at the The Pre-ICIS Workshop on Cross-Cultural Research in Information Systems, Washington D.C.

International-Telecommunication-Union. (2002). *ITU Internet Reports 2002: Internet for a Mobile Generation.*

Jarvenpaa, S. L., Lang, K. R., Takeda, Y., & Tuunainen, V. K. (2003). Mobile commerce at crossroads. *Communications of the ACM, 46*(12), 41–44. doi:10.1145/953460.953485

Jasperson, J., Carter, P. E., & Zmud, R. W. (2005). A comprehensive conceptualization of post-adoption behaviors associated with information technology enabled work systems. *Management Information Systems Quarterly, 29*(3), 525–557.

Karahanna, E., Straub, D. W., & Chervany, N. L. (1999). Information technology adoption across time: A cross-sectional comparison of pre-adoption and post-adoption beliefs. *Management Information Systems Quarterly, 23*(2), 183–213. doi:10.2307/249751

Khalifa, M., & Liu, V. (2001). Satisfaction with Internet-based services: a longitudinal study. *Journal of Global Information Management, 10*(3), 1–14.

Khalifa, M., & Liu, V. (2003). Determinant of satisfaction at different adoption stage of internet-based services. *Journal of the Association for Information Systems, 4*(5), 206–232.

Kim, B., Choi, M., & Han, I. (2009). User behaviors toward mobile data services: The role of perceived fee and prior experience. *Expert Systems with Applications, 36*(4), 8528–8536. doi:10.1016/j.eswa.2008.10.063

Kim, B., & Han, I. (2009). The role of trust belief and its antecedents in a community-driven knowledge environment. *Journal of the American Society for Information Science and Technology, 60*(5), 1012–1026. doi:10.1002/asi.21041

Kim, H. Y., Lee, I. S., & Kim, J. W. (2008). Maintaining continuers vs. converting discontinuers: Relative importance of post-adoption factors for mobile data services. *International Journal of Mobile Communications, 6*(1), 108–132. doi:10.1504/IJMC.2008.016007

Kim, S. S., & Malhotra, N. K. (2005). A longitudinal model of continued IS use: An integrative view of four mechanisms underlying post adoption phenomena. *Management Science, 51*(5), 741–755. doi:10.1287/mnsc.1040.0326

Lai, V. S., & Li, H. (2005). Technology acceptance model for internet banking: An invariance analysis. *Information & Management, 42*(2), 373–386. doi:10.1016/j.im.2004.01.007

Laroche, M., Bergeron, J., & Goutaland, C. (2003). How intangibility affects perceived risk: The moderating role of knowledge and involvement. *Journal of Services Marketing, 17*(2), 122–140. doi:10.1108/08876040310467907

Lee, I., Choi, B., Kim, J., & Hong, S. J. (2007). Culture-technology fit: Effects of cultural characteristics on the post-adoption beliefs of mobile internet users. *International Journal of Electronic Commerce, 11*(4), 11–51. doi:10.2753/JEC1086-4415110401

Lee, I., Kim, J. S., & Kim, J. W. (2005). Use contexts for the mobile data: A longitudinal study monitoring actual use of mobile data services. *International Journal of Human-Computer Interaction, 18*(3), 269–292. doi:10.1207/s15327590ijhc1803_2

Lee, I. S., Choi, B., Kim, J., Hong, S., & Tam, K. (2004). *Cross-cultural comparison for cultural aspects of mobile internet: Focusing on Korea and Hong Kong.* Paper presented at the Proceedings of the 2004 Americas Conference on Information Systems (AMCIS), New York.

Lee, M. K. O., Cheung, C. M. K., & Chen, Z. (2005). Acceptance of Internet-based learning medium: the role of extrinsic and intrinsic motivation. *Information & Management, 42*(8), 1095–1104. doi:10.1016/j.im.2003.10.007

Lee, Y. S., Kim, J. W., Lee, I. S., & Kim, H. Y. (2002). A cross-cultural study on the value structure of mobile internet usage: Comparison between Korea and Japan. *Journal of Electronic Commerce Research, 3*(4), 227–239.

Lefever, S., Dal, M., & Matthíasdóttir, Á. (2006). Online data collection in academic research: Advantages and limitations. *British Journal of Educational Technology, 38*(4), 574–582. doi:10.1111/j.1467-8535.2006.00638.x

Lippert, S. K., & Volkmar, J. A. (2007). Cultural effects on technology performance and utilization: A comparison of U.S. and Canadian users. *Journal of Global Information Management, 15*(2), 56–90.

Little, T. D. (1997). Mean and covariance structures (MACS) analyses of cross-cultural data: Practical and theoretical issues. *Multivariate Behavioral Research, 32*(1), 53–76. doi:10.1207/s15327906mbr3201_3

Liu, B. S. C., Olivier, F., & Subharshan, D. (2001). The relationships between culture and behavioral intentions toward services. *Journal of Service Research, 4*(2), 118–129. doi:10.1177/109467050142004

Lu, J., Yu, C. S., Liu, C., & Yao, J. E. (2003). Technology acceptance model for wireless Internet. *Internet Research: Electronic Networking Applications and Policy, 13*(3), 206–222. doi:10.1108/10662240310478222

Maitland, C. (1999). Global diffusion of interactive networks: The impact of culture. *AI & Society, 13*(4), 341–356. doi:10.1007/BF01205982

Marcus, A., & Emilie, W. G. (2000). Crosscurrents: Cultural dimensions and global web user-interface design. *Interaction, 7*(4), 32–46. doi:10.1145/345190.345238

Marsh, H. W., Balla, J. R., & McDonald, R. P. (1988). Goodness-of-fit indexes in confirmatory factor analysis: The effect of sample size. *Psychological Bulletin, 103*(3), 391–410. doi:10.1037/0033-2909.103.3.391

Mawhinney, C., & Lederer, C. (1990). A study of personal computer utilization by managers. *Information & Management, 18*(5), 243–253. doi:10.1016/0378-7206(90)90026-E

McCoy, S., Everard, A., & Jones, B. M. (2005). An examination of the technology acceptance model in Uruguay and the US: A focus on culture. *Journal of Global Information Management, 8*(2), 27–45.

McCoy, S., Galletta, D. F., & King, W. R. (2005). Integrating national culture into IS research: The need for current individual level measure. *Communications of the Association for Information Systems, 15*(12), 211–224.

McFarlin, D. B., & Sweeney, P. D. (1992). Distributive justice and procedural justice as predictors of satisfaction with personal and organizational outcomes. *Academy of Management Journal, 35,* 626–637. doi:10.2307/256489

Moores, T. T., & Gregory, F. H. (2000). Cultural Problems in Applying SSM for IS Development. *Journal of Global Information Management, 8*(1), 14–19.

Morris, M. G., & Dillon, A. (1997). How user perceptions influence software use. *IEEE Software, 14*(4), 58–65. doi:10.1109/52.595956

Nachum, G. (2003). *Owning the mobile content customer.* Retrieved June 11, 2005 from http://www.totaltele.com/interviews/display.asp?InterviewID=254

Nunnaly, J. (1978). *Psychometric theory.* New York: McGraw-Hill.

Nysveen, H., Pedersen, P. E., & Thorbjnsen, H. (2005). Intentions to use mobile services: Antecedents and cross-service comparisons. *Journal of the Academy of Marketing Science, 33*(3), 330–346. doi:10.1177/0092070305276149

Oh, H. (1999). Service quality, customer satisfaction, and customer value: A holistic perspective. *International Journal of Hospitality Management, 18*(1), 67–82. doi:10.1016/S0278-4319(98)00047-4

Oliver, R. L. (1980). A cognitive model of the antecedents and consequences of satisfaction decisions. *JMR, Journal of Marketing Research, 17*(3), 460–469. doi:10.2307/3150499

Ouellette, J. A., & Wood, W. (1998). Habit and intention in everyday life: The multiple processes by which past behavior predicts future behavior. *Psychological Bulletin, 124*(1), 54–74. doi:10.1037/0033-2909.124.1.54

Parthasarathy, M., & Bhattacherjee, A. (1998). Understanding post-adoption behavior in the context of online services. *Information Systems Research, 9*(4), 362–379. doi:10.1287/isre.9.4.362

Pedersen, P. E., & Ling, R. R. (2003). *Modifying adoption research for mobile Internet service adoption: Cross-disciplinary interactions.* Paper presented at the Proceedings of the 36th Hawaii International Conference on System Sciences (HICSS-36), Big Island, Hawaii.

Pedersen, P. E., Methlie, L. B., & Thorbjonsen, H. (2002). *Understanding mobile commerce end-user adoption: A triangulation perspective and suggestions for an exploratory service evaluation framework.* Paper presented at the Proceedings of the 36th Hawaii International Conference on System Sciences (HICSS-36), Big Island, Hawaii.

Pitkow, J. E., & Kehoe, C. M. (1996). Emerging trends in the WWW user population. *Communications of the ACM, 39*(6), 106–108. doi:10.1145/228503.228525

Png, I. P. L., Tan, B. C. Y., & Wee, K. L. (2001). Dimensions of National Culture and Corporate Adoption of IT Infrastructure. *IEEE Transactions on Engineering Management, 48*(1), 36–45. doi:10.1109/17.913164

Podsakoff, P. M., MacKenzie, S. B., Lee, J. Y., & Podsakoff, N. P. (2003). Common method biases in behavioral research: A critical review of the literature and recommended remedies. *The Journal of Applied Psychology, 88*(5), 879–903. doi:10.1037/0021-9010.88.5.879

Podsakoff, P. M., & Organ, D. W. (1986). Self-reports in organizational research: Problems and prospects. *Journal of Management, 12*(2), 531–544. doi:10.1177/014920638601200408

Preece, J., Rogers, Y., Sharp, H., Benyon, D., Holland, S., & Carey, T. (1994). *Human-computer interaction.* Boston, MA: Addison-Wesley.

Rahim, M. A., & Magner, N. R. (1995). Confirmatory factor analysis of the styles of handling interpersonal conflict: First-order factor model and its invariance across groups. *The Journal of Applied Psychology, 80*(1), 122–132. doi:10.1037/0021-9010.80.1.122

Rogers, E. M. (1995). *Diffusion of innovations* (4th ed.). New York: The Free Press.

Ryan, R. M., & Deci, E. L. (2000). Intrinsic and extrinsic motivations: Classic definitions and new directions. *Contemporary Educational Psychology, 25*(1), 54–67. doi:10.1006/ceps.1999.1020

Segars, A. H., & Grover, V. (1993). Re-examining perceived ease of use and usefulness: A confirmatory factor analysis. *Management Information Systems Quarterly, 17*(4), 517–525. doi:10.2307/249590

Shang, R. A., Chen, Y. C., & Shen, L. (2005). Extrinsic versus intrinsic motivations for consumers to shop on-line. *Information & Management, 42*(3), 401–413. doi:10.1016/j.im.2004.01.009

Simon, S. J. (2001). The impact of culture and gender on Web sites: An empirical study. *ACM SIGMIS Database, 32*(1), 18–37. doi:10.1145/506740.506744

Sir, J. C., Cho, C. H., Yang, H. J., An, I. H., & Kim, J. R. (2004). *2004 Survey on wireless internet use.* Seoul, Korea: Korea Network Information Center.

Spiller, J., Vlasic, A., & Yetton, P. (2007). Post-adoption behavior of users of internet service providers. *Information & Management, 44*(6), 513–523. doi:10.1016/j.im.2007.01.003

Straub, D. W. (1994). The effect of culture on IT diffusion: E-mail and fax in Japan and the United States. *Information Systems Research, 5*(1), 23–47. doi:10.1287/isre.5.1.23

Straub, D. W., Loch, K., Evaristo, R., Karahanna, E., & Srite, M. (2002). Toward a theory-based measurement of culture. *Journal of Global Information Management, 10*(1), 13–23.

Teo, T. S. H. (2001). Demographic and motivation variables associated with Internet usage activities. *Internet Research: Electronic Networking Applications and Policy, 11*(2), 125–137. doi:10.1108/10662240110695089

Teo, T. S. H., Lim, V. K. G., & Lai, R. Y. C. (1999). Intrinsic and extrinsic motivation in Internet usage. *International Journal of Management Science (Omega), 27*(1), 25–37. doi:10.1016/S0305-0483(98)00028-0

Teo, T. S. H., & Pok, S. H. (2003). Adoption of WAP-enabled mobile phones among internet users. *International Journal of Management Science (Omega), 31*(6), 483–498. doi:10.1016/j.omega.2003.08.005

Thang, J. Y. L., Hong, S. J., & Tam, K. Y. (2006). The effect of post-adoption beliefs on the expectation-confirmation model for information technology continuance. *International Journal of Human-Computer Studies, 64*(9), 799–810. doi:10.1016/j.ijhcs.2006.05.001

Tsikriktsis, N. (2002). Does cultural influence website quality expectation? An empirical study. *Journal of Service Research, 5*(2), 101–112. doi:10.1177/109467002237490

Van der Heijden, H. (2004). User acceptance of hedonic information systems. *Management Information Systems Quarterly, 28*(4), 695–704.

Van Slyke, C., Lou, H., Belanger, F., & Sridhar, V. (2004). *The influence of culture on consumer-oriented electronic commerce adoption.* Paper presented at the Proceedings of the 2004 Southern Association for Information Systems Conference.

Veiga, O. F., Floyd, S., & Dechant, K. (2001). Towards modeling the effects of national culture on IT implementation and acceptance. *Journal of Information Technology, 16*(3), 145–158. doi:10.1080/02683960110063654

Venkatesh, V., & Brown, S. A. (2001). A longitudinal investigation of personal computers in homes: adoption determinants and emerging challenges. *Management Information Systems Quarterly, 25*(1), 71–102. doi:10.2307/3250959

Venkatesh, V., Ramesh, V., & Massey, A. P. (2003). Understanding usability in mobile commerce. *Communications of the ACM, 46*(12), 53–56. doi:10.1145/953460.953488

Vlahos, G. E., & Ferratt, T. W. (1995). Information technology use by managers in Greece to support decision making: Amount, perceived value, and satisfaction. *Information & Management, 29*(6), 305–315. doi:10.1016/0378-7206(95)00037-1

Vlahos, G. E., Ferratt, T. W., & Knoepfle, G. (2000). *Use and perceived value of computer-based information systems in supporting the decision making of German managers.* Paper presented at the SIGCPR Evanston Illinois

Wright, B. K. (2005). Researching Internet-based populations: Advantages and disadvantages of online survey research, online questionnaire authoring software packages, and web survey services. *Journal of Computer-Mediated Communication, 10*(3).

Xiaowen, F., Chan, S., Brzezinski, J., & Xu, S. (2005). Moderating effects of task type on wireless technology acceptance. *Journal of Management Information Systems, 22*(3), 123–157.

Yi, M. Y., & Hwang, Y. (2003). Predicting the use of web-based information systems: self-efficacy, enjoyment, learning goal orientation, and the technology acceptance model. *International Journal of Human-Computer Studies, 59*(4), 431–449. doi:10.1016/S1071-5819(03)00114-9

Zhu, K., & Kraemer, K. L. (2005). Post-adoption variations in usage and value of e-business by organizations: Cross-country evidence from the retail industry. *Information Systems Research, 16*(1), 61–84. doi:10.1287/isre.1050.0045

ENDNOTES

1   This work was supported by the Korea Research Foundation Grant funded by the Korean Government, Basic Research Promotion Fund (KRF-2007-327-B00190). the authors also appreciate Dr. In Seong Lee for his help in data gathering and analysis.

2   Results of the pre-test and pilot test are available from the authors on request.

3   Instructions used in the online survey and survey anchor points are presented in Appendix 2.

4   In this confirmation process, checking the actual number of MDS uses in the past month was not possible. The phone companies' systems do not retain this information, out of concern for the privacy of their customers.

5   The results of exploratory factor analysis (EFA) are not shown in this paper, in accordance with Chin's (1998) insistence that the results of CFA have already involved the results of EFA.

6   Nunnaly (1978) recommends 0.7 as the threshold reliability value, but many researchers still consider 0.6 sufficient.

APPENDIX A: CONSTRUCTS AND CORRESPONDING QUESTIONS

Post-Usefulness (Parthasarathy & Bhattacherjee, 1998)

Instruction: Please indicate the extent to which you agree with each statement
Anchors: 1 *Strongly disagree... 7 Strongly agree*
  1. I find MDS useful in my daily life.
  2. Using MDS helps me accomplish things more quickly.
  3. Using MDS increases my productivity.
  4. Using MDS increases my chances of achieving things that are important to me (e.g., raise, promotion, advance in grade/evaluation/position, award/recognition).

Post–Monetary Value (Dodds & Monroe, 1985)

Instruction: Please indicate the extent to which you agree with each statement
Anchors: 1 *Strongly disagree... 7 Strongly agree*
  1. MDS is reasonably priced.
  2. MDS offers good value for the money.
  3. At the current price, MDS provides a good value.

Post–Ease of Use (Karahanna, Straub, & Chervany, 1999)

Instruction: Please indicate the extent to which you agree with each statement
Anchors: 1 *Strongly disagree... 7 Strongly agree*
  1. Learning how to use MDS was easy for me.
  2. My interaction with MDS was clear and understandable.
  3. I find MDS easy to use.
  4. It was easy for me to become skillful at using MDS.

Post-Enjoyment (Yi & Hwang, 2003)

Instruction: Please indicate the extent to which you agree with each statement
Anchors: 1 *Strongly disagree... 7 Strongly agree*
  1. Using MDS is enjoyable.
  2. Using MDS is pleasurable.
  3. I have fun using MDS.
  4. Using MDS is interesting.

Satisfaction (Parthasarathy & Bhattacherjee, 1998)

Instruction: How do you feel about your overall experience of using the MDS?
Anchors:
1. 1 Very dissatisfied ……………..……7 Very satisfied
2. 1 Very displeased ………………….7 Very pleased
3. 1 Very frustrated ………………..…7 Very contented
4. 1 Absolutely terrible …….............… 7 Absolutely delightful

Continuance Intention (Parthasarathy & Bhattacherjee, 1998)

Instruction: Please indicate the extent to which you agree with each statement
Anchors: 1 *Strongly disagree... 7 Strongly agree*
1. I intend to continue to use MDS in the future.
2. I intend to continue to use MDS in the future.
3. In the future, I plan to continue to use MDS frequently.

Uncertainty Avoidance (Lee, Choi, Kim, & Hong, 2007)

Instruction: Please indicate the extent to which you agree with each statement based on your overall experience with MDS.
Anchors: 1 *Strongly disagree... 7 Strongly agree*
1. I do not use MDS when I am not sure about its quality.
2. I get very upset when MDS does something strange.
3. I am reluctant to use MDS if the security of operations is compromised for any reason.

APPENDIX B: SURVEY INSTRUCTIONS AND ANCHOR POINTS

Survey Instructions

The WMIS (Worldwide Mobile Internet Survey) Project is conducting cross-cultural research on the influence of the Mobile Internet on people's lives in different countries. This survey is being conducted in Korea and Hong Kong simultaneously. The representativeness of the sample depends on **your response**, so please complete this questionnaire as accurately as you can. **All of your responses are completely confidential** and will be released only in aggregate form.

If you have any questions about this research, please email chlgns@cup.ac.kr. Thank you for your participation.

Step 1. Please enter your mobile phone number. Please be careful to enter the correct number. Otherwise, you can't be contacted if you win an award in the lottery.

Step 2. In this survey, "mobile Internet service" means wireless access to an assortment of data services using a mobile phone (e.g., SMS, e-mail, MMS, downloads, weather news, e-tickets, stock trading, etc.) available through mobile service providers (e.g., FOMA, i-mode, eZweb, J-sky, etc.).

This definition excludes devices such as laptops, small desktops, and PDA's that can connect to designated transmission stations through a local wireless connection (e.g., Bluetooth, Wireless LAN). Therefore, the following are **NOT** considered to be examples of MDS in our context:

- connecting to the Internet with your laptop in a Starbucks coffee-shop via wireless LAN (or Wi-Fi)
- downloading lecture notes onto your PDA via a campus wireless LAN
- downloading files onto your laptop via Bluetooth

Step 3. Your usage log will be checked with the help of your mobile phone service provider in this study. Do you agree to this? [Only those who permitted the authors to check their usage log data were allowed to participate.]

This work was previously published in International Journal of Global Information Management, Volume 18, Issue 2, edited by Felix B. Tan, pp. 1-29, copyright 2010 by IGI Publishing (an imprint of IGI Global).

Chapter 5

Untangling the Web of Relationships between Wealth, Culture, and Global Software Piracy Rates:
A Path Model

Trevor T. Moores
University of Nevada Las Vegas, USA

ABSTRACT

This article examines the relationship between Hofstede's national culture indices (IDV, PSI, MAS, and UAI), economic wealth (GNI), and national software piracy rates (SPR). Although a number of studies have already examined this relationship, the contribution of this article is two-fold. First, we develop a path model that highlights not only the key factors that promote software piracy, but also the inter-relationships between these factors. Second, most studies have used the dataset from the pre-2003 methodology which only accounted for business software and did not take into account local market conditions. Using the latest dataset and a large sample of countries (n=61) we find there is an important triadic relationship between PDI, IDV, and GNI that explains over 80% of the variance in software piracy rates. Implications for combating software piracy are discussed.

INTRODUCTION

Software piracy continues to be a major problem for the software industry. It is estimated that in 2007, 38% of all software in use was a pirated copy, accounting for US$47.1Bn in lost sales (BSA, 2008). In a number of recent studies, the national software piracy rate (SPR) has been related to socio-economic factors such as economic wealth (Gopal & Sanders, 2000; Ronkainen & Guerrero-Cusumano, 2001; Shin et al., 2004), trade with the USA (Depken & Simmons, 2004), income inequality (Andrés, 2006) and foreign direct investment (Robertson et al., 2008).

Culture is also seen as important (Husted, 2000; Moores, 2003), as well as socio-political factors such as the rule of law and levels of corruption (Banerjee et al., 2005; Marron & Steel, 2000).

DOI: 10.4018/978-1-61350-480-2.ch005

Given that wealth is a strong factor in almost all of these models, economic models have proposed combating software piracy by adjusting the cost of legal software to fit the market (Bae & Choi, 2006; Sundararajan, 2004), or by increasing the cost of pirating by some sort of version control, for instance, where only legitimate copies can benefit from online updates or service support (August & Tunca, 2008; Chiu et al., 2008; Wu & Chen, 2008).

In this article we argue that although wealth may appear to be the dominant factor in explaining national piracy rates, many of the factors used in these models are themselves inter-related. In particular, wealth, culture, and perceived levels of corruption are all significantly correlated (Davis & Ruhe, 2003; Getz & Volkema, 2001; Park, 2003). This suggests that software piracy is motivated by a web of factors.

We highlight these inter-relationships by developing a path model of the national software piracy rates using a measure of economic wealth and Hoftsede's four cultural dimensions. Using the largest available set of countries (n=61) over a 4-year period (2003 to 2006), we show that economic wealth is the most important direct effect, individualism (IDV) and power distance (PDI) interact to form key antecedents to levels of wealth, and hence, levels of software piracy. The model stresses the complexity of the positive and negative forces motivating software piracy, and can act as a blueprint for further research.

THEORETICAL BACKGROUND

Software is treated as an original work of authorship and, along with literary, musical, audio-visual, and artistic work, is protected by intellectual property rights (IPR) legislation in most countries around the world. Within the US, the main IPR legislation is the Copyright Act of 1976, which has been extended by the Digital Millennium Copyright Act (DMCA) of 1998. The DCMA implemented the provisions of the World Intellectual Property Organization (WIPO) Copyright Treaty of 1996, which now has more than 50 signatory countries. Software piracy, then, involves the copying, distribution, or sale of commercial software in violation of the end user license agreement (EULA) that comes with each piece of commercial software. In practice, this means making a copy of a commercial piece of software (say, Microsoft Office 2007), and giving or selling a copy to someone else.

An accurate measure of the level of software piracy in each country would be almost impossible to determine, given that it would require a count of the number of applications being packaged and sold by criminal gangs, as well as the number being shared amongst family and friends. As such, vendor organizations such as the Business Software Alliance (BSA) estimate the levels of piracy by assuming that for each personal computer sold, a certain amount of software will also be sold. Information about computer and software sales in provided by BSA member companies, such as Adobe, Dell, HP, IBM, and Microsoft

The software piracy rate (SPR) is then calculated as the ratio of missing sales to the estimated overall market, with some adjustments made for local conditions and for replacement hardware where software is being transferred from one machine to another. For instance, a SPR of 50% would suggest half of the software in use is a pirated copy. A global SPR is calculated by treating the world as a single market and is heavily influenced by the levels of piracy in the largest markets in the world, such as the U.S., China, and Russia. The SPR for a particular country is calculated by using country-specific data. Small markets with high SPRs would have little influence on the global SPR.

The BSA gathers data for more than 80 countries and estimates that between 1994 and 2002, the global software piracy rate averaged 41%, with a retail value averaging $11.9Bn a year (BSA, 2003). The average across each country, however, declined from 76% in 1994 to 55% in 2002. These declines coincided with new legislation and awareness programs by vendor organizations such as BSA and others. In 2003 the methodology changed to include operating systems and PC games, and using local analysts to determine computer sales and the "load" of software on each computer. Using this new methodology, the global piracy rate averaged 34% between 2003 and 2007, while the country average hovered around 60% (BSA, 2008). With the extra software included, lost sales jumped to an average of $36.7Bn a year.

Studies of the motivating factors for software piracy have ranged from demographic factors, such as social norms and age (Al-Rafee & Cronan, 2006; Kwong et al., 2003; Peace et al., 2003), to a belief that software piracy is justified because legal software is overpriced (Douglas et al., 2007; Introna, 2007; Moores & Chang, 2006). It has also been suggested that for many people, past behavior is often a good predictor of future behavior

(Cronan & Al-Rafee, 2008; Goles et al., 2008). Students tend to have high rates of piracy (e.g., Cronan et al., 2006; Gan & Koh, 2006), with an understanding of the nature and consequences of software piracy having little impact (Wooley & Eining, 2006). Indeed, there is often found to be a core of determined business students that see pirating almost as a right (Moores & Dhaliwal, 2004).

RESEARCH MODEL

At the national level, software piracy is often explained in terms of per capita wealth. Plotting the relationship between per capita gross national income (GNI) and national software piracy rate (SPR) for 2006 (see Figure 1), shows a strong negative relationship (r=-.87, p=.000), such that richer countries have lower piracy rates. Depending on the year chosen for analysis and the measures used, economic wealth has been found to account for more than 60% of the variance in

Figure 1. Relationship between GNI and software piracy rate for 2006 (n=94)

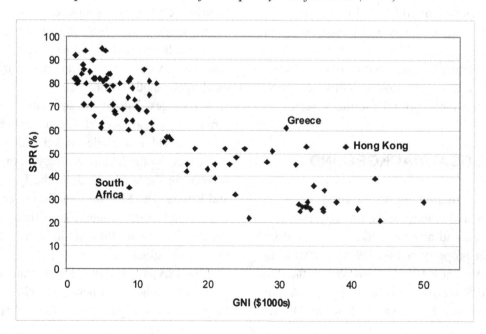

SPR (e.g., Gopal & Sanders, 2000; Ronkainen & Guerrero-Cusumano, 2001; Shin et al., 2004). Our first hypothesis, therefore, is:

H1: The higher the per capita economic wealth of a country, the lower the software piracy rate.

However, it is also noticeable that some countries are relatively poor but have a relatively low SPR (e.g., South Africa), while some relatively rich countries have higher than expected SPRs, such as Greece and Hong Kong. Although a number of different underlying factors are likely to be involved, a strong relationship has often been found between one or more of Hofstede's four cultural dimensions and SPR. The four dimensions are the power distance index (PDI), individualism-collectivism (IDV), masculinity-femininity (MAS), and uncertainty avoidance index (UAI)(Hofstede & Hofstede, 2005).

In brief, PDI refers to the extent to which the less powerful members of organizations and institutions accept and expect that power is distributed unequally, and those in power are beyond reproach. Countries with high PDI include Malaysia and Slovakia, while Canada and The Netherlands are low PDI countries. IDV refers to the extent to which individuality is prized above group (collectivist) ideals, and is a measure of the extent to which the individual is expected to look after themselves. High IDV countries include the US and Australia, while Ecuador and Guatemala are low IDV countries. MAS refers to expressions of traditional masculine ideals of achievement, power, and control, with Slovakia and Japan scoring high, and Norway and Sweden scoring low. Finally, UAI refers to the need for rules and regulations to avoid ambiguity and uncertainty in social settings. Greece and Portugal are high UAI countries, while Jamaica and Singapore score low.

Given that IDV (positively) and PDI (negatively) are significantly correlated with economic wealth, it is no surprise to find that these two dimensions have been related to SPR (Bagchi et al. 2006; Husted, 2000; Moores, 2003; Shin et al., 2004). Although MAS and UAI have not typically been related to SPR, both have been found to be positively related to levels of corruption (e.g., Davis & Ruhe, 2003; Park, 2003), with PDI and UAI also explaining rates of decline in SPR (Moores, 2008). As such, we include both MAS and UAI for completeness. Given that we are using a more up-to-date dataset in the analysis, it is worth confirming the relationship (or lack thereof) of these two dimensions to SPR. Therefore, we have:

H2a: The higher the individualism of a country, the lower the software piracy rate.

H2b: The higher the power distance of a country, the higher the software piracy rate.

H2c: The higher the masculinity of a country, the higher the software piracy rate.

H2d: The higher the uncertainty avoidance of a country, the higher the software piracy rate.

The link between IDV and PDI to economic wealth is confirmed not only as a direct effect on wealth (Hofstede & Hofstede, 2005), but also through corruption. In particular, PDI has been found to be significantly and positively correlated to levels of perceived national corruption (Davis & Ruhe, 2003; Getz & Volkema, 2001; Park, 2003). Corruption, in turn, has been significantly and positively correlated with levels of software piracy (Banerjee et al., 2005; Marron & Steel, 2000; Ronkainen & Guerrero-Cusumano, 2001).

The implication of these studies is that software piracy flourishes in countries that are also perceived as being corrupt, but corruption is an inefficient use of resources and stifles business, leading to overall economic hardship (Getz & Volkema, 2001; Robertson et al., 2008), which is itself another driving force behind software piracy. Therefore, we have:

H3a: The higher the individualism of a country, the higher the per capita economic wealth.
H3b: The higher the power distance of a country, the lower the per capita economic wealth.

Finally, we also take into account the relationship between PDI and IDV, which are significantly and negatively correlated (Hofstede & Hofstede, 2005). Although IDV tends to have the strongest relationship to SPR (Bagchi et al., 2006; Husted, 2000; Moores, 2003; Shin et al., 2004), we suggest the lack of significance for PDI may be due to PDI acting as an antecedent to wealth and IDV, rather than as a direct effect on SPR. This might explain why, although PDI has a significant correlation with SPR, it fails to be a significant variable in regression models (Husted, 2000; Moores, 2003).

Correlation does not imply causality, however, and it would be possible to generate an argument for PDI being an antecedent to IDV, or vice versa. We will take the more conservative approach, therefore, and suggest the relationship is reciprocal, where PDI and IDV determine each other to some extent. We complete our web of relationships, therefore, by suggesting:

H4: The higher the power distance of a country, the lower the individualism.

The research model, summarizing the web of relationships between wealth, culture, and software piracy rates, is illustrated in Figure 2.

METHOD

Given that BSA have increased the number of countries in their annual survey of global software piracy, a sample of 61 countries can now be derived for which wealth, culture, and SPR statistics are available. The latest survey includes SPR statistics for 114 countries and regions from 2002 to 2007 (BSA, 2008). Economic wealth is defined in terms of per capita gross national income (GNI). GNI is

Figure 2. Research model

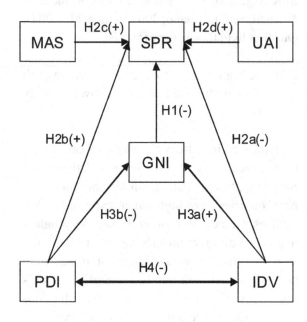

used instead of GDP because GNI includes GDP and net foreign earnings, which provides a better "bank balance" measure of overall wealth. GNI statistics for 227 countries and regions up to 2006 are available from the online World Development Indicators database (World Bank, 2008).

Culture scores are those provided by Hofstede (i.e., Hofstede & Hofstede, 2005). Although the dataset is thus biased by the availability of culture scores, comparing a plot of GNI against SPR for the countries used in the analysis (see Figure 3) to all the countries in the BSA survey (see again Figure 1), suggests the dataset used here is a good approximation of all countries in the BSA survey because the distribution of data points is similar. Finally, given that it is not clear whether the GNI for one year is responsible for the SPR of that particular year, we smooth the data by taking an average. Constrained by the availability of GNI statistics, this suggests SPR and GNI will be averaged from 2002 to 2006.

The path analysis is conducted, first, by applying multiple regression with SPR as the dependent variable and GNI, PDI, IDV, MAS, and UAI as the independent variables. This model

Figure 3. GNI versus software piracy rate for the dataset (n=61)

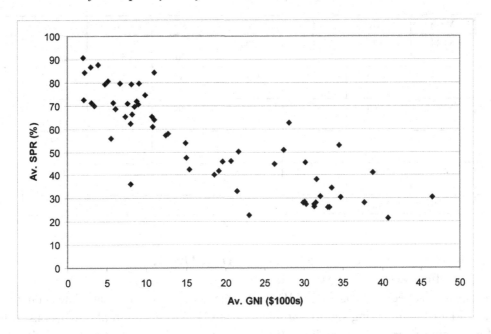

will test hypotheses H1 and H2. Second, multiple regression is again applied with GNI as the dependent variable and PDI and IDV as the independent variables. This model will test hypotheses H3. Finally, a test of hypothesis H4 is carried out by applying ordinary least squares regression with IDV as the dependent variable and PDI as the independent variable. Non-significant variables are trimmed from the analysis and the regression re-run to determine the final path coefficients.

RESULTS

Descriptive statistics for the six variables are given in Table 1. As expected, SPR is significantly correlated with PDI, IDV, and GNI (r>.7), with PDI, IDV, and GNI also having significant correlations (r>.6). The other correlations are much weaker or not significant. The results of the regression analysis are given in Table 2. As can be seen, even using the new methodology (post-2002), the dominance of the relationship between GNI and

SPR can be seen, with PDI and IDV also having a significant direct effect. Overall, the model explains more than 82% of the variance in SPR (see Table 3).

The relationship between PDI, IDV and GNI is further strengthened by the results of the second regression analysis (see Table 4), in which more than half of the variance in GNI is explained by PDI and IDV. The beta-coefficient suggests most of the variance is explained by IDV. Finally, the significant relationship between PDI and IDV is also confirmed by the third regression analysis (see Table 5), with PDI explaining more than 40% of the variance in IDV. These results provide support for all of the hypotheses except H2c and H2d (relating MAS and UAI to SPR). Removing MAS and UAI from the first regression analysis results in a final trimmed path model (see Figure 4).

Direct and indirect effects can be assessed by summing the product of pathways for each independent variable. For instance, PDI has a direct effect on SPR (.18), but also has indirect effects through GNI, and IDV. Summing these indirect effects (see Table 3), it can be seen that while PDI

Table 1. Descriptive statistics (mean, standard deviation, and correlation matrix)(n=61)

Var.	Mean	StdDev	1	2	3	4	5	6
1. SPR	53.88	20.33	1.000	.700 (.000)	-.781 (.000)	-.081 (.535)	.256 (.046)	-.850 (.000)
2. PDI	59.48	22.49		1.000	-.647 (.000)	.132 (.309)	.208 (.108)	-.612 (.000)
3. IDV	44.46	24.60			1.000	.133 (.309)	-.271 (.034)	.682 (.000)
4. MAS	49.85	19.89				1.000	-.034 (.795)	-.030 (.821)
5. UAI	68.52	24.02					1.000	-.293 (.022)
6. GNI	18.00	12.28						1.000

has a relatively weak direct effect on SPR, it has a strong indirect effect. In short, while GNI is the most important direct effect (r=-.85, β=-.55), IDV (β=-.78) and PDI (β=.70) having significant indirect effects, significantly influencing SPR through a reciprocal relationship between IDV and PDI (r=-.65).

Figure 4. Trimmed model

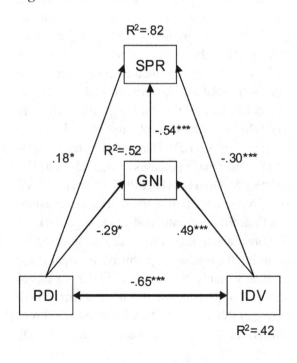

DISCUSSION

In almost all studies that relate economic wealth and Hofstede's cultural dimensions to levels of software piracy, multiple regression has been applied and wealth (per capita GNI, or GDP) has been found to be the most significant factor. We confirm this result using a larger and more up-to-date set of 61 countries over a four-year period (2003 to 2006). By also exploring the inter-relationships between these factors, however, we find that PDI has the highest overall effect, acting through IDV and GNI. This result has important implications when we consider the role of wealth and culture to understand the problem of software piracy, suggesting that any anti-piracy campaign must contend with opposing forces and the inter-relationship of wealth and culture.

The Problem of Wealth

The strength of the relationship between GNI and SPR suggests that software piracy is often a rational economic decision for poorer countries. In response, software vendors such as Microsoft have developed strategies to come to terms with this fact. For instance, establishing the Academic Alliance e-Academy in universities around the world, where Microsoft software is sold at a fraction of the retail price. In 2007, Microsoft also

Table 2. Results of regression analysis (n=61)

Var.	BETA	Std. Err. of BETA	B	Std. Err. of B	t-Test	p-level
(a) Dep. Var.=SPR (R²=.823; F(5,55)=51.081, p=.000)					t(57)	
Intercept			74.911	7.920	9.458	.000
PDI	.207	.081	.187	.073	2.560	.013
IDV	-.265	.089	-.219	.073	-2.987	.004
MAS	-.090	.060	-.092	.061	-1.505	.138
UAI	-.023	.060	-.020	.051	-.392	.697
GNI	-.552	.083	-.914	.137	-6.675	.000
(b) Dep. Var.=GNI (R²=.516; F(2,58)=30.864, p=.000)					t(58)	
Intercept			16.633	6.070	2.740	.008
PDI	-.294	.120	-.161	.065	-2.455	.017
IDV	.492	.120	.246	.060	4.106	.000
(c) Dep. Var.=IDV (R²=.418; F(1,59)=42.379, p=.000)					t(59)	
Intercept			86.529	6.902	12.537	.000
PDI	-.647	.099	-.707	.109	-6.510	.000

Table 3. Direct and indirect effects using the trimmed path model

Independent Variable	Direct Effect	Indirect Effects	Total
PDI	.18	.52	.70
IDV	-.30	-.48	-.78
GNI	-.54	---	-.54

NOTE: Dependent variable is SPR.

expanded the Unlimited Potential initiative, which offers a suite of products, including Windows and MS Office, for $3 to governments and students in emerging countries.

While these initiatives are laudable, the market-sensitive pricing strategy that might be expected to support developing countries has not been widely adopted. For instance, in the US, the retail price of a new copy of Windows Vista Ultimate is approximately $320, while in Thailand the retail price is $400. Thailand's per capita national income is one-sixth that of the US, and so it is no surprise to find Thailand's SPR in 2007 was 78%, with pirated copies of Vista selling for $3.50 soon after release (Cheung, 2006).

While it would be tempting to suggest that as countries become richer, software piracy rates will naturally decline, there appears to be resistance in many countries to switching from pirates to non-pirates. For instance, software vendors might hope the rate of software piracy will decline rapidly in countries that show dramatic increases in national wealth. However, the correlation between the rate of change in GNI and the rate of change in SPR for the 61 countries in our dataset is not significant (r=.14, p=.278). Indeed, of the five countries that saw a more than 40% increase in GNI from 2003 to 2006, only China (-11%) and Estonia (-4%) saw a drop in SPR, while Uruguay (+4%), Argentina (+6%), and Venezuela (+19%) saw increases. This result reinforces the point there is more to software piracy than the ability to pay.

Looking back at our plot of GNI against SPR (see Figure 1), it does appear there is a point of inflexion around GNI=$25K. For poorer countries below this threshold, the correlation between GNI and SPR is strong and negative (r=-.82, p=.000), while for richer countries the relationship is more of a scatter-plot, and the correlation is not significant (r=-.36, p=.105). This type of

Table 4. Results of 1-way ANOVA for the richest 22 countries, with level of SPR as the grouping variable

Var.	Mean			F-statistic	p-level
	All (n=22)	**High SPR (n=11)**	**Low SPR (n=11)**		
SPR	34.19	41.95	26.43	22.993	.000
PDI	42.14	49.91	34.36	5.300	.032
IDV	65.68	60.64	70.73	1.480	.238
MAS	50.77	47.27	54.27	.465	.503
UAI	56.95	57.55	56.36	.011	.919
GNI	32.54	33.06	32.02	.223	.642

point of inflexion has been noted before (e.g., Shin et al., 2004), and suggests that people are more willing to buy legal copies of software as their wealth increases, but a point comes in which other forces come into play, slowing the decline in SPR, and magnifying economic, political, and cultural differences.

The Problem of Culture

The problem with culture as a strategy against software piracy is twofold. First, as a general description of a large social group, a macro-measure of social thinking will not help obviate the deviant behavior of a sub-set of individuals. Given that most countries have IPR laws, people that engage in software piracy are often already aware that copying software is illegal (Wooley & Eining, 2006). Second, if the cultural dimensions represent deeply ingrained social programming, there is very little that software vendors from another country can do to alter that mindset. Indeed, any perception of political or economic colonialism may provoke a hostile response.

However, culture clearly has a significant part to play in explaining levels of software piracy, and can even provide a partial explanation for the scatter-plot of SPRs for the richer countries in our sample (see again, Figure 3). If we split the richest 22 countries (those around and above the observed $25K GNI inflexion point) into two equal groups according to high and low SPR, we find PDI is the only significant cultural or economic variable

(see Table 4). This result not only provides further support for the suggestion that PDI is an important antecedent factor to understanding levels of software piracy, but suggests the effect becomes important when a country reaches an inflexion point where poverty is no longer an excuse.

This result also suggests that economic models that posit a change in the behavior of software pirates based on the perceived value of the software (Bae & Choi, 2006), differences in the quality of real versus pirated software (Chiu et al., 2008), or restricting patches and updates to legal users only (Sundararajan, 2004; Wu & Chen, 2008) may be underestimating the motivating power of culture. In particular, copying software has virtually no cost, and there is no quality difference between the bits and bytes of the original and the copy. Furthermore, attempting to control access to patches and updates may appear a useful deterrent to piracy, but there is evidence to suggest that no matter what anti-piracy mechanisms are put in place, software crackers are motivated by the challenge of defeating those mechanisms (Goode & Cruise, 2006).

While software vendors, such as Microsoft, have attempted to combat software piracy with measures such as holographic disks to distinguish legitimate CDs from copies, and online activation codes to log each installation, flourishing arcades of software pirates exist in many countries. In Hong Kong, for instance, CDs containing thousands of dollars worth of software are being sold for the equivalent of a few US dollars. The CDs come with instructions on how to circumvent the activation process by creating validation codes using cracker programs such as Keygen (e.g., see Walls & Harvey, 2006). In spite of all the legislation and advertising campaigns over the last decade or more, no country has a piracy rate less than 20%, and most countries are above 60% (BSA, 2008). Clearly, software piracy will continue to be a difficult problem to combat.

According to the results of this study, however, a more successful anti-piracy campaign would blend a market-sensitive pricing policy with advertising that incorporates ideals of high IDV and low PDI. For instance, high IDV could be incorporated by selling the idea that having the latest software is an asset that can help individuals and organizations distinguish themselves from their peers, and by implication, sharing would reduce their ability to stay ahead. Similarly, the low PDI that accelerates the decline in software piracy for richer countries could be portrayed in terms of an "illegal is immoral" ethic, where it is shameful for any individual or organization to be caught using pirated software. For instance, Hong Kong's SPR dropped from 67% in 1997 to 56% in 1999 after the introduction of the Copyright Ordinance of 1997 that subsequently became law in 2000. "Losing face" can be a significant deterrent.

LIMITATIONS AND FURTHER RESEARCH

There are a number of limitations of this study that need to be addressed. In particular, Hofstede's cultural dimensions were developed over 30 years ago and it is likely that some, perhaps all, will have shifted. The relative position of each country, however, is likely to remain the same (Hofstede & Hofstede, 2005). Although Hofstede's cultural dimensions are the most popular to use, principally because statistics are available for a large number of countries, other value scales exist, such as Schwartz's Value Inventory (Schwartz, 1999), Trompenaars' four basic types of corporate culture

(Trompenaars & Hampton-Turner, 1998), the Global Leadership study (House et al., 2004), and the World Values Survey (Halman et al., 2007), although many of these value scores correlate significantly with those of Hofstede (Hofstede & Hofstede, 2005; Ng et al., 2007).

We also did not add further socio-political measures to the model because such measures often correlate very strongly with the GNI and culture variables, which could lead to problems of variance inflation, and obfuscate the true relationship between the variables in the model. For instance, most studies measure corruption in terms the corruption perceptions index (CPI) developed by Transparency International (see http://www. transparency.org/). Each country is rated on a scale of 1 to 10, with lower scores indicating more corruption. However, the average CPI score for the 61 countries in this study had a correlation of .88 and -.87 with GNI and SPR, respectively. Similarly, for those countries for which statistics are available (n=55), the correlation between wealth and expenditure on information and communication technology is almost identical (r=.93), and suggests that introducing technology into the model is repeating a measure of wealth.

We suggest, therefore, that the goal of further research will be to build on the path model presented here and develop measures that are theoretically distinct but significantly related to the original wealth and culture variables. Alternative economic measures include trade (Depken & Simmons, 2004), income inequality (Andrés, 2006), and foreign direct investment (Robertson et al., 2008). Expanding the culture variables will prove the most challenging, given that developing a measure that applies at the national level is extremely labor-intensive (Straub & Loch, 2006).

According to our model only factors related to PDI and IDV need to be developed, with the tension between PDI and IDV probably being the most important area of further research. For instance, PDI is strongly correlated with the standard of the educational system and the size of the middle class. Literacy rate and access to information has been correlated with levels of software piracy (Depken & Simmons, 2004; DiRienzo et al., 2007). The importance of education is that educational campaigns to raise awareness of intellectual property rights and the illegality of software piracy may help reduce the propensity to pirate software at the individual level (e.g., Kini et al., 2004; Kuo & Hsu, 2001).

CONCLUSION

We developed a path model that highlighted the inter-relationship of economic and cultural factors that influence levels of national software piracy. Using the latest dataset and a larger sample than was previously available, economic wealth, long understood to be a dominant direct effect, was again shown to be a key direct effect. However, our model also shows that PDI and IDV are important antecedent motivating factors, directly influencing economic wealth and having a larger total effect on SPR than economic wealth. On the basis of these results we suggest that software piracy may remain a largely economic issue. Attempts to reduce the overall rate must contend with the opposing forces of wealth and culture and the resistance in most countries to buying a product that is expensive and so easily copied. The most successful anti-piracy campaign is likely to be one that combines a market-sensitive pricing strategy with an advertising campaign that blends the ideals of high IDV and low PDI, especially for those countries making the transition from low-income to wealthy.

REFERENCES

Al-Rafee, S., & Cronan, T. P. (2006). Digital piracy: Factors that influence attitude towards behavior. *Journal of Business Ethics*, *63*(3), 237–259. doi:10.1007/s10551-005-1902-9

Andrés, A. R. (2006). Software piracy and income inequality. *Applied Economics Letters*, *13*(2), 101–105. doi:10.1080/13504850500390374

August, T., & Tunca, T. I. (2008). Let the pirates patch? An economic analysis of software security patch restrictions. *Information Systems Research*, *19*(1), 48–72. doi:10.1287/isre.1070.0142

Bae, S. H., & Choi, J. P. (2006). A model of piracy. *Information Economics and Policy*, *18*(3), 303–320. doi:10.1016/j.infoecopol.2006.02.002

Bagchi, K., Kirs, P., & Cerveny, R. (2006). Global software piracy: Can economic factors alone explain the trend? *Communications of the ACM*, *49*(6), 70–75. doi:10.1145/1132469.1132470

Banerjee, D., Khalid, A. M., & Sturm, J.-E. (2005). Socio-economic development and software piracy: An empirical assessment. *Applied Economics*, *37*(18), 2091–2097. doi:10.1080/00036840500293276

BSA. (2003). *Eighth Annual BSA Global Software Piracy Study* (available online at: http://global.bsa.org/globalstudy/2003_GSPS.pdf).

BSA. (2008). *Fifth Annual BSA and IDC Global Software Piracy Study* (available online at: http://global.bsa.org/idcglobalstudy2007/).

Cheung, H. (2006). *Windows Vista Ultimate for $3.50* (available online at: http:// www.tgdaily.com/ content/view/30080/118/).

Chiu, H.-C., Hsieh, Y.-C., & Wang, M.-C. (2008). How to encourage customers to use legal software. *Journal of Business Ethics*, *80*(3), 583–595. doi:10.1007/s10551-007-9456-7

Cronan, T. P., & Al-Rafee, S. (2008). Factors that influence the intention to pirate software and media. *Journal of Business Ethics*, *78*(4), 527–545. doi:10.1007/s10551-007-9366-8

Cronan, T. P., Foltz, C. B., & Jones, T. W. (2006). Piracy, computer crime, and IS misuse at the University. *Communications of the ACM*, *49*(6), 85–90. doi:10.1145/1132469.1132472

Davis, J. H., & Ruhe, J. A. (2003). Perceptions of country corruption: Antecedents and outcomes. *Journal of Business Ethics*, *43*(4), 275–288. doi:10.1023/A:1023038901080

Depken, C. A., & Simmons, L. C. (2004). Social construct and the propensity for software piracy. *Applied Economics Letters*, *11*(2), 97–100. doi:10.1080/1350485042000200187

DiRienzo, C. E., Das, J., Cort, K. T., & Burbridge, J. (2007). Corruption and the role of information. *Journal of International Business Studies*, *38*(2), 320–332. doi:10.1057/palgrave.jibs.8400262

Douglas, D. E., Cronan, T. P., & Behel, J. D. (2007). Equity perceptions as a deterrent to software piracy behavior. *Information & Management*, *44*(5), 503–512. doi:10.1016/j.im.2007.05.002

Gan, L. L., & Koh, H. C. (2006). An empirical study of software piracy among tertiary institutions in Singapore. *Information & Management*, *43*(5), 640–649. doi:10.1016/j.im.2006.03.005

Getz, K. A., & Volkema, R. J. (2001). Culture, perceived corruption, and economics: A model of predictors and outcomes. *Business & Society*, *40*(1), 7–30. doi:10.1177/000765030104000103

Goles, T., Jayatilaka, B., George, B., Parsons, L., Chambers, V., Taylor, D., & Brune, R. (2008). Softlifting: Exploring determinants of attitude. *Journal of Business Ethics*, *77*(4), 481–499. doi:10.1007/s10551-007-9361-0

Goode, S., & Cruise, S. (2006). What motivates software crackers? *Journal of Business Ethics*, *65*(2), 173–201. doi:10.1007/s10551-005-4709-9

Gopal, R. D., & Sanders, G. L. (2000). Global software piracy: You can't get blood out of a turnip. *Communications of the ACM*, *43*(9), 82–89. doi:10.1145/348941.349002

Halman, L., Inglehart, R., Díez-Medrano, J., Luijkx, R., Moreno, A., & Basáñez, M. (2007). *Changing Values and Beliefs in 85 Countries*. Boston (MA): Brill.

Hofstede, G., & Hofstede, G. J. (2005). *Cultures and Organizations: Software of the Mind* (2nd ed.). New York (NY): McGraw-Hill.

House, R. J., Hanges, P. J., Javidan, M., Dorfman, P. W., & Gupta, V. (2004). *Culture, Leadership, and Organizations: The GLOBE Study of 62 Societies*. Thousand Oaks (CA): Sage Publications.

Husted, B. W. (2000). The impact of national culture on software piracy. *Journal of Business Ethics*, *26*(3), 197–211. doi:10.1023/A:1006250203828

Introna, L. D. (2007). Singular justice and software piracy. *Business Ethics. European Review (Chichester, England)*, *16*(3), 264–277.

Kini, R. B., Ramakrishna, H. V., & Vijayaraman, V. (2004). Shaping moral intensity regarding software piracy: A comparison between Thailand and U.S. students. *Journal of Business Ethics*, *49*(1), 91–104. doi:10.1023/B:BUSI.0000013863.82522.98

Kuo, F., & Hsu, M. (2001). Development and validation of ethical computer self-efficacy measure: The case of softlifting. *Journal of Business Ethics*, *32*(4), 299–315. doi:10.1023/A:1010715504824

Kwong, K. K., Yau, O. H. M., Lee, J. S. Y., Sin, L. Y. M., & Tse, A. C. B. (2003). The effects of attitudinal and demographic factors on intention to buy pirated CDs: The case of Chinese consumers. *Journal of Business Ethics*, *47*(3), 223–235. doi:10.1023/A:1026269003472

Marron, D. B., & Steel, D. G. (2000). Which countries protect intellectual property? The case of software piracy. *Economic Enquiry*, *38*(2), 159–174. doi:10.1093/ei/38.2.159

Moores, T. T. (2003). The effect of national culture and economic wealth on global software piracy rates. *Communications of the ACM*, *46*(9), 207–215. doi:10.1145/903893.903939

Moores, T. T. (2008). An analysis of the impact of economic wealth and national culture on the rise and fall of software piracy rates. *Journal of Business Ethics*, *81*(1), 39–51. doi:10.1007/s10551-007-9479-0

Moores, T. T., & Chang, J. C. J. (2006). Ethical decision making in software piracy: Initial development and test of a four-component model. *MIS Quarterly*, *30*(1), 167–180.

Moores, T. T., & Dhaliwal, J. (2004). A reversed context analysis of software piracy issues in Singapore. *Information & Management*, *41*(8), 1037–1042. doi:10.1016/j.im.2003.10.005

Ng, S. I., Lee, J. A., & Soutar, G. N. (2007). Are Hofstede's and Schwartz's value frameworks congruent? *International Marketing Review*, *24*(2), 164–180. doi:10.1108/02651330710741802

Park, H. (2003). Determinants of corruption: A cross-national analysis. *Multinational Business Review*, *11*(2), 29–48.

Peace, A. G., Galletta, D. F., & Thong, J. Y. L. (2003). Software piracy in the workplace: A model and empirical test. *Journal of Management Information Systems*, *20*(1), 153–177.

Robertson, C. J., Gilley, K. M., Crittenden, V., & Crittenden, W. F. (2008). An analysis of the predictors of software piracy within Latin America. *Journal of Business Research*, *61*(6), 651–656. doi:10.1016/j.jbusres.2007.06.042

Ronkainen, I. A., & Guerrero-Cusumano, J. L. (2001). Correlates of intellectual property violation. *Multinational Business Review*, *9*(1), 59–65.

Schwartz, S. H. (1999). A theory of cultural values and some implications for work. *Applied Psychology: An International Review*, *48*(1), 12–47.

Shin, S. K., Gopal, R. D., Sanders, G. L., & Whinston, A. B. (2004). Global software piracy revisited. *Communications of the ACM*, *47*(1), 103–107. doi:10.1145/962081.962088

Straub, D. W., & Loch, K. D. (2006). Creating and developing a program of global research. *Journal of Global Information Management*, *14*(2), 1–28.

Sundararajan, A. (2004). Managing digital piracy: Pricing and protection. *Information Systems Research*, *15*(3), 287–308. doi:10.1287/isre.1040.0030

Trompenaars, F., & Hampden-Turner, C. (1998). *Riding the Waves of Culture: Understanding Diversity in Global Business* (2nd ed). New York (NY): McGraw-Hill.

Walls, W. D., & Harvey, P. J. (2006). Digital pirates in practice: Analysis of market transactions in Hong Kong's pirate software arcades. *International Journal of Management*, *23*(2), 207–215.

Wooley, D. J., & Eining, M. M. (2006). Software piracy among accounting students: A longitudinal comparison of changes and sensitivity. *Journal of Information Systems*, *20*(1), 49–63. doi:10.2308/jis.2006.20.1.49

World Bank. (2008). *World Development Indicators* (available online at http://go.worldbank.org/RVW6YTLQH0).

Wu, S.-Y., & Chen, P.-Y. (2008). Versioning and piracy control for digital information goods. *Operations Research*, *56*(1), 157–174. doi:10.1287/opre.1070.0414

This work was previously published in International Journal of Global Information Management, Volume 18, Issue 1, edited by Felix B. Tan, pp. 1-14, copyright 2010 by IGI Publishing (an imprint of IGI Global).

Section 2
International Perspectives

Chapter 6
ICT for Digital Inclusion:
A Study of Public Internet Kiosks in Mauritius

L.G. Pee
Tokyo Institute of Technology, Japan

A. Kankanhalli
National University of Singapore, Singapore

V.C.Y. On Show
National University of Singapore, Singapore

ABSTRACT

To bridge the digital divide, subsidized access to information and communications technology (ICT) is often provided in less-developed countries. While such efforts can be helpful, their effectiveness depends on targeted users' willingness to utilize the ICT provided. This study examines the factors influencing individuals' use of one such ICT, public internet kiosks, in Mauritius. Findings from a survey indicate that individuals' self-efficacy, perceived ease of use, perceived usefulness, subjective norm, and perceived behavioral control have significant effects. This study contributes to research by highlighting key factors influencing the use of public internet kiosks and discussing how the factors' perception and assessment differ from those in the developed world. Mauritius also provides an interesting context for this study, as her government has been actively promoting the diffusion of ICT in the country yet there have been limited empirical studies on Mauritian and sub-Saharan African users in the digital divide research. Suggestions for promoting the use public internet kiosks in less-developed countries are also provided.

INTRODUCTION

Driven by the rapid growth of the World Wide Web, information and communications technology (ICT) has been integrated into virtually every aspect of life, redefining the political, economic, social, and work environments. It is widely believed that universal access to ICT will promote economic development, global interaction, and learning that can in turn enhance standards of living and improve social welfare (Dewan & Riggins, 2005). However, a large gap still exists

DOI: 10.4018/978-1-61350-480-2.ch006

between ICT "haves" and "have-nots" in many parts of the world. "ICT haves" refer to those who can access ICT while "ICT have-nots" are socially disadvantaged individuals who have less opportunity to access and use ICT (Lam & Lee, 2006). The gap between these two groups is commonly termed as the "digital divide". Digital divide was first acknowledged by the United States (U.S.) Department of Commerce's National Telecommunications and Information Administration in a study that quantified ICT use by various socioeconomic groups (United States Department of Commerce, 1995). It has since been a topic of interest among researchers and practitioners in information system (IS), public administration, and sociology fields (e.g., Lam & Lee, 2006; Sipior et al., 2003), with studies spanning different levels of analysis (i.e., individual, organizational, national), subjects, and methodologies (e.g., case study, survey) (Dewan & Riggins, 2005).

To bridge the digital divide and enhance digital inclusion, it is important to provide the have-nots with access to computers and the internet (Dewan et al., 2010). Public internet kiosks are often set up for this purpose in regions where access to ICT is limited. Public internet kiosks are also known as telecenters, information kiosks, internet access points, community technology centers, and cyber-cafés and they aim to provide subsidized or free ICT access to rural populations (Schware, 2007). For example, in Kenya, the government launched the Pasha project in 2009 for setting up telecenters in rural areas to provide access to computers and the internet as well as computer training (Drury, 2011). In Peru, telecenters were deployed under the ERTIC project (Establecimientos Rurales de Tecnologías de la Información y Comunicación, i.e., rural ICT establishments) to provide farmers access to learning materials on agricultural practices and commerce portals through which they can sell their produce, in addition to basic access to computers and the internet (Heeks and Kanashiro, 2009). In the developed world as well, Canadian libraries provide public internet kiosks

that served as the main point of access for about 8% of Canadians (Umbach, 2004). Although providing ICT access is a necessary first step to narrow the digital divide, it is not sufficient to alleviate the problem as the benefits of ICT can only be reaped when it is accepted and utilized by the targeted users. Further, initiatives to provide ICT access can only be sustainable in the long term if user demand is strong enough for providers to charge fees to cover the costs. It is therefore important to understand the factors influencing individuals' use of public internet kiosks.

Although there have been many studies examining individuals' ICT acceptance and use (Schepers and Wetzels, 2007), it is necessary to study the use of public internet kiosks in its own right. Unlike personal computers, public internet kiosks are installed at public locations and shared among many users. For example, in Kenya's Pasha project, a basic telecentre has three computers while a standard telecentre has five computers and they are typically set up in areas with populations exceeding 5,000 (Drury, 2011). Therefore, issues such as convenience of location, wait time to use the computers, and cost of access are likely to be more prevalent and findings from studies of personal computers may not be directly applicable.

It is also interesting to examine ICT use in developing countries, which have been relatively unexplored in studies of ICT use compared to developed countries (Dwivedi et al., 2008; Mbarika & Byrd, 2009). Developing countries like Mauritius tend to face different constraints in ICT use compared to developed countries. For example, poverty, lack of infrastructure, and low information technology (IT) literacy have been found to be significant limitations that hinder the adoption and use of ICT in developing countries (Shih et al., 2008). These conditions are less prominent in developed countries and research findings and interventions designed in the developed world, therefore, may not be relevant to developing countries.

This study seeks to address the gaps identified above by examining factors influencing individuals' use of public internet kiosks in Mauritius. Prior studies on individuals' ICT use in the digital divide context have examined demographic characteristics of users such as gender, income, and level of education (e.g., Rice & Katz, 2003). Other studies have focused on analyzing patterns of use (e.g., Akhter, 2003) and identifying benefits of use (e.g., Locke, 2005). This study adds to existing research by examining technological, social, and situational factors influencing individuals' use of public internet kiosks. Specificities of the Mauritian population's ICT use compared to issues in developed countries are also highlighted.

This study aims to contribute to research by identifying and examining factors influencing individuals' use of public internet kiosks based on the technology acceptance model (Davis, 1989), theory of planned behavior (Fishbein & Ajzen, 1975), and social cognitive theory (Bandura, 2001). Data were collected from a survey of general Mauritian public who were the actual targeted users of public internet kiosks rather than from surrogates, such as employees in business organizations and university students, employed in previous studies. Construct measures that are more relevant to ICT users in less-developed countries are also developed and validated. The findings are expected to provide empirical evidence for factors influencing the use of public internet kiosks in developing countries and extend prior studies which have mostly been anecdotal in the form of case studies (e.g., Rangaswamy, 2007). Suggestions for promoting the use of public internet kiosks in developing countries are also discussed.

The following section describes the development of ICT and the initiative to set up public internet kiosks in Mauritius. Next, a review of existing literature on ICT use and its applicability to the digital divide context is discussed. The proposed research model and hypotheses are then presented, followed by details about data collection and analysis. The findings and their implications for research and practice are then considered.

INFORMATION AND COMMUNICATIONS TECHNOLOGY IN MAURITIUS

Mauritius is a small island nation in Sub-Saharan Africa with an estimated population of 1.3 million, constituted by descendants of the original immigrants from India, Europe, Madagascar, Africa, and China (United States Central Intelligence Agency, 2011). In 2011, it was estimated that 70.7% of her population is aged between 15 to 64 years old and 84.4% are literate. Although English is the official language, it is spoken by less than 1% of the population and the majority (80.5%) of the Mauritian population speak Creole. Mauritius is a middle-income developing country whose economy depends largely on manufacturing, sugar exports, tourism, and financial services. About 58% of the country's population lives in rural areas and only about 22.3% are internet users.

The Government of Mauritius recognized the potential of ICT in stimulating national development and proposed a five-year National ICT Strategic Plan in 2007 and a four-year follow-up plan in 2011. The plans aimed to transform Mauritius into a preferred centre of ICT skills, expertise, and employment in the region (Mauritius National Computer Board, 2007). Targets set in the plan include increasing personal computer ownership by at least 20,000 in households and 12,000 in primary schools, increasing broadband internet penetration by at least 250,000, and establishing 150 public internet kiosks across the island.

Even prior to the national plan in 2007, the Information and Communication Technologies Authority (ICTA) and Mauritius Post Limited had established public internet kiosks in 93 post offices since 2005 to facilitate the uptake of ICT (Mauritius Government Information Service, 2005). Through these kiosks, computer and internet access were provided to those who otherwise would not be able to afford personal computers. These kiosks offered facilities such as word processing, communication, and access to government and commercial information and web services. To ensure their sustainability, communities were encouraged to self-manage the kiosks and generate locally relevant content. Computer courses were also conducted regularly to train users.

This initiative is a suitable context for our study because the internet kiosks have been provided with the aim of enhancing digital inclusion in Mauritius. Mauritius has been witnessing rapid ICT growth in recent years since its participation in the United Nations Development Program (United Nations, 2007). The role of public internet kiosks in bridging the digital divide may therefore be more discernible here, allowing the identification of issues that may be relevant to other efforts for bridging the digital divide. The initiative is also interesting in that it provides an opportunity for us to understand individuals' perceptions with respect to ICT use in a developing Sub-Saharan Africa country, which is a rarely studied population in IS research (Mbarika & Byrd, 2009). This permits us to assess the applicability of technology acceptance theories (which have mostly been proposed and validated in developed countries)

in a less-developed context and understand how the considerations of public internet kiosk users in developing countries differ from those in developed countries.

CONCEPTUAL BACKGROUND

Technology Acceptance Theories

Individuals' ICT use has been studied from various theoretical perspectives, notably the theory of reasoned action (TRA), theory of planned behavior (TPB), and technology acceptance model (TAM) (Agarwal, 2000). Both TRA and TPB posit that behavior is affected by an individual's intention to perform it. *Intention* indicates how hard an individual is willing to try and how much effort the individual plans to exert to perform a behavior. It is a function of individuals' attitude towards the behavior and subjective norm about the behavior. *Attitude* is defined as the general feeling of favorableness towards the behavior, while *subjective norm* (SN) refers to the perceived extent to which important social referents would desire the performance of the behavior (Fishbein & Ajzen, 1975). TPB extends TRA by considering perceived behavioral control as a predictor of intention in addition to attitude and subjective norm (see Figure 1). *Perceived behavioral control* (PBC) refers to one's perception of control over

Figure 1. Theory of planned behavior (Ajzen, 1991)

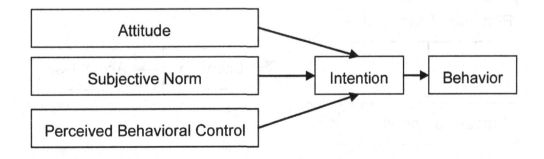

performing a behavior and recognizes the need to consider situational obstacles in predicting behavior (Ajzen, 1991).

Davis (1989) proposed TAM to explain and predict individuals' use of technology (see Figure 2). TAM posits that perceived usefulness and perceived ease of use predict one's intention to use technology, which in turn predicts the actual use behavior. *Perceived usefulness* (PU) is the degree to which one believes that using a particular technology would enhance performance, while *perceived ease of use* (PEOU) refers to the degree to which one believes that using the technology would be free of effort (Davis, 1989). Perceived ease of use is also expected to influence intention through perceived usefulness, in addition to its direct effect on intention to use technology.

Although TAM is a parsimonious and robust model that has been shown to be useful in explaining and predicting technology use behavior, the model clearly overlooks the social and situational contexts in which a technology is used (Venkatesh et al., 2003). This places a ceiling on the variance accountable by the theory and limits the variety of practical interventions available to policy makers in encouraging technology use. To address this gap, we extend the model to include subjective norm and perceived behavioral control proposed in TRA and TPB to study the use of public internet kiosk in Mauritius.

Other than TRA, TPB, and TAM, IS researchers have also gained additional insights into technology use through the Social Cognitive Theory (SCT). SCT is a rich and complex theory which posits that individuals' behavior results from complex interactions between their personal characteristics and situational factors (Bandura, 2001). Among the concepts proposed in the theory, self-efficacy is commonly studied with reference to technology acceptance (King & He, 2006). *Computer self-efficacy* refers to an individual's perceptions of his or her ability to use computers in the accomplishment of a task (Compeau & Higgins, 1995). It can influence one's decision about what behavior to undertake and the amount of effort and persistence put forth when faced with obstacles. In the context of this study, the IT literacy of the targeted users of public internet kiosks may be low. Computer self-efficacy is therefore likely to be a significant factor influencing their use of public internet kiosks. Accordingly, we consider the effect of self-efficacy in our proposed model.

Application of Technology Acceptance Theories in Developing Countries

TRA, TPB, TAM, and SCT have been applied in many studies involving a broad range of technologies and user populations. However, their application in developing countries is relatively

Figure 2. Technology acceptance model (Davis, 1989)

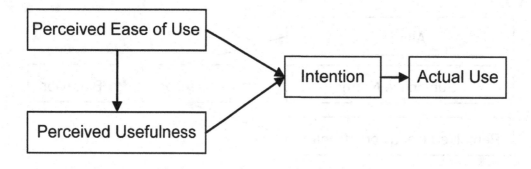

Table 1. Empirical studies applying TRA, TPB, TAM, and SCT in less-developed countries

Source	Theoretical Bases	Independent Variables	Dependent Variables	Key Findings	Methodology and Subject
Agbonlahor, 2006	- TAM - Diffusion of innovation (DOI)	- Access to IT - Image - Level of IT use - PEOU - PU - Result demonstrability - Training of IT - Visibility	- Number of applications used - Frequency of computer use	- Access to IT → number of applications used - Image → frequency of computer use - PEOU → frequency of computer use - PU → number of applications used - Result demonstrability → number of applications used - Training of IT → number of applications used	Survey of 780 lecturers from ten universities in Nigeria
Anandarajan et al., 2002	- TRA - TAM	- Computer skills - Organizational support - Organizational usage - PEOU - Perceived enjoyment - PU - Social pressure	- Micro computer usage - Job satisfaction	- Organizational support → perceived enjoyment - Organizational support → social pressure - Organizational usage → social pressure - PEOU → micro computer usage - PEOU → perceived enjoyment - PU → job satisfaction - Perceived enjoyment → job satisfaction - Social pressure → micro computer usage - System usage → job satisfaction	Survey of 143 individuals from nine organizations in Nigeria
Bankole et al., 2011	- TAM - TPB - TRA - Unified theory of use and acceptance of technology	- Convenience and cost - Effort expectancy - Individualism - Masculinity/ femininity - Power distance - Social factors - Trust and privacy - Uncertainty avoidance - User satisfaction - Utility expectancy	- Behavioral intention to use mobile banking - User behavior to use mobile banking	- Behavioral intention → user behavior - Effort expectancy → behavioral intention - Effort expectancy → utility expectancy - Individualism → effort expectancy - Masculinity/femininity → effort expectancy - Masculinity/femininity → utility expectancy - Power distance → behavioral intention - Uncertainty avoidance → effort expectancy - Uncertainty avoidance → trust and privacy - User satisfaction → effort expectancy - User satisfaction → utility expectancy - Utility expectancy → behavioral intention	Survey of 231 students and workers in Nigeria

continued on following page

Table 1. Continued

Brown, 2002	- TAM - SCT	- Computer anxiety - Ease of finding (navigation) - Ease of understanding - PU - Self-efficacy	- Usage of web-based learning technology	- Computer anxiety → PEOU - Ease of finding (navigation) → PEOU - Ease of understanding → PEOU - PEOU → PU - PEOU → usage - Self-efficacy → PEOU	Survey of 78 students using web-based learning technology in a South African University
Elbeltagi et al., 2005	- TAM	- Cultural characteristics - Decision maker characteristics - Decision support system characteristics - Environmental characteristics - External support - Internal support - Organizational characteristics - PEOU - PU - Task characteristics - Top management	- DSS usage	- PEOU → DSS usage - PU → DSS usage - Top management → PEOU	Survey of 294 chief executive officers and DSS unit managers within Egyptian governorates (administrative division of a country)
Furuholt et al., 2005	- TAM	- Financial capacity - Individual capability - Media exposure - Occupation	- Frequency of internet café use	- Individual capability → frequency of use - Media exposure → frequency of use	Survey of 270 users of internet cafes in Indonesia
Fusilier & Durlabhji, 2005	- TAM - TPB	- Attitude - Experience - PBC - PEOU - PU - SN	- Intention to use internet - Internet usage behavior	- Experience X PU* → intention to use internet - Experience X SN → intention to use internet - PBC → intention to use internet - PBC → internet use - PEOU → intention to use internet - PU → intention to use internet - SN → intention to use internet	Survey of 245 college students in Northwestern India
Gupta et al., 2008	- TAM - Unified theory of acceptance and use of technology	- Effort expectancy - Facilitating conditions - Performance expectancy - Social influence	- Behavioral intention to use internet - User behavior	- Effort expectancy → behavioral intention - Facilitating conditions → usage - Performance expectancy → behavioral intention - Social influence → behavioral intention	Survey of 102 employees of a government organization in India

continued on following page

Table 1. Continued

Ifinedo, 2006	- TAM - SCT	- Computer anxiety - Ease of finding - Ease of understanding - PEOU - PU - Self-efficacy	- Usage of computer - Continuance intention	- Computer anxiety → PEOU - Computer anxiety → PU - Ease of finding → PEOU - Ease of finding → PU - Ease of understanding → PEOU - Ease of understanding → PU - PEOU → PU - PU → continuance intention - Self-efficacy → PEOU - Self-efficacy → PU - Usage → continuance intention	Survey of 72 students from four tertiary institutions in Estonia
Kaba et al., 2006	TRA	- Familiarity within social group - Mobility required by task - Social influences	- Use of cellular phone	- Familiarity within social group → duration of calls - Mobility required by task → duration of calls - Mobility required by task → number of outgoing calls - Social influence → number of incoming calls	Survey of 463 cell phone subscribers in Guinea
Lin et al., 2011	- TAM - TPB	- Attitude towards using - Behavioral intention - Information quality - Information system quality - PEOU - PU	- Behavioral intention to use e-government system	- Attitude towards using → behavior intention - Information quality → PU - PEOU → attitude towards using - PEOU → PU	Survey of 167 e-government system users in Gambia
Meso et al., 2005	- TAM	- Accessibility of mobile ICTs - Age - Cultural influences - Gender - Level of education - PEOU - Perceived technology reliability - PU	- Socializing use of mobile ICTs - Business use of mobile ICTs	- Accessibility of mobile ICTs → business use of mobile ICTs - Accessibility of mobile ICTs → PEOU - Accessibility of mobile ICTs → PU - Cultural influences → PEOU - PEOU → PU - Perceived technology reliability → business use of mobile ICTs - Perceived technology reliability → PU - Perceived technology reliability → socializing use of mobile ICTs	Survey of 198 employees from universities, polytechnic institutes, and government ministries in Nigeria and Kenya

continued on following page

Table 1. Continued

Uzoka et al., 2007	- TPB	- Facilitating conditions - Internet accessibility - Internet and technological complexity - Management support - Perceived advantage of e-commerce - Perceived disadvantages of e-commerce	- Adoption of e-commerce	- Internet accessibility → adoption of e-commerce - Internet and technological complexity → adoption of e-commerce - Management support → adoption of e-commerce	Survey of 120 individuals from public and private organizations in Botswana
* X represents interaction effect between two variables					

limited compared to developed countries. In our review of major IS journals (e.g., Management Information Systems Quarterly, Information Systems Research, Journal of Management Information Systems, Communications of the AIS, Journal of the AIS) and journals dedicated to global IS research (e.g., Electronic Journal of Information Systems in Developing Countries, Journal of Global Information Management, and Journal of Global Information Technology Management), only 13 empirical studies were found to have applied these theories in less-developed countries (see Table 1). Although the review was not all-inclusive in that it did not consider all publications, it reflected the extent to which mainstream technology acceptance research has focused on this part of the world to date.

It can be observed from the review that except for the study by Furuholt et al. (2005), most studies were conducted in organizational contexts (i.e., university, government institution, private organization) where respondents tend to be better educated and the use of technology is often affected by organizational mandate. The use of public internet kiosks by the general population

has been largely overlooked in prior research and our study seeks to address this gap.

RESEARCH MODEL AND HYPOTHESES

Based on our review of the technology acceptance literature, an extended research model based on TAM is proposed (see Figure 3). Specifically, the model posits that individuals' use of public internet kiosks is predicted by intention, which is in turn determined by perceived ease of use and perceived usefulness as proposed in TAM. The model is extended by including subjective norm and perceived behavioral control from TPB. In addition, we hypothesize that self-efficacy from SCT influences perceived ease of use. The rationale for hypothesizing each relationship in the proposed model will be discussed in the following sections. Where relevant, specificities of the Mauritian context that may influence the hypothesis or construct conceptualization are also highlighted.

Perceived ease of use refers to the degree to which an individual believes that using a particular technology would be free of effort (Davis, 1989). Compared to a technology that is skill demanding, individuals are naturally more willing to use a more user-friendly technology that can achieve the same performance. Extensive empirical evidence accumulated in previous IS research has shown that perceived ease of use is significantly related to ICT use intention (e.g., Legris et al., 2003; Venkatesh et al., 2003). In Mauritius, only 22.3% of the population are internet users (United States Central Intelligence Agency, 2011). As most of the targeted users of public internet kiosks are likely to have little experience and knowledge of ICT, perceived ease of use is likely to be an important consideration when they decide whether to use the kiosks. Therefore, we hypothesize that:

H1: Perceived ease of use is positively related to individuals' intention to use public internet kiosks.

Perceived ease of using public internet kiosks may be influenced by individuals' computer self-efficacy (Venkatesh, 2000). *Computer self-efficacy* refers to an individual's perceptions of his or her ability to use computers in the accomplishment of a task (Compeau & Higgins, 1995). Individuals with a strong confidence in the use of computers are likely to perceive public internet kiosks as easy to use. In Mauritius, the targeted users of public internet kiosks generally lack experience with computers and the internet. In judging the ease of using the kiosks, they are likely to form an opinion based on their existing skills and past experience with technologies requiring a similar skill set (e.g. typewriter). Accordingly, we hypothesize that:

H2: Individuals' self-efficacy is positively related to perceived ease of using public internet kiosks.

Perceived usefulness refers to the degree to which an individual believes that using a particular technology would enhance performance (Davis, 1989). Outcome expectation is an important precursor to ICT use as individuals are likely to engage in behaviors that are expected to result in favorable consequences or solve problems. There is strong empirical support for the effect of perceived usefulness on intention to use ICT in the

Figure 3. Proposed model

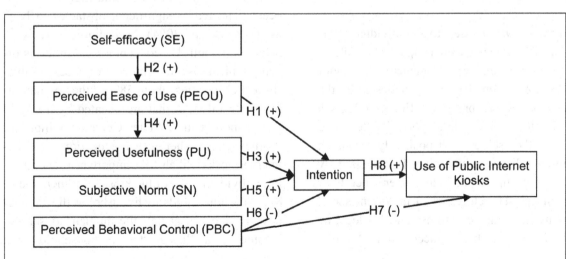

IS literature (Schepers & Wetzels, 2007). In the digital divide literature, studies on the sustainability of public internet kiosks have also emphasized the need to provide information and services that are relevant and valuable to attract users (Madon, 2005). In Mauritius, public internet kiosks offer a variety of services such as internet browsing, word processing, e-government, healthcare, email, and messaging. Some services are more efficient and cheaper than their offline equivalents (e.g., email versus snail mail) and these benefits are likely to increase individuals' intention to use public internet kiosks.

H3: Perceived usefulness is positively related to individuals' intention to use public internet kiosks.

TAM also proposes that perceived ease of use influences perceived usefulness (Davis, 1989). When public internet kiosks are perceived to be easy to use, one would be better able to utilize the functions available to achieve different purposes and complete desired tasks. This in turn may improve the perceived usefulness of public internet kiosks. Therefore, we hypothesize that:

H4: Perceived ease of use is positively related to perceived usefulness of public internet kiosks.

Subjective norm refers to an individual's perception of the extent to which important social referents would desire the performance of a behavior (Fishbein & Ajzen, 1975). As suggested by the concept of value-expressive influence (Deutsch and Harold, 1955), individuals will attempt to associate themselves with positively evaluated groups and distance themselves from negatively evaluated groups to maintain or enhance their self-concept. Therefore, when important social referents encourage one to use a technology, he or she is likely to behave accordingly. Further,

about 68% of the Mauritian population is of Indian descent (United States Central Intelligence Agency, 2011). This implies that the population can be characterized as having a collective culture according to Hofstede's dimensions of culture (Hofstede, 1984), where family and authoritarian values guide individuals' behavior (Lee-Ross, 2005; Thanacoody et al., 2006). Subjective norm may therefore be important in determining Mauritians' use of public internet kiosks as compared to countries with an individualistic culture. Hence, we hypothesize that:

H5: Subjective norm is positively related to individuals' intention to use public internet kiosks.

Perceived behavioral control refers to an individual's perception of control over performing a behavior and recognizes the need to consider situational obstacles in predicting behavior (Ajzen, 1991). TPB proposes that perceived behavioral control may influence an individuals' behavioral intention. When one believes that there exist significant situational hindrances related to a behavior, one's confidence in successfully carrying out the behavior weakens and this reduces one's intention to engage in the behavior to avoid disappointment. In the digital divide literature, geographical inaccessibility and high cost have been identified as significant obstacles to ICT use (Foster et al., 2004). Accordingly, perceived behavioral control is conceptualized in terms of geographical distance and cost of access in this study. In Mauritius, about 58% of the population lives in rural areas that are isolated by the lack of transport (United States Central Intelligence Agency, 2011). Therefore, the location of public internet kiosks is likely to be important in determining individuals' use of the kiosks. Further, about 8% of the households are living below the relative poverty line (set at half the median household income per adult, adjusted for household size, age composition, and economies of scale) (United States Central Intelligence Agency, 2011). As the targeted users of public internet kiosks are mainly individuals who can not afford personal computers, cost is likely to be a significant consideration in the use of public internet kiosks. In sum, we expect that perceived behavioral control in terms of geographical distance and cost of access is likely to have a significant effect on individuals' intention to use public internet kiosks.

H6: Perceived behavioral control is negatively related to individuals' intention to use public internet kiosks.

Perceived behavioral control may also directly influence actual behavior to the extent that it accurately reflects the controls that actually exist (Ajzen, 2002). This notion is similar to that of facilitating/inhibiting conditions proposed in Triandis' theory of interpersonal behavior (Triandis, 1977), which refers to the objective factors in individuals' environment that make a behavior easy or difficult to perform. In this study, we consider the inhibiting factors of distance and cost under perceived behavioral control. Accordingly, we hypothesize that:

H7: Perceived behavioral control is negatively related to individuals' use of public internet kiosks.

Intention is an indication of how hard one is willing to try and how much effort one plans to exert in order to perform a behavior. It is the subjective probability of performing a behavior. Intention has been found to be an accurate predictor of actual behavior in many IS and human behavior studies (Legris et al., 2003; Schepers & Wetzels, 2007). Similarly, we expect individuals' intention to use public internet kiosks to be related to their actual use behavior.

H8: Intention is positively related to individuals' use of public internet kiosks.

RESEARCH METHODOLOGY

To assess the proposed model, a field survey was conducted to collect data from users of public internet kiosks in Mauritius. Development of the survey instrument to suit the study's context is detailed below.

Construct Operationalization

Guidelines recommended by Churchill (1979) were adopted to develop the survey instrument. The first step involved specifying the domain of each construct by reviewing existing literature to establish a clear definition for each construct as discussed in the preceding sections. Suitable measurement items that could represent each construct were then gathered and selected from previously validated scales. Constructs in this study were categorized as either of fixed or variable content (Fishbein et al., 2001). For constructs with *fixed content*, the primary question is not what to measure, but how to measure. Among the constructs in this study, self-efficacy, perceived ease of use, perceived usefulness, intention, and use of public internet kiosks are constructs whose main content does not vary across contexts of study. In contrast, for constructs with *variable content*, their operationalization depends on the type of behavior and the population being examined. In this study, the content of subjective norm and perceived behavioral control are context specific as they need to be measured in terms of the social referents and behavioral controls relevant to Mauritian public internet kiosk users.

Specifically, items measuring self-efficacy were adapted from Taylor and Todd (1995). Items measuring perceived ease of use and perceived usefulness were adapted from Moore and Benbasat (1991). Intention and behavior of using public internet kiosks were assessed with items adapted from Davis et al. (1989). To determine the content of subjective norm and perceived behavioral control, we consulted existing literature on digital divide and interviewed public internet kiosk managers and users to identify salient referents and behavioral controls that are relevant to Mauritian users. In particular, subjective norm was measured in terms of influence of family members, friends, government, staff of public internet kiosks, and media sources. Perceived behavioral control was assessed in terms of individuals' geographical access to and costs involved in using public internet kiosks. Items measuring each construct are listed in the Appendix.

DATA ANALYSIS AND RESULTS

The proposed model was analyzed with Partial Least Squares (PLS), a structural equation modeling tool. PLS analysis concurrently tests the psychometric properties of measures (measurement model) and analyzes the strength and direction of relationships among constructs (structural model). It is also able to handle both formative and reflective constructs jointly occurring in a single model (Chin, 1998). In this study, all constructs are reflective except for subjective norm and perceived behavioral control. These constructs are considered formative because their items jointly define the construct and exclusion of an item could alter their conceptual meaning.

About 150 individuals at different public internet kiosks in Mauritius were approached for the survey. Among them, 78 completed the survey yielding a response rate of about 52%. The sample was sufficient to detect a small effect size of 0.30 at the alpha level of 0.05 and power level of 0.95 (Cohen, 1988). The respondents' demographic profile is shown in Table 2. As can be observed from the table, the majority of respondents were young (15-25 years old) and had little experience with computers (0-6 months). Most of them had little training in computers (0-6 months) and had secondary school education. The monthly income

Table 2. Demographic profile

Characteristic	Frequency	Percentage	Characteristic	Frequency	Percentage
Age			*Education*		
15-25	42	53.8%	Primary School or Below	5	6.4%
26-35	19	24.4%	Secondary School	41	52.6%
36-45	13	16.7%	Diploma	10	12.8%
>45	4	5.1%	Degree or Higher	22	28.2%
Experience with Computers (Months)			*Monthly Income*		
0-6	47	59.7%	<RS5,000 (~USD170)*	26	33.3%
			RS5,000 - RS9,999 (~USD340)	22	28.2%
7-12	25	32.5%	RS10,000 - RS19,999 (~USD690)	18	23.1%
>12	6	7.8%	RS20,000 - RS29,999 (~USD1,030)	3	3.8%
Training in Computers (Months)			>RS30,000 (~USD1,030)	6	7.7%
			Undisclosed	3	3.8%
0-6	56	71.8%	*Gender*		
7-12	21	26.9%	Male	53	67.9%
>12	1	1.3%	Female	25	32.1%
*RS: Rupee; USD: United States Dollar					

was low with 85% of respondents earning below USD690. The majority of respondents were male (67.9%).

Tests of Measurement Model

The measurement model was assessed by evaluating each scale's internal consistency, convergent validity, and discriminant validity. Reflective and formative constructs needed to be analyzed differently because unlike reflective constructs, different dimensions of formative constructs may not demonstrate internal consistency and correlations (Chin, 1998). Instead, the formative constructs were assessed by item weights to determine the relevance and contribution of each item.

For reflective constructs, internal consistency was assessed by calculating Cronbach's alpha. As shown in Table 3, all reflective constructs in our model had scores above the recommended level of 0.70 (Nunnally, 1978). Convergent validity was assessed by calculating: a) item reliability, b) composite reliability, and c) average variance extracted (AVE). Table 3 shows that all item and composite reliabilities were above the recom-

mended level of 0.70 (Nunnally, 1978) and all AVEs were above the recommended level of 0.50 (Chin, 1998).

Discriminant validity was assessed by examining: a) factor analysis with varimax rotation and b) item correlations. Results of factor analysis (see Table 4) showed that five factors corresponding to the five reflective constructs in our proposed model were extracted and Kaiser-Meyer-Olkin (which assesses whether the partial correlations among variables are small) measured 0.60, which was above the recommended value of 0.50. All item loadings on stipulated constructs were also greater than the required 0.50 (Hair et al., 1998) and all eigenvalues were greater than one. The item correlation matrix (see Table 5) showed that the non-diagonal entries (item correlation) do not exceed the corresponding diagonal entries (square root of AVE) for all constructs, indicating that measures of each construct correlated more highly with their own items than with items measuring other constructs. Based on these findings, we concluded that the discriminant validities of all scales were adequate.

Table 3. Psychometric properties of reflective and formative constructs

Reflective	Items	Item Loading	Formative Constructs	Items	Item Weight
Self-efficacy $\alpha = 0.73$; CR = 0.88; AVE = 0.79	SE1	0.90***	Subjective Norm	SN1	0.25
	SE2	0.88***		SN2	0.87*
				SN3	0.84
Perceived Ease of Use $\alpha = 0.84$; CR = 0.93; AVE = 0.86	PEOU1	0.93***		SN4	0.85*
	PEOU2	0.93***			
Perceived Usefulness $\alpha = 0.77$; CR = 0.77; AVE = 0.59	PU1	0.11*		SN5	0.19
	PU2	0.71*	Perceived Behavioral Control	PBC1	0.13
Intention $\alpha = 0.78$; CR = 0.89; AVE = 0.80	INT1	0.90***		PBC2	0.21
	INT2	0.88***		PBC3	0.52*
Use of Public Internet Kiosks $\alpha = 0.74$; CR = 0.80; AVE = 0.67	USE1	0.64***		PBC4	0.47*
	USE2	0.97***		PBC5	0.33*
α: Cronbach's Alpha; CR: Composite Reliability; AVE: Average Variance Extracted *Significant at p<0.05; **Significant at p<0.01; ***Significant at p<0.001					

Table 4. Factor analysis of reflective constructs

Construct Items	Components				
	1	2	3	4	5
Self-efficacy					
SE1	0.26	0.18	**0.87**	0.06	-0.03
SE2	0.29	0.28	**0.63**	0.12	0.37
Perceived Ease of Use					
PEOU1	**0.88**	-0.04	0.23	0.09	-0.06
PEOU2	**0.87**	0.06	0.17	0.13	0.06
Perceived Usefulness					
PU1	-0.04	-0.16	0.10	0.07	**0.89**
PU2	0.26	0.31	-0.39	-0.07	**0.75**
Intention					
INT1	0.03	**0.82**	0.13	0.20	-0.01
INT2	-0.02	**0.87**	0.23	-0.01	-0.16
Use of Public Internet Kiosks					
USE1	-0.03	0.14	0.16	**0.82**	0.08
USE2	0.23	0.02	-0.03	**0.83**	0.00
Eigenvalue	2.98	1.56	1.23	1.12	1.01
Variance (with Varimax rotation)	19.43	18.51	14.81	14.58	11.71
Cumulative Variance (%)	19.43	37.94	52.75	67.33	79.04
Variance (without rotation)	29.77	15.57	12.34	11.22	10.13

Table 5. Square root of AVE vs. correlation among reflective constructs

	Self-efficacy	Perceived Usefulness	Perceived Ease of Use	Intention	Use of Public Internet Kiosks
Self-efficacy	**0.89**				
Perceived Usefulness	0.16	**0.77**			
Perceived Ease Of Use	0.42	0.33	**0.93**		
Intention	0.32	0.27	0.13	**0.89**	
Use of Public Internet Kiosks	0.23	0.08	0.25	0.17	**0.82**

Since a single data collection method was used, the extent of common method bias was also examined with Harman's one-factor test. The test involves entering all constructs into an unrotated principal components factor analysis and examining the resultant variance (Podsakoff & Organ, 1986). As shown in Table 4, the first factor accounted for 29.77% of the variance while the last factor accounted for 10.13%. There was no sign of the first factor significantly dominating the variance and we therefore concluded that common method bias was unlikely.

For formative constructs, absolute value of item weights was examined to determine the relative contribution of items constituting each construct (Chin, 1998). Results indicated that friends and

Figure 4. Results of structural model analysis

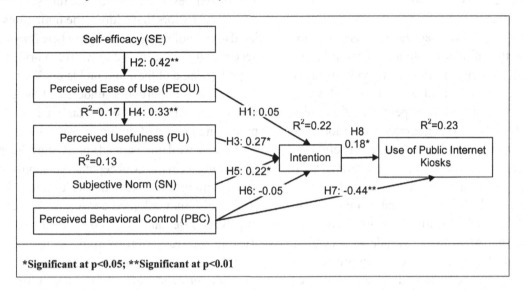

*Significant at p<0.05; **Significant at p<0.01

public internet kiosks' staff were considered the most significant social constituents enforcing subjective norm while cost was the primary perceived behavioral control that influenced individuals' use of public internet kiosks (see Table 3).

Tests of Structural Model

The structural model was analyzed by examining the significance of path coefficients and variance explained by the dependent variables. As shown in Figure 4, we found that perceived usefulness and subjective norm significantly influenced individuals' intention to use public internet kiosks and explained 22% of the variance in the construct. In addition, intention and perceived behavioral control significantly influenced individuals' use of public internet kiosks and explained 23% of its variance. We also found that self-efficacy significantly influenced perceived ease of use, which in turn increased perceived usefulness. Contrary to our hypotheses, perceived ease of use and perceived behavioral control did not have significant effect on intention. In sum, six out of our eight hypotheses were supported.

DISCUSSION AND IMPLICATIONS

Based on TAM, TPB, and SCT, this study identified and examined the factors influencing individuals' use of public internet kiosks in a developing country. Results of our survey in Mauritius showed that in addition to factors proposed in TAM (i.e., perceived ease of use and perceived usefulness), subjective norm and perceived behavioral control proposed in TPB and self-efficacy proposed in SCT had significant influences on individuals' use of public internet kiosks. Among them, perceived behavioral control had the most significant influence. Overall, the theories proposed to explain general technology use were also adequate in explaining individuals' ICT use in the digital divide context when they are appropriately operationalized to account for contextual specificities. Implications of our findings for research and practice in the use of public internet kiosks to bridge the digital divide are discussed next.

Implications for Research

We found that perceived ease of use did not significantly influence individuals' intention to use public internet kiosks as hypothesized in H1. However, perceived ease of use indirectly influenced intention through perceived usefulness (H3 and H4). This finding corroborates prior studies which often found that perceived usefulness is a stronger and more stable predictor of ICT use intention than perceived ease of use (e.g., Venkatesh et al., 2003). This finding may also be partly related to the Mauritian culture. It has been observed that Mauritians' attitude to education is largely utilitarian, where children are sent to schools to obtain certificates that are necessary to secure employment (Day-Hookoomsing, 2000). To the extent that proficiency in ICT can improve chances of employment, respondents may hold a similar utilitarian view about the use of ICT. They are therefore willing to use ICT if it is perceived as useful, even if the technology may not be easy to use. More research is needed to understand how culture influences the effect of perceived ease of use and perceived usefulness.

We also found that perceived behavioral control influences the use of public internet kiosks directly (H7) rather than through intention (H6). This suggests that in the digital divide context, individuals are still willing to use ICT despite expecting some situational obstacles. This finding might also be attributable to our conceptualization of perceived behavioral control in terms of geographical distance and cost of access, which are clearly observable and therefore stronger predictors of actual behavior (Ajzen, 2002). Future research may consider expanding the conceptualization of perceived behavior control by including other constraints such as individuals' language proficiency and waiting time for access.

This study contributes to research in several ways. First, based on TAM, TPB, and SCT, we identified the factors that have significant influence on individuals' use of public internet kiosks.

We looked beyond TAM to examine social and situational factors that address the limitations related to technology determinism, where providing access to ICT is seen as a quick fix to bridge the digital divide and facilitate inclusion. We recognize that it is important to understand the social and situational contexts of the targeted users for promoting the use of public internet kiosks.

Second, we demonstrated that existing theories explaining technology acceptance are applicable to developing countries with careful adaptations. While this study adopted constructs of theories largely proposed and validated in developed countries, the constructs were carefully operationalized to suit the context of developing countries. For example, prior studies on ICT use were often conducted in organizations in developed countries and they often operationalized subjective norm in terms of influence of colleagues (e.g., Hsu & Chiu, 2004). In our study of public internet kiosks in Mauritius, subjective norm was operationalized in terms of measures such as the influences of government and public internet kiosks' staff. Perceived behavioral control was also operationalized in terms of aspects more relevant to the context of developing countries (i.e., geographical distance and costs of access instead of personal computer ownership and organizational support). This highlights the importance of using context-relevant measures when applying existing theories to study phenomena in developing countries.

Third, we studied the actual users of public internet kiosks in Mauritius, which is an understudied population in technology acceptance research. The findings help to extend our understanding of technology acceptance outside the context of developed countries. The data collected are likely to be more valid than those from surrogates such as undergraduates and provide more relevant suggestions for promoting the use of public internet kiosks, as discussed next.

Implications for Practice

The finding that perceived ease of use influences intention indirectly through perceived usefulness (H3 and H4) rather than directly suggests that in practice, ensuring that ICT content and services are relevant and valued by the targeted users should take precedence over designing a friendly user interface. To improve relevance, the needs of the targeted users must be understood. In Mauritius, a bottom-up approach where users were encouraged to specify and build their own content was used. For example, university students were awarded academic credits for contributing content to web portals accessible via public internet kiosks. Significant content contributors were also publicly acknowledged through a system of annual rewards. To demonstrate the value of public internet kiosks, various functionalities of the kiosks could be publicized by demonstrating how the kiosks can be used to locate information regarding education and employment and for communicating with family and friends. Success stories of users could also be publicized through various media such as newsletters or television.

We found that self-efficacy significantly increases individuals' perceived ease of using public internet kiosks (H2). In Mauritius, the government has been active in increasing individuals' computer self-efficacy. For example, the Universal ICT Education Programme was introduced in 2006 to provide computer training to students, workers, and the population at large (Mauritius National Computer Board, 2006). Basic computing skill courses were offered in 59 schools and training centers across Mauritius. Participants who successfully completed the course were awarded the Internet and Computing Core Certification (IC3), which is a globally accepted certificate that recognizes one's knowledge level in using computer hardware, software, and the internet. In addition, ICT-related programmes were broadcasted on the Mauritian television network. The government also employed IT personnel to support the use of ICT in youth centers, women centers, and public internet kiosks (Soyjaudah et al., 2002). These initiatives can help to equip individuals with fundamental computer skills and alleviate their apprehension towards ICT. Providing positive and timely feedback in the form of certification also increases individuals' confidence in using computers (Sipior et al., 2003) and encourages them to continuously learn and develop new skills as ICT advances.

We also found that subjective norm is a significant factor influencing individuals' intention to use public internet kiosks (H5). This suggests that initiatives emphasizing peer influence (e.g., refer-a-friend program) are likely to be effective in promoting the use of public internet kiosks. Also, as mentioned earlier, the Mauritian culture can be characterized as largely collective, where authoritarian values guide individuals' behavior (Hofstede, 1984). This suggests that social referents who have rational and legitimate authority related to the use of public internet kiosks are likely to have significant influence on individuals' use of the kiosks. Indeed, results of our survey showed that staff members managing public internet kiosks are deemed to be one of the most important social referents related to the use of the kiosks. This suggests that roles such as "ambassadors" may be established to promote the use of public internet kiosks by publicizing how the kiosks may be used to accomplish various tasks.

Results of our study indicate that perceived behavioral control conceptualized in terms of geographical distance and cost of accessing public internet kiosks is the most significant factor determining individuals' use of public internet kiosks (H7). We also found that perceived behavioral control influences use of public internet kiosks directly rather than through intention (H6). This suggests that Mauritians' intention to use public internet kiosks is unlikely to be weakened by the perceived existence of behavioral controls. They are likely to be still willing to try despite expected obstacles. Therefore, efforts should be focused on eliminating these barriers to ICT use. At the time of our study, the use of public internet kiosks in Mauritius is free of charge. The government has been able to invest in infrastructure to provide access to ICT at affordable rates (Isaacs, 2007). This may be due to the government's overall objective of providing a supportive business environment that encourages foreign investment and economic development. Our study indicates that it is important to keep ICT access affordable to masses for the success of digital inclusion efforts. To overcome geographical barriers, the Mauritian government has also begun to operate Cyber Caravans (i.e., mobile public internet kiosks) to provide ICT access to more isolated areas.

Limitations and Suggestions for Future Research

The findings of this study should be interpreted in view of several limitations that also present opportunities for future research. First, although the sample size of 78 offered adequate power for our data analysis, future studies should examine a larger sample to achieve stronger statistical power and generalizability.

Second, our data were collected in a cross-sectional survey. We were therefore not able to draw conclusive evidence of causality, despite strong theoretical arguments and empirical support from prior studies. Future studies may consider collecting longitudinal data to better understand the causal relationships among factors influencing individuals' use of public internet kiosks.

Third, our data on the actual use of public internet kiosks were based on respondents' self-reports. Biases such as social desirability and self-presentation bias might be present. For example, a respondent might over report his/her usage to uphold personal image. To avoid these biases,

future research may consider using unobtrusive observations of ICT use via access logs. However, it is important to note that access logs may have limitations such as difficulty of identifying individual users as different users might share one account (Kurth, 1993). It may also be difficult to overcome ethical issues involving informed consent and invasion of privacy.

Fourth, as one of the first empirical studies applying TAM, TPB, and SCT to study the use of public internet kiosks in a developing country, the proposed model explained some of the variance in individuals' intention (22%) and actual behavior (23%) of using public internet kiosks. Future studies can further improve explanatory power by considering factors related to institutional context, personality traits, and other characteristics of technology. Institutional contexts such as workplace, school, community, and home may have a significant influence on individuals' use of public internet kiosks. Individuals who are members of institutions where ICT is an important functional element are likely to have more opportunities to learn about and explore different uses of public internet kiosks. This is particularly important for those who are disadvantaged by their socioeconomic status, as their institutional contexts provide them with opportunities to encounter ICT and see their personal relevance (Rogers, 1995). In line with this argument, a study of technology diffusion in rural communities showed that employment by a company that adopts a particular ICT is an important predictor of an individual's use of that technology beyond the workplace environment (Hollifield & Donnermeyer, 2003). This suggests that it may be important to consider the effects of institutional contexts.

Personality traits reflect the cognitive and affective structures maintained by individuals to facilitate adjustments to situations encountered. Researchers have suggested that traits such as personal innovativeness (Agarwal & Prasad,

1998) and risk-taking propensity (Kishore et al., 2001) may influence individuals' ICT use. Although these factors are difficult and sometimes impossible to manipulate directly, understanding their effects can help to identify enthusiastic users who may serve as champions to promote the use of public internet kiosks to other potential users.

Other than perceived ease of use and perceived usefulness considered in this study, characteristics of technology such as compatibility, trialability (Rogers, 1995), result demonstrability, and image (Moore & Benbasat, 1991) may offer additional insights into individuals' use of public internet kiosks. It has been found that individuals are better able to accept ICT that matches their habits. People also prefer ICT that can be tried before paying. Further, ICT with benefits and utility that are easily observable and that bestows its users with added prestige in their social community are also likely to attract more usage (Plouffe et al., 2001). It may be worthwhile to examine whether these observations apply to users in developing countries.

CONCLUSION

Following the National ICT Strategic Plan in 2007, Mauritius has demonstrated promising progress in her ICT development. For example, the country has been ranked 47 based on her network readiness index (composed of measures such as ICT infrastructure, ICT usage, and government prioritization of ICT) in the year 2011, which was an improvement by 6 ranks and about 4% in the index from the previous year (Mauritius ICT Indicators Portal, 2011). Among African countries, Mauritius has also been ranked second based on the United Nation's E-Government Readiness Index and first based on World Bank's Ease of Doing Business index (Mauritius ICT Indicators Portal, 2011). These show that bridging the digital divide also has the benefit of promoting economic development in developing countries.

While it is necessary to provide access to ICT in bridging the digital divide, it is also critical to promote its utilization by the targeted users. This study has shown that other than the technological characteristics of perceived ease of use and perceived usefulness, social referents such as public internet kiosk managers and situational characteristics such as geographical distance and cost of access also have significant influence on individuals' use of ICT in the developing country context. This suggests that merely providing ICT without understanding the social and situational contexts of the targeted users cannot adequately narrow the gap between ICT haves and have-nots. We hope that this study will encourage research to look beyond the notion of technology as a quick fix to tackle the digital divide and focus more on understanding users' contexts.

ACKNOWLEDGMENT

This chapter is an updated version of the article "Bridging the Digital Divide: Use of Public Internet Kiosks in Mauritius" published in Journal of Global Information Management (18:1), 2010.

REFERENCES

Agarwal, R. (2000). Individual acceptance of information technologies. In Zmud, R. W. (Ed.), *Framing the domains of IT management: Projecting the future...through the past*. Cincinnati, OH: Pinnaflex Educational Resources.

Agarwal, R., & Prasad, J. (1998). A conceptual and operational definition of personal innovativeness in the domain of information technology. *Information Systems Research*, 9(2), 204–215. doi:10.1287/isre.9.2.204

Agbonlahor, R. O. (2006). Motivation for use of information technology by university faculty: A developing country perspective. *Information Development*, *22*(4), 263–277. doi:10.1177/0266666906072955

Ajzen, I. (1991). The theory of planned behavior. *Organizational Behavior and Human Decision Processes*, *50*(2), 179–211. doi:10.1016/0749-5978(91)90020-T

Ajzen, I. (2002). Perceived behavioral control, self-efficacy, locus of control, and the theory of planned behavior. *Journal of Applied Social Psychology*, *32*(4), 665–683. doi:10.1111/j.1559-1816.2002.tb00236.x

Akhter, S. H. (2003). Digital divide and purchase intention: Why demographic psychology matters. *Journal of Economic Psychology*, *24*(3), 321–327. doi:10.1016/S0167-4870(02)00171-X

Anandarajan, M., Igbaria, M., & Uzoamaka, P. A. (2002). IT acceptance in a less-developed country: A motivational factor perspective. *International Journal of Information Management*, *22*(1), 47–65. doi:10.1016/S0268-4012(01)00040-8

Bandura, A. (2001). Social cognitive theory: An agentive perspective. *Annual Review of Psychology*, *52*(1), 1–26. doi:10.1146/annurev.psych.52.1.1

Bankole, F. O., Bankole, O. O., & Brown, I. (2011). Mobile banking adoption in Nigeria. *The Electronic Journal of Information Systems in Developing Countries*, *47*(2), 1–23.

Brown, I. T. J. (2002). Individual and technological factors affecting perceived ease of use of Web-based learning technologies in a developing country. *The Electronic Journal on Information Systems in Developing Countries*, *9*(5), 1–15.

Chin, W. W. (1998). Issues and opinion on structural equation modeling. *Management Information Systems Quarterly*, *22*(1), vii–xv.

Churchill, G. A. (1979). A paradigm for developing better measures of marketing constructs. *Journal of Marketing*, *16*(1), 64–73. doi:10.2307/3150876

Cohen, J. (1988). *Statistical power analysis for the behavioral sciences*. Hillsdale, NJ: Erlbaum.

Compeau, D. R., & Higgins, C. A. (1995). Computer self-efficacy: Development of a measure and initial test. *Management Information Systems Quarterly*, *19*(2), 189–211. doi:10.2307/249688

Davis, F. D. (1989). Perceived usefulness, perceived ease of use, and user acceptance of information technology. *Management Information Systems Quarterly*, *13*(3), 319–340. doi:10.2307/249008

Davis, F. D., Bagozzi, R. P., & Warshaw, P. R. (1989). User acceptance of computer technology: A comparison of two theoretical models. *Management Science*, *35*(8), 982–1002. doi:10.1287/mnsc.35.8.982

Day-Hookoomsing, P. N. (2000). Leadership training for improved quality in a post-colonial, multicultural society. *The International Journal of Sociology and Social Policy*, *20*(8), 23–32. doi:10.1108/01443330010789025

Deutsch, M., & Harold, G. B. (1955). A study of normative and informational social influences upon individual judgment. *Journal of Abnormal and Social Psychology*, *51*(3), 629–636. doi:10.1037/h0046408

Dewan, S., Ganley, D., & Kraemer, K. L. (2010). Complementarities in the diffusion of personal computers and the Internet: Implications for the global digital divide. *Information Systems Research*, *21*(4), 925–940. doi:10.1287/isre.1080.0219

Dewan, S., & Riggins, F. J. (2005). The digital divide: Current and future research directions. *Journal of the Association for Information Systems*, *6*(12), 298–337.

Drury, P. (2011). *Kenya's Pasha centres: Development ground for digital villages*. Cisco Internet Business Solutions Group. Retrieved from http://www.cisco.com/web/about/ac79/docs/case/Kenya-Pasha-Centres_Engagement_Overview_IBSG.pdf

Dwivedi, Y., Williams, M. D., Lal, B., & Schwarz, A. (2008). Profiling adoption, acceptance and diffusion research in the information systems discipline. *Proceedings of the European Conference on Information Systems*, Galway, Ireland.

Elbeltagi, I., McBride, N., & Hardaker, G. (2005). Evaluating the factors affecting DSS usage by senior managers in local authorities in Egypt. *Journal of Global Information Management*, *13*(2), 42–65. doi:10.4018/jgim.2005040103

Fishbein, M., & Ajzen, I. (1975). *Belief, attitude, intention and behavior: An introduction to theory and research*. Reading, MA: Addison-Wesley.

Fishbein, M., Triandis, H. C., Kanfer, F. H., Becker, M., Middlestadt, S. E., & Eichler, A. (2001). Factors influencing behavior and behavior change. In Baum, A., Revenson, T. A., & Singer, J. E. (Eds.), *Handbook of health psychology*. Mahwah, NJ: Lawrence Erlbaum.

Foster, W., Goodman, S., Osiakwan, E., & Bernstein, A. (2004). Global diffusion of the internet IV: The internet in Ghana. *Communications of the Association for Information Systems*, *13*(38), 654–670.

Furuholt, B., Kristiansen, S., & Wahid, F. (2005). Information dissemination in a developing society: Internet café users in Indonesia. *The Electronic Journal of Information Systems in Developing Countries*, *22*(3), 1–16.

Fusilier, M., & Durlabhji, S. (2005). An exploration of student internet use in India: The technology acceptance model and the theory of planned behavior. *Campus-Wide Information Systems*, *22*(4), 233–246. doi:10.1108/10650740510617539

Gupta, B., Dasgupta, S., & Gupta, A. (2008). Adoption of ICT in a government organization in a developing country: An empirical study. *The Journal of Strategic Information Systems*, *17*(2), 140–154. doi:10.1016/j.jsis.2007.12.004

Hair, J. F., Anderson, R. E., Tatham, R. L., & Black, W. C. (1998). *Multivariate data analysis*. Upper Saddle River, NJ: Prentice-Hall.

Heeks, R., & Kanashiro, L. (2009). Telecentres in mountain regions: A Peruvian case study of the impact of information and communication technologies on remoteness and exclusion. *Journal of Mountain Science*, *6*(4), 320–330. doi:10.1007/s11629-009-1070-y

Hofstede, G. (1984). *Culture's consequences: International differences in work related values*. Thousand Oaks, CA: Sage Publications.

Hollifield, C. A., & Donnermeyer, J. F. (2003). Creating demand: influencing information technology diffusion in rural communities. *Government Information Quarterly*, *20*(2), 135–150. doi:10.1016/S0740-624X(03)00035-2

Hsu, M., & Chiu, C. (2004). Internet self-efficacy and electronic service acceptance. *Decision Support Systems*, *38*(3), 369–381. doi:10.1016/j.dss.2003.08.001

Ifinedo, P. (2006). Acceptance and continuance intention of web-based learning technologies (WLT) use among university students in a Baltic country. *The Electronic Journal of Information Systems in Developing Countries*, *23*(6), 1–20.

Isaacs, S. (2007). *Survey of ICT and education in Mauritius. Survey of ICT and Education in Africa (volume 2): 53 country reports*. Washington, DC: infoDev / World Bank. Retrieved from http://www.infodev.org/en/Publication.354.html

Kaba, B., Diallo, A., Plaisent, M., Bernard, P., & N'Da, K. (2006). Explaining the factors influencing cellular phones use in Guinea. *The Electronic Journal of Information Systems in Developing Countries, 28*(6), 1–7.

King, W. R., & He, J. (2006). A meta-analysis of the technology acceptance model. *Information & Management, 43*(6), 740–755. doi:10.1016/j.im.2006.05.003

Kishore, R., Lee, J., & McLean, E. M. (2001). The role of personal innovativeness and self-efficacy in information technology acceptance: An extension of TAM with notions of risk. *Proceedings of the International Conference on Information Systems*. New Orleans, USA.

Kurth, M. (1993). The limits and limitations of transaction log analysis. *Library Hi Tech, 11*(2), 98–104. doi:10.1108/eb047888

Lam, J. C. Y., & Lee, M. K. O. (2006). Digital inclusiveness: Longitudinal study of internet adoption by older adults. *Journal of Management Information Systems, 22*(4), 177–206. doi:10.2753/MIS0742-1222220407

Lee-Ross, D. (2005). Perceived job characteristics and internal work motivation: An exploratory cross-cultural analysis of the motivational antecedents of hotel workers in Mauritius and Australia. *Journal of Management Development, 24*(3), 253–266. doi:10.1108/02621710510584062

Legris, P., Ingham, J., & Collerette, P. (2003). Why do people use information technology? A critical review of the technology acceptance model. *Information & Management, 40*(3), 191–204. doi:10.1016/S0378-7206(01)00143-4

Lin, F., Fofanah, S. S., & Liang, D. (2011). Assessing citizen adoption of e-Government initiatives in Gambia: A validation of the technology acceptance model in information systems success. *Government Information Quarterly, 28*(2), 271–279. doi:10.1016/j.giq.2010.09.004

Locke, S. (2005). Farmer adoption of ICT in New Zealand. *Proceedings of the Global Management & Information Technology Research Conference*, New York, USA.

Madon, S. (2005). Governance lessons from the experience of telecentres in Kerala. *European Journal of Information Systems, 14*(4), 401–416. doi:10.1057/palgrave.ejis.3000576

Mauritius Government Information Service. (2005). *ICT: Mauritius gearing up for a centre of excellence*. Retrieved from http://encounter-mauritius.gov.mu/ portal/site/ Mainhomepage/menuitem.a42b24128104d9845dabddd154508a0c/?content_id=e2fbb08e94d4b010VgnVCM1000000a04a8c0RCRD

Mauritius ICT Indicators Portal. (2011). *International indices*. Mauritius National Computer Board. Retrieved from http://www.gov.mu/ portal/sites/indicators/ International_Indices. html

Mauritius National Computer Board. (2006). *Universal ICT education programme*. Retrieved from http://www.gov.mu/portal/sites/uieptest/about.html

Mauritius National Computer Board. (2007). *Republic of Mauritius national ICT strategic plan 2007-2011*. Retrieved from http://www. gov.mu/portal/goc/ telecomit/ files/ NICTSP.pdf

Mbarika, V., & Byrd, T. A. (2009). An exploratory study of strategies to improve Africa's Least developed economies' telecommunications infrastructure: The stakeholders speak. *IEEE Transactions on Engineering Management, 56*(2), 312–328. doi:10.1109/TEM.2009.2013826

Meso, P., Musa, P., & Mbarika, V. (2005). Towards a model of consumer use of mobile information and communication technology in LDCs: The case of Sub-Saharan Africa. *Information Systems Journal, 15*(2), 119–146. doi:10.1111/j.1365-2575.2005.00190.x

Moore, G. C., & Benbasat, I. (1991). Development of an instrument to measure perceptions of adopting an information technology innovation. *Information Systems Research, 2*(3), 192–222. doi:10.1287/isre.2.3.192

Nunnally, J. C. (1978). *Psychometric theory* (2nd ed.). New York, NY: McGraw-Hill Book Company.

Plouffe, C. R., Hulland, J. S., & Vand, M. (2001). Research report: Richness versus parsimony in modeling technology adoption decisions: Understanding merchant adoption of a smart card-based payment system. *Information Systems Research, 12*(2), 208–222. doi:10.1287/isre.12.2.208.9697

Podsakoff, P., & Organ, M. (1986). Self-reports in organizational research: Problems and prospects. *Journal of Management, 12*(4), 531–544. doi:10.1177/014920638601200408

Rangaswamy, N. (2007). ICT for development and commerce: A case study of internet cafés in India. *Proceedings of the 9th International Conference on Social Implications of Computers in Developing Countries*, São Paulo, Brazil.

Rice, R. E., & Katz, J. E. (2003). Comparing internet and mobile phone usage: Digital divides of usage, adoption, and dropouts. *Telecommunications Policy, 27*(8/9), 597–623. doi:10.1016/S0308-5961(03)00068-5

Rogers, E. M. (1995). *Diffusion of innovations* (4th ed.). New York: The Free Press.

Schepers, J., & Wetzels, M. (2007). A meta-analysis of the technology acceptance model: Investigating subjective norm and moderation effects. *Information & Management, 44*(1), 90–103. doi:10.1016/j.im.2006.10.007

Schware, R. (2007). Scaling up rural electronic governance initiatives. *Proceedings of the 1st International Conference on Theory and Practice of Electronic Governance*, Macao, Hong Kong.

Shih, E., Kraemer, K. L., & Dedrick, J. (2008). IT diffusion in developing countries. *Communications of the ACM, 51*(2), 43–48. doi:10.1145/1314215.1340913

Sipior, J. C., Ward, B. T., Volonino, L., & Marzec, J. Z. (2003). A community initiative that diminished the digital divide. *Communications of the Association for Information Systems, 13*(1), 29–56.

Soyjaudah, K. M. S., Oolun, M. K., Jahmeerbacus, I., & Govinda, S. (2002). ICT development in Mauritius. *Proceedings of the 6th IEEE Africon Conference in Africa*, George, South Africa.

Taylor, S., & Todd, P. (1995). Understanding information technology use: a test of competing models. *Information Systems Research, 6*(2), 144–176. doi:10.1287/isre.6.2.144

Thanacoody, R., Bartram, T., Barker, M., & Jacobs, K. (2006). Career progression among female academics: A comparative study of Australia and Mauritius. *Women in Management Review, 21*(7), 536–553. doi:10.1108/09649420610692499

Triandis, H. C. (1977). *Interpersonal behavior*. Monterey, CA: Brooks/Cole Publishing Company.

Umbach, J. M. (2004). Libraries: Bridges across the digital divide. *Feliciter, 50*(2), 44.

United Nations. (2007). *Mauritius national ICT strategic plan final analysis report*. Retrieved from http://un.intnet.mu/undp/downloads/info/ social_development/nictsp/ report/ Final%20 Analysis%20Report%20Draft%20One1.zip

United States Central Intelligence Agency. (2011). *The world factbook of Mauritius*. Retrieved from https://www.cia.gov/library/publications/the-world-factbook/geos/mp.html

United States Department of Commerce. (1995). *Falling through the net: A survey of the "have nots" in rural and urban America*. Retrieved from http:// www.ntia.doc.gov/ ntiahome/ fallingthru.html

Uzoka, F. E., Shemi, A. P., & Seleka, G. G. (2007). Behavioral influences on e-commerce adoption in a developing country context. *The Electronic Journal of Information Systems in Developing Countries, 31*(4), 1–15.

Venkatesh, V. (2000). Determinants of perceived ease of use: Integrating control, intrinsic motivation, and emotion into the technology acceptance model. *Information Systems Research, 11*(4), 342–365. doi:10.1287/isre.11.4.342.11872

Venkatesh, V., Morris, M., Davis, G., & Davis, F. (2003). User acceptance of information technology: Toward a unified view. *Management Information Systems Quarterly, 27*(3), 425–478.

APPENDIX: CONSTRUCT OPERATIONALIZATION

Construct	Items	Source
Self-efficacy (SE)	SE1: I am very comfortable with using public internet kiosk on my own. SE2: Overall, I am able to use public internet kiosk even if there is no help.	Adapted from Taylor and Todd (1995)
Perceived Ease of Use (PEOU)	PEOU1: Using public internet kiosk is easy for me. PEOU2: Overall, public internet kiosk is easy to use.	Adapted from Moore and Benbasat (1991)
Perceived Usefulness (PU)	PU1: Using public internet kiosk significantly reduces time required to perform tasks. PU2: Using public internet kiosk significantly improves the quality of my work/ task outputs.	
Subjective Norm (SN) Product of likelihood that specific salient referents think one should perform the behavior (LSN) and the individuals' motivation to comply (CSN).	*LSN Scale* SN1a: My family members and relatives strongly suggest that I should use public internet kiosks. SN2a: My friends strongly suggest that I should use public internet kiosks. SN3a: The Government strongly suggests that we should use public internet kiosks to learn computers. SN4a: The staff of public internet kiosks strongly suggest that I should use public internet kiosks. SN5a: Public media strongly suggests that I should use public internet kiosks. *CSN Scale* SN1b: I am very willing to comply with my family members' and relatives' suggestions. SN2b: I am very willing to comply with my friends' suggestions. SN3b: I am strongly influenced by the Government's suggestions. SN4b: I am strongly influenced by the suggestions of public internet kiosk staff. SN5b: I am strongly influenced by public media (e.g., newspaper, magazine, television).	Developed based on Taylor and Todd (1995)
Perceived Behavioral Control (PBC)	*Geographical Access* PBC1: The location of public internet kiosk is very close to my workplace. PBC2: The location of public internet kiosk is very close to my home. *Cost* PBC3: It is very affordable to use public internet kiosk. PBC4: Using public internet kiosk involves very little cost. PBC5: It is very expensive for me to travel to use public internet kiosks (PBC1-PBC4 reverse coded).	
Intention (INT)	INT1: I intend to continue using public internet kiosk in future. INT2: I intend to use public internet kiosk more frequently.	Adapted from Davis et al. (1989)
ICT Use Behavior (USE)	USE1: I use public internet kiosks *(very rarely - sometimes - very frequently)*. USE2: I use public internet kiosks for *(very few - some - many)* tasks.	

Chapter 7

E–Business Assimilation in China's International Trade Firms:
The Technology–Organization–Environment Framework

Dahui Li
University of Minnesota Duluth, USA

Fujun Lai
University of Southern Mississippi, USA

Jian Wang
University of International Business and Economics, China

ABSTRACT

Despite the importance of international trade firms in China's economic development, there is only limited empirical evidence about how these firms assimilate Internet-based e-business in global supply chain operations. Using the Technology-Organization-Environment framework, this study investigates technological, organizational, and environmental factors which determine e-business assimilation in these firms. Based on survey data collected from 307 international trade firms in the Beijing area, we found that environmental uncertainty was negatively associated with e-business assimilation, while a firm's internal IT capability, relative advantage of e-business, learning orientation, and inter-organizational dependence were positive determinants of e-business assimilation. The effect of a firm's ownership type was also significant. Environmental uncertainty was the most important inhibitor, and IT capability and inter-organizational dependence were the most salient enablers of e-business assimilation.

DOI: 10.4018/978-1-61350-480-2.ch007

INTRODUCTION

China has become one of the world's top three international trade countries in recent years, together with the United States and Germany (MoC, 2007; US Commercial Service, 2005; US Department of State, 2007; WTO, 2007). The practice of international trade in China is different from that in developed countries. Export and import businesses in China are conducted exclusively by international trade firms which are granted trade rights by the Chinese government. Firms without licenses are not allowed to export and/or import goods and services. Therefore, international trade companies have been the middle men in the global supply chain relationship between foreign companies and Chinese domestic firms. In the past, only a few state-owned companies, such as those affiliated with the Ministry of Foreign Trade, were allowed to conduct international trade business. Since China joined the World Trade Organization (WTO) in December 2001, the Chinese government has eased restrictions on exports and imports. Various new policies have been implemented so that both domestic and foreign companies in China can trade directly—as long as they are licensed by the government (US Department of State, 2007). Both domestic and foreign Chinese firms no longer need to go through intermediary trade firms.

The Chinese international trade industry has been the most innovative industry in China for implementing information technology (IT) and information systems (IS). The industry has maintained the highest adoption rate of Internet technologies in China for some time (Guo & Chen, 2005). The China Ocean Shipping Company (COSCO) was the first Chinese company to adopt commercial computers in business operations (McFarlan & Rockart, 2004; McFarlan, Chen & Lane, 2005). In recent decades, the Chinese government provided significant support for deploying IT/IS in China's international trade industry. For example, the government initiated the "Golden Customs" project in 1993, which mandated that international trade firms inter-network with Chinese Customs via electronic data interchange (EDI). In 1998, the State Economic and Trade Commission, and the Ministry of Information Industry, jointly started the "Golden Trade" project, which provided general directions and guidelines for developing a national information infrastructure to promote online trading for the international trade industry. Since then, many infrastructures and platforms have been established to help the industry to do e-business. Following the previous definition of e-business (Zhu & Kraemer, 2005; Zhu, Kraemer, Xu & Dedrick, 2004), we define e-business as conducting business activities—such as sales, customer services, logistics, information sharing and coordination—on the Internet in order to connect to worldwide marketplaces and the global supply chain.

Although e-business empowered by Internet technologies is widely diffused in the global supply chain (Kirkman, 2002), there is only limited empirical research showing how Chinese firms have implemented e-business. Our current understanding of the international trade industry has been limited to either case studies (McFarlan & Rockart, 2004; McFarlan et al., 2005), or segmented analyses from a small scale survey (Guo & Chen, 2005). There are concerns that IT/IS adoption in these Chinese firms may be superficial, and that these firms may not be able to implement IT/IS to a deeper level, due to various internal and external challenges (Guo & Chen, 2005; McFarlan & Rockart, 2004).

Broadly speaking, how e-business is assimilated in Chinese firms remains a myth and is yet to be revealed. Except for limited empirical evidence (e.g., Dasgupta, Agarwal, Ioannidis & Gopalakrishnan, 1999; Molla & Licker, 2005a, 2005b; Xu, Zhu & Gibbs, 2004; Zhu et al., 2004), how e-business is conducted in developing countries is not well-examined. Unlike developed countries which have established e-business infrastructures, China and other developing countries do not have mature institutions and infrastructures supporting e-business operations (Tan & Ouyang, 2002). These countries also have some unique characteristics which are worth investigation. Therefore, we suggest that it will be interesting to examine whether theories which originated in the context of mature markets and developed countries can be applied in developing countries, despite different economic and regulatory environments (Dasgupta et al., 1999; Zhu & Kraemer, 2005).

To investigate e-business implementation in Chinese firms, we have conducted a large scale survey of international trade companies in the Beijing area. This paper reports the findings of the survey, and reveals technological, organizational, and environmental factors that drive these companies' e-business assimilation. In the paper, we develop a research model based on the well-established Technology-Organization-Environment (TOE) framework (Tornatzky & Fleischer, 1990). Most of the factors we include in our research model have been examined in the previous literature from Western countries. Using the same framework enables us to compare and contrast the unique patterns of e-business assimilation in Chinese firms with those found in the West. In the following, we present the theoretical background and develop our research hypotheses. We then explain our research methodology and report our data analyses and results. In the final section we provide discussion and implications of the study.

THEORETICAL BACKGROUND

E-Business Assimilation

In the management literature, assimilation means "an organizational process that (1) is set in motion when individual organization members first hear of an innovation's development, (2) can lead to the acquisition of the innovation, and (3) sometimes comes to fruition in the innovation's full acceptance, utilization, and institutionalization" (Meyer

& Goes, 1988, pp. 897). In the information systems literature, assimilation means "the process within organizations stretching from initial awareness of the innovation to potentially formal adoption and full-scale deployment" (Fichman, 2000, pp. 106). Several recent studies emphasize assimilation as a post-adoption stage (e.g., Chatterjee, Grewal & Sambamurthy, 2002; Purvis, Sambamurthy & Zmud, 2001). For example, assimilation is stated as "the extent to which the use of technology diffuses across the organizational projects or work processes and becomes routinized in the activities of those projects and processes" (Purvis et al., 2001, pp. 121). The definition suggests that the IT/IS being assimilated/implemented has become a routine part of an organization's business operation, and "is no longer perceived as something out of the ordinary" (Cooper & Zmud, 1990, pp. 124). Therefore, assimilation is similar to 'routinization' (Cooper & Zmud, 1990) and 'incorporation' (Kwon & Zmud, 1987).

Following a similar definition by Chatterjee et al. (2002) and Armstrong and Sambamurthy (1999), we define e-business assimilation as the extent to which a firm utilizes Internet technologies to facilitate the firm's e-business strategies and activities aimed at marketing, advertising, and buying and selling products and services in the firm's supply chain. Chatterjee et al. (2002) and Armstrong and Sambamurthy (1999) suggest that there are two distinct dimensions of e-business assimilation which are worthy of investigation. The first is strategy assimilation, which indicates the importance of Internet technologies emphasized by a firm when the firm makes strategic decisions. The second is activity assimilation, which is the extent to which the firm uses Internet technologies in its business activities and day-to-day operations. These two dimensions provide a unified understanding of assimilation at both the strategic and operational levels of a firm (Chatterjee et al., 2002). In this study we also conceptualize these two aspects as two dimensions of e-business assimilation.

Previous IT/IS Assimilation Research

Previous research on IT/IS assimilation is focused on investigating critical factors in the following five major categories: individual user, organization (or structure), technology (or innovation), task (or business process), and environment (Kwon & Zmud, 1987). Some studies apply theories such as knowledge-based view of firms, resource-based theories, and technology structuration theory and reveal factors inside the firm to explain the degree of IT/IS assimilation (e.g., Armstrong & Sambamurthy, 1999; Chatterjee et al., 2002; Purvis et al., 2001). Chatterjee et al. (2002), for example, examine top management championship, extent of coordination, and the rationale for strategic investment in assimilation, based on institutional theory and technology structuration theory. Drawing on resource-based theory and knowledge-based theory of the firm, Armstrong and Sambamurthy (1999) investigate the role of senior leaders' knowledge of IT/IS, CIOs' participation and interaction with the top management team, the vision of strategic IT, as well as the sophistication of IT infrastructure. Other studies rely on theories which examine the effect of external environment. These studies reveal the external factors rather than internal factors which shape the organization's behavior and the strategic leader's decision making. For example, Liang, Saraf, Hu, and Xue (2007) apply institutional theory in a study of the assimilation of Enterprise Resource Planning (ERP) systems. Hart and Sanders (1998) investigate the effects of power and dependence in the business relationship on the use of EDI.

Studies with theoretical focuses on either internal or external factors of IT/IS assimilation are useful for expanding our knowledge about the subject. However, these focuses may not provide a complementary view of the effects of both internal and external factors. In pursuit of a balanced view of IT/IS assimilation, we note several previous studies (e.g., Bolloju & Turban, 2007; Molla & Licker, 2005b; Ranganathan, Dhaliwal & Teo,

2004; Raymond, Bergeron & Blili, 2005; Zhu et al. 2004) adopt the Technology-Organization-Environment (TOE) framework (Tornatzky & Fleischer, 1990).

TOE is one of the most useful and popular frameworks for IT/IS implementation research. See Zhu et al. (2004) for a summary. Compared with a purely technological emphasis, TOE integrates contingent organizational and environmental factors that firms have to face (Kuan & Chau, 2001). Therefore, TOE provides a unified perspective from which we are able to examine a firm's internal and external factors, as well as technological features. These factors generally fall in three broad categories: the factors related to the technology to be implemented, the features of the organization which deploys the technology, and contingent factors in the macro-environment where the IT/IS is implemented. Previous literature has studied such technological factors as the innovation features (e.g., relative advantage, compatibility, and complexity) and an organization's existing technology features (Kwon & Zmud, 1987; Rogers, 1995). The most often examined organizational factors include firm size, firm age, centralization, formalization, and complexity of organization structure (Chau & Tam, 1997; Kwon & Zmud, 1987). The environmental context includes such factors as environmental uncertainty, heterogeneity, competition, and concentration (e.g., Chau & Tam, 1997; Kwon & Zmud, 1987).

The Relevancy of TOE Framework

The TOE framework is relevant to the study of e-business assimilation in Chinese firms. The advantage of the technological features of e-business over other business practices is pivotal. At the initial adoption of e-business practices, the relative advantage of e-business can be perceived or projected. As the assimilation process moves on, the relative advantage may have been achieved and demonstrated in the firm's performance. As the most innovative industry in China in terms of deploying IT/IS in business operations (Guo & Chen, 2005), Chinese international trade firms have to constantly evaluate and reevaluate the technological features in their supply chain operations.

However, simply focusing on the technological features of e-business may not be sufficient to explain a successful implementation of e-business. Such features may be able to explain the rate of diffusion of IT/IS, but not the pattern of an organization's assimilation of IT/IS over time (Fichman, 2000). Organizations that implement e-business may need to make corresponding adaptation to meet the technological requirements of e-business, such as streamlining business processes, so that the full potential and advantages of e-business may be achieved. Therefore, we have to consider whether the organization is able to facilitate such integration and adaptation, in order to understand the process of the firm's e-business assimilation.

An examination of an organization's internal factors may be especially important for studying firms such as Chinese international trade firms in developing countries, because these firms may not have the resources and capabilities necessary for survival and growth (Peng, 2003). The lack of internal resources and the capabilities needed to absorb e-business innovation in firm operations may impose many constraints on the successful implementation of e-business.

On the other hand, examining the organization's external environment is also important because firms in developing countries have to consider the uncertainties in their business environment. The rapid growth of Internet technologies and e-business practices in China is also accompanied by concerns about immature infrastructure and supporting industries such as banks and shipping (Tan & Ouyang, 2002). Finally, the assimilation of e-business is a joint effort, which means that the successful assimilation of e-business is not only achieved through a focal firm's operations, but also those of the business partners (Fichman, 2000). Therefore, we need to consider the business relationships between these trade firms and their partners.

With these concerns in mind, we believe that the TOE framework is appropriate for examining e-business assimilation issues in our study.

Figure 1. Research model

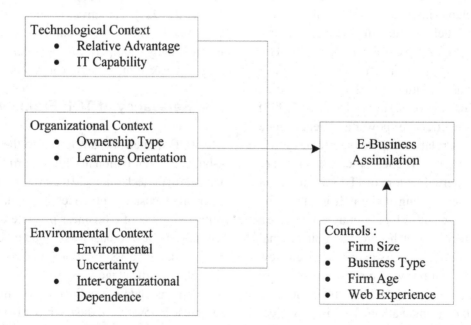

RESEARCH MODEL AND HYPOTHESES

Drawing on the TOE framework, we develop the research model as shown in Figure 1. The outcome of the model is e-business assimilation, which consists of two dimensions, as presented above. We examine relative advantage and IT capability as technological factors, ownership type and learning orientation as organizational factors, and environmental uncertainty and inter-organizational dependence as environmental factors. Research hypotheses are discussed as follows.

Technological Context

Relative Advantage

One of the most salient factors influencing e-business assimilation is the relative advantage of e-business. Kwon and Zmud (1987) and Rogers (1995) define relative advantage as the degree to which an innovation is perceived to be better than competing technologies. Realizing the potential benefits of a given technology over other existing technologies, a firm is more likely to implement such a technology in its business operations and strategies. Tornatzky and Klein (1982) report that relative advantage is one of the most widely studied variables in empirical studies which examine IT/IS adoption and diffusion at the firm level.

In the e-business context, the relative advantage of e-business over other technologies and business approaches is an attractive feature for firms. For example, travel agencies pay special attention to relative advantage of web technologies in their adoption and acceptance of these technologies (Wang & Cheung, 2004). Firms always evaluate the potential benefits and risks of implementing e-business in order to develop strategies to compete with rivals online (Molla & Licker, 2005b). Zhu, Kraemer, Gurbaxani, and Xu (2006) also find that expected benefits positively affect the

adoption of open-standard inter-organizational systems in firms.

The relative advantage of e-business may be an appealing factor to firms in China—a country which is undergoing the transition from a state-run centralized economy to free competition in the market economy. Chinese international trade firms, as intermediaries in the global supply chain, may be very knowledgeable about e-business through conducting business transactions with their business partners in the West or the market economy. Thus, we hypothesize

H1a: Relative advantage of e-business has a positive impact on e-business assimilation.

IT Capability

In addition to the appealing technological features of e-business, we have to examine the technological features of the firm which implements e-business. Whether the new technology can be assimilated to a firm's operation also depends on the firm's existing technological infrastructure. A firm's IT capability explains such a concern. According to Bharadwaj (2000), IT capability means a firm's ability to utilize and deploy IT-related resources, which include hardware, software, databases, IT manpower, and other IT-enabled intangibles. Companies with a higher level of IT capability have an advantage over competing firms, and have better business and financial performances (Bharadwaj, 2000).

Many studies have examined the effect of IT capability on the assimilation process of IT/IS in the business context (e.g., Cui, Zhang, Zhang & Huang, 2008). For example, the complexity of a company's IT infrastructure—an important indicator of the company's IT capability—has a positive effect on the adoption of open systems (Chau & Tam, 1997). Another aspect of IT capability, experienced IT/IS personnel, may usually have knowledge of innovative technologies and be aware of the risks and barriers in the adoption process. Therefore, they are likely to be active promoters and facilitators of the adoption of new technologies (e.g., Chau & Tam, 1997).

In the e-business assimilation context, firms with a higher level of IT capability may be more competent to deal with the complexity and potential risks of e-business operations. Advanced IT capability developed in the history of an organization is also an indicator of the importance of IT for the business' strategic committee in the long term plan (Fichman, 2000). As mentioned above, Chinese international trade firms are the most active players of IT/IS (McFarlan & Rockart, 2004; McFarlan et al., 2005), even before China started its economic transition to the market economy in the 1990s. These firms have accumulated a sufficient level of IT resources and capabilities which can enable the firms to migrate into the e-business platform. Therefore, we hypothesize

H1b: A firm's IT capability has a positive impact on e-business assimilation.

Organizational Context

Ownership Type

As an important organizational factor related to Chinese firms, ownership type draws our attention. Ownership type is a parsimonious organizational factor which can explain Chinese firms' strategy and performance (Mascarenhas, 1989; Peng, Tan & Tong, 2004). However, the effect of ownership type has been seldom examined in the IT/IS implementation literature because previous studies are for the most part based on the market economy, in which private ownership is the dominant ownership type. However, in China, there are several types of ownership among firms, such as state-owned, privately-owned, collectively-owned, and foreign-invested (Peng et al., 2004; Tan, 2002). The influence of ownership type on a firm's

behavior, as examined in previous organization and strategy literature (Mascarenhas, 1989; Peng et al., 2004; Tan, 2002), may be also manifested in Chinese firms' implementation of e-business.

Firms of different ownership types are believed to execute different business strategies and behaviors (Mascarenhas, 1989). For example, privately-owned companies usually pursue profit maximization, while state-owned firms have to emphasize public, municipal, and national preferences (Mascarenhas, 1989). Therefore, privately-owned companies are flexible and aggressive, while state-owned companies are characterized as risk averse, inflexible, bureaucratic, and subject to inertia.

From the innovation perspective, the source of technological investment for state-owned firms is limited to government sponsors and budget allocations, so that these firms do not have control over how to spend money. Privately-owned companies, on the other hand, are in full control of their resources and thus seem to be more innovative and are constantly looking for environmental opportunities (Peng et al., 2004; Tan, 2002). Accordingly, privately-owned companies are more likely to adopt proactive innovation strategies, while state-owned companies are more likely to adopt conservative strategies (Peng et al., 2004; Tan, 2002). Therefore, a firm's ownership type may be a clear factor to differentiate the level of e-business assimilation in Chinese international trade firms, which are of a variety of ownership types. In one of the few studies of the effect of ownership on information technology adoption, Dasgupta et al. (1999) point out the significance of ownership type in the study of developing countries. Martinsons (2004) finds clear differences between privately-owned firms and state-owned firms in terms of ERP implementation in China. Therefore, we hypothesize that

H2a: A firm's ownership type is a significant factor for e-business assimilation.

Learning Orientation

Chinese international trade firms are embedded with their business partners within the global supply chain. Learning the business practice from their business partners, especially international partners in the upstream or downstream of a firm's supply chain, is very critical to the success of Chinese international trade firms. While e-business has been widely adopted in developed countries, the practice may be new to Chinese firms and other firms in developing countries. Whether firms in developing countries can learn to apply modern business practice is critical to a firm's future.

A firm's learning orientation emphasizes that learning is vital to a firm's performance and survival (Sinkula, Baker & Noordewier, 1997). A firm with strong learning orientation is "an organization skilled at creating, acquiring, and transferring knowledge, and at modifying its behavior to reflect new knowledge and insights" (Garvin, 1993, pp. 80). In this study, we define learning orientation as the degree to which firms emphasize the value of learning as an inseparable segment of the organization's culture (Hult, Ketchen & Nichols, 2003).

Learning-oriented firms actively scan and respond to external innovations and environmental challenges (Calantone, Cavusgil & Zhao, 2002; Sinkula et al., 1997). Firms with higher level of learning orientation will be more active in exploring and introducing new business initiatives and state-of-the-art technologies (Calantone et al., 2002). A learning-oriented Chinese international trade firm may be proactive in gathering initial knowledge about what e-business is and how e-business can enhance business operations. Moreover, such a firm may be more active in implementing e-business in a very competitive manner to achieve the benefits of e-business. Therefore, we hypothesize that

H2b: A firm's learning orientation has a positive impact on e-business assimilation.

Environmental Context

Organizational theories have recognized the impact of environment on a firm's strategy and performance. However, theories developed in a market economy need to be justified or reconciled in developing countries like China (Molla & Licker, 2005a; Shenkar & von Glinow, 1994). This is because the infrastructures well-established in developed countries are not readily available in developing countries.

Environmental Uncertainty

Contingent environmental factors such as environmental uncertainty have been well-accepted as significant determinants of a firm's innovative behavior. In this study, we focus on the uncertainty of the e-business environment which is to provide various technological and institutional support for the operation of e-business. Technology assimilation is heavily dependent on sufficient supporting infrastructures from industries like logistics, transportation, and banking. A high level of maturity in the supporting infrastructure also helps a business to respond to external competition, enhance organizational learning, and seek and implement innovation strategies (e.g., Armstrong & Sambamurthy, 1999).

In a market economy, environmental uncertainty provides good opportunities for firms. Our conventional understanding of the effect of environmental uncertainty on innovation has been positive. Pfeffer and Leblebici (1977) point out that "Under conditions of relatively undifferentiated environments that are quite stable, organizations should be able to cope with the information processing requirements without elaborate information technology. It is when the organization faces a complex and rapidly changing environment that information technology is both necessary and justified" (p. 246). Empirical studies conducted in the West also confirm that when faced with higher uncertainty in the market, firms are more likely to adopt an innovation (e.g., Grover & Goslar, 1993).

However, this finding from the market economy may not apply to developing countries such as China, because China has been experiencing the transition to a market economy (Tan & Litschert, 1994). In such a transition, the rules and infrastructures required for doing business in the market economy are not established or are in the process of developing. Therefore, the basic infrastructures may not be available. In one of the first studies of a Chinese firm's innovation behavior during economic transition, Tan and Litschert (1994) find that Chinese companies are very concerned with environmental uncertainty in the development of business strategies. Chinese firms are usually less proactive and less innovative when they face dynamic and uncertain environments. Instead, they adopt risk-averse and defensive strategies (Tan & Litschert, 1994). This general pattern of Chinese firms' innovation behavior has also been tested in several other studies (e.g., Peng et al., 2004; Tan, 2002). Based on these previous findings of Chinese firms' innovation behavior, we posit that Chinese international trade firms are less likely to implement e-business without the support of mature infrastructures and institutions required for the operation of e-business.

H3a: Environmental uncertainty has a negative impact on e-business assimilation.

Inter-Organizational Dependence

Inter-organizational dependence is the extent to which an organization shares and exchanges resources with other organizations which may be customers, suppliers, and trade partners (Kwon & Zmud, 1987). The resource dependence theory (Pfeffer & Salancik, 1978) suggests that firms depend on a network of inter-organizational relationships to increase their competitive ability. This is especially important for firms in a transition economy, where firms are found to

rely on network-based strategies to obtain sufficient resources for growth (Peng, 2003; Peng & Heath, 1996).

Interdependence is a source of power that influences the innovation activities of a company and its business partners (Hart & Saunders, 1998). Implementation of inter-organizational information systems, such as electronic data interchange and supply-chain management, is a collective action where dependence among organizations can be the influential power implementing these systems. Empirical studies have consistently found a significant effect of interdependence on the adoption and implementation of inter-organizational information technology (e.g., Hart & Saunders, 1998) and e-business (e.g., Zhu, Kraemer & Xu, 2003; Zhu et al., 2004). Therefore, we hypothesize

H3b: Inter-organizational dependence has a positive impact on e-business assimilation.

Control Variables

Some variations in e-business assimilation can be explained only if controls are appropriately applied. Therefore, we include in this study four control variables, i.e., firm size, firm age, business type, and web experience, which have been examined in previous IT/IS assimilation literature.

Firm size, one of the most commonly studied factors in the innovation literature (Damanpour, 1992), is often used as a surrogate measure for total resources, slack resources, and organization structure (Rogers, 1995). Firm size may have a positive influence on adoption and assimilation behavior (Rogers, 1995; Tornatzky & Fleischer, 1990), since large organizations have economies of scale to leverage their investments in e-business assimilation. Therefore, we expect a positive relationship between firm size and e-business assimilation. *Firm age* could have a negative impact on e-business assimilation because older companies have more red tape, a lot of divisions, and are likely to favor the status quo (Chartterjee et al., 2002). Similar to Chartterjee et al. (2002), we expect a negative relationship between firm age and e-business assimilation. *Business type,* or the firm's primary industry, may also influence e-business assimilation. Because of the unique nature of Chinese international trade companies, these companies come from different primary industries, such as manufacturing and service. Some companies' only business activities are focused on importing and exporting products and services for other companies and we call these companies pure trading firms. Previous literature has found that service-oriented companies can have an advantage when they make extensive use of the Internet as a new business platform. Service-oriented companies, like financial services, are able to conduct a more extensive set of activities on the Internet than are manufacturers (Chatterjee et al., 2002). Therefore, we expect business type to significantly affect e-business assimilation. Finally, we include *web experience* (i.e. whether the company has a web site) as a control variable. Chatterjee et al. (2002) suggest that companies with web presence may have a high level of web assimilation. Thus, we also expect web experience to positively impact e-business assimilation.

RESEARCH METHOD

Data Collection

We used the survey method to collect data. The sample for the present study was selected from international trade companies listed in a database published by the Beijing Municipal Bureau of Commerce. All 2,075 companies in the database were contacted initially by phone to examine whether the company was interested in participating in a survey. 569 companies declined the invitation and 722 companies could not be reached. 784 companies stated their willingness to participate. We made follow-up phone calls to those 722 unreachable companies and reached 45

companies. 28 companies agreed to participate and 17 companies declined. So, our planned contact list included a total of 812 companies. During phone calls, we obtained the names and addresses of senior IT managers or operation managers of companies.

To help respondents better understand the survey questionnaire and to improve the survey response rate and data quality, we used a field interview, rather than a mail survey to collect data. Students who had completed e-business or e-commerce courses were hired to work as interviewers. The interviewers were trained in two one-hour sessions, one concentrated on interview skills and the other on questionnaire items.

Because of the budget constraints of our research, only 500 companies were randomly selected from our initial sample which included 812 reachable companies. Phone calls were made again to these 500 companies to set up interviews before the interviewers appeared on site. Because some companies had changed their mind about participating, field interviews were scheduled with 446 companies. The interviews with the senior IT managers or operations managers were conducted during a period of two months. Each interview lasted about 30 minutes. 139 were cancelled by the companies for various reasons and a total of 307 interviews were successfully completed.

Measures

Questions in the survey were developed based on a comprehensive review of the literature as well as expert opinions. This included successive stages of theoretical specification and refinement. Detailed definitions for all measurement items are shown in the appendix.

E-business assimilation can be measured using different approaches (Fichman, 2000). Some studies measure assimilation in terms of the extent of internal and external business functions and activities supported by IT/IS (e.g., Ranganathan et al., 2004; Raymond et al., 2005). Other studies, however, suggest including the extent of using IT/ IS in corporate strategies such as providing low cost products and services, providing production flexibility, and enhancing customer relationships (e.g., Armstrong & Sambamurthy, 1999; Chatterjee et al., 2002), in addition to using IT/ IS in business functions. Therefore, assimilation consists of both business strategies and business activities. Consistent with the conceptualization of e-business as mentioned in the previous section, we also treated e-business assimilation as a second-order factor including two first-order dimensions: strategy assimilation and activity assimilation. Although we followed the similar definitions of the two dimensions, the scales adopted in the questionnaire were different from those used in Armstrong and Sambamurthy (1999) and Chatterjee et al. (2002). Armstrong and Sambamurthy (1999) asked questions to capture the notion of using general IT to achieve a wide range of competitive advantages in different categories. Chatterjee et al. (2002), however, was focused on a specific IT, i.e., the company's web site, and therefore asked questions about using a company's web site to enhance company image, attract customers, and create a new advertising channel.

Considering the specific role of China's international trade firms as the middleman in the global supply chain, we emphasized how these firms viewed using Internet technologies to facilitate information and communication activities in the supply chain and global trade market. Therefore, to measure *strategy assimilation*, we asked how important Internet technologies were to shape the firm's business strategy in terms of using Internet technologies to attract customers, acquire information, share information, and maintain interactions with customers. We used a 5-point Likert scale for *strategy assimilation. Activity assimilation* was measured using six questions (binary scale) to capture whether or not a firm used Internet technologies in their business activities. The six

questions covered three categories of business activities which were information sharing within the firm and with customers and partners, customer interaction through online transactions, marketing and advertising, and business integration with ERP and customer relationships management systems.

Relative advantage of e-business usually includes reduced costs, more effective information processing, communication, data management, personnel management (MacKay, Oarent & Gemino, 2004; Wang & Cheung, 2004), and new business opportunities (Chatterjee et al., 2002). In our study, respondents were asked to indicate if they agreed that Internet technologies could bring more business opportunities, reduce costs, improve management, and improve efficiency. A binary scale was used for these items.

IT capability was measured by asking respondents to indicate if they had web sites, how they hosted their web sites (in-house server, rented virtual host, or rented dedicated host), and how they managed their web sites (full-time professional, part professional, or outsourcing). These three questions captured the basic resources available to the company (e.g., hardware, software, and IT human resources).

Ownership type was categorized into four types: state-owned, privately-owned, collectively-owned, and foreign-invested (Tan, 2002). *Learning orientation* was measured using the scale of commitment to learning (Sinkula et al., 1997). Adopted in several previous studies (e.g., Hult et al., 2003), the scale assessed a firm's basic belief about the value of learning, associated with a long-term strategic orientation (Calantone et al., 2002). The items were measured with a 5-point Likert scale.

Environmental uncertainty was measured by three items. Respondents were asked to rate the extent to which they agreed with the following three areas of supporting infrastructure within their industry: (1) the maturity of the virtual market as supported by Internet technologies, (2) the range of services that are supported by Internet technologies, and (3) the industry's interest in applying Internet technologies. These items followed a 5-point Likert-type scale.

Inter-organizational dependence was measured in terms of two items about the levels of implementation of Internet technologies among foreign business partners and among domestic business partners. The items were placed on a 5-point Likert scale.

Regarding the control variables, *firm size* was measured using the number of employees in a firm, while *firm age* was measured using the number of years the company operated in China. Both scales were continuous. In the following analyses, a natural logarithm transformation was applied on firm size and firm age to account for the distribution skewness and the diminishing effects of size and age. *Business type* was measured in terms of four types of industries: manufacturing, service, pure trading, and comprehensive. Comprehensive means that companies operated all kinds of businesses, including manufacturing products, providing services, trading products, and others. We measured *web experience* by asking the respondent whether the company had been running a web site or not, which was a binary scale.

Sample Statistics

Among the 307 respondents, 58% were pure trading companies that focused only on import and export, while 42% were other companies that had trade licenses to manage their own import and export activities, such as manufacturing, service, and comprehensive firms. More than 80% of these firms had at least 5 years of trade experience. Roughly half of the firms were medium-sized, with the number of employees ranging from 50 to 1,000. Products that were traded by these companies varied from machinery, electronics, and steel, to software, finance, and infrastructure.

ANALYSES AND RESULTS

We used Partial Least Squares (PLS) to analyze our data. PLS is preferred as a tool for theory development because the tool is concerned with relationships between research constructs (Gefen & Straub, 2005). PLS adopts components-based algorithms, and it is able to handle formative, single-item, and summated scales. PLS does not place a very high requirement for sample size and normal distribution on the source data (Gefen & Straub, 2005).

We used PLS-Graph in this study to examine the measurement model and structural model. A bootstrapping estimation procedure was used to assess the significance of the factor loading in the measurement model, and the path coefficient in the structural model (Gefen & Straub, 2005). Although the measurement and structural parameters were estimated together, the results are interpreted separately in two stages.

Measurement Model

Confirmatory factor analysis (CFA) was conducted using PLS. Based on the results of CFA, we analyzed convergent validity, discriminant validity, and reliability of all the multiple-item scales, following the guidelines from previous literature (Fornell & Larcker, 1981; Gefen & Straub, 2005). The measurement properties are reported in Tables 2 and 3.

Reliability was assessed in terms of composite reliability, which measures the degree to which items are free from random error and therefore yield consistent results. Composite reliabilities in our measurement model ranged from 0.79 to 0.97 (see Table 2), above the cutoff of 0.70. Convergent validity was assessed in terms of factor loadings and average variance extracted (AVE). Convergent validity requires a factor loading greater than 0.70 and an AVE no less than 0.50 (Fornell & Larcker,

Table 1. Respondent profile

Characteristic	N	Percent (%)
Business Type		
Pure Trading	177	58%
Manufacturing	68	22%
Service	32	10%
Comprehensive	30	10%
Ownership Type		
State-owned	132	43%
Foreign-invested	72	24%
Privately-owned	68	22%
Collectively-owned	29	9%
Missing	6	2%
Age		
Less than 5 years	56	18%
Between 5 and 15 years	139	45%
More than 15 years	112	37%
Number of Employees		
<= 49	88	29%
Between 50 and 199	77	25%
Between 200 and 999	84	27%
>=1000	58	19%
Trade Products		
Machinery and Electronic	86	28%
Chemical, Oil, Petrochemical, Pharmacy, Coal, Mine, and Steel	61	20%
Light Industrial Product, Craftwork, and Construction Material	51	17%
Software and Information Technology	40	13%
Textile and Garment	31	10%
Food, Grain, and Stock	19	6%
Service, Finance, and Infra-structure	11	4%
other	8	3%

1981). As shown in Table 2, except for four items, all the other items had significant factor loadings higher than 0.70 (p<0.01). AVEs ranged from 0.55 to 0.90. These suggested adequate convergent validity. Discriminant validity was assessed by comparing AVE of each individual construct with shared variances between this individual construct and all the other constructs. Higher AVE of the individual construct than shared variances suggests discriminant validity (Fornell & Larcker, 1981). Table 3 shows the inter-construct correlations off the diagonal of the matrix. Comparing all the correlations and square roots of AVEs shown on the diagonal, the results indicated adequate discriminant validity.

As mentioned above, e-business assimilation is a second-order factor. Following several previous studies (Chatterjee et al., 2002; Chin & Gopal, 1995), we specified a "molecular" model, in which e-business assimilation was measured by the factor scores of the two first-order factors. The loadings of the two factors were 0.929 for strategy assimilation and 0.692 for activity assimilation, respectively (Figure 2).

To evaluate the potential common method bias, we conducted a Harmon one-factor test (Podsakoff & Organ, 1986) on all first-order constructs. E-business assimilation was not included because it is a second-order construct. The test showed that the most covariance explained by one factor was only 27.92%. We also followed the approach of Liang et al. (2007) to test common method bias in PLS. The results indicated that loadings from the common method to indicators were insignificant and indicators' substantive variances were much greater than their method variances. In addition, we also checked the correlation matrix (see Table 3), which did not have any correlations of 0.9 or higher (Ettlie & Pavlou, 2006; Pavlou, Liang & Xue, 2007). The potential of common method bias was therefore not a major concern in this study.

Structural Model: Hypotheses Testing

The results of hypotheses testing are shown in Table 4 and Figure 2. To show the effects of control variables, we ran three models separately. The first model included control variables only, the second

Table 2. The measurement model

Construct	Item	Loading	Std. Error	T-Statistic	Composite Reliability	AVE
1. Relative Advantage	RA1	0.87	0.02	38.50	0.90	0.61
	RA2	0.65	0.04	17.00		
	RA3	0.54	0.05	11.92		
	RA4	0.82	0.03	29.66		
	RA5	0.88	0.02	35.71		
	RA6	0.88	0.02	41.63		
2. IT Capability	ITC1	0.97	0.00	228.78	0.97	0.90
	ITC2	0.94	0.01	130.49		
	ITC3	0.95	0.01	142.65		
3. Ownership Type	OT	1.00	NA	NA	NA	NA
4. Learning Orientation	LO1	0.73	0.04	18.72	0.88	0.66
	LO2	0.82	0.03	29.63		
	LO3	0.82	0.02	34.08		
	LO4	0.86	0.01	65.52		
5. Environmental Uncertainty	EU1	0.87	0.02	48.29	0.84	0.63
	EU2	0.80	0.03	26.14		
	EU3	0.71	0.04	17.92		
6. Inter-organizational Dependence	ID1	0.85	0.04	18.93	0.85	0.74
	ID2	0.87	0.03	25.39		
7. Strategy Assimilation	SA1	0.76	0.03	26.71	0.84	0.56
	SA2	0.70	0.04	17.69		
	SA3	0.76	0.02	30.70		
	SA4	0.76	0.02	30.96		
8. Activity Assimilation	AA1	0.74	0.03	22.20	0.79	0.55
	AA2	0.70	0.05	15.14		
	AA3	0.79	0.03	24.87		
9. Firm Size	FS	1.00	NA	NA	NA	NA
10. Business Type	BT	1.00	NA	NA	NA	NA
11. Firm Age	FA	1.00	NA	NA	NA	NA
12. Web Experience	WE	1.00	NA	NA	NA	NA

Note: All t-statistics are significant at 0.001 level; AVE=Average Variance Extracted; NA: not applicable to single-item measures

model included independent variables, and the third model (full model) included both control variables and independent variables. Overall, the full model in Figure 2 demonstrated high prediction power ($R^2 = 52.3\%$). The path from relative advantage to e-business assimilation was significant (b =0.139, p<0.05), suggesting that relative advantage of e-business was a significant driver of e-business assimilation in Chinese international trade firms. Therefore, H_{1a} was supported. The significant path loading of a firm's IT capability (b = 0.250, p<0.001) suggested that IT capability had a significant impact on e-business assimilation, supporting H_{1b}. The findings about H_{1a} and

Table 3. Inter-construct correlations

	1	2	3	4	5	6	7	8	9	10	11	12
1. Relative Advantage	0.78											
2. IT Capability	0.17	0.95										
3. Ownership Type	0.01	0.04	NA									
4. Learning Orientation	0.12	0.19	0.08	0.81								
5. Environ. Uncertainty	-0.07	-0.23	-0.14	-0.26	0.79							
6. Inter-org. Dependence	0.12	0.15	0.05	0.33	-0.40	0.86						
7. Strategy Assimilation	0.22	0.32	0.41	0.43	-0.56	0.34	0.75					
8. Activity Assimilation	0.27	0.27	0.19	0.23	-0.20	0.22	0.37	0.74				
9. Firm Size	0.10	0.11	0.01	0.11	-0.03	0.04	0.11	0.21	NA			
10. Business type	0.08	0.16	0.21	-0.01	-0.10	0.01	0.20	0.12	0.13	NA		
11. Firm Age	0.08	0.20	0.07	0.10	-0.08	0.10	0.08	0.12	0.27	-0.01	NA	
12. Web Experience	0.15	0.24	0.02	0.11	0.16	0.09	0.24	0.19	0.27	0.16	0.17	NA

Note: Square root of Average Variance Extracted (AVE) is shown on the diagonal of the matrix;
Inter-construct correlation is shown off the diagonal; NA: not applicable to single-item measures.

Figure 2. Research model with parameter estimates

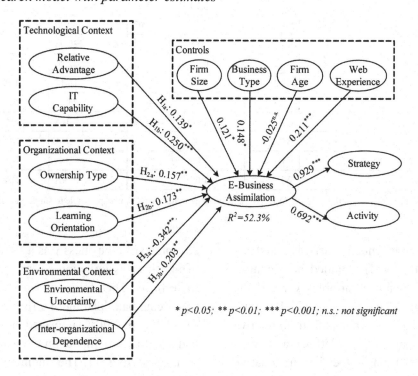

* $p<0.05$; ** $p<0.01$; *** $p<0.001$; n.s.: not significant

Table 4. Results of hypotheses testing

Variables	Model with Controls	Model without Controls	Model with All
Relative Advantage		0.141*	0.139*
IT Capability		0.258***	0.250***
Ownership Type		0.159**	0.157**
Learning Orientation		0.179**	0.173**
Environmental Uncertainty		-0.369***	-0.342***
Inter-org. Dependence		0.207**	0.203**
Firm Size (log)	0.124*		0.121*
Business Type	0.211***		0.148*
Firm Age(log)	-0.079[n.s.]		-0.025[n.s.]
Web Experience	0.277***		0.211***
R^2 *Explained*	*0.199*	*0.428*	*0.523*

*p<0.05; **p<0.01; ***p<0.001; n.s.: non-significant

H_{1b} indicated that technological factors were important for e-business assimilation.

Organizational factors investigated in our study were also important drivers of e-business assimilation. Both ownership type (b = 0.157, p<0.01) and learning orientation (b = 0.173, p<0.01) had significant influences on e-business assimilation, supporting H_{2a} and H_{2b}. The significant effect of ownership (H_{2a}) indicated that firms with different types of ownership had different levels of e-business assimilation. To further reveal how ownership type affected e-business assimilation, we investigated the differences of the two dimensions of e-business assimilation among firms of different types of ownership (Table 5). We found that state-owned companies had a lower level of strategy assimilation (mean=3.286) than those of privately-owned companies (mean=3.630, p<0.01) and foreign-invested companies (mean=3.669, p<0.01). Foreign-invested companies also had higher strategy assimilation than collectively-owned companies (mean=3.347, p<0.05). Furthermore, regarding activity assimilation, state-owned companies' level of activity assimilation (mean=1.811) was lower than those of privately-owned compa-

Table 5. Comparison of assimilation among firms of different ownership types

	(1) State-owned N = 132	(2) Privately-owned N = 68	(3) Collectively-owned N = 29	(4) Foreign-invested N = 72
Strategy	3.286	3.630	3.347	3.669
Activity	1.811	2.173	2.194	2.709

nies (mean=2.173, p<0.001), collectively-owned companies (mean=2.194, p<0.001), and foreign-invested companies (mean=2.709, p<0.001). Foreign-invested companies also had higher activity assimilation than those of privately-owned companies (p<0.001) and collectively-owned companies (p<0.001).

The path from environmental uncertainty to e-business assimilation was significant (b=-0.342, p<0.001), supporting H_{3a}. The negative coefficient indicated environmental uncertainty was a constraint factor for e-business assimilation. The path coefficient was the highest among all the paths investigated in this study, suggesting that environmental uncertainty was the most important determinant of e-business assimilation. Furthermore, the effect of inter-organizational dependence was also significant (b = 0.203, p<0.01), supporting H_{3b}.

Regarding control variables, firm age had no significant effect on e-business assimilation (b = -0.025; p>0.05). This suggests that international trade firms in China, whether old or young, were not different from one another in terms of e-business assimilation.

However, firm size, business type, and web experience were found significant (b = 0.121, 0.148 and 0.211; p<0.05, 0.05 and 0.001, respectively), suggesting that e-business assimilation was different among firms of different size and business type, and if they had web-site presence. Regarding business type, we conducted further analyses to reveal the differences. As shown in Table 6, comprehensive companies had a higher level of strategy assimilation (mean=3.742) than pure trading companies (mean=3.336, p<0.001), manufacturing companies (mean=3.493, p<0.05), and service companies (mean=3.452, p<0.01). The level of strategy assimilation of manufacturing companies was also higher than that of pure trading companies (p<0.05). Furthermore, pure trading companies' level of activity assimilation (mean=1.882) was lower than those of manufacturing companies (mean=2.375, p<0.05) and comprehensive companies (mean=2.674, p<0.01).

DISCUSSION

Our study has investigated technological, organizational, and environmental factors that account for the assimilation of e-business in Chinese international trade companies. The findings suggest that while these companies are enticed by the relative advantage of e-business and empowered by firms' organizational business capabilities (such as IT and learning), the assimilation of e-business is also constrained by the uncertainty of the e-business environment.

Environmental Factors

Contrary to previous findings based on the market economy (e.g., Grover & Goslar, 1993), we found that environmental uncertainty had a negative rather than positive impact on e-business assimilation in China's international trade industry. This finding is consistent with similar findings in research conducted in China, which suggests Chinese firms are risk-averse in the transitional process from planned economy to market economy (e.g., Tan & Litschert, 1994). The finding is also congruent with other e-commerce studies conducted in developing countries (e.g., Molla & Licker, 2005b), which find that environmental readiness or maturity is positively associated with e-business strategy and the adoption of e-business. Our finding also reveals that, unlike first movers and early adopters in the market economy, Chinese firms seem to be free-riders in the e-business diffusion process. That is, they take advantage of the established infrastructure in the industry and exploit the benefits generated from established e-business practices. If the infrastructure is uncertain and the level of maturity is low, Chinese international-trade firms are less likely to assimilate e-business, and more likely to follow a 'wait-and-see' strategy (Chau & Tam, 1997).

We also found that inter-organizational dependence had a significant impact on e-business assimilation. According to resource dependency theory (Pfeffer & Salancik, 1978), a firm is dependent on other organizations for necessary resources so that other organizations may exert essential external influences on the actions of the focal firm. Our finding suggests that inter-organizational connections and the dependency between suppliers, retailers, partners, and other supply chain members play very important roles in e-business assimilation. This is consistent with previous IT/IS implementation studies (e.g., Grover & Goslar, 1993; Iskander, Kurokawa & LeBlanc, 2001). A focal firm, together with its business stakeholders, seems to act collectively

Table 6. Comparison of assimilation among firms of different business types

	(1) Pure Trading N = 177	(2) Manufacturing N = 68	(3) Service N = 32	(4) Comprehensive N = 30
Strategy	3.336	3.493	3.452	3.742
Activity	1.882	2.375	2.192	2.674

in the assimilation of e-business. Because of the imbalance of dependence and power among these Chinese international trade firms (Hart & Saunders, 1998), more powerful companies may exert influences on other firms in the supply chain.

The highest coefficient of the path from environmental uncertainty to e-business assimilation suggests that environmental uncertainty was the most important constraining factor among the research variables we have included in the TOE framework. Furthermore, the relatively strong effects of environmental uncertainty and inter-organizational dependence suggest that environmental factors were more salient than other factors in the Chinese international trade firm's decision to assimilate e-business. The unpredictability and dynamics of the transitional process to the market economy, as well as the availability of formal and rule-based inter-organizational relationships, are major drivers of e-business assimilation.

Organizational Factors

We found that firms with different types of ownership had different levels of e-business assimilation. This is consistent with previous studies which find ownership type is a significant indicator of Chinese firms' strategy and behavior (e.g., Peng et al., 2004). Our study also confirms that firm ownership type is a clear measure of Chinese firms' IT innovation behavior. Examining the differences of e-business assimilation among firms of different ownership types, we found that state-owned companies were lagging behind other firms. It is noteworthy that firms with private ownership (the emerging ownership category

in the Chinese economy) stress the e-business assimilation strategy, as foreign-invested firms do, although the level of e-business activities is behind that of foreign-invested firms. Overall, foreign-invested companies were found to lead the assimilation of e-business among Chinese international trade firms.

The significant role of learning-orientation in e-business assimilation suggests that learning-oriented firms were likely to consider the implementation of e-business technologies as an approach to improve the learning ability of the firms. Chinese international trade firms are pioneers of adopting modern business concepts, strategies, and cultures which emphasize the role of learning in a firm's survival and sustained growth. Originating in the market economy, the practice of e-business has the potential to help a firm achieve a competitive position in the market (Zhu et al., 2004). However, the advantage of e-business could be contingent upon the firm's learning ability. If a firm is less learning-oriented, e-business may not be appealing to the firm. Learning orientation seems to be important for Chinese international trade firms. In order to cooperate and compete with firms in the global network, Chinese firms need to learn many lessons. This is impossible if Chinese firms do not cultivate learning ability and/or a learning culture.

Although we modeled firm size and business type as control variables, they can also be considered as organizational variables. The significant effect of business type on e-business assimilation also deserves discussion. To the best of our knowledge, there are no similar empirical studies done in the past; our findings are among the first. The Chinese government has enacted various policies to promote the development of certain industries while constraining others. For example, some industries are strictly controlled by the government (e.g., energy and transportation) while others (e.g., electronic and information technology) have been pushed to compete with foreign companies. Our findings reflect this. International trade firms, which are from different primary industries, have different levels of assimilating e-business.

Technological Factors

We found that a firm's IT capability was also a significant factor for e-business assimilation, which is consistent with previous studies (e.g., Chau & Tam, 1997; Cui et al., 2008; Grover & Goslar, 1993). This indicates that a firm's e-assimilation is affected by the firm's available IT capability and resources. If a firm does not possess sophisticated IT capability in terms of hardware, software, infrastructure, and other intangible capabilities such as employee skills, the firm is less likely to assimilate e-business. It is well-known that Chinese firms lack many modern resources for development and growth (Peng & Heath, 1996), such as accounting, finance, human resources, and logistics. The shortage of IT resources inside a firm seems also to limit the firm's innovation, in terms of the assimilation of e-business. Considering a firm's internal IT capability together with resources in the environment, such as Internet infrastructure and resources from other inter-organizations, we conclude that resource-related factors are critical in Chinese firms' e-business assimilation.

The relative advantage of e-business was also a significant determinant of e-business assimilation, which is consistent with previous literature (e.g., Chau & Tam, 1997). However, given the weakness of this effect, relative advantage was one of the least important factors. The assimilation of e-business in Chinese international trade firms was not as strongly driven by the purely technological feature of e-business as it was by other organizational and environmental factors.

Instead, firms are rational in their decision-making when incorporating Internet technologies into business operations. When e-business is tied to the firm's strategy and business activities to a much higher degree, the effect of technological features may decrease.

CONCLUSION

Limitations

Before we discuss the implications for research and practice, several limitations in the study are noteworthy, and our findings must be interpreted in light of these limitations. First, our data were collected from a specific area, Beijing. As a political and cultural center, Beijing is also one of China's most developed cities and regions; here, the Chinese central government exercises different policies than in other areas. Thus, generalization of the findings of the present study to other areas should be made with caution. Second, we relied on the response from a single informant of an international trade firm to measure our research constructs. We did not use any external and objective data sources, such as those from the national survey or statistical reports from the Chinese government. All the potential problems associated with single source data collection might be applied to this study. In the study, we examined common method bias and did not find the presence of this bias. Third, we adapted some measures, such as activities and relative advantage, using binary scales rather than 5-point or 7-point Likert scales. Thus, the depth of this questionnaire was not as detailed as those in previous studies. Some of the research constructs might not have been fully revealed by binary scales. However, the measurement model and the validation procedure reported in this study have shown that the quality of these measures was acceptable. Fourth, we did not include other social, cultural, and economical factors in the organizational and environmental contexts which may be relevant to Chinese firms. These factors may include a firm's strategic orientation toward technology and those toward the fundamental concepts of a market economy such as customer orientation and competitor orientation. Some technological factors which are not examined in this study, such as compatibility, complexity, cost, and image of e-business, may also be interesting to investigate in future studies.

Implications for Research

In this paper, we have examined the antecedent factors of e-business assimilation in Chinese international trade firms using the TOE framework. Our study represents one of the first studies of e-business assimilation in China. To date, the literature on IT/IS diffusion and assimilation in developing countries is relatively sparse, especially for China. As Walsham, Robey, and Sahay (2007) state in a special issue of IS in developing countries published by *MIS Quarterly*, IS in China is poorly represented in the literature. Our study responds to this call for research.

We have shown that the Technology-Organization-Environment framework (Tornatzky & Fleischer, 1990) is a suitable theoretical anchor for examining IT/IS assimilation in China. The framework has the advantage of integrating both internal and external factors that explain a firm's technology innovation behavior. Other theoretical perspectives (as we have pointed out earlier), such as institutional theory, knowledge-based theory, and resource-based theory of the firm, may also be feasible for studying IT/IS assimilation and diffusion in China. However, as our findings have indicated, these theories about IT/IS diffusion and assimilation, which have mainly developed from Western economies, may have different implications and different findings in developing countries. IT/IS innovations are "usually deeply intertwined with issues of power, politics, donor dependencies, institutional arrangement, and inequities of all sorts" (Walsham et al., 2007, pp.

324). The unique economic, social, and cultural challenges inherent in developing economies require attention from the research community in the study of IT/IS diffusion and assimilation. We concur with Fichman (2000) and suggest that some 'middle' theories tailored to the uniqueness of the Chinese economy and culture need be developed in future studies. Future studies based on these above-mentioned theoretical perspectives are suggested to incorporate these unique characteristics in the original theoretical frameworks, in order to better understand IT/IS diffusion and assimilation in developing countries.

A key finding of this study is that environmental uncertainty was an inhibitor, rather than an enabler, of Chinese international trade firms' e-business assimilation. This finding argues against the conventional wisdom rooted in Western economies, which often considers environmental uncertainty as an enabler of innovation behaviors. Future studies can continue this line of inquiry into the role of environmental uncertainty—an essential characteristic of China's economic transition today—before China is fully integrated into the global market economy.

This study also contributes to the IS literature by examining the role of ownership type in e-business assimilation. Although previous literature has found clear differences between firms of different ownership types (Mascarenhas, 1989; Peng & Heath, 1996), the effect of ownership type on IT/IS diffusion and assimilation has not been adequately examined in previous studies, especially in developing countries. While the dichotomy of public versus private ownership is dominant in the developed countries, developing countries have a diverse range of ownership types (Peng, 2003; Peng & Heath, 1996; Peng et al., 2004; Tan, 2002). Although the diversity of ownership types in developing countries may complicate this research, future studies of IT/IS diffusion and assimilation in developing countries should not ignore the effect of ownership type.

Although we found that ownership type had significant impact on e-business assimilation, the inherent mechanism is not clear. For example, it is not clear whether the strength of impacts of various antecedents on e-business assimilation is different among firms of different ownership types. That is, how ownership type moderates the effects of other antecedent factors on e-business assimilation deserves further investigation. Future studies can also integrate ownership type into frameworks and theories developed in previous IT/IS implementation literature to examine the mechanism of the effect of ownership type on IT/IS diffusion and assimilation.

In addition to the contextual factors examined in this study, other factors such as social networks are worthy of investigation. For example, "*guanxi*" is a special social factor in China. Future studies can investigate the effects of formal and informal business relationships (or *guan xi*) on e-business assimilation to answer questions such as how interpersonal *guan xi* between the managers and government agents can influence e-business assimilation.

Finally, from a methodology perspective, a longitudinal study can be conducted to compare the effects of different factors in the e-business assimilation process. Such a study would be especially useful to investigate how firms of different ownership (state-owned firms in particular) make changes during the transitional process in their implementation of IT/IS.

Implications for Practice

Our findings have some implications for the Chinese government, technology providers, and Chinese firms which plan to deploy e-business. First, the role of environmental uncertainty in e-business assimilation calls for the Chinese government to develop appropriate technological plans and policies to facilitate the development of a mature e-business environment and Internet infrastructure for Chinese firms. The government can rely on central coordination and can control the advantages of various resources to carry out such plans. With a high level of control over state-owned companies, governmental agencies need to push state-owned companies, which we found were lagging behind other firms in their e-business assimilation. Specific incentives and training need to be implemented for state-owned companies.

Dynamic changes in the environment also provide opportunities for providers to invest in technology and service areas such as third-party infrastructure providers, web hosting companies, online payment service providers, and third-party logistics. The establishment of these infrastructures may ease Chinese firms' concern about the uncertainties in the environment. These technology and service providers should be aware that Chinese firms which have better IT capabilities are more likely to assimilate e-business. These firms may be the initial targets of technology providers in order to develop customer relationships. Furthermore, firms of different ownership types have different levels of e-business assimilation. This is a reflection of different innovation behaviors among different Chinese firms. Technology providers may benefit more from developing business relationships with privately-owned companies first, which seem to be more innovative and thus more likely to assimilate e-business.

Our results indicated that firms with high learning-orientation are more likely to assimilate e-business. This finding suggests that Chinese firms need to develop the corporate culture which encourages organizational learning in order to facilitate e-business assimilation (Sinkula, 1994). These values and beliefs about organizational learning will act as clear signals to organizational members that the top management encourages them to learn e-business practices. The learning process can help organizational members better understand e-business technologies and business processes, which in turn may help e-business assimilation.

Our study has shown that international trade firms without high-level IT capabilities are less likely to assimilation e-business. In order to build superior IT capabilities, Chinese firms can seek to utilize the external IT capabilities of their inter-organizational partners, such as supply chain partners, market forces in their business operations, and the external IT infrastructure of the industry. These external IT resources can complement the lack of internal IT resources.

ACKNOWLEDGMENT

The research is supported by the 211 Project at University of International Business & Economics, Beijing, China.

REFERENCES

Armstrong, C. P., & Sambamurthy, V. (1999). Information technology assimilation in firms: The influence of senior leadership and IT infrastructures. *Information Systems Research, 10*(4), 304–327. doi:10.1287/isre.10.4.304

Bharadwaj, A. S. (2000). A resource-based perspective on information technology capability and firm performance: An empirical investigation. *MIS Quarterly, 24*(1), 169–197. doi:10.2307/3250983

Bolloju, N., & Turban, E. (2007). Organizational assimilation of web services technology: A research framework. *Journal of Organizational Computing and Electronic Commerce, 17*(1), 29–52.

Calantone, R. J., Cavusgil, S. T., & Zhao, Y. (2002). Learning orientation, firm innovation capability, and firm performance. *Industrial Marketing Management, 31*(6), 515–524. doi:10.1016/S0019-8501(01)00203-6

Chatterjee, D., Grewal, R., & Sambamurthy, V. (2002). Shaping up for e-commerce: Institutional enablers of the organizational assimilation of web technologies. *MIS Quarterly, 26*(2), 65–89. doi:10.2307/4132321

Chau, P. Y. K., & Tam, K. Y. (1997). Factors affecting the adoption of open systems: An exploratory study. *MIS Quarterly, 21*(1), 1–24. doi:10.2307/249740

Chin, W. W., & Gopal, A. (1995). Adoption intention in GSS: Relative importance of beliefs. *The Data Base for Advances in Information Systems, 26*(2&3), 42–64.

Cooper, R. B., & Zmud, R. W. (1990). Information technology implementation research: A technological diffusion approach. *Management Science, 36*(2), 123–139. doi:10.1287/mnsc.36.2.123

Cui, L., Zhang, C., Zhang, C., & Huang, L. (2008). Exploring IT adoption process in Shanghai firms: An empirical study. *Journal of Global Information Management, 16*(2), 1–17.

Damanpour, F. (1992). Organization size and innovation. *Organization Studies, 13*(3), 375–402. doi:10.1177/017084069201300304

Dasgupta, S., Agarwal, D., Ioannidis, A., & Gopalakrishnan, S. (1999). Determinants of information technology adoption: An extension of existing models to firms in a developing country. *Journal of Global Information Management, 7*(3), 30–40.

Ettlie, J. E., & Pavlou, P. A. (2006). Technology-based new product development partnerships. *Decision Sciences, 37*(2), 117–147. doi:10.1111/j.1540-5915.2006.00119.x

Fichman, R. G. (2000). The diffusion and assimilation of information technology innovations. In R. W. Zmud (Ed.), *Framing the domains of IT management: Projecting the future through the past* (pp.105-127). Cincinnati, OH: Pinnaflex Publishing.

Fornell, C., & Larcker, D. F. (1981). Evaluating structural equations with unobservable variables and measurement error. *JMR, Journal of Marketing Research, 18*(1), 39–50. doi:10.2307/3151312

Garvin, D. A. (1993). Building a learning organization. *Harvard Business Review, 71*(4), 78–91.

Gefen, D., & Straub, D. (2005). A practical guide to factorial validity using PLS-Graph: Tutorial and annotated example. *Communications of the Association for Information Systems, 16*, 91–109.

Grover, V., & Goslar, M. D. (1993). The initiation, adoption, and implementation of telecommunications technologies in U.S. organizations. *Journal of Management Information Systems, 10*(1), 141–163.

Guo, X., & Chen, G. (2005). Internet diffusion in Chinese companies. *Communications of the ACM, 48*(4), 54–58. doi:10.1145/1053291.1053318

Hart, P., & Saunders, C. S. (1998). Emerging electronic partnerships: Antecedents and dimensions of EDI use from the supplier's perspective. *Journal of Management Information Systems, 14*(4), 87–111.

Hult, G. T. M., Ketchen, D. J. Jr, & Nichols, E. L. Jr. (2003). Organizational learning as a strategic resource in supply management. *Journal of Operations Management, 21*(5), 541–556. doi:10.1016/j.jom.2003.02.001

Iskander, B. Y., Kurokawa, S., & LeBlanc, L. J. (2001). Adoption of EDI: The role of buyer-supplier relationships. *IEEE Transactions on Engineering Management, 48*(4), 505–517. doi:10.1109/17.969427

Kirkman, G. S. (2002). Executive summary. In G.S. Kirkman, P.K. Cornelius, J.D. Sachs, & K. Schwab (Eds.), *The global information technology report 2001-2002: Readiness for the networked world* (pp. xiii-xvi). Retrieved January 28, 2008, from http:// www.cid.harvard. edu/archive/ cr/ pdf/ gitrr2002_execsumm.pdf

Kuan, K. K. Y., & Chau, P. Y. K. (2001). A perception-based model for EDI adoption in small business using a technology-organization-environment framework. *Information & Management, 38*(8), 507–521. doi:10.1016/S0378-7206(01)00073-8

Kwon, T. H., & Zmud, R. W. (1997). Unifying the fragmented models information systems implementation. In R.J. Boland & R.A. Hirschheim (Eds.), *Critical issues in information systems research* (pp. 227-251). Chichester, England: John Wiley and Sons Ltd.

Liang, H., Saraf, N., Hu, Q., & Xue, Y. (2007). Assimilation of enterprise systems: The effect of institutional pressures and the mediating role of top management. *MIS Quarterly, 31*(1), 59–87.

MacKay, N., Oarent, M., & Gemino, A. (2004). A model of electronic commerce adoption by small voluntary organizations. *European Journal of Information Systems, 13*(2), 147–159. doi:10.1057/palgrave.ejis.3000491

Martinsons, M. G. (2004). ERP in China: One package, two profiles. *Communications of the ACM, 47*(7), 65–68. doi:10.1145/1005817.1005823

Mascarenhas, B. (1989). Domains of state-owned, privately held, and publicly traded companies in international competition. *Administrative Science Quarterly, 34*(4), 582–597. doi:10.2307/2393568

McFarlan, F.W, Chen, G., & Lane, D. (2005). Information technology at COSCO. *Harvard Business Review Case*, Product No.: 9-305-080.

McFarlan, F. W., & Rockart, J. (2004). China and information technology: An interview with Warren McFarlan from the Harvard Business School. *MIS Quarterly Executive, 3*(2), 83–88.

Meyer, A. D., & Goes, J. B. (1988). Organizational assimilation of innovations: A multilevel contextual analysis. *Academy of Management Journal, 31*(4), 897–923. doi:10.2307/256344

Ministry of Commerce of the People's Republic of China (MoC). (2007). *Statistics Data of Imports and Exports*. Retrieved Nov. 11, 2007, from http://zhs.mofcom.gov.cn/tongji.shtml

Molla, A., & Licker, P. S. (2005a). E-commerce adoption in development countries: A model and instrument. *Information & Management, 42*(6), 877–899. doi:10.1016/j.im.2004.09.002

Molla, A., & Licker, P. S. (2005b). Perceived e-readiness factors in e-commerce adoption: An empirical investigation in a developing country. *International Journal of Electronic Commerce, 10*(1), 83–110.

Pavlou, P. A., Liang, H., & Xue, Y. (2007). Understanding and mitigating uncertainty in online exchange relationships: A principal-agent perspective. *MIS Quarterly, 31*(1), 105–136.

Peng, M. W. (2003). Institutional transitions and strategic choices. *Academy of Management Review, 28*(2), 275–286.

Peng, M. W., & Heath, P. S. (1996). The growth of the firm in planned economies in transition: Institutions, organizations, and strategic choice. *Academy of Management Review, 21*(2), 492–528. doi:10.2307/258670

Peng, M. W., Tan, J., & Tong, T. (2004). Ownership types and strategic groups in an emerging economy. *Journal of Management Studies, 41*(7), 1105–1109. doi:10.1111/j.1467-6486.2004.00468.x

Pfeffer, J., & Leblebici, H. (1977). Information technology and organizational structure. *Pacific Sociological Review, 20*(2), 241–261.

Pfeffer, J., & Salancick, G. R. (1978). *The external control of organization: A resource dependency perspective*, New York, NY: Harper and Row Press.

Podsakoff, P. M., & Organ, D. W. (1986). Self-reports in organizational research: Problems and prospects. *Journal of Management, 12*(4), 531–544. doi:10.1177/014920638601200408

Purvis, R. L., Sambamurthy, V., & Zmud, R. W. (2001). The assimilation of knowledge platforms in organizations: An empirical investigation. *Organization Science, 12*(2), 117–135. doi:10.1287/orsc.12.2.117.10115

Ranganathan, C., Dhaliwal, J. S., & Teo, T. S. H. (2004). Assimilation and diffusion of web technologies in supply-chain management: An examination of key drivers and performance impacts. *International Journal of Electronic Commerce, 9*(1), 127–161.

Raymond, L., Bergeron, F., & Blili, S. (2005). The assimilation of e-business in manufacturing SMEs: Determinants and effects on growth and internationalization. *Electronic Markets, 15*(2), 106–118. doi:10.1080/10196780500083761

Rogers, E. M. (1995). *Diffusion of innovations*, New York, NY: Free Press.

Shenkar, O., & von Glinow, M. A. (1994). Paradoxes of organizational theory and research: Using the case of China to illustrate national contingency. *Management Science, 40*(1), 56–71. doi:10.1287/mnsc.40.1.56

Sinkula, J. M. (1994). Market information processing and organizational learning. *Journal of Marketing, 58*(1), 35–45. doi:10.2307/1252249

Sinkula, J. M., Baker, W. E., & Noordewier, T. (1997). A framework for market-based organizational learning: Linking values, knowledge, and behavior. *Journal of the Academy of Marketing Science, 25*(4), 305–318. doi:10.1177/0092070397254003

Tan, A., & Ouyang, W. (2002). *Global and national factors affecting e-commerce diffusion in China.* University of California, Irvine. Retrieved Nov. 11, 2007, from www.crito.uci.edu/publications/pdf/GEC2_China.pdf

Tan, J. (2002). Impact of ownership type on environment, strategy, and performance: Evidence from China. *Journal of Management Studies, 39*(3), 333–354. doi:10.1111/1467-6486.00295

Tan, J., & Litschert, R. J. (1994). Environment-strategy relationship and its performance implications: An empirical study of Chinese electronics industry. *Strategic Management Journal, 15*(1), 1–20. doi:10.1002/smj.4250150102

Tornatzky, L. G., & Fleischer, M. (1990). *The processes of technological innovation.* Lexington, MA: Lexington Books.

Tornatzky, L. G., & Klein, R. J. (1982). Innovation characteristics and innovation adoption-implementation: A meta-analysis of findings. *IEEE Transactions on Engineering Management, 29*, 28–45.

US Commercial Service. (2005). *Doing business in China: A country commercial guide for U.S. companies.* Retrieved Nov. 11, 2007, from www.buyusainfo.net/docs/x_7439525.pdf

US Department of State. (2007). *Background note: China.* Retrieved Nov. 11, 2007, from www.state.gov/r/pa/ei/bgn/18902.htm

Walsham, G., Robey, D., & Sahay, S. (2007). Foreword: Special issue on information systems in developing countries. *MIS Quarterly, 31*(2), 317–326.

Wang, S., & Cheung, W. (2004). E-business adoption by travel agencies: Prime candidates for mobile e-business. *International Journal of Electronic Commerce, 8*(3), 43–63.

World Trade Organization (WTO). (2007). *Risks lie ahead following stronger trade in 2006.* Retrieved Nov. 12, 2007, from http://www.wto.org/english/news_e/pres07_e/pr472_e.htm

Xu, S., Zhu, K., & Gibbs, J. (2004). Global technology, local adoption: A cross-country investigation of internet adoption by companies in the Unites States and China. *Electronic Markets, 14*(1), 13–24. doi:10.1080/10196780420000175261

Zhu, K., & Kraemer, K. L. (2005). Post-adoption variations in usage and value of e-business by organizations: Cross-country evidence from the retail industry. *Information Systems Research, 16*(1), 61–84. doi:10.1287/isre.1050.0045

Zhu, K., Kraemer, K. L., Gurbaxani, V., & Xu, S. (2006). Migration to open-standard interorganizational systems: Network effects, switching costs, and path dependency. *MIS Quarterly, 30*, 515–539.

Zhu, K., Kraemer, K. L., & Xu, S. (2003). E-business adoption by European firms: A cross-country assessment of the facilitators and inhibitors. *European Journal of Information Systems, 12*(4), 251–268. doi:10.1057/palgrave.ejis.3000475

Zhu, K., Kraemer, K. L., Xu, S., & Dedrick, J. (2004). Information technology payoff in e-business environments: An international perspective on value creation of e-business in the financial services industry. *Journal of Management Information Systems, 21*(1), 17–54.

APPENDIX: THE SURVEY INSTRUMENT

E-Business Assimilation

Strategy Assimilation: Please indicate the degree to which the following statements are related to the business strategies of your firm. Scale: 1-5. 1 very little; 2 little; 3 neutral; 4 much; 5 very much.

(SA1) Internet coverage is very important to your business, i.e., the number of global Internet users is very important for your firm to develop trade business.

(SA2) The application of Internet technologies makes it easier and quicker for your firm to get information about trade businesses, which helps to solve the problems of information asymmetry and information uncertainty.

(SA3) The application of Internet technologies in your firm enables information sharing between your firm and existing and potential customers.

(SA4) The application of Internet technologies in your firm enables interactions with your customers (i.e., Internet-based two-way and real-time communication with your existing and potential customers save costs).

Activity Assimilation: In what business activities does your firm use Internet technologies? Please check all of the following that apply. Scale: binary. 1 yes; 0 no.

(AA1) Information sharing within the firm

(AA1) Information sharing with customers and partners

(AA2) Online transactions

(AA2) Online marketing and advertising

(AA3) Enterprise resources planning

(AA3) Customer relationship management

Technological Context

Relative Advantage: What advantage do you think your company can achieve through assimilating Internet technologies. Please check all of the following that apply. Scale: binary. 1 yes; 0 no.

(RA1) Bring more business opportunities

(RA2) Improve efficiency

(RA3) Save cost

(RA4) Improve customer relationship

(RA5) Improve management (coordination, sharing)

(RA6) Meet regulatory requirements imposed by the government.

IT Capability

(ITC1) Did your company have a web site? 0: No 1: Yes

(ITC2) How did your company host a web site?

1: Rent ISP virtual host 2: Rent ISP dedicated host 3: In-house server

(ITC3) Who was in charge of your company's web site?

1: No professional 2: In-house part-time professional 3: Outsourcing to the third-party 4: In-house full-time professional

Organizational Context

Ownership Type: (OT) What is your firm's ownership type?
1: state-owned 2: collectively-owned 3: privately-owned 4: foreign-invested
*Learning Orientation:*Please indicate the degree to which you agree the following statements. Scale: 1-5. 1 very little; 2 little; 3 neutral; 4 much; 5 very much.
(LO1) Managers basically agree that our organization's ability to learn is the key to our competitive advantage.
(LO2) The basic values of this organization include learning as key to improvement.
(LO3) The sense around here is that employee learning is an investment, not an expense.
(LO4) Learning in my organization is seen as a key commodity necessary to guarantee organizational survival.

Environmental Context

Environmental Uncertainty: Please indicate the degree to which you agree the following statements. Scale: 1-5. 1 very little; 2 little; 3 neutral; 4 much; 5 very much.
(EU1) In our industry, the Internet-based virtual market is mature, and most businesses are conducted to the Internet (reversed).
(EU 2) In our industry, Internet technologies provide support for various services, such as payment, logistics, and credit reporting, and the technology platform is very mature (reversed).
(EU 3) The whole industry is very interested in the application of Internet technologies (reversed).
Inter-organizational Dependence: Please indicate the degree to which you agree the following statements. Scale: 1-5. 1 very little 2 little 3 neutral 4 much 5 very much
(ID1) Our foreign customers, suppliers, and trade partners use Internet technologies.
(ID2) Our domestic customers, suppliers, and trade partners use Internet technologies.

Control Variables

Firm Size: (FS) How many employees does your company have?
Business Type: (BT) What is your firm's business type?
1: pure trading 2: manufacturing 3: service 4: comprehensive (all business)
Firm Age: (FA) How many years has your company been operating in China?
Web Experience: (WE) Please indicate whether your company has a website. Yes No

This work was previously published in International Journal of Global Information Management, Volume 18, Issue 1, edited by Felix B. Tan, pp. 39-65, copyright 2010 by IGI Publishing (an imprint of IGI Global).

Chapter 8
An International Comparative Study of the Roles, Rules, Norms, and Values that Predict Email Use

Mark F. Peterson
Florida Atlantic University, USA

Robert Konopaske
Texas State University, USA

Stephanie J. Thomason
University of Tampa, USA

Tomasz Lenartowicz
Florida Atlantic University, USA

Norm Althouse
University of Calgary, Canada

Mark Meckler
University of Portland, USA

Nicholas Athanassiou
Northeastern University, USA

Mark E. Mendenhall
University of Tennessee-Chattanooga, USA

Gudrun Curri
Dalhousie University, Canada

Andrew A. Mogaji
Benue State University, Nigeria

Julie I. A. Rowney
University of Calgary, Canada

ABSTRACT

This chapter extends communication and technology use theories about factors that predict e-mail use by explaining the reasons for cultural contingencies in the effects of managers' personal values and the social structures (roles, rules and norms) that are most used in their work context. Results from a survey of 576 managers from Canada, the English-speaking Caribbean, Nigeria, and the United States indicate that e-mail use may support participative and lateral decision making, as it is positively associated with work contexts that show high reliance on staff specialists especially in the U.S., subordinates, and unwritten rules especially in Nigeria and Canada. The personal value of self-direction is positively related to e-mail use in Canada, while security is negatively related to e-mail use in the United States. The results have implications for further development of TAM and media characteristic theories as well as for training about media use in different cultural contexts.

DOI: 10.4018/978-1-61350-480-2.ch008

INTRODUCTION

Managers everywhere have come to consider e-mail to be as ordinary a communication medium for discussing work events as are the telephone, fax, written documents, and face-to-face conversations. Despite its common use, the distinctive informal, explicitly written, easily transmitted qualities of e-mail make it a more comfortable medium for managers with some cultural backgrounds, values, and work contexts to use than for others (Carlson & Zmud, 1999). Managers of organizations that have divisions in multiple nations need to be trained to become aware of the typical use of e-mail in different parts of their organization to effectively manage cross-border interactions (MacDuffie, 2008) and disseminate company information to large audiences in nations other than their own. More generally, managers in all organizations need to be trained about possible differences between their own preference about e-mail use and the preference of their communication partners. Differences in e-mail use preferences, whether rooted in cultural differences, personal differences, or differences in organizational practices, have the potential to create misunderstandings of messages and interpersonal stresses. Organizational training to keep managers aware of such differences has the potential to promote cross-cultural communication in buyer-seller relationships, inter-organizational collaborative arrangements, and the many interpersonal contacts that occur in multinational organizations. A major contribution of projects like the present one is to provide information that can be used to adapt such training to particular cultural settings. In particular, we consider which personal and organizational characteristics associated with e-mail use preferences are unique to particular nations and which ones are relatively universal among English-speaking nations that differ substantially in their cultural and socio-economic situation.

The present study contributes to the international communication and technology literatures about factors predicting the tendency of managers to use e-mail by drawing from cross-cultural and psychology research about personal values and social structures. One of these cross-cultural literatures deals with cultural characteristics of the nation in which one lives (e.g., Hofstede, 2001). These are characteristics that one may personally value or not value, but that a nation's members inevitably know well, through long experience have come to find normal, and are more likely to prefer than are members of other nations (Peterson & Wood, 2008). A second is the literature about personal values (Rokeach, 1968; Schwartz, 1994). While knowing a nation very well for all of one's life and regularly facing influence to conform to its norms can affect one's personal values, expressed values can still show considerable within-nation variability (Au, 1999). The third is the social structure literature rooted in role theory about the links of individual managers to members of their role set and to impersonal social forces like organizational and societal norms (Smith, Peterson & Schwartz, 2002; Peterson & Smith, 2008). This line of research draws upon an earlier literature (Tannenbaum et al., 1974; Weber, 1947; Kluckhohn & Strodtbeck, 1961) suggesting that societies differ in typical role relationships (for a broader discussion of these three types of cross-cultural research, refer to Peterson & Soendergaard, in press). Specifically, the social structures we consider include the self, role senders such as superiors, subordinates, coworkers, colleagues, specialists, friends, family, formal organizational rules and procedures, and norms including both unwritten organizational rules, and beliefs that are widespread in one's nation (Peterson & Smith, 2000). Interest in social structures is represented to a limited degree in the larger body of international comparative research built on surveys of values. For example, studies of national culture (e.g., Hofstede, 2001; House et al.,

2004) and personal values (e.g., Schwartz, 1994) typically include variables like power distance that includes reliance on people in authority as an element, collectivism that includes relying on friends and colleagues, and uncertainty avoidance that includes reliance on rules. The comparative research we follow in our treatment of social structures draws from role theory (Biddle & Thomas, 1966; Kahn et al., 1964) by directly representing reliance on different categories of roles, rules, and norms in organizations rather than inferring it from national cultural values (e.g., Smith, Peterson & Schwartz, 2002).

The present study integrates these three generic cross-cultural literatures with two major lines of theory about e-mail use: theories of media richness and social presence (Daft & Lengel, 1986) and the Technology Acceptance Model (TAM: Davis, 1989; Venkatesh & Davis, 2000) to explain variations in the use of e-mail across nations. We consider three kinds of variables that are likely to predict managers' use of e-mail: (1) nationality, (2) personal values, and (3) the social structures that managers report to be used most heavily in their workplace. Each of these three kinds of variables has either been studied before or has provided a basis for speculative theory development about technology use. These variables have been chosen because they distinguish the cultural contexts embedded in nations that serve to influence individual decisions to use e-mail. Instead of relying on national patterns of technology acceptance and internet usage, or the "tip of the iceberg," these variables bring attention to the underlying cultural mechanisms that influence the acceptance and use of e-mail by individuals within nations. National culture and personal values that are likely to have been influenced by national and subcultural norms are especially relevant to extensions of TAM that include norms.

As will become apparent as the hypotheses are developed, no one element included in our project has been thoroughly studied and no prior study has dealt with all of these factors in combination

as predictors of technology use (Martins, Gilson & Maynard, 2004). In particular, possible national moderating effects on each of the other predictors need to be considered rather than assuming that relationships found in the most typically studied nations apply to all. Despite limitations in the available research and theory base about interactions between nation and the other predictors, we will propose tentative hypotheses which are tested using data obtained from surveys of managers in four nations in which English is widely spoken – the United States, the English-speaking islands of the Caribbean, Nigeria, and English-speaking sections of Canada. From a research standpoint, these nations have the advantage of providing considerable variability in socioeconomic characteristics and culture dimensions especially likely to affect the use of e-mail, yet the similarity in language reduces translation difficulties typical of multiple-nation research.

LITERATURE REVIEW

We draw from the three generic cross-cultural literatures on cultural characteristics, personal values, and social structures to contribute to two major lines of theory about the use of e-mail as a communication medium. One is theory about media richness and social presence (Daft & Lengel, 1986). This line of theory includes characteristics of technologies, social context, and both the deliberate and non-conscious choices that people make when communicating. The second is the technology acceptance model (Davis, 1989; Venkatesh & Davis, 2000), which is an adaptation of the Theory of Reasoned Action (TRA; Fishbein & Ajzen, 1975) applied to technology. TRA was broadly conceived to predict the behaviors of people in specific situations. TAM more narrowly focuses on technology usage, positing that perceived usefulness and ease of use predict intentions to use technology which in turn predict actual usage (Davis, 1989; Venkatesh & Davis, 2000). As noted in a meta-analysis by Legris, Ingham, and Collerette (2003), TAM has evolved to include other factors related to human and social change processes, such as subjective norms like those typical of a given nation and work situation and experiences like those that shape personal values. The importance of each factor varies by the stage of IT implementation (Legris, Ingham, & Collerette, 2003).

Dual processing theories of cognition (Kahneman, 2003) imply that TAM will be most helpful when individuals process information consciously and that media richness and social presence theories will be most helpful when individuals process information automatically or unconsciously. Research drawing from TAM focuses mainly on the deliberate, planned choices by individuals and organizations that are especially relevant to the introduction and early acceptance of a new technology, whereas media richness and social presence theories are more closely linked to nonconscious cognitive processes that have more influence once a technology is fully adopted. As we will detail in the following, despite its focus on conscious choices, we find TAM helpful when considering the difference between how e-mail is used in nations like Nigeria where it is likely to be a relatively recent part of work life as compared to how it is used in more technologically developed nations.

National Differences in E-Mail Use

Relative acceptance of information technology in the nations studied here differs substantially. The United Nations Development Program (2001) provides an index which measures technology acceptance across nations based on technology creation, technology diffusion and human skills (Desai, Fukuda-Parr, Johansson, & Sagasti (2002). The United States ranks #2 and Canada ranks #9 on this index, as nations on the cutting edge of technology. Jamaica ranks #49 and is considered a "dynamic adopter." Barbados and Nigeria fall

much lower on the index and are unranked, included with nations considered "marginalized." This overall technology difference is reflected in the World Bank Development Indicators Database (World Bank 2006) of internet use.

National culture is linked to technology use indirectly through its implications for national economic development (Inglehart & Baker, 2000), and directly through its implications for cultural norms affecting technology use. National culture is also likely to affect norm-based preferences for the sort of relatively explicit, spontaneous, moderately informal communication that e-mail promotes. Yet while culture is known to affect how people communicate and how they handle work events, culture's implications for information technology use are less well demonstrated (Avolio & Dodge, 2001). Only a few empirical studies have addressed this issue (e.g. Baba, Gluesing, Ratner & Wagner, 2004; Jarvenpaa, Knoll & Leidner, 1998; Tan, Wei, Watson, Clapper, & McLean, 1998). Despite this literature's limited empirical depth, it does provide some conceptual grounds for theorizing about culture's implications for e-mail use (e.g., Baba et al., 2004).

Most cultural analyses of e-mail use focus on implications of individualism and collectivism. Constructs related to individualism and collectivism have been argued to influence e-mail use through the mechanism of societal norms and personal values having to do with maintaining harmony. The basis for this explanation is that collectivism arises from historical or ongoing situations in which societies come to be organized around stable patterns of relationship, and that social norms and sanctions develop to support the harmonious continuation of these relationships. For example, Tan and colleagues (1998) found that people from individualist cultures tend to be more frank and outspoken and are more likely to use e-mail to challenge opinions, including those of the majority. People from collectivist cultures, they explain, are likely to value harmony over confrontation and are less likely to use e-mail to challenge the opinions of others. These findings align with those from Venkatesh and Zhang (2010), who analyzed technology adoption in a collectivist (China) and an individualist (U.S.) country. In the U.S., the role of "social influence" in predicting technology adoption was tied to multiple contingencies, while in China, the role was tied to a single contingency. In the United States, when adoption of a technology was mandatory within an organization, older women with less experience using a technology were more likely to seek the advice and help of others in using the technology. In China, those with less experience were more likely seek guidance from others, regardless of gender, age, or the voluntariness of adoption.

The other argument for the influence of individualism and collectivism is based on the dependence of communication on contextual information in collectivist societies. The basis for this dependence is that even more than requiring harmony, the quality of stability and multifaceted nature of groups characterizing collective societies means that their members come to understand one another very well. Interaction, then, does not require explication since shared contextual understanding means that words are often superfluous. Chen, Chen, and Meindl (1998) evoke the contextual information argument when they note that the norms of collective cultures support face-to-face communication over written or technology-mediated communication, since the former provides better contextual information about social meanings, intentions, and feelings. The norms of individualist cultures, in contrast, support instrumental, direct communication and are concerned with the efficiency of communication to save time and avoid hassles. They posit that cooperation can better be fostered in collective societies with face-to-face communication.

Most cross-cultural discussion of reliance on context in communication draws from an anthropological distinction between high context and low context societies. Hall (1976) and Hall and Hall (1990) find cultural variation in preferences

for explicit or "low context" messages compared to "high context" messages that assume greater implicit understanding. For example, based on an ethnographic study of large American, German, and French companies, Hall and Hall (1990) argue that Anglo-Americans and Germans prefer low context communication and lack extensive, well-developed information networks, while the French prefer high context communication. Although Hall does not emphasize the individualism-collectivism dimension by name, Gudykunst and Ting-Toomey (1988) link it to dependence on context cues by suggesting that instrumental communication is favored in individualistic, low context cultures, while indirect, affective communication is favored by collective, high context cultures. Their language of "instrumental" versus "affective" also hints at the harmony explanation. Bhagat and colleagues (2002) combine elements of the harmony and social context explanations more fully by suggesting that people from individualistic cultures prefer explicit messages and message credibility, while those from collectivist societies prefer high context, tacit messages, and consensual decision making. Since e-mail is relatively explicit and low in context compared to phone and personal conversations, e-mail is likely to be more popular in individualist/low context than in collectivist/high context societies. Among the nations represented in the present study, the United States and Canada are higher in individualism than most other nations, while the culture of Jamaica is intermediate, and West Africa (including Nigeria) is very collective (Hofstede, 2001; Schwartz, personal communication).

For a combination of economic, general technology acceptance, and more proximally cultural reasons, we anticipate national mean differences in e-mail use. Specifically, e-mail is likely to be used most by respondents in Canada and the United States, next most by those in the Caribbean, and least by those in Nigeria. These predictions are consistent with findings from prior studies of internet diffusion across multiple cultural contexts (e.g., Zhao, Kim, Suh, and Du, 2007). However, since the nations we study here differ in both socioeconomic and cultural characteristics, the theoretical implications of nation differences in average levels of e-mail use are impossible to disentangle. Consequently, we will treat nation as a control variable rather than try to give theoretical meaning to mean differences between nations in the use of e-mail. As we will describe, the basis for predicting moderating effects of nation on other predictors are far more interpretable from a cultural rationale than an economic rationale. Consequently, we will develop theories of individualism/collectivism and high/low context communication to state and test hypotheses about the moderating effects of nation.

Social Structures as Sources of Guidance Predicting E-Mail Use

Since e-mail has been argued to promote informal communication among peers over more formally structured communication channels, we expect that the extent to which managers report that they rely on social structures that suggest informal communication will be linked to their e-mail use. Studies of national culture dimensions show some interest in aspects of social structure like leaders (reflected in power distance) and rules (reflected in uncertainty avoidance) of nations (Hofstede, 2001; House et al., 2004) that provide the larger cultural context for the social structures emphasized in particular organizations within nations. Peterson and Smith (2000) and Smith, Peterson and Schwartz (2002) deal with national differences in social structures more directly. They begin from role theory to suggest that organization cultures develop over time in which particular types of organization events are influenced by people who occupy particular roles. Among these are roles like subordinates, colleagues, and superiors, as well as non-organizational roles like family and friends. Formal rules, organizational norms, and societal norms also guide behavior in a way similar to the

guidance actively provided by people who occupy organizational and non-organizational roles. DeSanctis and Monge (1999) take a similar role structure perspective when they suggest that computer mediated communication promotes lateral interaction and broader, more diverse participation while restricting domination and hierarchy. They suggest that electronic communication media also promote access to experts. Staff experts in organizations, though, are divided between people who provide guidance (e.g., finance managers to senior corporate leaders), others who provide direct service to other organization members (e.g., information systems support staff), and others who combine advisor and controlling functions (e.g., human resources and accounting officers; Mintzberg, 1979). Computer mediated communication also promotes use of expert information by providing relatively easy access to people in the know anywhere in a manager's role set (Ahuja & Carley, 1999).

Given this variety of functions, we will state and test the DeSanctis and Monge (1999) hypothesis, while also considering other possible implications of the taxonomy of social structures listed above from Peterson and Smith (2000; Smith, Peterson & Schwartz, 2002). Strong reliance on several of these social structures that provide managers with sources of guidance for handling work events is likely to be negatively associated with e-mail use. The informal quality of the typical use of e-mail makes it a relatively less appropriate medium in settings where people rely heavily on formal rules than in less formal work settings. Reliance on national norms to make decisions implies that managers already know what these norms are, negating the necessity of consulting with a superior, colleague, etc. Consequently, use of e-mail should be less important in settings where national norms are heavily used. Relying on one's own judgment rather than other sources to make decisions also implies relatively little consultation with other parties through e-mail.

H1a: Reliance on formal rules, superiors, national norms, or oneself will be negatively associated with the use of e-mail.

The argument that e-mail promotes participation and lateral relationships is noted above (De-Sanctis & Monge, 1999). In the present project, this can be tested both for lateral relationships within one's organization and relationships extending to family and business friends in other organizations. Reliance on the unwritten rules or informal norms of an organization is also expected to be facilitated by e-mail exchanges about what exactly are the organization's norms and how they apply in particular situations.

H1b: Reliance on unwritten rules, colleagues at one's own level, staff specialists, subordinates, family, and friends in other organizations will be positively associated with the use of e-mail.

In addition to the largely U.S. literature on which Hypotheses 1a and 1b are based, there are reasons to expect national contingencies. One basis for predicting national contingencies follows from a technology acceptance model logic about stages in technology diffusion. Specifically, the use of e-mail may be acceptable and possible, but not normative or easy, in societies at the early and intermediate stages of e-mail technology acceptance. People whose social context means that they need to make the effort to use e-mail will do so more than will others. This line of logic means that all of the main effects predictions for social structures will be more likely to hold in national contexts at the earlier stages of technology acceptance than in others. For the nations included in the present study, this line of reasoning suggests that e-mail availability and established norms for its use will not only promote its overall use in the United States and Canada, but will also limit differences in use among individuals in these two nations due to their local social context. The prevailing social

structures in a manager's context will have stronger effects in the Caribbean where its use is acceptable and possible, but where its use is somewhat impeded by technology resource limitations and by values of traditionalism. We anticipate that the even greater resource limitations and collectivist cultural norms of Nigeria supporting high context communication may even restrain those managers whose contexts most strongly support e-mail use from doing so.

Another kind of contingency hypothesis relates the distinctive informal, explicitly written, easily distributed qualities of e-mail to particular configurations of national culture and social structure context. For example, one such configuration is that the explicit quality of e-mail makes it appropriate for use in societies that rely heavily on bureaucracy to manage work contexts where individuals indicate that formal rules are especially important. Developing hypotheses based on all such configurations, however, would be too complex and uncertain given the current state of the literature. Instead, we rely on the TAM based logic provided above to predict stronger effects of social structures in the Caribbean, but will treat other interactions with nation as exploratory and return to their implications for future research when discussing the results.

H1c: The predicted relationships between relying on social structures and the use of e-mail will differ by nation. In particular, these relationships will be stronger in the Caribbean than in the other nations studied.

Personal Values Predicting E-Mail Use

Considering the values that people express in surveys is helpful because making predictions about individuals based on national and demographic characteristics has limitations. Shared nationality and the more physically observable demographic characteristics certainly reflect commonalities in background and context, and they provide an important basis for personal and social identity. Studying socioeconomic and cultural characteristics of nationality and demographics is useful for making predictions about individuals, because people's ability to consciously and accurately report values that actually reflect the way they think and behave is limited (Nisbett & Wilson, 1977). Nevertheless, despite a shared context of societal norms and values, the values of individuals vary within a nation (Au, 1999), and the link of demographics to particular values or patterns of cognition is often illusive. Asking people explicit questions using interviews or paper and pencil surveys is especially useful for identifying individual differences in explicit thought processes, although it is less useful for identifying differences in automatic cognitive processes that affect so much of behavior (Peterson & Wood, 2008). A substantial body of research suggests that survey reports of personal values are often predictive (Schwartz & Sagiv, 1995).

Our solution here is to continue studying national and demographic characteristics as noted above, but also to study respondents' expressed values. Schwartz (1992; Schwartz & Sagiv, 1995) developed a measure of values that has been used to provide a set of dimensions applicable to individuals and a somewhat different set of dimensions applicable to nations. In so doing, he drew from established measures of values as well as values suggested by collaborators in many nations. The individual level values are:

1. **Self-Direction:** independent thought and action derived from needs for control and mastery.
2. **Stimulation:** need for change, novelty, variety, arousal, and thrill-seeking.
3. **Hedonism:** need for pleasure or sensuous gratification for oneself.
4. **Achievement:** need for personal success through demonstrating competence according to prevailing cultural standards.

5. **Power:** need for attainment of social status and prestige, along with control over people and resources.
6. **Security:** need for safety, harmony, stability of society, relationships, and the self.
7. **Conformity:** restraint of actions and impulses likely to upset or harm others and violate social norms and expectations.
8. **Benevolence:** preservation and enhancement of the welfare of people with whom one is in frequent personal contact.
9. **Universalism:** understanding, appreciation, tolerance, and protection for the welfare of all people and for nature.
10. **Tradition:** respect, commitment, and acceptance of the customs and ideas that traditional culture or religion provide to the self.

In order to ascertain which of these values might predict individual differences in e-mail use, we considered studies that had previously linked personality or value characteristics to e-mail use and also considered the basic theories that have been used to predict e-mail use to consider whether any values reflected in the SVS might be theoretically relevant to predicting e-mail use. We were able to find sufficient empirical and theoretical support for offering predictions for four of the SVS measures, but not for the other six measures. Two values, conservatism and security, were considered relevant because of their relationship to theories about conservative values that have been included in the TAM literature, while two other values self-direction and power, were considered relevant because of their relationship to theories about hierarchical orientation in the literature about e-mail use.

TAM may be used to help explain a conscious connection between personal values and e-mail use in various stages of e-mail adoption. During the early stages, it is likely that innovators and early adopters will embrace e-mail, while late adopters and laggards will not. Late adopters and laggards who likely to hold conservative values may consciously choose to use more traditional forms of communication over e-mail. Conservative values within the Schwartz taxonomy include tradition and security, two frequently investigated themes in the literature about e-mail use. Notably, an emphasis on tradition includes preference for established technologies as compared to new ones. Although in many parts of the world, e-mail has been used long enough that it is now well established, it is still appropriate to test the hypothesis that people holding traditional values will tend more than others to avoid electronic communication technology. For example, people holding traditional values may reside in rural areas, as opposed to urban areas. One study has found that people residing in rural areas use e-mail less than their metropolitan or non-metropolitan city counterparts (Spennemann & Atkinson, 2003). Similarly, security has been a significant issue in electronic communication, both as communications have the potential to be monitored unbeknownst to the communicating parties and because e-mail once sent can be distributed broadly by the recipient to third parties without the sender's knowledge. People who value security may prefer face-to-face or telephone conversations, where the messages are generally not recorded. People who value security and/or tradition are less open to change (Schwartz, 1992; Caprara, Vecchione, & Schwartz, 2009) and therefore may be less innovative, which is associated with greater risk, uncertainty, and imprecision (Agarwal & Prasad, 1998). Personal innovativeness relates to technology acceptance (Agarwal & Prasad, 1998), so those who oppose innovation may negatively relate to e-mail usage.

As reported by DeSanctis and Monge (1999), numerous studies have found that parties communicating via e-mail are less prone to hierarchy and domination and are more accepting of equality of participation and exchange. The link of e-mail to an emphasis on lateral communication as distinct from either a preference to work and think individually or a preference to rely on rules

and hierarchy suggests other hypotheses about implications that the values related to a need for control or domination will have for e-mail use. People who are self directed prefer to rely on their own judgment when interpreting work events and therefore are less likely to use e-mail to contact others to make sense of events. People who value power tend to have higher levels of social cynicism and lower levels of religiosity (Leung, Au, Huang, Kurman, Niit, & Niit, 2007). They also tend to rely on rules and hierarchy, and are therefore less likely to use forms of communication that promote lateral relationships, such as e-mail.

We could find no theoretical basis for predicting that the SVS value dimensions of stimulation, hedonism, achievement, conformity, benevolence, or universalism would predict use of e-mail based on the conceptualizations of these six dimensions as described above. That is, we identified no literature that would suggest that e-mail is likely to be more stimulating than are other media, that it will contribute to more hedonic pleasure than will other media, that it will promote achievement goals or reflect competence more than will other media, that its use is more or less consistent with conformity values of personal restraint than are other media, that its use reflects greater benevolence in the sense of enhancing the well-being of others than do other media, or that it shows greater or lesser universalism in the sense of tolerance for others and for nature than do other media.

We propose the following hypothesis.

H2a: Tradition, security, self-direction and power values will be negatively associated with the use of e-mail.

The rationales for predicting national differences in the relationships between personal values and e-mail use follow the same two principles as those noted for national contingencies in the implications of social structures. That is, e-mail use will be more strongly linked to personal values in nations at an intermediate level in the technology acceptance process where e-mail use is not yet fully normative, but where sufficient infrastructure is present to make it available for those who are motivated to use it. Specifically, we note above that the Caribbean lags the U.S. and Canada in e-mail adoption, such that it is more possible for individuals to successfully do their jobs without using this technology in the Caribbean than it is in the U.S. or Canada. In general, we expect that the personal values noted in Hypothesis 2a will be more highly associated with e-mail use in the Caribbean than in the U.S. or Canada. Other reasons to expect national differences in the implications of values come from local issues, such as the particular concern for security in the U.S. since 9/11. However, the argument that this concern is present only in the U.S. and not in Canada or the Caribbean is sufficiently speculative that implications of such idiosyncratic aspects of nationality need to be treated in an exploratory way. As noted in the description of methods below, measures of personal values were added to the project after the data collection was completed in Nigeria.

H2b: The predicted relationships of values to the use of e-mail will vary by nation. In particular, these relationships will be stronger in the Caribbean than in the other nations studied.

The hypotheses have been proposed to answer the questions of whether: (1) theories of e-mail use and acceptance developed in the United States can be extended to less developed nations, such as Nigeria and the English-speaking Caribbean, (2) whether psychological and social characteristics concerning certain social structures and values relate to the use of e-mail, and (3) whether cross-cultural differences in the use of e-mail exist, despite the increasing prevalence of e-mail use in global communication. Figure 1 presents a summary of the hypotheses.

METHOD

The data analyzed here are from working people who have subordinates – 168 from the United States, 98 from Canada, 78 from the Caribbean,

Figure 1. Proposed model of the social structures and personal values relating to the use of e-mail

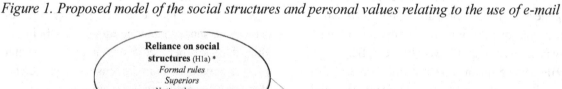

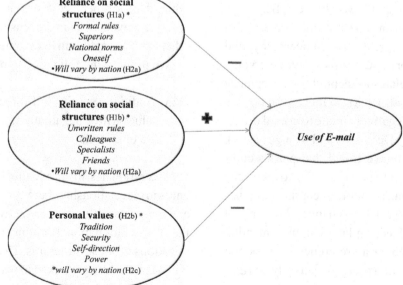

and 232 from Nigeria. Respondents were either attending managerial training programs or were enrolled in MBA programs designed for working managers. Data were collected as part of a training exercise during class sessions in which the surveys were administered and debriefed. Since the survey was part of an ordinary class exercise, although those who preferred not to turn in their surveys were free to keep them, participation rates were very high. A true response ratio of eligible respondents to completed surveys cannot be calculated precisely, since it is not possible to know whether a given person in the audience was eligible to participate in the project (i.e., was indeed a working person with subordinates who was raised in the nation in which the survey was administered) apart from the answers provided in the surveys themselves. For example, many people from the U.S. and Canada were raised outside of those nations (in many cases, in Asia) and immigrated later in life. Also, a number of respondents were self employed as consultants, had professional positions (e.g., lawyers), or for some other reason were not currently working or had no subordinates. Some of these ineligible individuals completed surveys that were then deleted (not included in the sample reported here) while others who were ineligible chose not to turn in their survey. To confirm that all respondents currently were not only working but had managerial responsibilities, respondents were omitted who did not complete the questions about subordinates, or who indicated that they had encountered the 8 events (see "e-mail use" below) less than "rarely" on average over the past few months. (Additional data collection and participation details are available from the authors.)

In order to limit the possibility that results might be due to regional subcultures within nations, care was taken to represent several geographical subcultures of each nation. In Canada, surveys were administered in Vancouver (19), Calgary (56), and Toronto (23). In the United States, surveys were administered in Fort Lauderdale (46), Wilmington, North Carolina (24), Chattanooga (16), Portland, Oregon (62), and Boston (20). In the Caribbean, surveys were administered in Jamaica (55) and Barbados (23). The surveys in Nigeria were administered in Lagos and included respondents from the nation's three major ethnic groups: Hausa (71), Igbo (69), and Yoruba (92). The survey used in Nigeria did not include the measures of values.

Sample demographic characteristics are provided in Table 1 and Table 2. Demographic characteristics that predict e-mail use are controlled in the analyses testing the hypotheses. Descriptive statistics and reliabilities for the multiple-item measures are provided in Tables 3, 4, 5, and 6.

E-Mail Use

The e-mail use criterion is an eight-item scale captured by the following question asked separately for each of eight work events: "When you do face a situation of this type, how often do you use e-mail to discuss it with others?" The eight kinds of events were: "When it is necessary to appoint a new subordinate in your department," "When one of your subordinates is doing consistently good work," "When one of your subordinates is doing consistently poor work," "When some of

Table 1. Demographic Characteristics

Nations	N	% Male	Age
Canada	98	55.10%	42.4
United States	168	53.90%	32
Caribbean	78	37.20%	32.7
Nigeria	232	72.30%	38.5
Tests of nation differences			
F statistic			39.59
Chi square		34.7	
p value		p <.001	p <.001
Degrees of freedom		3	3,557
Sample size		574	

Table 2. Demographic Characteristics

	Organization	Organization	Organization	Department
Nations	Size	Ownership	Industry[1]	Specialty[2]
Canada	1000+	MNC	Finance	Service
	56.10%	32.60%	25.00%	24.40%
United States	1000+	Priv	Prof	Finance
	42.80%	45.20%	11.70%	25.20%
Caribbean	100-1000	Gov	Finance	Service
	42.10%	41.90%	23.70%	31.10%
Nigeria	1000+	Gov	Service	Service
	67.30%	62.80%	76.10%	29.30%
Tests of nation differences				
F statistic				
Chi square	58.6	153.9	117.6	88.7
p value	p <.001	p <.001		
Degrees of freedom	6	12	36	30
Sample size	566	566	334	451

[1] The industry and percentage noted are for the one most mentioned in each nation. The industry codes made fewer distinctions in the form used in Nigeria. Organization Ownership response alternatives were Private, Government, MNC or Other. Organization industry was coded into 18 industry types for Canada, the US and Jamaica. The most frequent, as noted, were finance (Fin) and professional (Prof). For Nigeria, a simpler coding system was used that included Service, Manufacturing, Automated Manufacturing and Other.

[2] Department type codes had 12 categories including Direct Service Delivery, Financial and Accounting, and Sales. The specialty and percentage noted are for the one most mentioned in each nation.

the equipment or machinery used in your department seems to need replacement," "When another department does not provide the resources or support you require," "When there are differing opinions within your own department," "When you see the need to INTRODUCE new work procedures into your department, " and "When the time comes to EVALUATE THE SUCCESS of new work procedures." Responses were provided on a five-point Likert scale ranging from 1 – "Never" to 5 – "Very often." The descriptive statistics for the use of e-mail as a criterion are listed in Table 3.

Table 3. Descriptive Statistics for E-Mail Use Criterion

	Canada		United States		Caribbean		Nigeria	
	Mean	Alpha	Mean	Alpha	Mean	Alpha	Mean	Alpha
	(S.D.)		(S.D.)		(S.D.)		(S.D.)	
Use of E-mail	2.91	.89	2.78	.88	2.34	.92	2.21	.93
	(.93)		(.91)		(.98)		(.93)	

Responses ranged were coded from 1 to 5 with a higher value corresponding to more reported use of a given source.

Note: Alpha is an assessment of reliability, ranging from 0 to 1. Higher levels correspond to greater reliability, or consistency, in the measure.

Table 4. Descriptive Statistics for Social Structures Criterion: US and Canada

	Canada		United States	
	Mean	Alpha	Mean	Alpha
SOCIAL STRUCTURES	(S.D.)		(S.D.)	
Formal Rules	3.46	.84	3.34	.85
	(.78)		(.79)	
Unwritten Rules	3.18	.84	3.41	.83
	(.75)		(.72)	
Subordinates	3.06	.77	2.91	.83
	(.66)		(.76)	
Specialists	2.65	.81	2.53	.82
	(.75)		(.85)	
Colleagues	3.02	.84	3.18	.85
	(.71)		(.74)	
Superiors	3.58	.88	3.87	.85
	(.85)		(.76)	
Self	3.67	.91	3.63	.87
	(.82)		(.75)	
Widespread Beliefs	2.34	.92	2.47	.91
	(.92)		(.87)	
Family	1.33	.93	1.46	.95
	(.61)		(.74)	
Friends	1.33	.91	1.45	.93
	(.57)		(.68)	

Nationality

Nationality was coded based on the location where data were collected, and respondents who indicated that they were from other locations were deleted. All respondents in Nigeria and almost all in the Caribbean and Canada were born in the area where the data were collected. A substantial number of respondents surveyed in the United States were deleted as being born in other nations, however no single non-U.S. nationality was represented by a sufficient number of respondents to do meaningful analyses of particular immigrant groups.

Demographics

Demographic information was coded as indicated in Table 1. Consistent with education demographics throughout the Caribbean, there are more women than men in the Caribbean sample of managers attending MBA programs. The average ages exceeding 30 years in each nation are consistent with the focus on working managers. The Canadian respondents are somewhat more senior than the others. The department types and organization types of respondents ranged broadly in each nation studied. Table 1 reports only the most frequently mentioned department and organization types. More complete distributions for these demographic variables are available from

Table 5. Descriptive Statistics for Schwartz Value Survey Measures

	Canada		United States		Caribbean	
	Mean	**Alpha**	**Mean**	**Alpha**	**Mean**	**Alpha**
	(S.D.)		(S.D.)		(S.D.)	
Benevolence	6.75	.68	6.80	.77	7.23	.71
	(.75)		(.89)		(.68)	
Traditionalism	4.45	.49	4.77	.70	5.39	.60
	(.98)		(1.24)		(1.1)	
Power	4.85	.62	4.84	.78	5.42	.70
	(1.02)		(1.23)		(1.2)	
Achievement	6.95	.65	6.82	.60	7.48	.69
	(.78)		(.79)		(.71)	
Stimulation	5.89	.67	5.69	.64	5.79	.45
	(1.19)		(1.20)		(1.09)	
Self-Direction	6.89	.70	6.80	.64	7.17	.71
	(.84)		(.83)		(.81)	
Universalism	6.24	.73	6.13	.79	6.59	.77
	(.84)		(.96)		(.91)	
Hedonism	6.52	.69	6.56	.66	6.16	.42
	(1.32)		(1.22)		(1.31)	
Conformity	6.15	.66	6.16	.70	7.16	.75
	(1.05)		(1.06)		(.98)	
Security	6.26	.64	6.25	.69	6.84	.57
	(.90)		(.97)		(.83)	

Data for the Schwartz Value Survey were not collected in Nigeria. Higher values indicate a higher rating on a dimension. Scores add 2 to eliminate negatives and zeros.

Note: Alpha is an assessment of internal consistency, with ranges from 0 to 1. Higher levels correspond to greater internal consistency, or reliability, in the measure.

the authors. Demographic effects of gender, age, education, organization size, and organization ownership on the use of e-mail were tested and controlled where significant.

Social Structures Providing Sources of Meaning

The measures of ten sources that respondents report drawing from to handle the eight work events about which e-mail use was asked (see "E-mail use" above) were based on measures described by Smith, Peterson and Schwartz (2002). The ten resulting eight-item scales are averages of respondents' answers to questions that ask "to what extent are the actions taken affected by each of the following?" -- "Formal rules and procedures," "Unwritten rules about 'How we do things around here,'" "My subordinates," "Specialists outside my department," "Other people at my level," "My superior," "Opinions based on my own experience and training," "Beliefs which are widely accepted in my country about what is right," "Members of my family," and "Friends outside this organization." Responses were provided on a 5-point Likert scale ranging from "To a very small extent" (coded 1) to "To a very large extent" (coded 5). As indicated in Tables 4 and 5, all show adequate psychometrics in all nations studied.

Schwartz Value Survey (SVS)

The conceptual origins of the Schwartz Value Survey measures are described in the section leading to Hypothesis 2. Schwartz (1992, 1994; Schwartz & Sagiv, 1995) proposed measures of 10 individual level values based on their conceptual meaningfulness informed by the results of a smallest space analysis. The smallest space analysis was performed on a large database consisting of college students and secondary school teachers from many nations. The labels of each construct, the number of items used to represent it, and an example value item included in each measure are as follows: Benevolence (9 items; e.g., helpful), Traditionalism (6 items; e.g.,, humble), Power (5 items; e.g., authority), Achievement (6 items; e.g., successful), Stimulation (3 items; e.g., varied life), Self-Direction (6 items; e.g., own goals), Universalism (9 items; e.g., social justice), Hedonism (2 items; e.g., pleasure), Conformity (4 items; e.g., politeness), and Security (6 items; e.g., social order). These have become standard measures for studying differences in values in the comparative management (Egri & Ralston, 2004) and cross cultural psychology literatures (Schwartz & Boenke, 2004). Respondents were presented two lists containing sixty value items (such as "respect for tradition," "family security," or "humble) that included 56 items from the SVS and four additional values considered by Lenartowicz and Roth (2001). They are asked to rank the importance of each of the value items as guiding principles in their lives. Responses range from -1 – "opposed to the principles that guide you" to 7 –"of supreme importance as a guiding principle in your life." In order to eliminate the negative and zero values to facilitate interpretation, the scores are transformed by adding a constant of "2" so that the resulting possible range is from 1

to 9. The Cronbach alpha reliability coefficients shown in Table 5 for the values measures vary with at least one alpha below .70 for each of the four predictors for which hypotheses are proposed. The reliability of security is below .60 in the Caribbean. Reliability coefficients are generally lower when a subset of the items are used to form the scales as is typically done in cross cultural research (Schwartz, Verkasalo, Antonovsky, & Sagiv, 1997). Differences in reliability for these measures are taken into account in interpreting the results. Values data were not collected in Nigeria.

DATA ANALYSIS

The effects of the control variables on the use of e-mail were assessed using ANOVA for the categorical control variables of organizational ownership type (government, MNC, private, or other) and gender. Pearson correlations were used for the continuous variables of age, years of education, and organization size. Control variables found to be significant were included in subsequent hypotheses testing. Since a control variable consisting of the average of all SVS measures had no relationship to the email use criterion (r =.01) and did not increase variance explained in any of the regressions, it was not included in the regressions reported. The hypotheses were tested using hierarchical moderated regression. Categorical variables were dummy coded for use in the regressions. Interaction terms were created by multiplying dummy codes for nation with each predictor. Hypotheses were tested using multiple regressions in four hierarchical steps. The first step entered the control variables. The second step entered the nation variables. The third step added the hypothesized predictors. The fourth step added interaction terms consisting of the dummy coded nation variables multiplied by the predictors. Because of the large number of social structure measures (10) and interaction terms, the fourth step for the measures use of social structures was done with only those that showed significant main effects in the third step. As an exploratory analysis, the Schwartz Values Survey measures that were not predicted to have effects were used in a separate set of hierarchical regressions. This exploratory analysis also serves as a way to evaluate whether significant results for the tests of the predicted relationships are likely to be simply due to same-source response bias. That is, if non-predicted relationships are significant and similar in magnitude to predicted relationships, response bias would be a possible explanation.

RESULTS

Table 6 provides a summary of the hypotheses and results.

The categorical control variable, organization ownership, showed a significant main effect on the use of e-mail (F=19.88, df=3, 546, p <.001). One of the two continuous control variables showed a significant correlation with the use of e-mail -- age (r=-.15, p <.001), but the other, organization size, did not (r = -.03, n.s.). In contrast to a few IT studies (e.g., Gefen & Straub, 1997; Venkatesh & Morris, 2000; Lippert & Volkmar, 2007), gender effects were not found in the present sample. The variables showing significant effects, organization ownership and age, are controlled in the hypothesis tests that follow.

Table 3 shows significant mean differences in e-mail use between nations. We also tested whether nation was an important control variable after considering other controls by using a hierarchical regression that first entered a block containing the demographic controls followed by a block entering dummy codes for nations. The results show that the mean differences among nations in the use of e-mail that appear in Table 3 are significant after controlling for the effects of the demographics (R² change =.04, df 3, 478, p <.001). As indicated in Table 3, e-mail is most used in Canada, then the U.S., then the Caribbean, and least in Nigeria.

Hierarchical moderated regressions were used to Hypotheses 1a, 1b and 1c about the relationship of social structures to use of e-mail. The first step in the regression controlled for the significant effects of the demographic controls in the use of e-mail. Results indicated that the variance in the use of e-mail explained by the demographic controls (R²) was .11 (p <.001). The second step controlled for the effects of nations, resulting in a significant change in incremental variance (R² change =.04, p <.001). The third step shows significant main effects for a number of the sources (as detailed in Table 7), supporting parts of Hypotheses 1a and 1b (R² change =.17, p<.001). The fourth step assessed the moderating effect of nation on those sources that showed significant main effects of the regression in the third step or that

Table 6. Summary of Hypotheses Testing

Research Variables	Hypotheses	Findings
Formal rules, superiors, national norms, oneself **(H1a)**	Negative relationship	Not supported: Positive relationship for reliance on formal rules
Unwritten rules, colleagues, specialists, friends **(H1b)**	Positive relationship	Partial support: Positive relationship for reliance on unwritten rules and specialists
Relationships predicted in H1a and H1b will vary by nation **(H2a)**		Supported
Tradition, security, self-direction, power **(H2b)**	Negative relationship	Not supported
Relationships predicted in H2b will vary by nation **(H2c)**		Supported

Table 7. Variance Explained in Regressions Predicting the Use of E-mail from Sources of Meaning Overall

Model	R^2	R^2 change	F change	Df
1	.12***	.12***	17.57***	4,503
2	.16***	.04***	7.29***	3,500
3	.33***	.17***	24.46***	5,495
4	.37***	.04**	2.24**	15,480

Model 1 predictors: age, type of organization (government, MNC, other)

Model 2 predictors: age, type of organization, country (Jamaica, U.S., Nigeria)

Model 3 predictors: age, type of organization, country, sources of meaning (formal rules, unwritten rules, subordinates, specialists, family)

Model 4 predictors: age, type of organization, country, sources of meaning, sources of meaning x country interactions

The F test is an overall test of the significance of the regression equation, whether the independent variables in the regression equation together predict the dependent variable. The R^2 refers to the variance explained in the dependent variable by the independent variables in each regression equation. The R^2 change refers to the incremental variance explained by the study variables over the control variables.

*** p <.001

showed significant effects in at least one of the regressions conducted separately in each nation. This regression shows significant interactions of nation and source, offering some weak support for Hypothesis 1c (R^2 change =.04, p <.01).

In order to readily interpret the effects most directly related to the hypotheses, Table 8 shows standardized betas for regressions conducted separately in each nation. The figures in the left hand column provide the standardized betas associated with use of different social structures for the overall sample. This column corresponds to the main effects in step 3 of the hierarchical regression. These betas indicate that reliance on unwritten rules, subordinates and staff specialists are, as predicted in Hypothesis 1b, positively associated with use of e-mail. Other parts of Hypotheses 1a and 1b are not supported.

The test of the interactions between nation and sources showed a significant increase in variance explained as predicted in Hypothesis 1c. In general, most of the significant relationships shown in Table 8 indicate that the main effects predictions are supported more frequently than are other predictions. In several instances such as use of formal rules and unwritten rules in Canada, the betas are very consistent with the overall betas, but fall slightly short of the usual p <.05 signifi-

cance level. Several of the predicted relationships are most evident in Nigeria. Consistent with the main effects predictions, Nigerians who indicate high use of unwritten rules, subordinates and, especially, family report using e-mail more often than do other Nigerians. Although an overall negative effect of relying on formal rules was predicted, a positive relationship is found. For the U.S. respondents, the main effects prediction that relying on specialists is associated with using e-mail is supported.

Results testing Hypotheses 2a and 2b about the implications of personal values for e-mail use are provided in Tables 9 and 10. These measures are available for Canada, the United States and the Caribbean, but not for Nigeria. The first hierarchical step including the demographic controls showed an R^2 of.06 (p <.001). The second step showed a significant difference among these three nations (R^2 change =.03, p <.05.) The third step showed no significant main effect of the four values measures (R^2 change =.02, n.s). Thus, hypothesis 2a is not supported for the combined sample. The fourth step shows a significant interaction (details in Table 9) of values and nations (R^2 change =.05, p <.05) supporting hypothesis 2b. The regression for the combined sample shown in Table 10 is generally consistent with the find-

Table 8. Standardized Betas for Regressions Predicting the Use of E-Mail from Sources of Meaning: Overall and for Nations

	Overall	Nigeria	Canada	U.S.	Caribbean
Age	-0.08	-0.1	-0.18	-0.09	0.03
Govt.	-.12*	-0.14	-0.06	-0.02	-.43**
MNC	.09*	0.06	0.11	0.03	0.21
Other	0.04	-0.07	0.06	0.09	-0.04
Nigeria	-.39***				
U.S.	-0.06				
Caribbean	-.14**				
Formal rules	.10*	0.07	0.21	0.06	0.04
Unwritten rules	.13**	.19**	0.21	-0.02	0.13
Subordinates	.10*	.17*	0.12	0.14	0.03
Specialists	.18***	0.11	0.16	.23*	0.15
Family	0.09	.34***	0.02	-0.16	-0.09
F test	19.94***	17.58***	2.57**	2.14*	3.84***
R^2	.33***	.47***	.33**	.11*	.37***
Degrees freedom	12,495	9,177	14,72	9,153	9,58

Since nation and organizational ownership are dummy codes of categorical predictors, one category is dropped. The dummy variable representing Canada and private organizational ownership are the drop variables for nation and organizational ownership in the overall regression equations. The dummy variable representing private organizational ownership is the drop variable for the set of organizational ownership dummy variables in the separate nation regressions.

* $p < .05$

** $p < .01$

*** $p < .001$

ing of non-significant overall effects. The other three columns presenting betas for each nation help interpret the specific variables that produce the value by nation interactions. Self-direction is positively related to e-mail use in Canada. Security is negatively related to e-mail use in the United States. Since the reliability of the values measures differs among the three nations for each of the values studied, the possibility needs to be considered that significant effects are disproportionately found for more reliable measures. Referring to

the reliability coefficients shown in Table 3, this does not seem to have been the case, however. The values most highly associated with e-mail have alpha coefficients that are near the median for the values measures in each nation.

As an exploratory analysis, the six value measures not predicted to affect the use of e-mail were included in a second similar hierarchical regression (details available from the authors). No significant main effects of these other values or interactions with nation were found.

Table 9. Variance Explained in Regressions Predicting the Use of E-mail from the Schwartz Value Survey Overall

Model	R²	R² change	F test	Df
1	.06***	.06	5.28***	4,315
2	.09***	.03**	5.06***	6,313
3	.11***	.02	3.85***	10,309
4	.16***	.05*	3.16***	18,301

Model 1 predictors: age, type of organization (government, MNC, other)

Model 2 predictors: age, type of organization, country (Jamaica, U.S.)

Model 3 predictors: age, type of organization, country, Schwartz Value Survey variables (self direction, power, tradition, security)

Model 4 predictors: age, type of organization, country, SVS variables, SVS variables x country interactions

R^2 refers to the variance explained in the dependent variable by the independent variables in each regression equation. The R^2 change refers to the incremental variance explained by the study variables over the control variables

* $p < .05$

** $p < .01$

*** $p < .001$

DISCUSSION AND IMPLICATIONS

The research contributes to the international literature about e-mail use in several ways. It provides a field test in several nations, including Nigeria and the Caribbean, locales that have rarely been studied, to evaluate the generalization of a number of prior results from U.S. samples and theorized relationships about variables associated with the use of e-mail. It also develops and tests new theory that answers the need noted in a recent review for additional research into psychological characteristics and social characteristics that affect the use of e-mail (Martins, Gilson & Maynard, 2004). It does so in two nations, Canada and the United States, that have the socioeconomic and cultural characteristics that have been implicitly assumed as the theoretical context for prior research, and also in the English-speaking Caribbean and Nigeria, nations that have the potential to show quite different results from the U.S. and Canada.

Apart from the statistics, we were reminded when doing the project that while e-mail is indeed frequently used in each of the locales that we considered, global availability of e-mail is not to be taken entirely for granted. We learned

that in the business center of Nigeria, Lagos, not only is internet access somewhat sporadic, but personal electric generators are necessary for the frequent times each week when public electricity is unavailable for businesses and private homes of even affluent people. If e-mail is interrupted, as is more likely in Nigeria than in the other locales considered here, a manager is faced with the necessity of thinking through whether it is best to wait for it to resume or to pick up a phone or make a personal visit to have a conversation.

Our results contribute to the discussion of the way in which personal values and work contexts are associated with e-mail use (Carlson & Zmud, 1999). The results suggest that the use of particular kinds of roles, rules and norms in managers' organizational context is more closely related to their e-mail use than are a manager's personal values. The research also suggests that theoretical formulations based on a combination of functional and institutional theories of media use are plausible. Some significant e-mail predictors that are unique to particular nations suggest that models of the uniqueness of social constructions about media use in particular nations may be appropriate as well. In this last respect, in addition to finding support

Table 10. Standardized Betas for Regressions Predicting E-Mail Use from Schwartz Value Survey: Overall and for Nations

	Overall	Canada	U.S.	Caribbean
Age	-.09	-.26**	-.03	-.02
Govt.	-.1	.04	.03	-.33*
MNC	.16**	.14	.07	.29*
Other	.06	.08	.07	.02
U.S.	-.07			
Caribbean	-.19**			
Traditionalism	-.05	.17	-.06	-.24
Security	-.11	.02	-.22*	.17
Power	.12	.15	.12	-.08
Self direction	.08	.21*	.02	-.07
F test	3.85***	2.65**	1.31	3.68**
R²	.11***	.20**	.07	.33**
Degrees freedom	10,309	8,83	8,151	8,59

The dummy variables representing Canada and private organizational ownership were dropped from the regression equations.
* p <.05
** p <.01
*** p <.001

for prior theory and research, the present results contribute some unanticipated or only partially anticipated findings that can provide a beginning for further theory development and research.

Implications of Confirmed Main-Effects Hypotheses

The findings suggest a refinement to theory that posits that the use of e-mail promotes participative processes and lateral relationships over vertical ones (DeSanctis & Monge, 1999). The results suggest that e-mail may support lateral decision making, as e-mail use is positively associated with reliance on unwritten rules, subordinates, and staff specialists, without working against control by rules. As a direction for possible future research, the unexpected positive relationship between reliance on rules and e-mail use raises the question of whether requests from managers

for clarifications about the application of organizational rules to particular situations is especially common in organizations that are prone to rely on rules. From the standpoint of organizational politics, situations where e-mail does not work against corporate authority are reminiscent of the classic notion that control is not fixed, but expandable (Tannenbaum et al., 1974). Using e-mail to increase influence from additional parties need not work against following organizational rules and policies. Instead, it may increase the quality of decisions by increasing many parties' influence to anticipate a broader range of implementation issues before conclusions are reached. However, the national contingencies found suggest that there may be some national boundaries to this conclusion. For example, the relationship between relying on unwritten rules and e-mail use appears to be negligible in the U.S. and the relationship between relying on subordinates and e-mail use appears to be negligible in the Caribbean. In contrast to the finding of substantial main effects of the use of roles, rules and norms on e-mail use, the lack of links between values and the use of e-mail in the overall results is somewhat surprising.

Implications of Confirmed Nation Interactions

We started with two basic kinds of rationales for the hypotheses about interactions of social structures and personal values with nation. A theoretical rationale based on the Technology Acceptance Model would suggest that people in locations where technology is relatively new or less used would still be thinking consciously about technology use choices. Although there are indeed a larger number of significant betas for reliance on particular social structures in the Nigeria results than in the results for other nations, the differences can be accounted for in part from the larger sample size in Nigeria. Consequently, this result is not readily interpretable using a TAM rationale which suggests that there will be more effects of organizational context in nations in which one would expect more conscious attention to media use choices.

The results suggesting significant interactions of nation with either organizational context or personal values show three substantial effects that can be interpreted using both TAM ideas and theory based on media richness and social presence. Reliance on family has a particularly strong relationship to use of e-mail in Nigeria, while security values are negatively related to e-mail use in the U.S. and self-direction is positively related to e-mail use in Canada. The largest difference among nations in the relationship of roles, rules and norms to e-mail use suggests that degree of reliance on family members may have more influence on e-mail use in Nigeria than in the other nations studied. Among the possible explanations are that for those individual mangers or in those

work settings in Nigeria where others at the work site are not suitable sources of guidance about what to do, managers rely on e-mail to communicate with an extended family network for advice. Perhaps paradoxically, this interpretation would suggest that interactions with family require less social presence in Nigeria than elsewhere. The interactions of nation with personal values provide evidence for the stronger effects of personal values in the U.S. and Canada where the use of e-mail is more normative than in the Caribbean. As the use of e-mail becomes more ubiquitous and accepted in business communications and processes in a developing nation, a positive relationship between its use and reliance on unwritten rules, or the "way we do things around here" seems plausible. The largest two betas shown in Table 10 are for the negative relationship in the U.S. between security and the use of e-mail and the positive relationship in Canada between self-direction and the use of e-mail. The security effect in the U.S. could reflect the strong sensitizing effect of the 9/11 incident to security issues that is likely to be more evident among some U.S. respondents than it is to respondents from the Caribbean and Canada. That sort of explanation might be best explained by a modification of TAM which would suggest that technology acceptance can be reduced when a national crisis reduces general acceptance of a particular technology. To the extent that self-direction values are related to individualism and a preference for low context communication, this result provides an individual level analogy to theory of culture and communication that is typically posed at the societal level. That is, social presence is less necessary for people holding extremely individualistic values, at least for the Canadian respondents.

Another base for interpreting the interactions is to consider the reasons for particular configurations of effects. One is to recognize the nature of e-mail as a device that can be used to provide explicit documentation. Prior research (Smith,

Peterson & Schwartz, 2002) as well as the present data (Tables 3, 4, and 5) suggests that the social structure of Nigeria is based more substantially on explicit, formal rules than are the other nations studied. That does not necessarily mean that rules are followed, but it does mean that debates about propriety are more likely to rest on explication. Perhaps the positive relationship between reliance on subordinates and the use of e-mail in Nigeria may be a function of superiors' use of e-mail to give explicit assignments to subordinates and for subordinates to document compliance. Evaluating this speculation would benefit from replication that we have underway in other cultural contexts, but experimental evidence and qualitative accounts of the explanations that people in these nations provided for their e-mail use would also be helpful. The explanations for all of these interaction effects require follow up in further research.

Implications of Demographic Analyses

We treated the demographic factors as control variables rather than posing hypotheses for them mainly due to lack of a strong theoretical basis for forming hypotheses. Although applying un-predicted findings without further confirmation risks overemphasizing chance results, several are strong enough to stimulate further theory and research. The results showing main effects of two control variables, age and organization ownership suggest further study. The findings that suggest that older respondents overall are less likely to use e-mail than their younger counterparts is consistent with the literature relating technology acceptance to experience (e.g., Legris et al., 2003), yet the size and mean age of the Nigerian sample may have influenced these results. Given that the means indicate low use of e-mail in governments, our sense is that these effects are largely due to the difference between governments and other organizations.

Implications for Practice

The findings have implications for what managers should expect who do business across multiple nations, particularly the ones studied here. Understanding the cultural, personal and contextual factors behind the use of e-mail is important to multinational organizations both in their internal cross-border contacts and in the business transacted across borders with other organizations. Kundu, Niederman, and Boggs (2003) note that telecommunication advances are an underlying reason for the phenomenal growth of international business. They also note that firms that develop, maintain, and improve their information technology management skills are best able to sustain and enhance their core competencies and ultimately achieve a competitive advantage. Understanding the IT usage of individuals within firms is a necessary condition for ensuring the effective deployment of firm resources (Taylor & Todd, 1995). For example, firms employing global boundary spanners may assume that the primary form of communication between individuals will be through e-mail. Investments in an e-mail system, personal computers and hand-held communication devices may be made to benefit from e-mail's long-term cost efficiencies. The cost advantage of e-mail over other forms of communication in global communication may mean little in societies where individuals are simply less familiar and less comfortable with e-mail and perhaps more comfortable with face-to-face conversations or telephone calls. Offering training programs for these less comfortable users that are designed to promote the use of e-mail in varying business situations may benefit organizations with employees who communicate across borders. Offering training programs for all employees who communicate across borders on the cross-cultural preferences and uses of e-mail could additionally be beneficial to organizations in the promotion and management of e-mail to communicate across borders.

Since over half of the present Canadian data comes from Calgary, Canada's oil center, and that Nigeria's economy is heavily oil-dependent, interactions between managers in these two nations has practical significance. The greater use of e-mail in Canada than in Nigeria suggests that the Canadian managers might find it quite normal and comfortable to use e-mail when interacting with Nigerians, but Nigerians might find it less comfortable. Caribbean managers deal with both Canada and the United States as major trading partners. Since the amount of e-mail communication appears to be greater in the United States and Canada than in the Caribbean, Caribbeans might make less use of e-mail as a form of communication with managers in the U.S. or Canada. However, intercultural interactions also are well recognized to be quite complex, because once managers become aware of particular kinds of differences in cultural norms between nations, each party tends to adapt in ways that are difficult to anticipate. In conducting the present project, the researchers struggled with just these sorts of differences in communication media preferences. Comparative projects like this one should be considered as only part of understanding intercultural interaction. The way managers adapt to intercultural differences when using e-mail is an area for further research.

One recent study distinguished organization cultures by those that emphasize *rules versus risks* and also those that emphasize *rewards versus relations* (Borg, Groenen, Jehn, Bilsky, & Schwartz, 2011). The authors found that a better person-organization fit can be achieved when the personal values of workers align closely to these types of organization culture. They note that a person who values tradition and security is better served in an organization that values rules. In contrast, a person who values achievement may be better served in a rewards-focused organization. An area for future research would be to examine whether these types of organization cultures correspond to the use of email by workers within the firms. Current theories of technology, coupled

with findings from the present study suggest that email use would be more prevalent in organizations that value rules and relations, yet a further examination is warranted.

The results should also be considered when training managers in different nations about the utility of e-mail and when setting policy about use of e-mail within each nation's facilities of organizations operating in multiple nations. Like most social science research, university education, corporate training, and corporate policy setting throughout the world tends to be based on theory developed and results obtained in U.S. universities. Training or policies that encourage the use of e-mail in the expectation that it will support lateral communication, for example, may need to be reconsidered in light of the present results. In contrast, communicating formal or informal company policies, guidelines, or regulations via e-mail is likely to be well-received, even in varying cultural contexts. Follow up analysis by scholars in each nation is needed to determine whether or not the nation differences found here replicate and, if so, to provide explanations. Perhaps the most general implication of the present study is that e-mail should not be treated as something that works against the application of formal rules.

Application requires an expectation that an intervention, cultural training about the use of e-mail in this case, will have an effect. Application using this sort of causal reasoning is always tentative when based on cross-sectional research. The direction of influence is in most cases conceptually more likely to be from the variables treated as predictors to the use of e-mail rather than the reverse. That is, e-mail cannot possibly affect nationality or the demographic predictors. The argument that e-mail use affects values seems unlikely. The theory suggesting links between the use of e-mail and sources, however, rests on a reciprocal causal logic. That is, when sources amenable to e-mail are heavily used, e-mail use will be promote their use; when e-mail is heavily used, sources amenable to the use of e-mail will be drawn from. Of course, none of this causal logic is by any means certain given the present research design. Experimental research should continue to sort out these reciprocal effects in multiple nations and organization types.

LIMITATIONS

Considering the project's contributions as well as limitations suggests additional directions for future research. The measures are all obtained from a single survey collected at a single administration in each locale. While the results linking nationality and demographic characteristics to the use of e-mail and those treating nationality as a moderator cannot be explained by response bias, the results linking social structures and personal values to the use of e-mail may be affected by response biases. The varying strengths of relationships with the use of e-mail for the social structure measures and the finding of a larger proportion of significant effects for those for which positive relationships were predicted than those for which negative re-lationships were predicted suggest that response bias is not a plausible overall explanation for the results. Similarly, the lack of significant effects for the values measures for which effects on e-mail are unlikely suggests that survey-wide differences in individuals' propensity to answer positively or negatively do not produce the significant results obtained. Use of an alternative research method to overcome the project's same-source limitation in a field setting would be challenging. In principle, one might actually count the number of e-mails sent by respondents having different demographic characteristics in different nations who express dif-ferent values and work in different social contexts. Carrying out this sort of design is less likely to be practicable than, say, for studies that predict turnover or supervisor evaluations of performance rather than e-mail use as separate-source criteria.

Experimental studies, small-scale confirmation of the present results in individual nations, and ethnographic projects would be helpful, however.

Another limitation suggesting a need for further research is to consider additional demo-graphics. Although the data draw from people in many parts of the United States and Canada, the regional subsamples are not large enough to consider within-nation regional differences in culture. However, our data were more adequate for checking for such effects in Nigeria. Ad-ditional analyses not reported here indicate that the results are quite similar for the three Nigerian tribal groups studied. For Nigeria, we also have a smaller preliminary data base for Muslim Hausa that may ultimately provide the basis for further evaluation of a group at an even earlier stage in the information technology diffusion process.

The lack of personal values data for Nigeria is another limitation that awaits a future opportunity to do further data collection in this understudied nation.

We used a relatively early 56 item version of the Schwartz Value Survey. A more recent 57 item version (Schwartz, 2007) that revises and adds an item to strengthen the hedonism scale would be especially useful for research based on hedonism hypotheses.

CONCLUSION

Theories of media richness and social presence or social context can be combined with theories of individual characteristics to provide the conceptual basis for understanding the influence of national culture, personal values, and social structures on a manager's use of e-mail. Interpretive and constructivist theories that emphasize human autonomy and agency in using media should supplement rather than replace theories focusing on physical properties of communication media.

The potential richness and social presence of e-mail as a medium for communicating in a personal, emotional, nuanced way is bounded by its technical characteristics. Nevertheless, the national context, work context, and individual characteristics of e-mail users that are likely to affect TAM "ease of use" variables like comfort and skill in using e-mail affect a person's preferences and ability to take full advantage of its technical potential.

ACKNOWLEDGMENT

The authors would like to thank and Lawrence Nicholson, David C. Thomas, Betty Jane Punnett, and the faculty and staff of the University of West Indies for assistance with data collection in the Caribbean. This research was supported in part by the InternetCoast Adams Professorship and the Lynn Chair in International Business at Florida Atlantic University. A previous version of the chapter appear as Peterson, M.F., Thomason, S.J., Althouse, N., Athanassiou, N., Curri, G., Konopaske, R., Lenartowicz, T., Meckler, M.R, Mendenhall, M.E., Mogaji, A., & Rowney, J.I.A. (2010). Social structures and personal values that predict e-mail use: An international comparative study. Journal of Global Information Management, 18(2), 57-84.

REFERENCES

Agarwal, R., & Prasad, J. (1998). A conceptual and operational definition of personal innovativeness in the domain of information technology. *Information Systems Research, 9*(2), 204–215. doi:10.1287/isre.9.2.204

Ahuja, M. K., & Carley, K. M. (1999). Network structure in virtual organizations. *Organization Science, 10*(6), 741–757. doi:10.1287/orsc.10.6.741

Au, K. Y. (1999). Intra-cultural variation: Evidence and implications for international business. *Journal of International Business Studies, 30*(4), 799–812. doi:10.1057/palgrave.jibs.8490840

Avolio, B. J., & Dodge, G. E. (2001). E-leadership: Implications for Theory, Research, and Practice. *The Leadership Quarterly, 11*(4), 615–668. doi:10.1016/S1048-9843(00)00062-X

Baba, M. L., Gluesing, J., Ratner, H., & Wagner, K. H. (2004). The contexts of knowing: natural history of a globally distributed team. *Journal of Organizational Behavior, 25*(5), 547–587. doi:10.1002/job.259

Bhagat, R. G., Kedia, B. L., Harveston, P. D., & Triandis, H. C. (2002). Cultural variations in the cross-border transfer of organizational knowledge: An integrative framework. *Academy of Management Review, 27*(2), 204–239.

Biddle, B. J., & Thomas, E. J. (1966). *Role theory: Concepts and research.* New York, NY: John Wiley and Sons.

Borg, I., Groenen, P. J. F., Jehn, K. A., Bilsky, W., & Schwartz, S. H. (2011). Embedding the organization culture profile into Schwartz's theory of universals in values. *Journal of Personnel Psychology, 10*(1), 1–12. doi:10.1027/1866-5888/a000028

Caprara, G., Vecchione, M., & Schwartz, S. H. (2009). Mediational role of values in linking personality traits to political orientation. *Asian Journal of Social Psychology, 12*, 82–94. doi:10.1111/j.1467-839X.2009.01274.x

Carlson, J. R., & Zmud, R. (1999). Channel expansion theory and the experiential nature of media richness perceptions. *Academy of Management Journal, 42*(2), 153–170. doi:10.2307/257090

Chen, C. C., Chen, X., & Meindl, J. R. (1998). How can cooperation be fostered? The cultural effects of individualism-collectivism. *Academy of Management Review, 23*(2), 285–304.

Daft, R. L., & Lengel, R. H. (1986). Organizational information requirements, media richness and structural design. *Management Science, 32*(5), 554–571. doi:10.1287/mnsc.32.5.554

Davis, F. D. (1989). Perceived usefulness, perceived ease of use, and user acceptance of Information Technology. *Management Information Systems Quarterly, 13*(3), 319–339. doi:10.2307/249008

Desai, M., Fukuda-Parr, S., Johansson, C., & Sagasti, F. (2002). Measuring the technology achievement of nations and the capacity to participate in the network age. *Journal of Human Development, 3*(1), 95–122. doi:10.1080/14649880120105399

DeSanctis, G., & Monge, P. (1999). Introduction to the special issue: Communication processes for virtual organizations. *Organization Science, 10*(6), 693–703. doi:10.1287/orsc.10.6.693

Egri, C. P., & Ralston, D. A. (2004). Generation cohorts and personal values: A comparison of China and the United States. *Organization Science, 15*(2), 210–220. doi:10.1287/orsc.1030.0048

Fishbein, M., & Ajzen, I. (1975). *Belief, attitude, intention, and behavior: An introduction to theory and research.* Reading, PA: Addison-Wesley.

Gefen, D., & Straub, D. W. (1975). Gender differences in the perception and use of e-mail: An extension to the technology acceptance model. *Management Information Systems Quarterly, 21*(3), 389–400.

Gudykunst, W. B., & Ting-Toomey, S. (1988). *Culture and interpersonal communication.* Newbury Park, CA: Sage Publications.

Hall, E. T. (1976). *Beyond culture.* Garden City, NY: Anchor Books, Doubleday and Company.

Hall, E. T., & Hall, M. R. (1990). *Understanding cultural differences.* Yarmouth, ME: Intercultural Press.

Hofstede, G. (2001). *Culture's consequences.* Thousand Oaks, CA: Sage Publications.

Jarvenpaa, S., Knoll, & Leidner, D. (1998). Is anybody out there? Antecedents of trust in global virtual teams. *Journal of Management Information Systems, 14*(4), 29–64.

Kahn, R. L., Wolfe, D. M., Quinn, R. P., Snoek, J. D., & Rosenthal, R. A. (1964). *Organizational stress: Studies in role conflict and ambiguity.* New York, NY: Wiley.

Kahneman, D. (2003). A perspective on judgment and choice: Mapping bounded rationality. *The American Psychologist, 58*(9), 696–720. doi:10.1037/0003-066X.58.9.697

Kluckhohn, F. R., & Strodtbeck, F. L. (1961). *Variations in value orientations.* Evanston, IL: Row, Peterson.

Kundu, S. K., Niederman, F., & Boggs, D. J. (2003). The relevance of global management Information Systems to international business. *Journal of Global Information Management, 11*(1), 1–4.

Legris, P., Ingham, J., & Collerette, P. (2003). Why do people use Information Technology? A critical review of the technology acceptance model. *Information & Management, 40*(3), 191–204. doi:10.1016/S0378-7206(01)00143-4

Lenartowicz, T., & Roth, K. (2001). Does subculture within a country matter? A cross-cultural study of motivational domains and business performance in Brazil. *Journal of International Business Studies, 32*(2), 305–325. doi:10.1057/palgrave.jibs.8490954

Leung, K., Au, A., Huang, X., Kurman, J., Niit, T., & Nii, K. (2007). Social axioms and values: A cross-cultural examination. *European Journal of Personality*, *21*, 91–111. doi:10.1002/per.615

Lippert, S. K., & Volkmar, J. A. (2007). Cultural effects on technology performance and utilization. *Journal of Global Information Management*, *15*(2), 56–90. doi:10.4018/jgim.2007040103

MacDuffie, J. P. (2008). HRM and distributed work: Managing people across distances. In Walsh, J. P., & Brief, A. P. (Eds.), *The Academy of Management Annals* (*Vol. 1*, pp. 549–616). New York, NY: Lawrence Erlbaum.

Martins, L. L., Gilson, L. L., & Maynard, M. T. (2004). Virtual teams: What do we know and where do we go from here? *Journal of Management*, *30*(6), 805–835. doi:10.1016/j.jm.2004.05.002

Nisbett, R. E., & Wilson, T. D. (1977). Telling more than we can know: Verbal reports on mental processes. *Psychological Review*, *84*(3), 231–259. doi:10.1037/0033-295X.84.3.231

Peterson, M. F., & Smith, P. B. (2000). Sources of meaning, organizations, and culture: Making sense of organizational events. In Ashkanasy, N., Wilderom, C. P. M., & Peterson, M. F. (Eds.), *Handbook of organizational culture and climate* (pp. 101–116). Thousand Oaks, CA: Sage.

Peterson, M. F., & Smith, P. B. (2008). Social structures and processes in cross cultural management. In Smith, P. B., Peterson, M. F., & Thomas, D. C. (Eds.), *Handbook of cross-cultural management research* (pp. 35–58). Thousand Oaks, CA: Sage Press.

Peterson, M. F., & Soendergaard, M. (in press). Traditions and transitions in quantitative societal culture research in organization studies. *Organization Studies*.

Peterson, M. F., & Wood, R. (2008). Cognitive structures and processes in cross cultural management. In Smith, P. B., Peterson, M. F., & Thomas, D. C. (Eds.), *Handbook of cross-cultural management research*. Thousand Oaks, CA: Sage.

Rokeach, M. (1968). *Beliefs, attitudes, and values*. San Francisco, CA: Jossey-Bass.

Schwartz, S. H. (1992). Universals in the content and structure of values: Theoretical advances and empirical tests in 20 countries. In M. Zanna (Ed.), *Advances in Experimental Social Psychology, 25*, 1-65. Orlando, FL: Academic Press.

Schwartz, S. H. (2007). Universalism values and the inclusiveness of our moral universe. *Journal of Cross-Cultural Psychology*, *38*(6), 711–728. doi:10.1177/0022022107308992

Schwartz, S. H., & Boehnke, K. (2004). Evaluating the structure of human values with confirmatory factor analysis. *Journal of Research in Personality*, *38*(3), 230–255. doi:10.1016/S0092-6566(03)00069-2

Schwartz, S. H., & Sagiv, L. (1995). Identifying culture-specifics in the content and structure of values. *Journal of Cross-Cultural Psychology*, *26*(1), 92–116. doi:10.1177/0022022195261007

Schwartz, S. H., Verkasalo, M., Antonovsky, A., & Sagiv, L. (1997). Value priorities and social desirability: Much substance, some style. *The British Journal of Social Psychology*, *36*(1), 3–18. doi:10.1111/j.2044-8309.1997.tb01115.x

Smith, P. B., Peterson, M. F., & Schwartz, S. H. (2002). Cultural values, sources of guidance, and their relevance to managerial behavior: A 47-nation study. *Journal of Cross-Cultural Psychology*, *33*(2), 188–208. doi:10.1177/0022022102033002005

Spennemann, D. H. R., & Atkinson, J. S. (2003). A longitudinal study of the uptake of and confidence in using e-mail among parks management students. *Campus-Wide Information Systems, 23*(2), 55–67. doi:10.1108/10650740310467754

Tan, B. C. Y., Wei, K. K., Watson, R. T., Clapper, D. L., & McLean, E. (1998). Computer-mediated communication and majority influence: Assessing the impact in an individualistic and a collectivist culture. *Management Science, 44*, 1263–1278. doi:10.1287/mnsc.44.9.1263

Tannenbaum, A. S., Kavcic, B., Rosner, M., Vianello, M., & Wieser, G. (1974). *Hierarchy in organizations*. San Francisco, CA: Jossey-Bass.

Taylor, S., & Todd, P. A. (1995). Understanding Information Technology usage: A test of competing models. *Information Systems Research, 6*(2), 144–176. doi:10.1287/isre.6.2.144

Venkatesh, V., & Davis, F. D. (2000). A theoretical extension of the technology acceptance model: Four longitudinal field studies. *Management Science, 46*(2), 186–205. doi:10.1287/mnsc.46.2.186.11926

Venkatesh, V., & Morris, M. G. (2000). Why don't men ever stop to ask for directions? Gender, social influence, and their role in technology acceptance and usage behavior. *Management Information Systems Quarterly, 24*(1), 115–139. doi:10.2307/3250981

Venkatesh, V., & Zhang, X. (2010). Unified theory of acceptance and use of technology: U.S. Vs. China. *Journal of Global Information Management, 13*(1), 5–28.

Weber, M. (1947). *The theory of social and economic organization*. New York, NY: Free Press.

World Bank. (2006). *World development indicators database*. Retrieved October 1, 2008, from http://web.worldbank.org

Zhao, H., Kim, S., Suh, T., & Du, J. (2007). Social institutional explanations of global internet diffusion: A cross-country analysis. *Journal of Global Information Management, 15*(2), 28–55. doi:10.4018/jgim.2007040102

Chapter 9
Critical Factors of ERP Adoption for Small- and Medium- Sized Enterprises:
An Empirical Study

She-I Chang
National Chung Cheng University, Taiwan

Shin-Yuan Hung
National Chung Cheng University, Taiwan

David C. Yen
Miami University, USA

Pei-Ju Lee
National Chung Cheng University, Taiwan

ABSTRACT

Small and Medium-sized Enterprises (SMEs) play a vital and pervasive role in the current development of Taiwan's economy. Recently, the application of Enterprise Resource Planning (ERP) systems have enabled large enterprises to have direct contact with their clients via e-commerce technology, which has led to even fiercer competition among the SMEs. This study develops and tests a theoretical model including critical factors which influence ERP adoption in Taiwan's SMEs. Specifically, four dimensions, including CEO characteristics, innovative technology characteristics, organizational characteristics, and environmental characteristics, are empirically examined. The results of a mail survey indicate that the CEO's attitude towards information technology (IT) adoption, the CEO's IT knowledge, the employees' IT skills, business size, competitive pressure, cost, complexity, and compatibility are all important determinants in ERP adoption for SMEs. The authors' results are compared with research on IT adoption in SMEs based in Singapore and the United States, while implications of the results are also discussed.

DOI: 10.4018/978-1-61350-480-2.ch009

INTRODUCTION

In recent years, the evolution and application of information technology (IT) have enabled large enterprises to gain the capability of contacting clients directly via e-Commerce technology. In order to cooperate with large firms, SMEs need to incorporate information systems into their operations. Consequently, SMEs should improve their organizational structure and operational business processes by constantly strengthening and aligning their business partners within fast changing markets, while also being keenly responsive to external challenges (Iacovoui et al., 1995; Thong & Yap, 1995).

In order to cope with these aforementioned future challenges, SMEs must aggressively adopt newer market ideas, be sensitive to advanced technology, equip themselves with better, innovative solutions to maintain their core competencies, values and norms, and focus on future research and development issues (Adams et al., 1992; Ballantine et al., 1998; Grandon, 2004; Thong, 2001). SMEs not only face a highly competitive business environment, but they are also restricted by such factors as existing financial constraints, lack of access to expert knowledge, and an exposure to external pressures (Chwelos et al., 2001; Houben et al., 1999; Welsh & White, 1981). SMEs usually lack sufficient experience in the management of IT and information system professionals (Cragg & King, 1993; Iacovou et al., 1995). Even the SMEs that are willing to recruit experienced information system professionals have trouble retaining them on a full time basis (Gable, 1991). Furthermore, although factors such as technology, time, and human resources may not be major constraints for large organizations, this remain the main resource issue in SMEs implementation and management of their business plans and development. As a result, organizational theories that apply to large enterprises are not likely to be applicable to SMEs'. Prior literature also supports that influencing factors for large enterprises to adopt new information

systems may not be applicable to SMEs (Grandon, 2004; Raymond, 1985; Riemenschneider et al., 2003; Thong, 2001; Yap et al., 1992).

Enterprise Resource Planning (ERP) is an information systems technology and concept that was developed and initiated in the 1990's. ERP systems have proven to be useful solutions in integrating business processes and resources with the enterprises' operational and management strategies (Amoako-Gyampah & Salam, 2004; Davenport, 1998; Holsapple & Sena, 2005; Klaus, Rosemann, & Gable, 2000; Markus et al., 2000; Sarker & Lee, 2003). Several potential tangible and intangible benefits of applying ERP technology to businesses include such items as: improving organizational performance; understanding a range of functionalities not catered to in the existing system; increasing returns from the implemented ERP systems; streamlining an organization's internal processes; lowering costs for developing an ERP that more accurately reflects business needs; improved capability to react to a changing environment; and improved customer satisfaction related to services rendered or products manufactured (Chang & Gable, 2002; Davenport, 2000; Li, 1999; Rajagopal, 2002).

Although many prior studies focused on ERP, few have examined the effects of ERP adoption on SMEs (Bingi et al., 1999; Chen & William, 1998; Grandon, 2004; Holland & Light, 1999; Levy et al., 1999; Mabert et al., 2003; Rajagopal, 2002; Sarker & Lee, 2003; Scott & Kaindl, 2000). There are several existing studies focused on organizational characteristic for ERP adoption within SME such as the studies of Chan (1999); Kagan et al. (1990); Lang et al. (1997); Levy and Powell (2000); and Zinatelli et al. (1996). On the other hand, some characteristics about SMEs mentioned by previous studies include only CEO characteristics, Environmental characteristics and Innovation characteristics (i.e., SMEs were not only face a highly competitive business environment). In addition, they are also restricted by not studying such factors as existing financial constraints, lack of access to expert knowledge, and an exposure to external pressures (Chwelos et al., 2001; Houben et al., 1999; Welsh & White, 1981). SMEs usually lack sufficient experience in the management of IT and information system professionals (Cragg & King, 1993; Iacovou et al., 1995). Even the SMEs that are willing to recruit experienced information system professionals may have trouble retaining them on a full time basis (Gable, 1991). To this end, there are, being different from prior studies, this study aims to develop and test a comprehensive framework containing internal and external characteristics of SMEs in order to examine which factors affect SMEs' ERP adoption decisions.

Similar to many other emerging technologies, organizations typically exhibit a significant time gap between the adoption stage and the realization of benefits from effective management of ERP systems (Stratman & Roth, 2002). Although the technology for ERP systems has been available for a number of years, its full potential is yet to be realized. ERP systems have grown in popularity and with promising potential, there appears to be insufficient insight into the critical factors of ERP adoption for SMEs. This sustained interest in adopting and realizing the benefits of ERP systems in general, and the effect on ERP adoption by SMEs in particular, indeed provide the rationale for this study. In addition, a better understanding of the critical factors of ERP adoption by SMEs will also help direct the ERP research agenda.

This study will employ Taiwanese SMEs as an empirical target to discuss which factors influence them to adopt ERP systems. There are several reasons why this study chose SMEs in Taiwan. During the past three decades, economic development in Taiwan has depended heavily upon the development of small- and medium-sized enterprises (SMEs). According to the data from an annual white paper released by the Ministry of Economic Affairs in 2008, there were approximately one million and two hundred thousand SMEs in

Taiwan, equivalent to nearly ninety-eight percent of all businesses operating in Taiwan. Moreover, the sales of SMEs are approaching ten trillion NT dollars (28.34% of all enterprises). In terms of the domestic sales ratio, there are 32.49% and 67.51% in the SMEs and Large enterprises, respectively. In terms of the export sales ratio, there are 17.02% and 82.98% in the SMEs and Large enterprises, respectively. Based on the above results, the domestic sales ratio of SMEs is larger than the export sales ratio. Due to their flexibility, agility, and efficiency, SMEs have become the backbone of Taiwan's economic miracles. According to the survey from the Market Intelligence and Consulting Institute (MIC, 2007), most large enterprises have implemented ERP systems, especially in the manufacturing sector where a 90% implementation rate has been reached. Further, information technology investment in SMEs had 3% growth in 2006, with the top three adopted systems including customer relationship management (CRM), enterprise resource planning (ERP), and Business Intelligence (BI). In addition, the trade liberalization trend since Taiwan's entry into the World Trade Organization (WTO), the globalization of economic development, the rapid dispersion of industrial knowledge, and the impact of the application of latest technologies on industries are all major challenges for the continuous survival of SMEs in Taiwan.

This study has two main contributions. Firstly, In practice, to meet challenges in the highly competitive business environment, SMEs must aggressively adopt advanced information technology for survival. The adoption of ERP by SMEs is a crucial issue in today's business environment, while most available literatures are mostly focus on the study of large firms. Secondly, prior literature still lack of a comprehensive framework about critical factors of ERP Adoption for SMEs. Therefore, this study investigates some important characteristics and develops and tests an comprehensive model to discuss would enhance the theoretical value in this issue. Further, this comprehensive framework can shed a light as SMEs would like to adopt ERP and speed up the efficiency of decision.

LITERATURE REVIEW

The Adoption of ERP Systems

As discovered by Beard and Sumner (2004), Chang and Gable (2002) and Davenport (1998), although ERP systems promise better resource planning and execution, in addition to improved product quality and reliable service delivery (Hunton et al., 2003), most organizations have not fully realized the expected benefits from the adoption of these aforementioned systems. In their discovery they found that: (1) more than 40% of large software projects are unsuccessful (Davenport, 1998); (2) 90% of ERP implementations end up late and/or over budget (Davenport, 1998); (3) continuous shortages and concomitant turnover of ERP staff contribute to high ERP costs (Beard & Sumner, 2004); (4) the growth of ERP consulting services has led to a proliferation of methods, techniques and tools, causing additional implementation difficulties (Chang & Gable, 2002); (5) 67% of enterprise application integration initiatives could be considered either negative or unsuccessful (Davenport, 1998). These difficulties associated with packaged software implementations actually suggest that many organizations underestimate the implementation/adoption issues and problems often encountered throughout the ERP life cycle. As the number of organizations implementing ERP systems increase and ERP applications within organizations proliferate (Bancroft, 1998; Beard & Sumner, 2004; Davenport, 2000; Kwahk & Lee, 2008; Xu & Ma, 2008), critical factors for improved understanding of ERP adoptions by SMEs are required, so that development and management functions/activities can be effectively monitored.

For example, the adoption of ERP systems have prevailed throughout most enterprises since 1997, when SAP set up its first branch office in Taiwan. Following the entry of Taiwan into the WTO, the interaction between Mainland China and Taiwan became closer. According to the white paper released by the Ministry of Economic Affairs (2008), the investment amount of SMEs to China is about USD 10 billion. On the other hand, due to the large business market and decreased language obstacle, the first option for SMEs' investment is China. Overall, Taiwan's SMEs have been deeply and heavily dependent on China. In order to handle cross-region transactions, the need for information sharing and faster transmission capabilities became requirements in business operations wherever the head office was located. Further, more and more large-sized enterprises in Taiwan had pioneered to adopt ERP systems in order to appreciate its advantages such as the effective handling of transactions, and the constant reduction of repeated work and redundancy inside the company (MIC, 2008). In 2002, the directorate general of Budget Accounting and Statistics, Executive Yuan of Taiwan announced that the prevailing rate of the ERP market had reached a saturation level of 53% of the ERP adoptions in domestic, large-sized enterprises. With the steady popularity of ERP adoption in Taiwan large companies, small firms may be able to learn faster and more effectively from their peers. Meanwhile, more and more ERP providers value the potential of the SME market and thus increase have increased their penetration into the market with appropriate promotional strategies and better-fit functional alternatives (Chen et al., 2008). Secondly, many leading system providers around the world have implemented the suggestion to cooperate with manufacturers in Taiwan to install ERP systems to help increase production efficiency in the supply chain. Thirdly, the effort to transfer investments from Taiwan's SMEs to mainland China presents the urgent need to have

ERP systems that provide assistance in cross-strait business management issues.

Though the ERP market has slowly increased over the past years, the adoption of ERP systems in Taiwan's SMEs appears to be sluggish (Hung & Liang, 2001). Thus far, ERP adoption has been favored by larger organizations. Features and business process flows developed to facilitate the adoption of ERP systems are largely based on the operational practices in large organizations. Further, the consulting and project management methodologies normally utilized are simply based on such experiences to adopt ERP systems in larger enterprises. The needs, operating requirements, logistics fulfillment, and financial capabilities of SMEs are however, vastly different from those of the large organizations (MIC, 2007). The experience in the management of ERP projects that are focused on the adoption of information technology by SMEs is also limited. According to the study by Tsai et al. (2005), ERP adoption by SMEs in Taiwan has generally been in a "wait-and-see" situation, because ERP adoption has to match the reengineering of business processes, an issue with which the SMEs did not have much experience. Hence, Hung and Liang (2001) concluded that most surveyed enterprises still hesitated to move forward with ERP adoption. While the aforementioned practitioner- and academic-oriented research provides valuable insights concerning the effective use of ERP technology and the implementation process issues related to the adoption of ERP systems, there is a need for more systematic, empirical analysis of the critical factors that lead to successful adoption and management of ERP systems.

Critical Factors to IT Innovation Adoption

ERP systems can be viewed as a unique IT application, so it is an acceptable practice to use IT-related literature as the theoretical basis and reference to explore ERP adoption. The Innovation Diffusion Theory of Rogers (Rogers, 1995) has been one of the most widely accepted theories in the innovation technology area. Subsequently, researchers began to combine it with other factors to become more comprehensive and explanatory. Rogers further defined innovation as a new concept with great influences on the enhancement of personal and organizational goals and measures. According to the definition of Poutsma et al. (1987), innovation has four major characteristics, which include: (1) process innovation or production innovation; (2) innovation beginning from the basis or incremental innovation; (3) innovation happening in a technology push or market pull environment; and (4) planned or incidental innovation.

Based on Poutsma's definition, ERP belongs to the category of process innovation through the employment of new methods, machineries, and/or equipment that enhance data processing and distribution logic functions, and service standards. ERP can also be categorized as an innovation from which it utilizes information technology to realize the benefits of implementation, which is a type of major reform. The potential reason for ERP adoption is in accordance with the strong trend of the technology push or pull market characteristic introduced earlier and matches Poutsma's definition to characterize ERP as a planned innovation.

Enterprise systems are different from many previously studied IT innovations, because they involve all areas and functions of management (Bradford & Florin, 2003). In addition to similarities, there are also significant differences between IT and ERP. For example, ERP systems are integrative and require significant changes to processes across the entire organization to ensure successful implementation. Furthermore, ERP systems include a number of functionalities, such as human resources planning, decision support applications, regulatory control, quality, elements of supply chain management, and maintenance support that are beyond the traditional focuses of IT (Mabert et al., 2003). The benefits of ERP systems are not only recognized by large firms;

SMEs can also benefit from ERP systems through the reduction of data and application redundancy by accomplishing systems and applications integration. In addition, ERP systems can "level the playing field" for larger firms by providing location and time independence, and by reducing the need for frequent communication (Amoako-Gyampah & Salam, 2004; Beard & Sumner, 2004). However, in spite of the many potential advantages of ERP systems addressed so far, the benefits from its adoption by SMEs remain limited (Tsai et al., 2005).

Raymond (1985) and Buonanno et al. (2005) found that organizational theory and practical operations applicable to the domain of large enterprises might not be suitable for smaller ones. Welsh and White (1981) and Chwelos et al. (2001) showed that small enterprises operate their businesses in a highly competitive environment and thus, are limited to resources such as insufficient financial capital, shortage of professionals, and easy exposure to external pressure. Similarly, Yap et al. (1992) also found that compared to large businesses, SMEs lack competent human, financial, and material resources. In gathering the variables concerned, this study will focus on the literature applicable to technology innovation in the domain of SMEs.

Considering that the adoption of new technology in small enterprises is different from that of large enterprises, the factors from these aforementioned studies that were found to be critical to the adoption of IT by different researchers are summarized in Table 1.

There are several factors that affect SMEs' decisions to adopt and implement ERP systems. Financial capability, considering the high cost of ERP adoption, is the main burden for smaller companies' ability to adopt and implement ERP systems (Gartner Group & Dataquest, 1999). Furthermore, Lang et al. (1997) indicate that information-seeking practices can influence SMEs to adopt ERP; also, Kagan et al. (1990) found that the main difference between SMEs and larger enterprises in the decision to adopt ERP systems is the lack of information system management in SMEs. On the other hand, SMEs do not have enough resources to commit an effective portion towards generating the information needed to successfully adopt and implement ERP systems (Chan, 1999). Buonanno et al. (2005) summarize some factors may affect ERP adoption in SMEs, including the scarcity of resource and strategic planning in IS, and the limited expertise in information technology (Levy & Powell, 2000; Zinatelli et al., 1996).

Table 1. Factors to adoption of information technology

Source	Dimensions and Exogenous Variables
Cragg and King (1993)	Motivators of IT growth: (1) Relative advantage, (2) Competitive pressure, (3) Consultant support, (4) Managerial enthusiasm. Inhibitors of IT growth: (1) IT education, (2) Managerial time, (3) Economic factors, (5) Technical factors.
Iacovou et al. (1995)	Perceived benefits: (1) Awareness of direct benefits, (2) Awareness of indirect benefits. Organizational readiness: (1) Financial resources, (2) Technical resources. External pressure: (1) Competitive pressure, (2) Imposition by partners.
Thong and Yap (1995)	CEO characteristics: (1) CEO's innovativeness, (2) CEO's attitude toward IT adoption, (3) CEO's IT knowledge. Organizational characteristics: (1) Business size, (2) Competitiveness of environment, (3) Information intensity.
Thong (1999)	CEO characteristics: (1) CEO's innovativeness, (2) CEO's IT knowledge. IS characteristics: (1) Relative advantage, (2) Complexity, (3) Compatibility. Organizational characteristics: (1) Business size, (2) Employee's IT knowledge, (3) Information intensity. Environmental characteristics: (1) Competition.
Premkumar and Roberts (1999)	Innovation characteristics: (1) Relative advantage, (2) Cost, (3) Complexity, (4) Compatibility. Organizational characteristics: (1) Top management support, (2) Business size, (3) IT expertise. Environmental characteristics: (1) Competitive pressure, (2) External support, (3) Vertical linkages.

RESEARCH MODEL AND HYPOTHESES

To investigate crucial factors that influence ERP adoption, these four categories --- "CEO's characteristics", "innovative technology characteristics", "organizational characteristics", and "environmental characteristics," set forth by Thong's study (Thong, 1999), are utilized as the basic foundation in this study. There are two points that why we adopt Thong's model. First, this model or constructs conducted by Thong have broadly been verified in the small business or IS adoption issues (Al-Qirim, 2006; de Guinea et al., 2005; Lee, 2004; Quaddus & Hofmeyer, 2007). Second, this model or constructs conducted by Thong have less been verified in the ERP adoption issues. Due to the specific characteristic of SME such as the structure of decision making, resource and so on, there should verify whether model of IS adoption exist in the ERP adoption of SME. In addition, the justification for the proposed research model and the associated hypotheses are presented in Figure 1 and discussed below.

CEO Characteristics

The CEO is an important entrepreneur figure who is crucial in determining the innovative attitude of a particular small business (Raymond, 1985). Thong's study (Thong, 1999) found that CEO innovativeness is positively associated with the decision to adopt IS in small businesses. ERP is a complex, technological innovation that requires large outlays of financial resources; the large amount of financial resources needed is especially difficult for SME's with scarce resources. If ERP implementation is not successful, an SME can suffer unavoidable damages, and consequently may not be able to recover and compete with other businesses in the future; hence, ERP adoption is a risky venture. Chang et al. (2008) indicate that CEOs of Taiwan SMEs demonstrate some specific characteristics such as flexibility and excellent responsibility in operation, and a SME with an innovative CEO who is willing to take the risk and associated challenge can help to increase the likelihood for such an adoption.

H1: Businesses with more innovative CEOs are more likely to adopt ERP systems.

In an SME, the chief decision-maker is the CEO. The CEO's perception of the adoption of IT is critically important. A degree of uncertainty exists since adoption of IT is risky, and each CEO perceives the degree of risk and uncertainty associated with IT in a different manner. If the CEO perceives the benefits of IT adoption to outweigh the risks, then the business is more likely to adopt IT. A study by Thong and Yap (1995) indicated that CEO attitude towards IT adoption significantly affects IT adoption in Singaporean SMEs.

According to Hung and Liang (2001), since Taiwan's cultural characteristics are more fearful of risk and apprehensive to seek adoption of innovations, CEOs from Taiwan's SMEs usually take a "wait-and-see" attitude/approach to finding successful IT innovations. Based on the above discussion, the CEOs' attitude towards IT adoption should play a role in ERP adoption in SMEs.

H2: Businesses with CEOs who have a more positive attitude towards IT adoption are more likely to adopt ERP systems.

In an earlier study, Niedleman (1979) attributed a lack of IT knowledge as the reason for European SMEs failure to recognize the full benefits of IT utilization. Thong's study (Thong, 1999) also indicated that SMEs with CEOs who are more knowledgeable about IS/IT are more likely to adopt them. Gable and Raman (1992) found that CEOs in small businesses tend to lack the basic knowledge and awareness needed to successfully adopt IS. If, in turn, these CEOs could realize the benefits of IS, they will be more willing to adopt such technology. Therefore, this study proposes the following hypotheses.

Figure 1. Research model

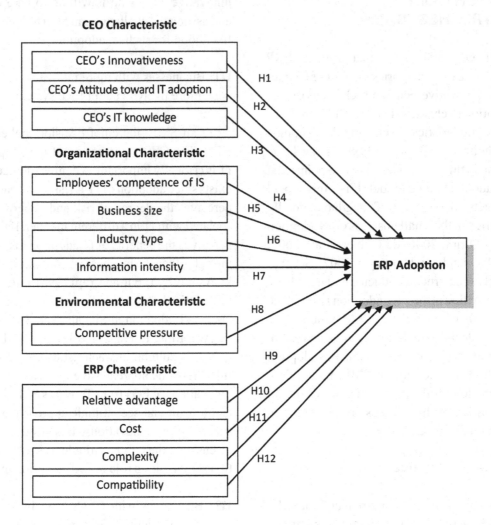

H3: Businesses with CEOs who are more knowledgeable about IT are more likely to adopt ERP systems.

Organizational Characteristics

DeLone's study (DeLone, 1988) illustrates that SMEs may lack the appropriate amount of professional knowledge about information systems. Thus, the lack of financial resources and knowledge in SMEs will delay the adoption of new IT. Amoako-Gyampah and Salam (2004) reported that the effective understanding of IT and the resulting participation of all levels of employees are signifi-

cant factors in successful ERP adoption within the SME segment. Yap et al. (1992) revealed that IT experience and user involvement in the organization also affect the adoption of new information technology. According to Thong's study (Thong, 1999), the competence of employees in the field of information systems has a positive influence on the adoption of information technology in small businesses. Pai et al. (2007) in their study of large companies in Taiwan, point out the critical factor of adopting ERP systems in Taiwan is employee rather than system, in particular, employee's IT capability.

H4: Businesses with employees who are more knowledgeable about IT are more likely to adopt ERP systems.

Larger organizations are found to have greater slack in resources and therefore are more capable of experimenting with new innovations. Grover's study (Grover, 1993) noted that business size was the major factor that affected the adoption of new IT. Even among small businesses, the larger SMEs are more comfortable to take risks with new technologies (Palvia et al., 1994). Thong's study (Thong, 1999) also supported that business size was influential to IT adoption, especially in small businesses. Additional studies also indicated that business size is one of the key factors in the acceptance of IT (Lees, 1987; Premkumar & King, 1992). Pan and Jang (2008) also support this variable as a crucial factor for adopting ERP in Taiwan.

H5: Businesses that are larger in size are more likely to adopt ERP systems.

Information systems adoption varies with the nature of industry types. The empirical study conducted by Premkumar and King (1992) revealed that a number of associated planning characteristics, such as information system planning, information system resources, the role of information systems in organizations, enterprise's positioning analysis, contribution to organizational performance, and implementation success, varied significantly among different industry groups. Prescriptive studies indicate that some industries have greater potential for exploiting IT than others and therefore differ in ERP adoption. Thong (1999) indicates that industry category affects ERP adoption. Moreover, firms in industries with short product life cycles and rapid product innovation are regarded as having high uncertainty, thus leading to the greater diffusion of IT innovation (Melville & Ramirez, 2008).

H6: The likelihood of ERP adoption varies among different industries.

Information intensity refers to the timely supply of supporting information required by a product or service to maintain the enterprise's competitiveness. The requirements of information processing vary among enterprises. Yap's study (Yap et al., 1992) noted that businesses in different sectors have different information processing needs and requirements and those in more information-intensive sectors are more likely to adopt IT than those in less information-intensive sectors. Further, Thong's study (Thong, 1999) indicated that while information intensity may not have a direct positive relationship to the adoption of IT in small businesses, information intensity may also have an indirect influence on IS adoption via the characteristics of information systems. This study thus proposes the following hypothesis.

H7: Businesses located in more information intensive environments are more likely to adopt ERP systems.

Environmental Characteristics

A business in an environment that is more competitive would feel a greater need to use IS to gain a competitive advantage. Premkumar et al. (1997) indicated that EDI has become a strategic necessity to compete in the market place. Premkumar et al. (1997) also discovered that pressures from competitors exert considerable influence on the adoption of EDI in an enterprise. Although Pan and Jang (2008) found little influence on the decision whether or not to adopt ERP in Taiwan's context, the studies from Cragg and King (1993), Iacovou et al. (1995), Premkumar and Ramamurthy (1995), and Premkumar and Roberts (1999) support that competitive pressures contribute to a portion of critical factors that influence the adoption of new strategies in SMEs. As such, we contend that:

H8: Businesses that are in a more competitive environment are more likely to adopt ERP systems.

The organization's perception of the value of the innovation adoption (ERP systems) whether globally or locally in Taiwan in terms of the associated advantages and disadvantages compared to existing solutions is an important consideration in making the adoption decision (Wang et al., 2006; Waarts et al., 2002). In this study, the authors address organizations' value assessment through a number of variables related to the frequently cited advantages of ERP adoption. These advantages include reduced turnaround time, increased customer service, reduced costs, and prompt information for organizational decisions. This study thus suggests that the relative advantages of ERP systems will facilitate the adoption of ERP for an organization.

H9: The greater the perceived relative advantage of ERP systems the more likely they will be adopted.

For small businesses, the costs of hardware and software remain a sizeable deterrent to the adoption of technology; therefore firms need to evaluate the relevant costs before adopting a new technology. According to the study of Tornatzky and Klein (1982), technologies that are perceived to be low in cost are more likely to be adopted. In the case of Taiwanese manufacturing industry, Chiu et al. (2001) found that most SMEs are concerned about the limited financial resource available for new technology adoption. Premkumar and Roberts (1999) also indicated that costs are an important determinant of Internet adoption in rural small businesses. Furthermore, Bingi et al. (1999) found that costs are crucial to ERP adoption. Therefore, this study proposes the following hypotheses.

H10: The lower the perceived costs of ERP systems the more likely they will be adopted.

Complexity is the degree of difficulty associated with understanding and learning how to use an innovation (Rogers, 1995). This perceived complexity of IT will negatively influence the adoption decision (Grover, 1993). Companies around the world that perceive the ERP system adoption to be a complex business solution will tend to diffuse it slowly and in limited capacity, and consequently may not realize its full benefits in the adoption and future stages (Wang et al., 2006; Bradford & Florin, 2003).

H11: The lower the perceived complexity of ERP systems the more likely they will be adopted.

Compatibility is referred to as the degree to which an innovation is perceived as being consistent with the existing values, needs, and past experiences of the potential adopter (Rogers, 1995). If IS are compatible with existing work practices, the SMEs will be more likely to adopt them. In an ERP environment, it is likely that certain software will be retained and integrated with the newer ERP system. A company that attempts to adopt a new ERP technology that is incompatible to its existing platform, is more likely to face a delay in adoption. This is mainly the result of the organization's need to replace its hardware and operating system, which entails high switching costs. The compatibility between ERP software and retained systems has a positive influence on ERP adoption (Waarts et al., 2002). Wang et al. (2006) and Chiu et al. (2001) also concluded compatibility is a critical factor for Taiwan manufacturing companies in their study. Based on the above discussion, this study proposes the following hypothesis.

H12: The greater the perceived compatibility of existing IT and new ERP systems in an organization, the more likely it will adopt the new ERP technology.

RESEARCH METHOD

The data was collected using a questionnaire survey. First, this study created a sampling frame with the assistance of the Hand Tool Manufacturers' Association, the Small and Medium Enterprises Association, and the China Credit Information Publication. In this study, the criteria for defining an SME were adopted from the Small and Medium Enterprise Administration (SMEA) and the Ministry of Economic Affairs (MOEA) in Taiwan. Then, the 139 usable responses out of 800 possible SMEs that matched the aforementioned SMEA criteria were collected using a systematic sampling method and analyzed using discriminant analysis. To verify the construction of the research model and measurement instruments, several professors and users were also interviewed to modify the research model and the construction of the questionnaire.

Instruments Development and Pretest

The collection and measurement operations of the research variables are summarized in Table 2. Items were adapted from previously used scales, while all perceptual items were measured by a five-point Likert scale representing a range from "strongly disagree" to "strongly agree." To ensure suitability and correctness of the questionnaire content, a group of information management experts were employed to inspect whether or not there were any incorrect use of words or sentences that needed revision or improvement. Finally, experienced supervisors and academic experts from the IS field were invited to conduct a pretest in order to verify that all questions were clear enough to avoid confusion and misunderstanding.

Data Collection

The actual data collection process was carried out through the following three steps. First, information, such as the company name and company contact person, was collected. Since not all directories of the three organizations provided the same contact information for the CEO of each SME, the contact person may have been the SME owner, a supervisor of the IS department, or a top-level manager. Second, a package was mailed to the contact person of each of the SMEs in the survey sample. The respondents also were assured of the confidentiality of their responses. Finally, a follow-up telephone call was made to the contact person of the remaining non-respondents after a two-week waiting period. With an initial response rate of 18.5%, a total of 148 copies of the questionnaires were received. However, nine questionnaires out of 148 were incomplete and were hence excluded from the final sample, leaving a total of 139 effective responses. Consequently, the effective response rate was measured at 17.4%.

Reliability and Validity

Table 3 shows the values for Cronbach's α range from 0.65 to 0.83. All the reliability coefficients met Nunnally's (Nunnally, 1978) guideline of 0.60 and hence, the research variables proved to be reliable.

In this study, the validity test was assessed through content and construct validity. Content validity was established through the extensive process of item selection and refinement in the development of the instrument. The items used for construct measurement were derived from prior empirical studies, and were adapted to suit the context of this research. Upon finishing the questionnaire design, information management experts scrutinized all items to ensure that there

Table 2. Definition and measurement of the variables

Variable	Definition	Measurement	Items in the Questionnaire
CEO's innovativeness (CI)	The degree to which CEO risks to accept the latest IT	Thong (1995)	CI1-CI3 in Section 1
CEO's attitude toward IT adoption (CA)	CEO's attitude and viewpoint toward the latest IT	Thong (1995)	CA1-CA3 in Section 1
CEO's IT knowledge (CK)	IT-related knowledge possessed by the CEO	Thong (1995)	CK1-CK2 in Section 1
Employees' IT knowledge (EK)	Employees' know-how, competence, and experience concerned with information systems	Thong (1999)	EK1-EK3 in Section 2
Business size (BS)	The measure according to the number of employees	Thong (1995)	BS1 in Section 2
Industry type (IT)	The industry type to which the enterprise belonged		IT1 in Section 2
Information intensity (II)	The degree of an enterprise's demand for information	Thong (1995)	II1-II3 in Section 2
Competitive pressure (CP)	The pressure borne by the enterprise under the sharply competitive environment	Thong (1995)	CP1-CP2 in Section 3
Relative advantage (RA)	The degree to which ERP systems are recognized by the user to improve the status quo	Premkumar and Roberts (1999)	RA1-RA4 in Section 4
Cost (CO)	The cost incurred in the ERP adoption	Premkumar and Roberts (1999)	CO1-CO3 in Section 4
Complexity (CL)	The degree to which ERP systems are difficult to be understood and used	Premkumar and Roberts (1999)	CL1-CL2 in Section 4
Compatibility (CT)	The degree to which ERP is recognized to correspond to the existing value, past experience, and potential adopter's demand	Premkumar and Roberts (1999)	CT1 in Section 4

Table 3. Reliability analysis

Construct	Cronbach's α
CEO's innovativeness	0.73
CEO's attitude towards IT adoption	0.74
CEO's IT knowledge	0.65
Employees' IT knowledge	0.73
Information intensity	0.76
Business size	N/A
Industry type	N/A
Competitive pressure	0.72
Relative advantage	0.83
Cost	0.73
Complexity	0.74
Compatibility	N/A

were no errors, omissions, or misuse of words that might confuse the respondents. In addition, extensive pilot testing of the instruments involving members with practical experience, such as academic scholars and supervisors of the IS department from various industries, were employed to ensure that the items were relevant from a practitioner's perspective. The construct validity of the research variables was examined using the principal component factor analysis with varimax rotation. The results are provided below in Table 4. The results of the factor analysis indicated that all the factor loadings were greater than the cutoff point of 0.50, as recommended by Nunnally (Nunnally, 1978). Nonetheless, one item (II1) of the information intensity construct was eliminated due to significant cross-loading of employees' IT knowledge. Other items are all loaded on their hypothesized variables. As a result, construct validity was demonstrated.

RESULTS

Sample Characteristics and Sample Representativeness

The sample characteristics of the 139 businesses are shown in Table 5. Of the businesses sampled, 29 were in the electronic information sector (20.9%), 89 were in the manufacturing sector (64.0%), and 21 were in the service sector (15.1%). Out of the responding sample, 50 SMEs (36.0%) had already adopted ERP systems while 89 respondents (64.0%) had not at the time of the study. Among the 50 ERP adopters, 43 businesses (86.0%) were in the manufacturing and electronic information sectors. This figure corresponds closely with the profile of ERP adoption in the Taiwanese SME environment. The chi-square goodness-of-fit test was used to test whether the sample data ratio, including ERP adopters and non-adopters, comes from the population with a specific distribution. The results indicated that this sample is representative of a population having adopted ERP. Moreover, this study also tests for response biases between the responses and non-responses using the independent sample t-test. The results revealed no significant differences in terms of the age of the firm, total capital, or number of employees. Thus, no response bias existed in this study.

Correlation Matrix

The Pearson correlation matrix was examined to determine the extent of multicollinearity problems (Table 6). The greatest squared correlation among the independent variables was 0.23 between the aggregated measure of competitive pressure and complexity. None of the squared correlations was close to 0.80 to suggest a problem with multicollinearity among the research variables (Hair et al., 2006). The results reveal that there was no evidence of significant multicollinearity among the research variables proposed in this study.

Hypotheses Testing

Discriminant analysis was used to test the research hypotheses proposed. The analysis provided a statistical procedure to identify the research variables that best discriminated between adopters and non-adopters. Since industry type was a categorical variable among the three groups, two dummy variables were used to represent it. Table 7 shows the results of the discriminant analysis. The overall model is significant (Wilks' Lambda=0.563, $\chi 2$=75.296, p=0.000), which means ERP adopters and non-adopters can be effectively discriminated by the discriminant function. Further, the results indicate that eight variables, including CEO's attitude toward IT adoption (p=0.002), CEO's IT knowledge (p<0.001), employees' IT knowledge (p<0.001), business size (p<0.001), competitive pressure (p=0.005), costs (p<0.001), complexity (p<0.001), and compatibility (p=0.001), have significant effects on ERP adoption. The percentage of these correctly classified was 82.0%, which exceeds the hit ratio of 53.9% that would have been expected due to chance (Hair et al., 2006). The individual correct classification rate for adopters and non-adopters was 84.0% and 80.9%, respectively.

DISCUSSION

After testing the aforementioned hypotheses, the testing results were summarized in Table 8.

CEO Characteristics

Through data analysis, both characteristics, including CEOs' attitude toward IT adoption and CEOs' IT knowledge, have a positive influence on whether or not ERP systems are adopted. Particularly, a CEO's IT knowledge is more significant than a CEO's attitude toward IT adoption in discriminating between adopters and non-adopters

Table 4. Factor analysis

Construct Items	RA	CI	EK	CA	II	CO	CL	CP	CK	CT
RA1	.733 [a]									
RA2	.762									
RA3	.715									
RA4	.622									
CI1		.805								
CI2		.784								
CI3		.768								
EK1			.693							
EK2			.756							
EK3			.746							
CA1				.694						
CA2				.600						
CA3				.774						
II2 [b]					.889					
II3					.911					
CO1						.727				
CO2						.709				
CO3						.710				
CL1							.651			
CL2							.859			
CP1								.852		
CP2								.866		
CK1									.699	
CK2									.915	
CT1										.803
Eigenvalue	2.535	2.252	2.150	2.121	2.094	2.040	1.912	1.736	1.626	1.370

CI: CEO's innovativeness; CA: CEO's attitude toward IT adoption; CK: CEO's IT knowledge;

EK: Employees' IT knowledge; II: Information intensity; CP: Competitive pressure;

RA: Relative advantage; CO: Cost; CL: Complexity; CT: Compatibility

[a]: Factor loading.

[b]: II1 deleted due to significant cross-loading with EK.

of ERP systems. Due to the lack of IT knowledge, most small businesses always place a priority on selecting hardware rather than software, and thus, often fail to achieve their desired goals (Lees, 1987). The application of IT in Taiwan's SMEs is deeply affected by the CEOs. If a CEO has clear, IT-orientated ideas, the level and sophistication of information application will become greater.

Otherwise, the business operations remain only in lower organizational levels such as a traditional transaction processing system. Moreover, a CEO's favorable attitude toward management is proven to be beneficial in expanding the level and scope of the application of information systems. A CEO's positive attitude and solid IT knowledge are closely related to the adoption of ERP.

Table 5. Sample characteristics

Item	Characteristics	Frequency	Percent
Industry type	Electronic information	29	20.9
	Manufacturing	89	64.0
	Service	21	15.1
ERP adoption	Adopter	50	36.0
	Non-adopter	89	64.0
Age of firm (years)	< 4	5	3.6
	5-9	22	15.8
	10-14	20	14.4
	15-19	25	18.0
	20-24	29	20.9
	> 25	38	27.3
Total capital (NT$ million) [a]	< 12	32	23.0
	13-29	31	22.3
	30-46	11	7.9
	47-63	5	3.6
	64-80	23	16.5
	> 81	37	26.6
Number of employees	< 20	35	25.2
	21-80	48	34.5
	81-140	21	15.1
	141-200	25	18.0
	> 201	10	7.2
Number of IS personnel	No formal IS Department	45	32.4
	1-3	69	49.6
	4-6	12	8.6
	7-10	9	6.5
	> 11	4	2.9
Computer expenditure (NT$ thousand)	< 100	36	25.9
	101-500	48	34.5
	501-1000	30	21.6
	1001-1500	10	7.2
	> 1501	15	10.8

[a]: US$1 = NT$32 approximately.

In viewing the significant characteristics of CEOs, this study differs considerably from the views of Singaporean researchers such as Thong and Yap (1995) and Thong (1999), in terms of the "CEO's innovativeness" variable. According to the study by Hung and Liang (2001), a significant cultural characteristic of SMEs in Taiwan is the fear of risk and less adventurism towards the discovery of innovations, whereas Singaporean CEOs think exactly the opposite. Thus, the CEO's

Table 6. Pearson correlation matrix

Variable	(1)	(2)	(3)	(4)	(5)	(6)	(7)	(8)	(9)	(10)
(1) CEO's innovativeness										
(2) CEO's attitude towards IT adoption	.365*									
(3) CEO's IT knowledge	.303*	.203*								
(4) Employees' IT knowledge	.459*	.433*	.309*							
(5) Information intensity	.269*	.232*	.167*	.355*						
(6) Business size	.134	-.023	-.068	-.146	-.036					
(7) Competitive pressure	.075	.139	.081	.218*	.242*	-.101				
(8) Relative advantage	.241*	.423*	.153	.234*	.442*	-.088	.314*			
(9) Cost	.008	.184*	.026	.088	.026	-.204*	.093	.179*		
(10) Complexity	.143	.349*	.166	.231*	.051	-.221*	-.048	.176*	.520*	
(11) Compatibility	.039	.304*	.143	.170*	.165	-.174*	.135	.272*	.385*	.453*

*: significant at $p < 0.05$; **: significant at $p < 0.01$.

innovativeness is the outstanding variable that supports the adoption of new information technology in Singapore's SMEs, whereas the CEO's innovativeness in Taiwan shows no strong influence on the adoption of new information technology.

Organizational Characteristics

The results reveal that employees' IT knowledge, and business size positively affect the adoption of ERP systems. SMEs usually lack experts who are proficient in information systems and IT applications. This argument is supported by the fact that 37.9% of ERP non-adopters lack IT/IS professionals and 26.4% of them claim a shortage of development resources. Due to cost considerations for hiring individuals with specialized skills like IT knowledge, almost all of Taiwan's SMEs employ fewer than five IS personnel; in fact some businesses do not even have a formal IS department. Nevertheless, when the scale of business operations increases to a certain degree, an urgent need for IS specialists tends to rise. Many businesses recruit employees with a great deal of experience in the field of information systems to lower the risks created from ERP adoption (Thong,

1999). It has also been found that the larger the business size, the greater the possibility of ERP adoption. Some small businesses may not have the need for ERP systems to disseminate and integrate information within their enterprise. In addition, only those businesses that have adequate financial and organizational resources actually consider ERP adoption as a viable option. These findings are similar to Thong's study (Thong, 1999) that found that larger businesses with employees who have more knowledge of technological innovation are more likely to utilize a greater amount of IS innovations.

The levels of ERP adoption do not vary among industry groups. Although current ERP promotion is rated highest in the manufacturing and electronic information industries (these two industries actually account for the majority of sampled businesses), the three industries have no significant differences in terms of ERP adoption. The results also reveal that information intensity does not significantly affect ERP adoption. The reason for this finding can be explained by the sample that was used in this study, since the sample in this study was concentrated in the information-intensive (regardless of the presence or lack of

Table 7. Discriminant analysis of ERP adoption

Variable	Discriminant Loading	Adopter (n=50)		Non-adopter (n=89)		Significance
		Mean	S.D.	Mean	S.D.	
CEO's innovativeness	0.095	2.258	0.678	2.140	0.559	0.295
CEO's attitude toward IT adoption	0.284	2.537	0.704	2.161	0.616	0.002**
CEO's IT knowledge	0.384	2.787	0.785	2.200	0.768	0.000**
Employees' IT knowledge	0.448	2.474	0.699	1.910	0.522	0.000**
Information intensity	-0.044	1.848	0.729	1.910	0.683	0.624
Business size	0.413	3.080	1.158	2.135	1.170	0.000**
Industry Type[a]	0.024	0.157	0.366	0.140	0.351	0.786
	0.133	0.685	0.467	0.560	0.501	0.141
Competitive pressure	0.256	2.625	0.830	2.232	0.695	0.005**
Relative advantage	0.098	2.348	0.608	2.234	0.575	0.280
Cost	0.395	3.484	0.643	2.965	0.715	0.000**
Complexity	0.448	3.462	0.684	2.853	0.705	0.000**
Compatibility	0.304	2.978	0.768	2.520	0.762	0.001**

Wilks' Lambda = 0.526, χ^2 = 83.751, d.f. = 8, sig. = 0.000

Classification accuracy

	Total	ERP non-adopter	ERP adopter
ERP non-adopter	89	72(80.9%)	17(19.1%)
ERP adopter	50	8(16.0%)	42(84.0%)

Overall accuracy 82.0%; Chance accuracy 53.9%

**: significant at p < 0.01.

[a]: Since industry type is a categorical variable with the three groups, two dummy variables were used to represent it.

ERP adoption) manufacturing and electronic information industries. In terms of organizational characteristics, there are no remarkable differences between this study and those of Thong and Yap (1995) and Thong (1999).

Environmental Characteristics

After examining the environmental characteristics, this study concludes that competitive pressure is positively crucial to ERP adoption. The rapid development of information technology has altered the behavior of many business operations, and as a result, SMEs now face more competitive challenges. These competitive pressures may signify that other SMEs have begun to use ERP systems to improve their operational efficiency and maintain their competitive advantage. According to the International Monetary Fund (IMF), the costs of logistics incurred in the purchasing, sales, and inventory functions comprise 10% of the total production costs in the U.S.A., 13% in Taiwan, and 26% in China. Competence in information processing is critical to cost reduction, because it

Table 8. Results of the hypotheses testing in Taiwan and Singapore

Hypothesis	Description	Support or not
H1	Businesses with more innovative CEOs are more likely to adopt ERP systems.	No
H2	Businesses with CEOs who have a more positive attitude towards IT adoption are more likely to adopt ERP systems.	Yes
H3	Businesses with CEOs who are more knowledgeable about IT are more likely to adopt ERP systems.	Yes
H4	Businesses with employees who are more knowledgeable about IT are more likely to adopt ERP systems.	Yes
H5	Businesses that are larger in size are more likely to adopt ERP systems.	Yes
H6	The levels of ERP adoption vary among industry groups.	No
H7	Businesses that are in more information intensive environments are more likely to adopt ERP systems.	No
H8	Businesses that are in a more competitive environment are more likely to adopt ERP systems.	Yes
H9	The greater the perceived relative advantage of ERP systems, the more likely they will be adopted.	No
H10	The lower the perceived cost of ERP systems, the more likely they will be adopted.	Yes
H11	The lower the perceived complexity of ERP systems, the more likely they will be adopted.	Yes
H12	The greater the perceived compatibility of ERP systems, the more likely they will be adopted.	Yes

can promote production and operational efficiency. In particular, when numerous enterprises gradually transfer their production sectors to China to take advantage of the cheap labor cost, the competence of information processing will directly influence the resulting competitive advantage. Enterprises not using ERP will fail to compete with others, and will ultimately face future difficulties at the time of ERP adoption. Thus, sharp competition is positively related to the utilization of ERP systems.

The results of this study are remarkably different from Thong and Yap (Thong et al., 1996) and Thong (1999) in terms of the variable, "competitive pressure". Thong's study (Thong, 1999) noted that competitive pressure has no direct influence on the IT adoption in Singaporean SMEs. In contrast, Premkumar and Roberts (1999) indicated that competitive pressure is a highly significant variable of IT adoption in American SMEs. Generally speaking, competitive pressure is a significant factor in IT adoption for SMEs in Taiwan and the U.S.A., but not significant in Singaporean SMEs.

ERP Characteristics

ERP characteristics, including costs, complexity, and compatibility, are variables that significantly affect ERP adoption. According to sample characteristics, 92% of the sampled businesses spend less than NT$1 million on computer expenditures. Due to the financial constraints among SMEs, costs are a crucial factor for IS/IT adoption. Some ERP packages that are suitable for small businesses, such as the "Super Assistant" and the "Small Special Assistant," are priced at around NT$0.3 million, while "Workflow ERP," for middle-sized companies, is priced at about NT$1.5 million and are estimated to generate annual sales between NT$.2 and 2billion. All of these software packages are offered by local ERP vendors; however Taiwan SMEs still encounter the problem of high costs. Thus, costs are a predictive variable with a remarkable influence on ERP adoption.

In terms of the variable, "complexity," the results indicate that "no understanding of ERP" was the major reason that for ERP non-adoption (44.8%); thus concluding that many SMEs are still not familiar with the ERP technology. When ERP systems are more capable of integrating the information system with the operation information from all departments within an organization, the organization will tend to adopt ERP to gain a competitive advantage. However, the complexity of ERP should be greatly reduced to decrease the risk for SMEs in ERP adoption.

Compatibility of ERP is also crucial to ERP adoption in SMEs. Indefinite goals and high employee turnover rates are problems that SMEs may face in adopting ERP systems. Some enterprises are often unaware of their own needs or expected outcomes and some may even mistake ERP systems for a panacea. If ERP adoption is compatible with existing working practices, the business will be more likely to adopt.

Among ERP characteristics, there is a remarkable difference between this study and that of Thong(1999), in terms of the variable, "relative advantages," possibly due to the fact that Thong(1999) studied IS adoption, while this study focused on ERP adoption. Generally, IS has had a long history and has reach its maturity in the application domain. Thus, the relative advantages of IS can easily be identified by most CEOs of SMEs. However, the ERP system is still viewed as an early development and technology for SMEs and has not yet reached its maturity stage in the application domain. It is not surprising that relative advantages do not significantly affect ERP adoption.

CONCLUSION

Recommendations and Implications for the Practice

The results reveal that the compatibility of ERP is a crucial factor to ERP adoption in SMEs. Due to insufficient consideration, enterprises are often unaware of their own expectations and expected outcomes, which in turn, cause them to greatly overestimate the capability of ERP systems. Moreover, before adopting ERP systems, it is important for enterprises to understand their internal demands and associated operation processes, and then select an ERP system that is compatible with their existing work practices.

Costs are one of the critical factors of ERP adoption in SMEs. Numerous ERP software vendors have begun to launch ERP systems at different prices to suit the budgets and needs of various SMEs. Therefore, SMEs with limited resources may consider adopting more suitable ERP systems based on their own financial and organizational resources.

The complexity of ERP technologies causes installation difficulty and thus most SMEs have to rely on vendors and consultants. Many failure cases of ERP systems can be attributed to the inappropriate selection of vendors and consultants. With regard to the selection of vendors and consultants, SMEs should ensure that those selected have the capability to address the complexity issue. In addition, an easy-to-understand and usable ERP system is a better choice than a powerful and complicated one for SMEs.

During the last several years, the Taiwan government has injected a great deal of effort and capital into promoting an e-program for SMEs in order to educate and help businesses adopt ERP systems. The government's e-service team offers free services such as consultation, technical guidance, and counseling. It suggests that SMEs make the best use of these external resources offered by the Taiwan government to complement ERP adoption. Further, due to financial resource limitation, more SMEs would encounter several obstacles to adopt information technology, especially as ERP that need higher adoption cost. Therefore, while government set up policy to assist SME to adopt IT, it should consider more about resource allocation like subsidy allocation. On the other hand, in order to enhance the knowledge base of SMEs'

CEOs, this study suggests that governments hold several conferences or workshops to promote the ERP adoption benefits recognized by peer firms in the same industry. Moreover, some small enterprise don't clearly and fully understand which weakness of their IT capability, and this reason often decrease their willing to adopt information technology. The government can plan and execute digital-divide project to mitigate these problems.

Based on this survey, 44.8% of the respondents referred to the "no understanding of ERP" factor as the main reason for ERP non-adoption. This implies that SMEs are still not familiar with ERP technologies and related concepts. Since the survey findings support that a CEO's attitude has a positive influence on the ERP adoption decision, ERP software vendors and consultants are advised to spend more time and effort in convincing and educating business owners and managers about the advantages of ERP adoption. In addition, since SMEs comprised of CEOs and employees who are more knowledgeable about ERP technologies are more likely to adopt them, ERP software vendors and consultants ought to educate CEOs and employees to create greater ERP awareness throughout the entire organization.

The results also reveal that costs and complexity affect ERP adoption in SMEs. Therefore, ERP software vendors and consultants ought to provide various ERP solutions that are affordable to SMEs. Moreover, instead of sophisticated multi-functional systems, options that are easier to use should be marketed to SMEs in the early stages of ERP adoption.

Finally, another suggestion is related to the fact that business size and competitive pressures significantly affect the ERP adoption decision. ERP software vendors and consultants ought to set their sights on larger and more competitive business segments, because they are more likely to adopt ERP systems.

Recommendations and Implications for the Government

During the last several years, the Taiwan government has invested a great deal of effort and capital into the promotion of an e-program for SMEs in order to educate and help businesses to adopt ERP systems. The government's e-service team offers free services such as consultation, technical guidance, and counseling. It actively encourages the SMEs to make the best use of these aforementioned resources offered by the Taiwan government to complement their ERP adoption. Further, due to the financial resources limitation, more SMEs would encounter some obstacles to adopt information technology as ERP that need higher adoption cost which SME cannot afford to. Therefore, while government set up policy to assist SME to adopt IT, it should also take into consideration about the resources allocation such as subsidy allocation. On the other hand, in order to enhance the knowledge base of SMEs' CEOs, this study suggests that governments to hold related conferences or workshops to promote the ERP adoption benefits recognized by peer firms in the same industry. Moreover, for these some small enterprises which don't clearly and fully understand the weakness of their IT capability and thus, decrease their willingness to adopt information technology, the government usually provides the assistance to plan and execute digital-divide project to mitigate these problems.

Recommendations and Implications for the Academics

This study also provides some implications for academics. Although there are a great deal of literatures that examine the adoption factors and successful critical factors of IS adoption, few studies test those on SMEs. This study employed several relevant research studies with different perspectives to develop and test a comprehensive model (internal and external environment).

Further, this study employed rigorous sample selection and empirical testing via discriminant analysis to determine the critical factors of ERP adoption in SMEs. Compared with prior studies that were focused on the adoption of information technologies such as Electronic Data Interchange (EDI), Inter-Organization Systems (IOS), and ERP systems, this study finding is advantage to larger companies to evaluate in terms of the value chain of business (Premkumar & Ramamurthy, 1995; Soliman & Janz, 2004). However, the results of this study show that ERP systems are still viewed as an early development and technology in SMEs and have still not yet attained the maturity stage in the application domain. In addition, although present literatures state that innovative CEOs are more likely to adopt ERP systems in big companies (Premkumar & Ramamurthy, 1995; van Everdingen & Wierenga, 2002), this study cannot support this hypothesis, because the main characteristics of CEOs that were studied were the fear of risk and lack of adventurism needed to create innovation (Hung & Liang, 2001).

There are two avenues that can be provided for future research. First, this study focuses on discussing the critical factors of ERP adoption, so future research can study the extent to which these factors also affect IS adoption. Second, future studies can extend this model to include several additional contextual variables from organizational theory, such as the capability of software vendors and cultural influence, to study their affects on SMEs' ERP adoption decision.

REFERENCES

Adams, D. A., Nelson, R. R., & Todd, P. A. (1992). Perceived usefulness, ease of use, and usage of information technology: a replication. *Management Information Systems Quarterly, 16*(2), 227–247. doi:10.2307/249577

Al-Qirim, N. (2006). Personas of E-Commerce adoption in small businesses in New Zealand. *Journal of Electronic Commerce in Organizations, 4*(3), 18–45.

Amoako-Gyampah, K., & Salam, A. F. (2004). An extension of the technology acceptance model in an ERP implementation environment. *Information & Management, 41*(6), 731–745. doi:10.1016/j.im.2003.08.010

Ballantine, J., Levy, M., & Powel, P. (1998). Evaluating information systems in small and medium-sized enterprises: issues and evidence. *European Journal of Information Systems, 7*(4), 241–251. doi:10.1057/palgrave.ejis.3000307

Bancroft, N. H. (1998). *Implementing SAP R/3: How to introduce a large system into a large organization*. London: Manning/Prentice Hall.

Beard, J. W., & Sumner, M. (2004). Seeking strategic advantage in the post-net era: viewing ERP systems from the resource-based perspective. *Strategic Information Systems, 13*(2), 129–150. doi:10.1016/j.jsis.2004.02.003

Bingi, P., Godla, J. K., & Sharma, M. K. (1999). Critical issues affecting an ERP implementation. *Information Systems Management, 16*(3), 7–14. doi:10.1201/1078/43197.16.3.19990601/31310.2

Bradford, M., & Florin, J. (2003). Examining the role of innovation diffusion factors on the implementation success of enterprise resource planning systems. *International Journal of Accounting Information Systems, 4*(3), 205–225. doi:10.1016/S1467-0895(03)00026-5

Buonanno, G., Faverio, P., Pigni, F., Ravarini, A., Sciuto, D., & Tagliavini, M. (2005). Factors affecting ERP system adoption. *Journal of Enterprise Information Management, 18*(4), 384–426. doi:10.1108/17410390510609572

Chan, R. (1999). Knowledge management for implementing ERP in SMEs. In *Proceedings of the 3rd Annual SAP Asia Pacific*. Singapore: Institute of Higher Learning Forum.

Chang, S. I., Chang, H. C., & Chang, I. C. (2008). The key factors of innovation ability and industry's high value in small and medium-sized enterprises: Evidence from information services industry. *Taiwan Business Performance Journal*, *1*(2), 175–201.

Chang, S. I., & Gable, G. G. (2002). A comparative analysis of major ERP lifecycle implementation, management and support issues in Queensland Government. *Journal of Global Information Management*, *10*(3), 36–54.

Chen, J. C., & Williams, B. C. (1998). The impact of electronic data interchange on SMEs: Summary of eight British case studies. *Journal of Small Business Management*, *36*(4), 68–72.

Chen, R. S., Sun, C. M., Helms, M. M., & Jin, W. J. (2008). Role negotiation and interaction: An exploratory case study of the impact of management consultants on ERP system implementation in SMEs in Taiwan. *Information Systems Management*, *25*(2), 159–173. doi:10.1080/10580530801941371

Chiu, F. Y., Wu, H. C., & Ho, T. F. (2001). *The strategy framework of ERP implementation for small and medium enterprise in Taiwan- A case study of manufacturing industry*. Paper presented at the National Computer Symposium, Taipei, Taiwan.

Chwelos, P., Benbasat, I., & Dexter, A. S. (2001). A Dexter research report: empirical test of an EDI adoption model. *Information Systems Research*, *12*(3), 304–321. doi:10.1287/isre.12.3.304.9708

Cragg, P. B., & King, M. (1993). Small-Firm computing: Motivators and inhibitors. *Management Information Systems Quarterly*, *17*(1), 47–60. doi:10.2307/249509

Davenport, T. H. (1998). Putting the enterprise into the enterprise system. *Harvard Business Review*, *76*(4), 121–131.

Davenport, T. H. (2000). *Mission critical: Realizing the promise of enterprise systems*. Boston: Harvard Business School Press.

De Guinea, A., Kelley, H., & Hunter, M. G. (2005). Information systems effectiveness in small businesses: Extending a Singaporean model in Canada. *Journal of Global Information Management*, *13*(3), 55–79.

Delone, W. H. (1988). Determinants of success for computer usage in small business. *Management Information Systems Quarterly*, *12*(1), 51–61. doi:10.2307/248803

Fink, D. (1998). Guidelines for the successful adoption of information technology in small and medium enterprises. *International Journal of Information Management*, *18*(4), 243–253. doi:10.1016/S0268-4012(98)00013-9

Gable, G. G. (1991). Consultant engagement for first time computerization: a pro-action client role in small business. *Information & Management*, *20*(2), 83–93. doi:10.1016/0378-7206(91)90046-5

Gable, G. G., & Raman, K. S. (1992). Government initiatives for IT adoption in small businesses: experiences of the Singapore Small Enterprise Computerization Programme. *International Information Systems*, *1*(1), 68–93.

Gartner Group and Dataquest. (1999). *ERP software publishers: service strategies and capabilities*. New York: Gartner Group and Dataquest.

Grandon, E. E., & Pearson, J. M. (2004). Electronic commerce adoption: An empirical study of small and medium US businesses. *Information & Management*, *42*(1), 197–216.

Grover, V. (1993). An empirically derived model for the adoption of customer-based interorganizational systems. *Decision Sciences, 24*(3), 603–640. doi:10.1111/j.1540-5915.1993.tb01295.x

Hair, J. F., Anderson, R. E., Tatham, R. L., & Black, W. C. (2006). *Multivariate data analysis.* Upper Saddle River, NJ: Prentice-Hall.

Holland, C., & Light, B. (1999). A critical success factors model for ERP implementation. *IEEE Software, 16*(3), 30–36. doi:10.1109/52.765784

Holsapple, C. W., & Sena, M. P. (2005). ERP plans and decision-support benefits. *Decision Support Systems, 38*(4), 575–590. doi:10.1016/j.dss.2003.07.001

Houben, G., Lenie, K., & Vanhoof, K. (1999). A knowledge-based SWOT-Analysis System as an Instrument for Strategic Planning in Small and Medium Sized Enterprises. *Decision Support Systems, 26*(2), 125–135. doi:10.1016/S0167-9236(99)00024-X

Hung, S. Y., & Liang, T. P. (2001). Cross-Cultural Applicability of group decision support systems. *International Journal of Management Theory and Practices, 2*(1), 1–12.

Hunton, J. E., Lippincott, B., & Reck, J. L. (2003). Enterprise resource planning systems: comparing firm performance of adopters and nonadopters. *International Journal of Accounting Information Systems, 4*(3), 165–184. doi:10.1016/S1467-0895(03)00008-3

Iacovou, C. L., Benbasat, I., & Dexter, A. S. (1995). Electronic Data interchange and small organizations: adoption and impact of technology. *Management Information Systems Quarterly, 19*(4), 465–485. doi:10.2307/249629

Kagan, A., Lau, K., & Nusgart, K. R. (1990). Information system usage within small business firms. *Entrepreneurship: Theory and Practice, 14*(3), 25–37.

Klaus, H., Rosemann, M., & Gable, G. G. (2000). What is ERP? *Information Systems Frontiers, 2*(2), 141–162. doi:10.1023/A:1026543906354

Kumar, V., Maheshwari, B., & Kumar, U. (2003). An investigation of critical management issues in ERP implementation: evidence from Canadian organizations. *Technovation, 23*(10), 793–807. doi:10.1016/S0166-4972(02)00015-9

Kwahk, K. Y., & Lee, J. N. (2008). The role of readiness for change in ERP implementation: theoretical bases and empirical validation. *Information & Management, 45*(7), 474–481. doi:10.1016/j.im.2008.07.002

Lang, J. R., Calantone, R. J., & Gudmundson, D. (1997). Small firm information seeking as a response to environmental threats and opportunities. *Journal of Small Business Management, 35*(1), 11–23.

Laughlin, S. P. (1999). An ERP game plan. *The Journal of Business Strategy, 20*(1), 32–37. doi:10.1108/eb039981

Lee, J. (2004). Discriminant analysis of technology adoption behavior: A case of internet technologies in small businesses. *Journal of Computer Information Systems, 44*(4), 57–66.

Lees, J. D. (1987). Successful development of small business information systems. *Journal of Systems Management, 38*(8), 32–39.

Levy, M., & Powell, P. (2000). Information systems strategy for small- medium-sized enterprises: an organizational perspective. *The Journal of Strategic Information Systems, 9*(1), 63–84. doi:10.1016/S0963-8687(00)00028-7

Levy, M., Powell, P., & Galliers, R. (1999). Assessing information systems strategy development frameworks in SMEs. *Information & Management, 36*(5), 247–261. doi:10.1016/S0378-7206(99)00020-8

Li, C. (1999). ERP packages: what's next? *Information Systems Management, 16*(3), 31–35. doi:10.1201/1078/43197.16.3.19990601/31313.5

Mabert, V. A., Soni, A., & Venkataramanan, M. A. (2003). Enterprise resource planning: Managing the implementation process. *European Journal of Operational Research, 146*(2), 302–314. doi:10.1016/S0377-2217(02)00551-9

Markus, M. L., Axline, S., Petrie, D., & Tanis, C. (2000). Learning from adopters' experiences with ERP: problems encountered and success achieved. *Journal of Information Technology, 15*(4), 245–265. doi:10.1080/02683960010008944

Melville, N., & Ramirez, R. (2008). Information technology innovation diffusion: An information requirements paradigm. *Information Systems Journal, 18*(3), 247–273. doi:10.1111/j.1365-2575.2007.00260.x

MIC (Market Intelligence & Consulting Institute). (2007). *Current and tendency of information technology investment in Taiwan SMEs during 2006-2007.* Taiwan: MIC.

MIC (Market Intelligence & Consulting Institute). (2008). *Current and tendency of information technology investment in Taiwan large company during 2007-2009.* Taiwan: MIC.

Niedleman, L. D. (1979). Computer usage by small and medium sized European firms: an empirical study. *Information & Management, 2*(2), 67–77. doi:10.1016/0378-7206(79)90008-9

Nunnally, J. (1978). *Psychometric theory.* New York: McGraw-Hill.

Pai, J. C., Lee, G. G., Tseng, W. G., & Chang, Y. L. (2007). Organizational, technological and environmental factors affecting the implementation of ERP systems: Multiple–case study in Taiwan. *Electronic Commerce Studies, 5*(2), 175–196.

Palvia, P., Means, D. B., & Jackson, W. M. (1994). Determinants of computing in very small businesses. *Information & Management, 27*(3), 161–174. doi:10.1016/0378-7206(94)90044-2

Pan, M. J., & Jang, W. Y. (2008). Determinants of the adoption of enterprise resource planning within the technology organization environment framework: Taiwan's communications industry. *Journal of Computer Information Systems, 48*(3), 94–102.

Poutsma, E. F., Van Uxem, F. W., & Walravens, A. H. C. M. (1987). *Process innovation and automation in small and medium sized business.* Delft, The Netherlands: Delft University Press.

Premkumar, G., & King, W. R. (1992). An empirical assessment of information systems planning and the role of information systems in organizations. *Journal of Management Information Systems, 9*(2), 99–125.

Premkumar, G., & Ramamurthy, K. (1995). The role of interorgainzational and organizational factors on the decision mode for adoption of interorganizational systems. *Decision Sciences, 26*(3), 303–336. doi:10.1111/j.1540-5915.1995.tb01431.x

Premkumar, G., Ramamurthy, K., & Crum, M. (1997). Determinants of EDI adoption in the transportation industry. *European Journal of Information Systems, 6*(2), 107–121. doi:10.1057/palgrave.ejis.3000260

Premkumar, G., & Roberts, M. (1999). Adoption of new information technologies in rural small business. *Omega, 27*(4), 467–484. doi:10.1016/S0305-0483(98)00071-1

Quaddus, M., & Hofmeyer, G. (2007). An investigation into the factors influencing the adoption of B2B trading exchanges in small businesses. *European Journal of Information Systems, 16*(3), 202–215. doi:10.1057/palgrave.ejis.3000671

Rajagopal, P. (2002). An innovation-diffusion view of implementation of enterprise resource planning (ERP) systems and development of a research model. *Information & Management, 40*(2), 87–114. doi:10.1016/S0378-7206(01)00135-5

Raymond, L. (1985). Organizational characteristics and MIS success in the context of small business. *Management Information Systems Quarterly, 9*(1), 37–60. doi:10.2307/249272

Riemenschneider, C. K., Harrison, D. A., & Mykytyn, P. P. (2003). Understanding IT adoption decisions in small business: Integrating current theories. *Information & Management, 40*(4), 269–285. doi:10.1016/S0378-7206(02)00010-1

Rogers, E. M. (1995). *Diffusion of Innovations.* New York: The Free Press.

Sarker, S., & Lee, A. (2003). Using a case study to test the role of three key social enablers in ERP implementation. *Information & Management, 40*(8), 813–829. doi:10.1016/S0378-7206(02)00103-9

Scott, J. E., & Kaindl, L. (2000). Enhancing functionality in an enterprise software package. *Information & Management, 37*(3), 111–122. doi:10.1016/S0378-7206(99)00040-3

Soliman, K. S., & Janz, B. D. (2004). An exploratory study to identify the critical factors affecting the decision to establish internet-based interorganizational information systems. *Information & Management, 41*(6), 697–706. doi:10.1016/j.im.2003.06.001

Stratman, J. K., & Roth, A. V. (2002). Enterprise resource planning (ERP) competence constructs: Two-stage multi-item scale development and validation. *Decision Sciences, 33*(4), 601–628. doi:10.1111/j.1540-5915.2002.tb01658.x

Thong, J. Y. L. (1999). An integrated model of information systems adoption in small business. *Journal of Management Information Systems, 15*(4), 187–214.

Thong, J. Y. L. (2001). Resource constrains and information systems implementation in Singaporean small business. *Omega, 29*(2), 143–156. doi:10.1016/S0305-0483(00)00035-9

Thong, J. Y. L., & Yap, C. S. (1995). CEO characteristics, organizational characteristics and information technology adoption in small businesses. *Omega, 23*(4), 429–442. doi:10.1016/0305-0483(95)00017-I

Thong, J. Y. L., Yap, C. S., & Raman, K. S. (1996). Top management support, external expertise and information systems implementation in small business. *Information Systems Research, 7*(2), 248–267. doi:10.1287/isre.7.2.248

Tornatzky, L. G., & Klein, K. J. (1982). Innovation characteristics and innovation adoption-implementation: a meta analysis of findings. *IEEE Transactions on Engineering Management, 29*(11), 28–45.

Tsai, W. H., Chein, S. W., Fan, Y. W., & Cheng, J. M. (2005). Critical management issues in implementing ERP: Empirical evidence from Taiwanese firms. *International Journal of Services and Standard, 1*(3), 299–318. doi:10.1504/IJSS.2005.005802

Umble, E. J., Haft, P. R., & Umble, M. M. (2003). Enterprise resource planning: Implementation procedures and critical success factors. *European Journal of Operational Research, 146*(2), 241–257. doi:10.1016/S0377-2217(02)00547-7

Van Everdingen, Y., & Wierenga, B. (2002). Intra-firm adoption decision: Role of inter-firm and intra-firm variables. *European Management Journal, 20*(6), 649–663. doi:10.1016/S0263-2373(02)00116-0

Waarts, E., Everdingen, Y. M., & Hillegersberg, J. (2002). The dynamics of factors affecting the adoption of innovations. *Journal of Product Innovation Management, 19*(6), 412–423. doi:10.1016/S0737-6782(02)00175-3

Wang, T. G. E., Klein, G., & Jiang, J. J. (2006). ERP misfit: Country of origin and organizational factors. *Journal of Management Information Systems, 23*(1), 263–292. doi:10.2753/MIS0742-1222230109

Welsh, J. A., & White, J. F. (1981). A small business is not a little big business. *Harvard Business Review, 59*(4), 18–32.

Xu, Q., & Ma, Q. (2008). Determinants of ERP implementation knowledge transfer. *Information & Management, 45*(8), 528–539. doi:10.1016/j.im.2008.08.004

Yap, C. S. (1990). Distinguishing characteristics of organizations using computers. *Information & Management, 18*(2), 97–107. doi:10.1016/0378-7206(90)90056-N

Yap, C. S., Soh, C. P. P., & Raman, K. S. (1992). Information systems success factors in small business. *Omega, 20*(5-6), 597–609. doi:10.1016/0305-0483(92)90005-R

Zinatelli, N., Cragg, P. B., & Cavaye, A. L. M. (1996). End user computing sophistication and success in small firms. *European Journal of Information Systems, 5*(3), 172–181. doi:10.1057/ejis.1996.23

This work was previously published in International Journal of Global Information Management, Volume 18, Issue 3, edited by Felix B. Tan, pp. 82-106, copyright 2010 by IGI Publishing (an imprint of IGI Global).

Chapter 10
Key Issues in Information Systems Management:
An Empirical Investigation from a Developing Country's Perspective

D. Li
Peking University, China

W.W. Huang
Ohio University, USA & Harvard University, USA

J. Luftman
Stevens Institute of Technology, USA

W. Sha
Pittsburg State University, USA

ABSTRACT

There have been periodical studies on key IS management issues facing the IT industry in North America; however, an empirical investigation on key IS management issues in developing countries has been largely ad hoc and inadequate. This paper identifies and analyzes important issues faced by CIOs in the developing country of China. The results of this study are based on two national wide CIO surveys in China, where the first was conducted in 2004 and followed by a more recent survey in 2008. The authors provide insight for both IS practitioners and researchers who have interests in developing countries. Data analysis indentified key IS management issues and demonstrated similarities as well as differences between the two rounds of surveys. Although some strategic IS issues were still within the top 10 on both the 2004 and 2008 lists, their importance ratings were different. Implications of the findings are also discussed.

DOI: 10.4018/978-1-61350-480-2.ch010

INTRODUCTION

Research on key issues in IS management has been published by MIS Quarterly (MISQ) or MISQ Executive (MISQE) in the last three decades (e.g., Ball & Harrison, 1982; Brancheau & Wetherbe, 1986; Watson, 1989; Niederman, Brancheau, & Wetherbe, 1991; Brancheau, Janz, & Wetherbe, 1996; Hovar, 2003; Luftman, Kempaih, & Nash, 2006; Luftman & Kempaiah, 2008). Since these studies were conducted by surveying U.S. senior managers, the top IS issues from the findings largely reflected perspectives and views of senior managers in North America. Additionally, key IS management issues could be influenced by political, economical, cultural, and technological infrastructure factors in different countries (e.g., Deans et al., 1991). Therefore, key issues in IS management identified from the perspective of North America may not be the exact ones faced by senior managers in other countries, especially when such countries have different technological infrastructure and cultural background.

Although there were a few prior studies examined key IS management issues from outside of North America's perspectives, this research stream has not been systematically studied in other countries periodically. Particularly, no such study has been done for the last eight years from the perspective of an Asian developing country. Asia has maintained relatively faster economic development speed in the last decade, especially in its IT industry, even during the current world economic downturn period. A study on key IS management issues facing an Asian country is important and timely.

The rest of the paper is organized as follows: first, existing research literature on the top 10 key IS management issues is reviewed; second, this study's survey method is presented; third, data gathered through the survey are analyzed and presented. Discussion and implications were provided in the final section of the paper.

PREVIOUS FINDINGS REGARDING TOP IS MANAGEMENT ISSUES

The top IS management issues have been systematically examined by IS researchers. Before 1990, IS strategic planning was clearly the most important issue. It was consistently ranked as the number one issue among U.S. private sectors (Ball & Harrison, 1982; Dickson et al., 1984; Hartog & Herbert, 1985; Brancheau & Wetherbe, 1986). During this time, there were rapidly and complex changes in the application of technology in the business environment. Business organizations began to be more dependent on Information Technology. The urgent need for the integration of technology into business missions makes IS strategic planning the top priority for IS managers. This top priority was strengthened by the imperative need for end-user training because of the proliferation of end-user computing technology. Lack of support from top management also make the strategic IS planning a priority. During the 1980s, IS managers were trying to position themselves within their organizations. The issues about the role of IS managers, particularly how to measure the effective of IS in terms of the alignment with organization strategic goals, i.e., the contribution of the IS organization frequently turned up on the top issue list. At the beginning of the 1980s, technology issues such as communication protocols, network layers, system development methodologies were also among the priority list of IS managers.

During the middle of the 1980s, the issues of software development, database administration, information architecture development and integration of technologies gained more attention. End-user computing continued to receive a lot of attention. IS managers tend to focus on end-user computing training and satisfaction. The management of IS human resources only made the top 10 list twice (Ball & Harrison, 1982; Dickson et al., 1984). Communication with the top management also received insignificant attention among IS managers (Number 8 on Hartog & Herbert's 1985 list).

At the end of 1980s, Caudle, Gorr, and Newcomer (1991) conduct a national survey about the key information system management issues in the public sectors of the U.S. They found that although IS strategic planning was consistently ranked number one in previous studies, long term planning is only ranked number seven on the public sector IS managers' list. The comparison of U.S. public sector with the private sector may indicate an interesting difference: while U.S. private sector has consistently seen the importance of the IS planning process, this issue is not even in the top 5 management issues list in the U.S. public sector. One possible reason could be that the public sector tended to focus more on operational issues than on strategic direction. The importance of integration of the current and future technologies was ranked number one in the public sector. The importance of this issue is on the decline in the private sector, i.e., number two on the Dickson et al.'s (1984) list, number seven on the Hartog and Herbert's (1985) list and number 10 on the Brancheau and Wetherbe's (1986) list. Two software development issues made it to the top 10 list, i.e., information requirements identification and software maintenance. More specific system development issues, such as end-user computing, office automation, data security, database management and distributed data processing, were also among the top 10 list.

During the 1990s, the focus of IS managers shifted considerably. The center of the focus evolved to the developing and building of a responsive IT infrastructure. The number one issue at the beginning of the 1990s was "developing an information architecture" (Niederman, Brancheau, & Wetherbe, 1991), and the number one issue in the middle of 1990s was "building a responsive IT infrastructure" (Brancheau, Janz, & Wetherbe, 1996). The importance of end-user computing issues such as end-user training and satisfaction declined dramatically. The facilitation of organizational learning and usage of IT was ranked number five in 1991, but dropped out of the top 10 in 1996. IS strategic planning

was no longer considered very important (number three on Niederman, Brancheau, & Wetherbe's, 1991 list and number nine on Brancheau, Janz, & Wetherbe's, 1996 list). The importance of the issue of improving the competitive advantage of IS dropped from number two (Brancheau & Wetherbe, 1986) to number eight (Niederman, Brancheau, & Wetherbe, 1991). The same trend also held for the need to align IS goals with the goals or missions of the organization. This may reflect on the fact that IT in the U.S. was becoming a more mature discipline and the position of the IS within a company had been strengthened. One interesting development in the early 1990s was the importance issues of IS employee recruiting, training, and retentions. The issues of specifying, recruiting and developing IS human resources was ranked number four at the beginning of 1990s (Niederman, Brancheau, & Wetherbe, 1991), but completely dropped out of top 20 issues several years later (Brancheau, Janz, & Wetherbe, 1996). The comparison of the top 10 U.S. IS management issues in the last three decades is listed in Table 1.

While most top ten IS management issues studies were conducted in North America, some researchers examined the same topic in other regions and countries, such as Australia (Watson, 1989), Taiwan (Harrison & Farn, 1990) Slovenia (Dekleva & Zupancic, 1996). Although these studies were conducted in different regions of the world at different times, their results shows resemblance with the U.S. studies on issues such as IS strategic planning, end-user computing, alignment of IS goals with the missions of the organization, the position of IS in the organization, etc. This might be due to the fact that IT infrastructures in terms of the main software (like Microsoft office suite) and hardware (like various types of computers and network products), are quite standard in the world, and the knowledge and skill of using those software and hardware are similar as well. As a result, some top IS management issues were similar across geographical borders.

Table 1. Top 10 IS management issues in US

Rank	Ball and Harrison, 1982	Dickson, et al, 1984	Hartog and Herbert, 1985	Brancheau and Wetherbe, 1986	Caudle, Gorr and Newcomer, 1991	Niederman, Brancheau and Wetherbe, 1991	Brancheau, Janz and Wetherbe, 1996
1	IS planning	IS planning	IS planning	IS planning	Integration of Technologies	Developing an Information Architecture	Building a responsive IT infrastructure
2	Measuring IS effectiveness	End-user computing	Aligning IS with Business Goals	IS for competitive advantage	Comprehensive Planning Integration	Making Effective Use of the Data Resource	Facilitating and managing business process redesign
3	Impact of communications on IS	Integration of Technologies	Software development	Educating all managers	Information Requirements Identification	Improving IS Strategic Planning	Developing and managing distributed systems
4	Role of information resource manager	Software development	Database administration	IS's role and contribution	End-user Computing	Sepcifying, Recruiting, and Developing IS Human Resources	Developing and implementing an information architecture
5	Decision support systems	Measuring IS effectiveness	End-user computing	Aligning IS in the organization	Office Automation	Facilitating Organizational Learning and Use of IS Technologies	Planning and managing communication networks
6	Office automation	Educating all managers	Data security	End-user computing	Data Security	Building a Responsive IT Infrastructure	Improving the effectiveness of software development
7	IS human resources	Aligning IS with business goals	Integration of technologies	Database administration	Long-term planning mechanisms	Aligning the IS Organization With That of the Enterprise	Making effective use of the data resources
8	Educating non-IS management	IS human resources	Educating senior personnel	Info. Architecture development	Database Management System Impact	Using Information Systems for Competitive Advantage	Aligning the IS organization within the enterprise
9	Centralization vs. decentral of Is	Database administration	Software quality assurance	Measuring IS productivity	Distributed Data Processing	Improving the Quality of Software Development	Improving IS strategic planning
10	Employee job satisfaction	Decision support systems	Telecommunication technology	Integration of technologies	Software Maintenance	Planning and Implementing a Telecommunications Systems	Implementing and managing collaborative support systems

234

An interesting finding is the commonly identified issue of the training and retention of IS professionals, which was rated number two on Watson's list, number four on Harrison and Farn's list, and number two on Dekleva and Zupancic's list. This may be due to overall shortage of IS professionals, inadequate training of IS professionals, and IS professionals' need for further improvement in areas outside of computing. Another interesting finding was that the IS managers in these three countries consistently gave higher ratings on the issues of improving communication with the top management. Better communication can improve the involvement of top management, and IS professionals can have more chances to show the value and contribution of the IS department to the top management.

Chen, Wu, and Guo (2007) did a survey on the top 12 IS key management issues of China in 1999. Their top IS issues list was independently derived by directly asking IS department managers and scholars of China in 1999. As a result, there is no basis for comparative analysis of their findings with similar studies conducted in other countries available in the research literature. For example, some commonly recognized key IS management issues in prior studies conducted in North American and other countries like IS strategic planning, alignment between IS strategy and business strategy, and building of enterprise-wide IS infrastructure, were not in their top 12 IS issues list, which was acknowledged as one main research limitation in their published paper.

The top 10 IS issues in the above studies are listed in Table 2.

IT investment in China increased tremendously during the last decade. According to China's State Economic and Trade Commission, the average investment in IT reached 10 million U.S. dollars, and the speed of the adoption of IT among industries such as retailing, banking, e-business, telecommunications, and e-government has accelerated as well. For example, ERP adoption in China was at 45% for the top 1,000 corporations in 1999 and reached 83% in 2004 (CNET, 2004). The top IS issues list of Chen, Wu, and Guo (2007) was studied in 1999 and their survey subjects were largely IT department managers, not CIOs. Given the fast development pace of the IT industry, past research has shown that it is necessary and important to conduct surveys on key IS issues periodically, especially in less studied developing countries.

RESEARCH METHODOLOGY

This study examined top IS issues in China through two surveys on CIOs of China's enterprises. An initial list of key IS issues was first determined by reviewing relevant research literature as discussed above. A preliminary study was then made to identify initial key issues faced by Chief Information Officers (CIOs) of China's large enterprises. The CIO concept was introduced into China in the 1980's, but only recently has become widely accepted by most of big Chinese enterprises. The China CIO Association (CCIOA) was established in late 2003. There are about 250 members in CCIOA, representing dominant IT corporations or organizations in China. One co-author of this paper was invited to be an advisor for this newly created CIO community. Through this connection, the CIO association issued a letter to support and urge its members to participate in this survey. A national wide survey was thus possible for the first time in China.

The survey initially started in late 2003 when the national CIO Club's first annual meeting was held, where all CIO members were present. During the meeting, designed preliminary questionnaire was sent out to CIOs for their feedbacks on the key issues. Some interviews were also conducted to solicit the CIOs' feedback on the preliminary questionnaire. Based on their feedback, the questionnaire was revised and the final version of the questionnaire was completed (Appendix A shows the survey questionnaires).

Table 2. Top 10 IS management issues in other countries outside North America

Issues	Watson, 1989	Harrison and Farn, 1990	Dekleva and Zupanncic, 1996	Chen, Wu and Guo, 2007
1	Improving IS strategic planning	Communications with end users	Inadequate appreciation of IS by executives and other users and their lack of involvement in IS development	Status and power of persons in charge of IS departments in enterprises
2	Specifying, recruiting, and developing human resources	Communications with top management	Education of IS professionals	Organizational mechanisms with which enterprises manage the IS department
3	Developing an information architecture	Improving the productivity of information systems professionals in the development of applications	Lack of IS strategic planning	Internal managerial and organizational level of IS departments
4	Improving the effectiveness of software development	IS staff development and the maintenance of attractive career paths	Management of IS function	Investments in the IS department
5	Aligning the IS organization with that of the enterprise	Improving the productivity of maintenance activities	Organizational problems	Support of and acquaintance with IT of high-level managers
6	Increasing understanding of the role and contribution of Is	Creating and promoting information management activities which provide or enhance competitive advantages for the firm	Education of IS users	Support of middle-level manager
7	Using information systems for competitive advantage	Maintaining close agreement between the goals of the organization and the goals of the information service group	Integration of subsystems into the comprehensive information architecture	Ability to utilize computers of the staff other than those of the IS department
8	Facilitating and managing end-user computing	Training end users to be effective participants in the development of applications	Telecommunication infrastructure in Slovenia and its links to the world	Technology competence of staff in IS departments
9	Promoting effective use of data as a resource	Fostering more effective use of work stations by professionals	Executive IS	Applications of advanced IT in enterprise
10	Facilitating organizational learning and the use of IS technologies	Accomplishing more complete integration of systems through better interface and interconnectivity standards	National and ISO-compatible IS standards	Staff training on IT

The finalized questionnaires were sent out to all the 250 CIO members of the association, and within one month, 50 responses were returned. A research assistant was hired to make long-distance calls (or through fax or email) to contact those CIOs who didn't fill in the questionnaire. By early 2004, a total of 92 responses were received, which represented a 37% response rate. No significant differences were found between the first 50 responses and subsequent 43 responses in the top 10 key IS issues list. As a result, non-response bias was not a major issue in the survey. 61% of the participants have undergraduate degrees, 34% have master degrees and 2% have doctorate degree.

A follow-up survey was sent in 2008, four years after our first CIO survey. After China's CCIOA was set up in 2003, its members have increased over time. In 2008 survey, CIOs from 500 corporations were surveyed. 150 usable surveys were returned, which represents a response rate of 30%. 60% of the participants have undergraduate degrees, 20% have master degrees, 9% have MBA

Figure 1. Distribution of Survey Corporations/Organizations in Industry

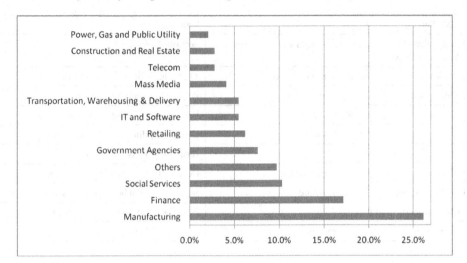

degrees, and 5% have doctorate degrees. Most CIOs are relatively young. 42% of surveyed CIOs are between 36-45 years old and 38% are in 26-35 years old. Only 19% are over 46 years old and 2% are below 26 years old. More than 40% CIOs have over 10 years' IT working experience, 29% have 5-10 years' IT working experience and 20% have less than 5 years' experience. In addition, 22% of CIOs work in public sector, 14% in private corporations, 47% in state-run corporations, 8% in foreign-owned corporations and 6% in joint-ventures. 23.3% are listed corporations. 21% are large-scale corporations/organizations, 53.8% are medium-sized and 25.2% are small-sized. Figure 1 below shows the distribution of participation corporations in various industries.

RESEARCH RESULTS AND FINDINGS

The results are listed in Table 3. While several traditional issues are included, there are also new issues more related to China's particular IT industry environment. These10 issues were classified into 3 categories: (1) Strategic issues; (2) Operational management issues; (3) Technology issues.

Strategic issues

Strategic issues are related to enterprises' vision, mission, and direction of future development. These are main concerns for enterprise top management. In the 2004 survey, this category include three issues: (#1) integration of information technologies with enterprise business practices; (#3) enterprise information systems strategies; and (#8) improve the responsiveness to business environment.

In the past, the focus of the IS department in most Chinese organizations was to support daily operations. But, many organizations realize that their success depends on the effective and efficient integration of information technologies with their business objectives. It is no surprise that the integration of information technologies with enterprise business practices is ranked as the number one issue. To most Chinese enterprises CIOs how to align information technologies and their business is a big challenge. The rapid changing business environment, evolving new technologies, fluid IT human resources, and the need to develop and maintain information systems all require CIOs to focus on the alignment of IS strategies and their business strategies. This issue is related to enterprise information systems

Table 3. Key IS management issues in China

Ranking	2004 Survey	2008 Survey
1	Integration of Information technologies with enterprise business practices	Security & Privacy
2	Improving enterprise information system security	The building of enterprise IS infrastructures
3	Enterprise Information Systems strategies	Applications' adoption and use
4	Making effective use of data resources	Making effective use of the data resources
5	The influence of CIO and IS departments	Enterprise Information Systems strategies
6	Business Process Reengineering	Business Process Management
7	The building of enterprise IS infrastructures	Developing and installing Information System
8	Building a responsive IT infrastructure	Network System building and its management
9	The evaluation of ROI of IS	Business Performance Management
10	Integration of different suppliers' open systems	Knowledge Management

strategies issue (#3). Lack of mature strategic IS planning methodologies is a common problem recognized by Chinese CIOs.

Improving enterprise information system security (#2) is taken very seriously by most CIOs. Chinese Internet users have increased 94 million in the beginning of 2005 to 338 million by June 2009 (CNNIC, 2005; Ye, 2009). As more and more organizations are connected through Internet, security risks increase dramatically. An intrusion could cause heavy monetary and valuable data losses, in addition to reputation damages. Because of this, firewall and virus-protection software have tremendous market popularity. The virus-protection software package "Rui Xing" was sold 10 million licenses in a short time period (Shi, 2004).

How to improve the responsiveness of an enterprise to the complex business environment is rated eighth. It should be noted that China's enterprise business environment is complex: many big enterprises are state-owned, with a significant percentage of private-owned and foreign-owned enterprises. Currently, state-owned enterprises are facing strong competition from private-owned and foreign-owned enterprises. This complex structure

coupled with the quickly changing information technologies makes Chinese enterprises' business environment less certain and less predictable. How to help organizations quickly respond to changes in the business environment is an important problem that Chinese CIOs are facing.

After four years, some strategic issues were more or less resolved. In 2008, the importance of the strategic issues in the 2004 top 10 list was decreased, except one issue. The #3 issue in the 2004 list, enterprise information systems strategies, was now rated as #5 in the 2008 list. This indicates that this issue, though still important, is not as important to CIOs as it was four years ago. The #1 strategic issue on the 2004 list, integration of information technologies with enterprise activities, was not on the 2008 top 10 list. In addition, one new strategic issue, business performance management, has made it onto the 2008 top 10 list, which was not in the 2004 list. This may show that more CIOs now realize the importance of evaluating IT/IS from the perspective of business performance, not just from a technical perspective. Finally, the #2 rated strategic issue on the 2004 list, security & privacy, was increased to the top issue in the 2008 list.

Operational Management Issues

The first issue in this category in our 2004 survey is making effective use of data resources (#4). This shows that as the penetration of IT increases, the collection of large data repositories is fundamental to many enterprises. However, these data are usually unrecognized, unorganized, or inaccessible. How to effectively extract useful information and knowledge from this data is a daunting task to CIOs. The influence of the CIO and IS departments (#5) is another issue in this category. It has reported that there are at least about 10,000 CIOs in China's enterprises (CNIC, 2003). The CIO is a relative new managerial position in China's corporate structures, and the wide spread of IS adoption only happened in the last 20 years. Therefore, CIOs may feel that IS importance to the success of the enterprise may still be underestimated.

Another issue in this category is business process reengineering (#6). China's economic system used to be a planned system before the 1980s. The way most Chinese big state-run enterprises used to operate their business complies with this system. As China's economy changed to a market-based economy, many enterprises, particularly state-owned medium to large size enterprises, have restructured themselves to adapt to the new environment. IS can be a business enabler in this transformation process. The last one in this category is the evaluation of Return on Investment (ROI) for IS (#9). During the last decade many major state-owned enterprises have invested heavily in IT, but they still lack good evaluation methods for estimating the return of the investment. How to justify the IS investment concerns China's CIOs, since they have to prove the value of IT and the IS department.

In this category, two issues were not changed in the top 10 list during the previous four years period; #4 (Making effective use of the data resources) and #6 (Business process management). Two prior issues in the 2004 list, #5 (the influence of CIO

and IS departments) and #9 (the evolution of ROI of IS), were not in the top 10 list in the 2008 list. Two new issues not in the 2004 list, #8 (Network systems building and its management) and #10 (Knowledge management), made into the 2008 list, indicating that new IS important issues and challenges would appear as time goes on.

Technology Issues

Building enterprise IS infrastructure (#7) is a traditional issue, but it is still considered important by most Chinese CIOs. An effective enterprise IS infrastructure is still lacking in many Chinese organizations. A large number of enterprises have IS infrastructures with different kinds of hardware, databases, critical applications, and operating system platforms. Different computing needs, massive network construction, and the alignment of IS functions and business functions make the building of a solid integrated IS infrastructure a complex situation. Leveraging them into one highly productive integrated infrastructure is a difficult but fundamental task.

The last issue is the integration of different suppliers' system platforms (#10). The traditional relationship between China's enterprises has been restricted to buyer-supplier relationships. With the development of IT industry and business environment, deeper cooperation between companies and subsequent establishment of supply chain become common. However, many companies have their own unique IT platforms and infrastructure. This makes integration a daunting task for all companies involved.

While the #10 issue of the 2004 list, integration of different suppliers' open systems, was not in the 2008 list. The #7 issue, the building of enterprise IS infrastructures, in the 2004 list has increased its importance to become the #2 issue in the 2008 list. This may signal that after four years' with the development in ERP adoption and implementation, this issue has not been well resolved and has even become more important and challenging. Survey debriefings with CIOs also revealed that after years of successful or failed experience, more and more CIOs realize that successful ERP implementation was not just an issue of funding and technical expertise. ERP's effective functioning and potential business value realization would rely more on business process reengineering and IS strategic planning/alignment.

DISCUSSION AND IMPLICATIONS

To compare the findings from this study with the results from earlier studies, the top 10 IS management issues from this study are compared with other lists from Australia, Republic of China (Taiwan), Slovenia and US in earlier studies. The results are listed in Table 4.

The comparison results from Table 4 show some similar key IS issues between China and other countries. IS communities around the world seem to share common emphasis on the building of IS infrastructure and the alignment of IS departments and organization missions, business practices and planning. The results from this study are more comparable with the results generated from the study of Australian IS managers. One issue that concerns all China, Australia, and Slovenia CIOs, is the appreciation of the contribution of CIOs and their IS departments. This might be due to the early development of IT industry in these countries, as compared to their US counterpart where in US, this concern in terms of the role of CIOs and IS department to corporation is not a major issue anymore.

While the commonalities of the top IS management issues may reflect the relative homogenous of the IS communities around the world, there are some significant differences between China and other countries. The first difference is about end-user computing. Issues related to end-user computing were rated number eight and number ten in Australia (Watson, 1989), number nine in Republic of China (Harrison & Farn, 1990), number

four in US (Caudle, Gorr, & Newcomer, 1991), and number sixth in Slovenia (Dekleva & Zupanncic, 1996). There is no issue in China's top 10 list that are related to end-user computing in both 2004 and 2008 survey. This result may indicate that end-user computing might not be as important in China's corporations yet.

The second major difference between this study and previous studies is the importance of IS professional recruiting, training and retention. This issue is rated number one in a more recent US survey (Luftman & Kempaiah, 2008), number two in Australia, number four in Republic of China, and number two in Slovenia. In China, this issue is rated number eleven in our 2004 survey, out of the top 10 list in both 2004 and 2008 surveys. This could be due to the fact that China, as the biggest developing country in the world, has not only big pool for its labor force, but also a bigger pool of talents because of its tradition of emphasizing on education. Therefore, China doesn't have the problem of lack of both labor force and IT professionals. This may also help explain that end-user computing is not a major issue for China's CIOs because they have enough IT professionals to handle IT projects without the need to ask end-users to do extra computing works. Furthermore, the issues of ROI, the integration of different suppliers' system platform, knowledge management, and the need to ensure quick reaction to the dynamics of the business environment are unique to China's IS CIOs.

While periodical study on key IS issues is conducted in North America every two years, similar studies in other countries, especially in developing countries, are not. Although there are similar key IS issues like IS/IT security across different countries, noticeable differences exist as well. More research should be conducted in other countries, especially in developing countries. This may become more important as the world economy becomes more global and developing countries will have more potential for their economic and IT/IS development than developed countries.

Table 4. Key IS Management Issues Comparison between China and other countries

Issues	Watson (1989)	Harrison and Farn (1990)	Dekleva and Zupanncic (1996)	Luftman and Kempaiah (2008)	This Study
1	Improving IS strategic planning	Communications with end users	Inadequate appreciation of IS by executives and other users and their lack of involvement in IS development	Important1 Attracting, developing, and retaining IT professionals	Security & Privacy (2008) Integration of Information technologies with enterprise activities (2004)
2	Specifying, recruiting, and developing human resources	Communications with top management	Education of IS professionals	IT and business alignment	The building of enterprise IS infrastructures (2008) Improving enterprise information system security (2004)
3	Developing an information architecture	Improving the productivity of information systems professionals in the development of applications	Lack of IS strategic planning	Build business skills in IT	Applications' adoption and use (2008) Enterprise Information Systems strategies (2004)
4	Improving the effectivenss of software development	IS staff development and the maintenance of attractive career paths	Management of IS function	Reduce the cost of doing business	Making effective use of the data resources (2008 & 2004)
5	Aligning the IS organization with that of the enterprise	Improving the productivity of maintenance activities	Organizational problems	Improve IT quality	Enterprise Information Systems strategies (2008) The influence of CIO and IS departments (2004)
6	Increasing understanding of the role and contribution of Is	Creating and promoting information management activities which provide or enhance competitive advantages for the firm	Education of IS users	Security and privacy	Business Process Management (2008 & 2004)
7	Using information systems for competitive advantage	Maintaining close agreement between the goals of the organization and the goals of the information service group	Integration of subsystems into the comprehensive information architecture	Manage change	Developing and installing Information System (2008) The building of enterprise IS infrastructures (2004)
8	Facilitating and managing end-user computing	Training end users to be effective participants in the development of applications	Telecommunication infrastructure in Slovenia and its links to the world	IT strategic planning	Network System building and its management (2008) Improve the responsiveness to the business environment (2004)
9	Promoting effective use of data as a resource	Fostering more effective use of work stations by professionals	Executive IS	Making better use of information	Business Performance Management (2008) The evaluation of ROI of IS (2004)
10	Facilitating organizational learning and the use of IS technologies	Accomplishing more complete integration of systems through better interface and interconnectivity standards	National and ISO-compatible IS standards	Evolving CIO leadership role	Knowledge Management (2008) Integration of different suppliers' open systems (2004)

More specifically, future research should study on whether and how factors such as national economic development level, IT infrastructure condition and culture affect key IS issues between developed and developing countries. Although we know those factors being main differences between developed and developing countries, we know little about whether or how those factors affect CIOs in the two categories of countries in terms of their main IS/IT concerns, and whether such correlations could be validated in empirical investigations. Although the nature of this study is largely exploratory, it represents a useful step toward the determination of the critical IS management issues and subsequent theory-guided research on these issues in developing countries in the future.

Another research limitation is that the survey was conducted within the members of China CIO Club. About 70% of the members are from large or medium-sized companies and organizations. This sample may not be a good representative sample of all businesses, particularly small businesses. Therefore, the interpretation of these results should be done with caution.

For CIOs or executives who have business in China or are considering to do so in the near future, they should not think that because information technology is the same everywhere, key IS issues should be quite similar so best practices or experience of using IS/IT in US corporations can be easily implemented and used in other developing countries like China. In fact, key IS issues faced by CIOs of US and China are quite different. For example, in US, issues such as the building of enterprise IS infrastructure and enterprise information systems strategy have already been considered and thus are not major issues anymore to the US CIOs; but they are key issues for China's CIOs (the number two and number 5 key IS issues respectively). As a result, they should not under-estimate the big task of adopting and implementing major IS infrastructure

and application systems in China, because these issues may not only incur huge investment, but also have a high risk of implementation failure. The successful past experience and best practices from US corporations may not work well in China.

China is in the process of modernizing its IT infrastructure in corporations. Consequently, China CIOs are mainly concerned about how to work effectively to meet the business development pace and requirements. This concern is reflected in the top 10 key IS issues list in terms of the building of enterprise IS infrastructure (#2 key IS issue), applications' adoption and use (#3 key IS issue), enterprise information systems strategies (#5 key IS issue), developing and installing information systems (#7 key IS issue) and network system building and its management (#8 key IS issue). All these key IS issues can only be effectively resolved with careful planning, sufficient investment, and specialized knowledge/skills. As a result, this will bring in many business opportunities to IT corporations of US and other developed countries where advanced IT/IS knowledge and skills are more available. Even developing countries like India, InfoSys, TCS, and Wipro, the three big India IT corporations, identified big business opportunities to set up their branches in China. Our findings in the top 10 IS issues list can help those corporations to understand more about what the main IS/IT business opportunities are in China, as well as provide directions for them to work out effective market-penetration strategies. For example, business strategies should be centered on those areas such as how to help China's business and corporations to build up enterprise infrastructure, and how to add values to business by adopting application systems and enterprise information systems.

Understanding of those key IS issues can also provide Chinese CIOs with guidance and insights in setting current and long term business and IS development goals, as well as future priorities for their organizations.

CONCLUSION

This study is the first national wide survey of top IS issues facing CIOs in the biggest developing country, China. Our 2004 and 2008 survey results indicate that first, there were some similar key IS management issues as well as some interesting differences between two rounds of surveys. While some strategic IS issues are still within top 10 in both 2004 and 2008 lists, their importance ratings are Enterprise information systems strategies, rated as #3 issue in the 2004 list, is rated as #5 in the 2008 list. The #1 strategic issue in the 2004 list, integration of information technologies with enterprise activities, is not in the 2008 top 10 list. In addition, one new strategic issue, business performance management, makes it onto the 2008 top 10 list, which is not in the 2004 list. Second, there are similar IS key issues that are concerned by CIOs of both developed countries like U.S. and developing countries like China. There are still some IS key issues that are unique to developed countries such as U.S., and vice versa. Future research can further investigate what factors lead to those similarities and differences, and how those factors contribute to the theory-building in the research field.

For IS researchers, this study identifies a specific list of IS key issues faced by China CIOs, which might be relevant and useful to those who are interested in studying IS issues in developing countries like China, as well as to those Multi-National Companies that have business interests in China.

REFERENCES

Ball, L., & Harris, R. (1982). SMIS Membership Analysis. *Management Information Systems Quarterly*, *6*(1), 19–38. doi:10.2307/248752

Benbasat, I., & Zmud, R. W. (1999). Empirical research in Information Systems: the practice of relevance. *Management Information Systems Quarterly*, *23*(1), 3–16. doi:10.2307/249403

Brancheau, J. C., Janz, B. D., & Wetherbe, J. C. (1996). Key Issues in Information Systems Management: 1994-1995 SIM Delphi Results. *Management Information Systems Quarterly*, 225–242. doi:10.2307/249479

Brancheau, J. C., & Wetherbe, J. C. (1987). Key Issues in Information Systems Management. *Management Information Systems Quarterly*, *11*(1), 23–45. doi:10.2307/248822

Caudle, S. L., Gorr, W. L., & Newcomer, K. E. (1991). Key Information System Management Issues for the Public Sector. *Management Information Systems Quarterly*, 171–188. doi:10.2307/249378

Chen, G., Wu, R., & Guo, X. (2007). Key issues in information systems management in China. *Journal of Enterprise Information Management*, *20*(2), 198–208. doi:10.1108/17410390710725779

China Enterprise IT application development report. (2003). China National Information Center. Retrieved from http://www.cirm.net.cn

CNNIC report-Top News. (2005). *Chinese Information System Engineering, 16*.

Davenport, T. H., & Markus, M. L. (1999). Rigor vs. relevance revisited: response to Benbasat and Zmud. *Management Information Systems Quarterly*, *23*(1), 19–23. doi:10.2307/249405

Deans, P. C., Karwan, K. R., Goslar, M. D., Ricks, D. A., & Toyne, B. (1991). Identification of key international information systems issues in U.S.-based multinational corporations. *Journal of Management Information Systems*, *7*(4), 27–50.

Dekleva, S., & Zupancic, J. (1996). Key Issues in Information Systems Management: a Delphi Study in Slovenia. *Information & Management, 31*, 1–11. doi:10.1016/S0378-7206(96)01066-X

Dickson, G. W., Leitheiser, R. L., Wetherbe, J. C., & Nechis, M. (1984). Key information Systems Issues for the 1980's. *Management Information Systems Quarterly, 8*(3), 135–159. doi:10.2307/248662

Frenzel, C. W. (1991). *Information Technology Management*. Boston: Boyd & Fraser.

Harrison, W. L., & Farn, C. K. (1990). A Comparison of Information Management Issues in the United States of America and the Republic of China. *Information & Management, 18*(4), 177–188. doi:10.1016/0378-7206(90)90038-J

Hartog, C., & Herbert, M. (1986). 1985 Opinion Survey of MIS Managers: Key Issues. *Management Information Systems Quarterly, 10*(4), 351–361. doi:10.2307/249189

Lee, A. S. (1999). Rigor and relevance in MIS research: beyond the approach of positivism alone. *Management Information Systems Quarterly, 23*(1), 29–34. doi:10.2307/249407

Luftman, Kempaiah, & Nash. (2006). Key Issues For IT Executives 2005. *MIS Quarterly Executive, 5*(2), 81-99.

Luftman & Kempaiah. (2008). Key Issues For IT Executives 2007. *MIS Quarterly Executive, 7*(2), 99–112.

Moynihan, T. (1990). What Chief Executives and Senior Managers Want from Their IT Departments. *Management Information Systems Quarterly, 14*(1), 15–25. doi:10.2307/249303

Niederman, F., Brancheau, J. C., & Wetherbe, J. C. (1991). Information Systems Management Issues for the 1990s. *Management Information Systems Quarterly, 15*(4), 475–502. doi:10.2307/249452

Olikowski, W. J., & Iacono, C. S. (2001). Desperately seeking the 'IT' in IT research – a call to theorizing the IT artifact. *Information Systems Research, 12*(2), 121–134. doi:10.1287/isre.12.2.121.9700

Shan, W. (2001). The IT work force in China. *Communications of the ACM, 44*(7), 76. doi:10.1145/379300.379319

Shi, J. L. (2004). Contract of Competitors. *Chinese Information System Engineering*, 74-78.

Watson, R. T. (1989). Key Issues in Information System Management: an Australian Perspective. *Australian Computer Journal, 21*(3), 118–129.

Ye, J. (2009). China's Internet Population Hits 338 Million. *The Wall Street Journal*. Retrieved from http://blogs.wsj.com/chinarealtime/2009/07/17/chinas-internet-population-hits-338-million/

Zhang, Q. H. (2002). ERP Crisis, ZDNet China. *eWeek, 22*. Retrieved from http://www.zdnet.com.cn/news/software/story/0,3800004741,39038580,00.htm

APPENDIX A. SURVEY QUESTIONNAIRE

Questionnaire for 2003

Please rate the following subjects regarding their importance to your company's development: (Follow the scoring from 1 to 5, where 1 = unimportant, 3 = medium, 5 = very important)

1. Influence of CIO and IT sector on your company
2. How to improve the security of the information systems in your company
3. How to protect the privacy of employees in the enterprise informatization
4. The approach to manage the enterprise information infrastructure
5. Integration of IT and your enterprise business
6. How to improve the speed of your company's response to environmental change
7. How to improve the company's capacity to respond to complex conditions
8. How to introduce business solutions based on IT
9. Enterprise informatization strategy
10. The business solutions based on wireless and mobile information technology
11. Knowledge management
12. Construction of your enterprise information infrastructure
13. Business process reengineering (BPR)
14. Human resource management in IT sector
15. How to improve internal users' satisfaction with the information sector
16. The evaluation of returns on investment on enterprise informatization
17. How to train employees in IT awareness and the relative abilities
18. How to use new technologies in IT innovation
19. Effective use of data resources
20. Information system outsourcing
21. Planning and management of communications networks
22. Management and use of the original application system
23. Planning and integration of various open systems from different vendors
     Questionnaire for 2008

Please rate the following subjects regarding their importance to your company's development: (Follow the scoring from 1 to 5, where 1 = unimportant, 3 = medium, 5 = very important)

1. Construction and use of all kinds of application systems (such as ERP, CRM)
2. The approaches to develop and introduce information systems
3. The approaches to integrate middleware and application systems
4. Construction of your enterprise information infrastructure
5. The approaches to access to and accumulate data resources
6. The assurance of the enterprise information system's security
7. Business intelligence and decision support system
8. Wireless and mobile information technology

9. The approaches of knowledge management and document management
10. Enterprise information strategic planning methods
11. Service oriented architecture (SOA)
12. Green computing
13. Business process management (BPM)
14. Service science, management and engineering (SSME)
15. IT governance standards such as ITIL, COBIT
16. ROI evaluation model of enterprise informatization investment
17. Data mining and visualization
18. IT outsourcing (such as SAAS)
19. Enterprise performance management
20. The use of new technologies for business innovation
21. Construction and management of communications networks
22. IT applications in supply chain
23. Artificial intelligence (AI)
24. E-commerce and network marketing

This work was previously published in International Journal of Global Information Management, Volume 18, Issue 4, edited by Felix B. Tan, pp. 19-35, copyright 2010 by IGI Publishing (an imprint of IGI Global).

Chapter 11
Outsourcing of Community Source:
The Case of Kuali

Manlu Liu
Zhejiang University, China & Rochester Institute of Technology, USA

Xiaobo Wu
ZheJiang University, China

J. Leon Zhao
City University of Hong Kong, China

Ling Zhu
Long Island University, USA

ABSTRACT

Community source (a community-based open source) has emerged as an innovative approach to developing enterprise application software. Different from the conventional model of in-house development, community source creates a virtual community that pools human, financial, and technological resources from member organizations to develop custom software. Products of community source are available as open source software to all members. To better understand community source, the authors studied the Kuali project through interviewing its participants. The interview analysis revealed that community source faced a number of challenges in project management, particularly in the areas of staffing management and project sustainability. A viable solution to these issues, as supported by the findings in the interview and the literature review on the drivers and expected benefits of outsourcing, is outsourcing software development in community source projects. The authors accordingly proposed a research framework and seven propositions that warrant future investigation into the relationship between community source and software outsourcing.

DOI: 10.4018/978-1-61350-480-2.ch011

INTRODUCTION

Due to an increasingly competitive business environment, organizations demand customized application software that can meet their specialized and strategic requirements. If they cannot find suitable one in a commercial market, organizations feel compelled to develop the software in-house. According to the study of Perry & Quirk (2007), however, the overall cost of in-house development is so much higher than that of buying commercial software that many organizations cannot afford the former approach. Furthermore, it is impractical for many organizations to achieve all the necessary competence for building software themselves. As a result, they seek out strategic alliance to jointly develop their desired software. Strategic alliance is a formal relationship between two or more parties to pursue a set of agreed-upon-goals or to meet critical business needs while the parties remaining independent organizations (Yoshino & Rangan, 1995). The concept has recently been applied to the development of open-source enterprise application software (EAS) (Agerfalk & Fitzgerald, 2008). The new model of software development is referred to as community-base open source, or "community source," in which member organizations invest and collaborate with each other to develop custom, open-source EAS. A community source project is "an open source project that requires significant investments from institutional partners in both human resources and cash contributions" (Liu, Wang, & Zhao, 2007).

Community source is a unique form of open source: First, rather than relying on commercial software vendors, the member organizations of a community source project pool their monetary and human resources together in order to develop EAS; second, comparing with an institution developing software independently, the organizations in a community source project can have better control over the software development; and third, developers in a community source project are employees of the member organizations, who are designated to work on the project. Due to these unique features, however, community source projects face a couple of challenges. On one hand, it is optimal for a community source project to attract as many participating organizations as possible in order to share resources, reduce costs, and minimize risks. On the other hand, the management of the community source project becomes increasingly complicated and difficult with the increasing number of participating organizations. In order to understand these challenges well, we studied a real-world ongoing community source project Kuali (www.kuali.org). The Kuali case offered us a great opportunity to look into the project management issues of community source in a higher education setting. Upon completion of this study, we were able to propose that outsourcing the software development in a community source project could be a viable solution to those issues.

The primary research question of this study is: What are the motivations and potential benefits of outsourcing software development in community source? In the following sections, we first introduce the background of community source and the case of Kuali; we then present the literature review on open source/community source and outsourcing of software development; after we present the research methodology of the study and the interview results from the Kuali case, we propose a research framework and seven propositions derived from the findings of the Kuali case study. The chapter concludes with the implications of the study and suggestions of future research directions.

BACKGROUND

Community Source and the Case of Kuali

Community source model requires formal collaboration among member institutions via a virtual mechanism. It also requires significant investment contributed by the participating institutions. Among all the resources contributed, the IT employees of participating institutions are designated as the software developers of a community source project.

For this research we study a particular community source project: Kuali. The Kuali project was initially established to develop a community source financial system, Kuali Financial System (KFS), for partner colleges and universities in the United States. The development of the KFS was based on an existing financial information system at Indiana University. For the project, a community source organization, the Kuali organization, was founded in late 2004 by the University of Hawaii, the rSmart group, NACUBO, and Indiana University. The project received a $2.5 million start-up grant from the Andrew W. Mellon Foundation and follow-up funding from the participating institutions. Those institutions that invested in a significant amount at the development stage are referred as the development partners of the project. Currently, the Kuali project is funded primarily by its development partners, the number of which has grown to more than twenty.

The development partners of the Kuali project collaborate with each other on the platform of the Kuali organization, which works in a hierarchical organizational structure (see Figure 1). The Kuali Board is the final decision maker of the Kuali organization. The functional council, project manager, and technical council report to the Kuali Board and supervise the software development teams. The development partners contribute 130 software developers from their own institutions to work on the Kuali project.

Since the establishment of the Kuali organization, two more software development projects have been added in addition to the original KFS project: Kuali Coeus (KC), a research administration system, and Kuali Student (KS), a student service system. The KFS has completed by (which year?) and the KC and KS are still under development as of 2011.

Proprietary Software, Open Source, and Community Source

Table 1 shows the similarities and differences with regard to developer, end user, and characteristic among proprietary software, open source and community source, which are three primary models for software development. A software company designs and develops a piece of proprietary software and the organizations and individuals that use the software pay for it. In open source model,

Figure 1. The organizational structure of the Kuali project (www.kuali.org)

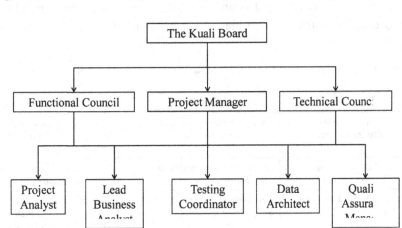

volunteer developers develop software that is free to use. Community source model is different from the other two models in that software developers are employees from partner organizations. The partner institutions invest significantly and coordinate with each other in the software development project. The final product of the project is open source and available to each partner.

While proprietary software was described as a cathedrals and open source as a bazaar by Raymond (1999), community source can be seen as a shopping mall with a virtual organization consisting of partner institutions, each of which commands its own employees. There are many stores in a shopping mall, and each store has its own employees. These stores are under the leadership of an organization that manages the shopping mall. Similarly in community source, a community source organization manages partner institutions and facilitates their collaboration on the development of community source software. Effective collaboration among partner institutions is the key to the success of community source development.

LITERATURE REVIEW

Open Source and Community Source

In the past decade, open source software (OSS) development has often been characterized as an innovative way to develop software (DiBona, Ockman, & Stone, 1999; Raymond 1999). OSS has become a significant competitor with commercial software businesses; it even dominates the software markets today (Vixie, 1999).

The concept of community source was first discussed by Wheeler (2004). Since then community source has emerged as an innovative model for the development of scalable and flexible information systems collaborated by multiple organizations (Liu, Wang, & Zhao, 2007). Several issues of community-based development of enterprise applications have been studied from different perspectives. On one hand, it is optimal for the community source project to attract as many participating institutions as possible in order to share resources, reduce costs, and minimize risks (Liu, Zeng, & Zhao, 2008). On the other hand, management of community source project becomes increasingly complicated and difficult with the increasing number of participating institutions (Liu, Wheeler, & Zhao, 2008). Managing community source projects poses new challenges that are not found in the development of commercial software or conventional open source software (Liu, Wu, Zhao, & Zhu, 2010).

Outsourcing Software Development

Outsourcing may be viewed as a natural step in the evolution of a business (Davis, Ein-Dor, King, & Torkzadeh, 2006). An organization can outsource a wide selection of business activities, particularly IT-related activities such as software

Table 1. Comparison of proprietary software, open source, and community source

	Developer	End User	Characteristics
Proprietary Software	Employees in third-party software companies	Organizations or individuals who pay for the software	End users purchase the software. The initial investment required for the software development is made by the software company.
Open Source	Volunteer software developers	Organizations or individuals who use the software for free	The software is free for end users. The developers are volunteers.
Community Source	Employees from development partners	Development partners and other participating institutions	The initial investment required for the software development is shared by the partners. The final product is open source.

development. A typical process of software outsourcing is that a software is developed by external vendors, purchased by clients, and then customized and integrated with other internal systems (Davis et al., 2006). Significant portions of a firm's software development activities have been routinely outsourced for years (McFarlan & Nolan, 1995). There are no signs that this trend of IS outsourcing will cease (Beverakis, Dick, & Cecez-Kecmanovic, 2009). In fact, programming and software development continues to move out of internal IT department into outsourced companies (King, 2008). According to a recent report, programming, software developing and testing, and software maintenance is among the most outsourced works of firms (Aspray, Mayadas, & Vardi, 2006).

The sourcing decision of software development is basically a make-or-buy issue. Prior to the commencement of a customized development project, the organization must decide whether to develop the software internally or to outsource it to external developers (Wang, Barron, & Seidmann, 1997). Aside from the obvious benefit of cost savings, previous studies have identified two primary motivations for outsourcing software development: to leverage the IS expertise of external vendors and to focus more attention on the organization's core competencies (Beverakis et al., 2009). Subsequently, the expected strategic benefits from outsourcing are: (a) improved IS development and (b) increased operational efficiency and effectiveness and thus enhanced overall performance (Beverakis et al., 2009; DiRomualdo & Gurbaxani, 1998). The following two sub-sections explore further the staffing and core-focus factors in outsourcing.

IS Staffing Factor as an Outsourcing Driver

With regard to IS development, there has been an industry-wide concern about the insufficiency and inferior quality of internal developers and their lagging performance. Many organizations confront a wide array of disparities between the required capabilities and skills to deliver the software and the actual technology capabilities and skills of in-house IS staff (DiRomualdo & Gurbaxani, 1998). Subsequently, research has found a positive relationship between IS outsourcing and the discrepancies between the desired and actual levels of in-house IS staff support and quality (Dibbern, Goles, Hirschheim, & Jayatilaka, 2004). When an organization recognizes that it does not have the skills and resources necessary to achieve its goal in software development, outsourcing then becomes a viable solution. Through outsourcing, the organization can close the gap strategically and obtain cost-effective access to specialized software development skills and capabilities (DiRomualdo & Gurbaxani, 1998; Finlay & King, 1999). In fact, this was one of the primary reasons at the beginning of IS outsourcing (McFarlan & Nolan, 1995). In a review of the outsourcing history, Davis, Ein-Dor, King, & Torkzadeh (2006) explained that the outsourcing of IT began with the hiring of external professionals to work in areas where organizations did not have sufficient skills and/or enough people to accomplish the software development jobs.

Even if the insufficiency of in-house software development skill is not a primary concern, some companies still outsource their IS development in order to do better those things that they are already doing, utilizing state-of-the-art skills, tools, and competencies from external sources, and improving the quality of software development (DiRomualdo & Gurbaxani, 1998). As they enter into outsourcing arrangements, organizations discover that external vendors could provide high-quality and skilled technical developers and more flexible and responsive design and development of IT systems (Khan & Fitzgerald, 2004). In general, the technical expertise and specialization of outsourcers result in better and more effective systems at a significantly lower cost than those developed by the internal IS department (Wang et al., 1997).

In addition, some organizations may have trouble attracting, managing, and retaining their in-house IS developers. Outsourcing offers a way to gain those software development skills without getting involved in the complex staff management issues those organizations are not skilled in and even do not want to manage (McFarlan & Nolan, 1995). Outsourcing vendors have the specialty and proficiency at recruiting and managing software developers, which makes them better than internal IS functions/groups (DiRomualdo & Gurbaxani, 1998). Furthermore, leveraging external developers provides organizations the flexibility to hire IS staff only when necessary and for only as long as needed (Jiang, Frazier, & Prater, 2006). In this way, outsourcing simplifies IS staff management and saves related human resource costs.

Core Consideration in Outsourcing

After delegating software development management to external vendors, an organization, especially when under financial and human resource pressures, may find itself able to direct its key resources to and focus its management time and energy on its core activities (DiRomualdo & Gurbaxani, 1998; McFarlan & Nolan, 1995). Actually, the motivation for outsourcing has evolved from a primary focus on cost reduction to an emphasis on improving overall organization performance. This latter objective can be realized by off-loading responsibilities to an experienced external partner and gaining better internal resource deployment and management (Davis et al., 2006; DiRomualdo & Gurbaxani, 1998).

According to a survey and analysis conducted by Dibbern, Hirschheim, & Jayatilaka (2004), consideration of IS outsourcing is also a consequence of a shift in business strategy. Many companies have abandoned their diversification strategies to focus on core competencies. Senior executives have come to believe that in order to sustain their companies' competitive advantages they should concentrate on what they do best and

outsource the rest (Dibbern et al., 2004). Similarly, after reviewing the history and prospects of IT outsourcing, Davis et al. (2006) argue that an organization should not engage in activities that are outside its core competencies, because others can generally do such functions better, less expensively, and/or faster. Therefore, many organizations have "upgraded" their in-house jobs to focus on "core" activities, with "commodity" activities being outsourced to specialists (Davis et al., 2006; King 2008). For example, while many programming and software development jobs are being outsourcing, new in-house jobs are being created that focus on management and customizing of externally developed software (Davis et al., 2006).

The "job upgrading" concept is also supported by King's (2008) observations of a number of post-outsourcing organizations. The author argues that the need for some traditional capabilities in an organization will virtually disappear. Among the capabilities that will clearly need to be downgraded are those in systems development, since much of an organization's software will be developed on an outsourced basis and customized to fit the organization's unique needs. On the other hand, a new internal capability is needed, the capability reflecting a deep understanding of the organization's business, operations, goals, priorities, and strategies (King, 2008).

The Motivations of Offshoring

In today's globalized economy, outsourcing IT services globally is simply too compelling to be ignored. Globalization has resulted in billions of people joining the free-market world, and millions of companies involving in world trade of products and services. This trend has produced a world where not only goods are globally tradable, but so is labor and service, which can be sent over a wire rather than physically relocated (Aspray et al., 2006). Offshore outsourcing software development is a well paradigm for globalization of

IT service. The main incentive for offshore outsourcing is quality criteria. The ideal outsourcing partner assures high-quality work at low prices and a modern IT infrastructure, and guarantees international quality standards. Another incentive for offshore outsourcing is core competence focus. By offshore outsourcing processes outside its core business, an enterprise can devote itself entirely into value-added activities within its core competencies. This can help to unlock internal resources (Erber & Sayed-Ahmed, 2005). Gupta, Seshasai, & Mukherji (2007) pointed out that the key drivers for offshoring will be strategic, not economic. They argued that to reap the full benefits from offshoring and to develop sustainable models, one need to treat offshore vendors as strategic partners rather than as mere low cost service providers. One can employ professionals in multiple parts of the world, perform tasks at all times of the day, and bring new products and services quicker to the market by offshoring. The creation of new globally distributed workforces and global partnerships can lead to major strategic advantages for companies and countries alike.

Given that offshore outsourcing is simply a specific form of outsourcing (Erber & Sayed-Ahmed, 2005), we use outsourcing to mean software development through either domestic or foreign third-party vendors. As we discussed in the previous sub-section of outsourcing, many in-house IS departments lacked the necessary technical talent, management skills, and financial resources for software development projects (DiRomualdo & Gurbaxani, 1998). On the other hand, there was no shortage of computer skills and competencies either from domestic outsourcing or international offshoring (Davis et al., 2006). Furthermore, most third-party software development firms in the U.S. as well as other countries have adopted Six Sigma or other approaches to quality management (Davis et al., 2006). As a result, outsourcing of software development to supplement or even replace permanent in-house IS staff has increased dramati-

cally in the real world (Ang & Slaughter, 2001). Organizations outsource their software development to the external vendors which can provide not only the lowest-cost but also the best-qualified software developers worldwide (A. Kakabadse & Kakabadse, 2005). The direct benefits expected from such practices are improvement of software development and acquisition of better and cheaper technical skills and competencies at a global scale (DiRomualdo & Gurbaxani, 1998). Furthermore, companies will benefit from outsourcing not just by accessing to better technical skills and expertise to software development, but also by freeing up limited financial and human resources and easing internal management attention on developers (DiRomualdo & Gurbaxani, 1998). Specifically speaking, outsourcing allows an organization to rely on management in external vendors to oversee software development tasks at which it might be at a relative disadvantage, so that the organization can increase managerial attention and resource allocation to those tasks that it does best (Jiang et al., 2006).

In summary, we find from our literature review that (a) many organizations regard outsourcing software development as a solution for the problems of poor performance of their in-house IS staff and lack of skilled professionals and a solution for the issue of in-house IS staff management; and (b) released financial, human and management resources by outsourcing can in turn give organizations an opportunity to concentrate on business analysis, performance improvement, new initiatives, and strategic planning. These two motivations of outsourcing in turn will help organizations gain sustainable strategic advantage. We further find that the main incentive and driver for offshore outsourcing software development are consistent with the ones of general outsourcing. Accordingly, Figure 2 illustrates the motivations and expected benefits of outsourcing software development.

RESEARCH METHODOLOGY AND FINDINGS

Research Methodology

In order to achieve an in-depth understanding of the motivations and expected benefits of outsourcing in community source project, we adopted the qualitative approach of interviewing Kuali development partners on the KFS development. Thirteen interviews were conducted during a biannual conference of Kuali Days.

All interviewees were senior executives, project managers, developers, or other IS staff members from Kuali's partner institutions. Most of them had extensive involvement in the development of the KFS project. Since Andrew Mellon Foundation, rSmart Group, and NACUBO played important roles as did other partner universities in the Kuali project, we also interviewed managers from these three institutions. Most interviews ran around 30 minutes. Three of them lasted between one to two hours.

The interviews were "structured open-response interviews," (King, 1994) in which the formats of interview guideline and schedule were similar to a structured survey. However, most of the questions were open-ended and non-directive. There was flexibility during the interview to focus on particular questions, factual information and evaluative comments, and to skip others, based on the interviewee's personal experience and expertise. For example, we focused more on technical issues when we interviewed developers. As a result, the interview provided a complete descriptive report of interviewees' personal perceptions, beliefs, views, and feelings about the KFS development project.

The interview procedure worked as follows. First, we explained to the participants the purpose of this study. Second, we asked the participants to provide their general impressions about the KFS development. Third, we asked about the problems during the KFS development. Finally, we asked for suggestions on improving the later development of the KRA and KS. We did not ask directly about the benefit of outsourcing in community source development since most respondents had no experience with outsourcing. Instead, we asked about problems arising in the KFS development and invited suggestions for resolving the problems. Some respondents in deed mentioned outsourcing as a solution to the staffing problem in community source and talked about the related benefits of outsourcing.

Figure 2. Motivations and expected benefits of outsourcing software development

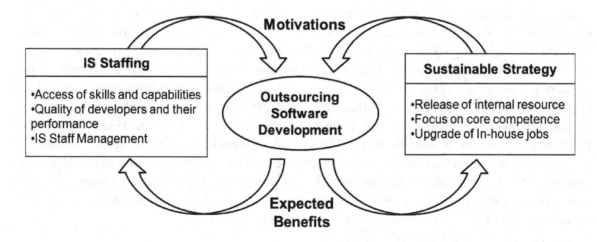

Interview Findings

Table 2 presents the categories and codes that emerged from the content analysis of the interview notes. The detailed explanations and findings are reported as follows in a descriptive, narrative, and succinct form, categorized into five sub-sections and supported by interviewees' quotations (in *italics* in the following texts). In order to ensure anonymity all respondents' last names were removed in the report.

Managing Developers

There are nine counts in the interview transcripts that interviewees mentioned the challenge of managing software developers in the community source project. This was the most-mentioned problem during the interview. At the beginning of the project, "*the developers are actually not there; they need to be recruited later on*" (Steve, project manager). Later on, partner universities contributed their own employees as the software developers for the project. However, there was an "*unbalance problem*" of such human resource contributions between different partner institutions. Tim, a member of the Kuali organization, pointed out the problem that "*some large universities over-contribute to project development, while some small universities cannot contribute enough.*" Some participating colleges even had a "*difficult time to get enough developers [to work on Kuali's projects]*" (Tim, member of the Kuali organization).

Another problem about software developers emerges as every participating university was required to contribute developers to the project. When the number of development partners grew, the number of developers also grew. "*The increasing number of development partners put pressure to the Kuali organization for project management*" (Mike, member of the Kuali organization). "*The number of developers is [also growing] too large [for such a small organization] to manage effectively*" (Steve, project manager). All developers worked in different locations at different time, and reported to both their home institutions and the Kuali organization. This was essentially a virtual team that the Kuali organization had to coordinate and manage, but managing such "*a virtual team is not easy*" (Steve, project manager). It was even more "*challengeable to make [developers from] multiple universities which have different culture to work in a team. How to balance their commitment becomes a key issue*" (Alex, member of the Kuali organization).

Finally, every time "*when the new development partner joins Kuali, new developers will join the development team... [I]t is not easy [for those new developers] to come into the team quickly and it takes lots of time and effort [for both the developers and the development team]*" (Bob, developer). This point is supported by a statement made by a developer from a new participating university: "*We join Kuali in the late stage. We have difficult time to involve in the project development*" (John, developer).

Table 2. Summary of the interview analysis results

Category	Codes	Ranking by quote counts
Staffing problem of the community source project	Managing / management	1
	Quality / performance	2
	Turn-over rate / recruiting	4
Sustainability problem of the community source project	Sustainability / focus shift	4
Outsourcing as a solution for the problems	Outsourcing	3

The interviews at this point led us to find that while software developers were the most basic human resource for a community source project to perform its job, the creation, the growth, and the change of this developer team were out of the control of the project. Managing such a virtual team therefore became an indefinite and intractable problem.

Quality Issue of Developers

The second most-mentioned challenge was the quality of developers. It "*is a big issue. In the future project, I wish we could select high-quality developers*" (Steve, project manager). George, a senior executive in the Kuali organization, suggested that "*we should hire good quality developers to work for us.*" The quality of the developers working for the project depended largely on the contributing institutions and thus there was a large variance on the developers' quality. One interviewee gave an example, "*it is hard for X University to get good developers since the living expense there is very high*" (Alex, member of the Kuali organization). Another example was that "*[the developers contributed by] X College ... cannot do the work they promised*" (Tim, members of the Kuali organization).

Another quality-related issue identified in the interviews was the training program for the developers. "*We need to better educate [our] developers... Our training program in the future needs to be more organized*" (Alex, member of the Kuali organization). Moreover, "*we need to assign somebody responsible for training program [for our developers]. We do not give this specific role in KFS right now*" (Steve, project manager).

High Turnover Rate of Developers

The high turn-over rate of the developers was also identified as a challenge for the project. According to Alex and Mike, members of the Kuali organization, "*the current problem for KFS is the high turn-over rate of developers*"; "*we have high turn-over rate of developers in KFS. In some universities, the turn-over rate of developers is very high.*" As a sequence, "*the high turn-over rate of developers delays our progress*" (Steve, project manager). Steve further stated that "*we could be much more effective if we don't need to deal with this issue.*"

Sustainability of the Community Source Project

When we conducted the interviews, the KFS was almost completed and the KRA and KS were being launched. After learning lessons from developing the KFS and in looking ahead to the future of the Kuali organization, several interviewees identified the sustainability requirement for the organization. At the strategic level, George, a senior executive of the Kuali organization, indicated that "*sustainability for Kuali is our behavior of volition... Right now, the Kuali organization should focus on sustainability.*" At the project level, Steve, a project manager, also identified sustainability as "*a big issue.*" Kate, a developer, suggested that "*right now how to sustain the system is very important for us.*" However, "*there is not enough resource for sustaining the system*" (Steve, project manager).

Outsourcing as a Solution for the Problems

As shown above, we revealed several problems of software development in the community source project. As the Kuali organization moved on to its next software project, Eric, a member of the Kuali organization from its external sponsor, indicated that "*Kuali needs to spend more time to design in KS than that in KFS and KRA. The development problems now will become more serious in KS if the problems are not solved properly.*" Since in-house software development had been shown to have a number of issues with this community source project, the interviewees suggested that outsourcing be a potential solution. "*I think out-sourcing developers might be an option for future projects… Outsourcing developers will help the Kuali organization [concentrate its energy to] manage more development partner [institutions]*" (Mike, member of the Kuali organization).

Tim, another member of the Kuali organization, disclosed another motivation of outsourcing—focus shift—"*[from now on] the Kuali organization should focus less on technical issues and more on community issues.*" George, the senior executive of the Kuali organization, also indicated the shift of the organization's strategic focus could lead to outsourcing. "*During KFS development, the Kuali organization focused on managing the develop-ment project. Right now, the Kuali organization should focus on sustainability*" (George, the senior executive).

DISCUSSION: OUTSOURCING OF COMMUNITY SOURCE

At the beginning of his interview, George, the senior executive of the Kuali organization, ex-plained that software development experienced three eras: "*[During] the first era, each of us built it; [in] the second era, we bought it…then hammered it…sent more money; [in] the third era, we now built it together [e.g. community source].*" He then posed the question of "*what next*". According to our literature review on the motivations and expected benefits of outsourcing and our interview analysis on the challenges of the community source project, we proposed that outsourcing of community source development be an answer to George's question. The results in the Kuali interview analysis specifically led us to construct a research framework for outsourcing of community source development, as shown in Figure 3. The framework describes the rela-tionship between in-house IS staffing problems and outsourcing and the relationship between outsourcing and the sustainable strategy of orga-

nizations. Based on the research framework that maps with the concepts in Figure 2 and evidenced by the interview results, seven propositions were developed and discussed as follows.

THE STAFFING PROBLEM OF DEVELOPER

Proposition H1a: The higher the turn-over rate of developers, the lower the quality of develop-ers working for a community source project.

Proposition H1b: The higher the turn-over rate of developers, the more difficult in managing the in-house developer team in a community source project.

In traditional open source projects most develop-ers participate voluntarily and are motivated by their own interests. The developers in the com-munity source project are employees assigned by the development partners. The problem of developers' turn-over rate in community source is therefore contributed by three facts. First, the turn-over rate of IS staff in the home institution itself is high. Due to the high cost of living and/or relatively low salary, some developer might leave the home institution and thus not work for the community source project any more. Second, the development partners might re-assign some developers to other projects and replace them with new developers. And last, when the com-munity source organization grows, new partners join the project and a few old partners drop out. Subsequently, developers come and go over time.

The high turn-over rate makes it impossible for the community source project to maintain a consistent quality within the developer team and to manage a relatively stable developer team. However, there is no such a reliable IS human resource in hand for the community source organization to access. The organization has to devote an enormous amount of attention, time,

Figure 3. The proposed research framework of outsourcing community source projects

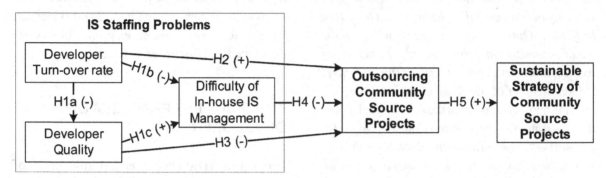

effort, and resources to deal with this developer issue and consequently the development progress is slowed down.

Proposition H1c: The lower the quality of the developers, the more difficult in managing the in-house developer team in a community source project.

The quality of the developers had been a serious management problem in our community source case. The quality of the developers is not within the control of the community source organization. Instead, it depends on the IS human resource that individual partner can access. As the development partners "contribute" their own developers to the project, the quality of the developers varies greatly. In the Kuali case, some institutions, especially those small colleges, were even unable to provide competent developers to perform their jobs for the community source project. Moreover, the overall quality of software developers working in educational institutions generally cannot compare with the quality of developers at professional software companies.

The poor quality of the developers not only negatively affects the quality of the software development but also makes the management of such a diverse team very complicated and time- and effort-consuming. In the Kuali case, in order to deal with the quality issue, the community

source organization had to allocate a great deal of human and financial resources, especially for a developer training program.

Outsourcing as a Solution to the Developer Staffing Problem

Proposition H2: The high turn-over rate of the developers is a driver to outsource software development of a community source project to third-party vendors.

Proposition H3: The low quality of the in-house developers is a driver to outsource software development of a community source project to third-party vendors.

The developers designated by partner institutions might have private interests or incentives from the home institution that conflict with those of the community source project. These developers differ greatly in experience, skills, and business and technical knowledge. The turn-over rate and quality problems of developers are essentially the issue of accessibility of IS skills and capabilities in the community source project.

According to the motivations of outsourcing identified in Figure 2, the needs to access sufficient and reliable IS skills and competencies and to ensure the performance of software development drive the consideration of outsourcing. Senior executives in many organizations (including the community source organization in our case) believe that external IS vendors possess better technical and business expertise in providing IS development and provide the development more efficiently and effectively than internal IS departments. Globalization further makes it possible to access the best external vendors worldwide. DiRomualdo & Gurbaxani (1998) described two company cases in the early era of outsourcing. When Dow Chemical realized it was losing IS staff with critical technical skills, it created unique outsourcing joint ventures both domestically and internationally, to gain access to a broader talent pool worldwide. In a similar case, Philips Electronics restructured its IS organization and outsourced system delivery to both domestic and foreign joint ventures with professional software and system integration companies. Therefore, outsourcing could be a viable solution for the problems of lack of IS skills and capacities and poor performance of in-house IS staff.

Proposition H4: The difficulty of managing the in-house developers is a driver to outsource software development of a community source project to third-party vendors.

The costs associated with communication among and train of developers, identification of system requirements, and project management can be significant for in-house software development. In the case of Kuali, for a small community source organization, managing such a virtual team in dispersed locations, with a large and ever-growing number, high turn-over rate, and varied skills of developers, was particularly difficult, intractable, and thus resource-consuming.

Outsourcing makes it possible not to get involved in the complex IS staff management issues that the community source organization is unable to handle or not skilled in. Moreover, by outsourcing the software development, the community source organization can accommodate

more partner institutions without worrying about managing and coordinating the developers "contributed" by them. Outsourcing will fundamentally change the way how a community source system is developed since it eliminates the need of programmer contribution and simplifies in-house IS staff management.

A primary reason for some organizations retaining their software development in-house is trade secrets or critical key processes embedded in their software and systems that they would not wish to be made available to outsiders (King, 2008). This is, however, not a big concern for a community source project in the case of outsourcing because the community source software is fully accessible to all development partners and becomes open source once the development is completed. Thanks to the open source feature, community source partners are able to customize the software and integrate it into their own organizations.

Sustainable Strategy

Proposition H5: Outsourcing of software development will help community source management focus more on project sustainability.

Our study of Kuali shows that the root problem of community source is inadequate and ineffective coordination among its partner institutions. The management team of the Kuai project has to invest time and efforts on dealing with uneven contributions of human and financial resources from its partner institutions, combating the problem of uneven quality of developers from different partners, and making new arrangements whenever a developer (or a partner) joins or leaves the project. This problem of ineffective coordination has distracted the project from its long-term management goals and made it harder for the project to develop systems to sustain and strengthen the strategic alliance among its participating institutions. Therefore, a number of executive members of the Kuali project in our interviews indicated that outsourcing could help them shift the strategic focus of the project and achieve the goal of sustainability.

Outsourcing of software development can ease internal IS staff management and thus release the human, financial as well as other resources needed for the community source project. According to Davis et al. (2006), software outsourcing means that a much smaller proportion of the effort involves in-house IS staff, while a higher proportion lies in the software developers of the external vendor. The organization that outsources software development will not need large numbers of developers and programmers and its in-house IS team will shrink significant in size. The smaller function of the in-house team will involve project management, technology architects, and system sustainability. Thanks to the easy and low-cost access to developers worldwide, the organization will thus save the investment in manpower, facilities, and equipment for software development.

Moreover, by outsourcing software development, the community source organization, under financial and human resource pressures, is able to allocate and concentrate its resource, time, and effort more efficiently on overall project management and performance, on system deployment, customization, integration and sustainability, and on organizational planning and strategies. In this way, the in-house jobs will be "upgraded" to focus on core activities and priorities, and thereby able to help the organization gain sustainable advantages, survive, and achieve healthy growth.

CONCLUSION

Implications for Practice

Community source has just emerged recently as an innovative and economic way to develop enterprise information systems, not only in educational settings like the case in our study, but potentially in commercial settings as well. By pooling the resources of partner institutions, this new development model allows the sharing of costs, risks, and responsibilities and the sharing and control of the application/system which the partner institutions develop together. More important, most development partners of a community source project are themselves the users of the final product. Community source model offers the development partners an open development platform to share their common values, strategic goals, concepts, ideas, and knowledge with each other. The application/system developed out of this community source model could, therefore, be better shaped by the strategic requirements of the organizations and better tailored to different organizational environments and business needs than those generic commercial software/systems can.

Despite the advantages promised by community source, our study of the Kuali case shows that managing system developers in such a virtual environment and under limited resources is quite challenging, due to unbalanced development partners' contributions, decentralized locations, high turn-over rate, and poor quality and/or performance of the developers. The experience and issues of the Kuali project depicted in this research provide other organizations insights into community source development.

The findings in our study further suggest that organizations making a similar decision to join or create a community source project and expecting a similar challenge of IS staff management consider outsourcing as a practical alternative for internal software development. Outsourcing will allow the project management of community source to deal with a single software contractor/vendor rather than a large number of individual developers from multiple partners, while still keeping control of software design and quality of the final delivery which is guaranteed by international industry standards. Subsequently, outsourcing can help community source management focus on its core activities: coordinating resources and services of the project, designing new systems and applications, sustaining the project, and strategic planning.

A shift from in-house development to outsourcing might lead to new issues of outsourcing relationship management and the problem of unequal monetary commitments from different participating institutions might persist. Even though they are out of the scope of this study, these issues merit an investigation during a post-outsourcing study of community source.

Research Contributions, Limitations, and Future Study

To the best of our knowledge, our research is among the first ones focusing on the community source phenomenon and its interrelationship with outsourcing. Both open source software development and outsourcing have gained momentum in IS research in recent years. This research takes one step forward to incorporate the novel concept of community source with outsourcing. The rich information and in-depth understanding gained from the extensive survey of previous literature and an interview analysis of a real-world case of the Kuali community source project allow us to propose a research framework to study the drivers of outsourcing in community source and expected values of outsourcing for community source.

Since community source as a new model of software development is still at its infant stage, limited cases are available for us. Nevertheless, in a future study of post-outsourcing community source, we plan to conduct a larger scale quantitative survey, which is based on the research framework we propose in this research. We expect to validate the findings in the current qualitative study and test the seven propositions in our research framework. We will also investigate the new issues and challenges of outsourcing in community source, and tradeoffs and alternatives of outsourcing. In this way, our research has the potential to develop a general and solid theory for outsourcing of community source development.

Furthermore, we hope this study will inspire other researchers to use the research framework we propose to investigate further the interrelationships among community source, open source, and outsourcing.

REFERENCES

Agerfalk, P. J., & Fitzgerald, B. (2008). Outsourcing to an unknown workforce: Exploring opensourcing as a global sourcing strategy. *Management Information Systems Quarterly, 32*(2), 385–409.

Ang, S., & Slaughter, S. A. (2001). Work outcomes and job design for contract versus permanent information systems professionals on software development teams. *Management Information Systems Quarterly, 25*(3), 321–350. doi:10.2307/3250920

Aspray, W., Mayadas, F., & Vardi, M. Y. (2006). *Globalization and offshoring of Software: A report of the ACM Job Migration Task Force.*

Beverakis, G., Dick, G. N., & Cecez-Kecmanovic, D. (2009). Taking information systems business process outsourcing offshore: The conflict of competition and risk. *Journal of Global Information Management, 17*(1), 32–48. doi:10.4018/jgim.2009010102

Davis, G. B., Ein-Dor, P., King, W. R., & Torkzadeh, R. (2006). IT offshoring: History, prospects and challenges. *Journal of the Association for Information Systems, 7*(11), 770–795.

Dibbern, J., Goles, T., Hirschheim, R., & Jayatilaka, B. (2004). Information systems outsourcing: A survey and analysis of the literature. *The Data Base for Advances in Information Systems, 35*(4), 6–97.

DiBona, C., Ockman, S., & Stone, M. (1999). *Open sources: Voices from the open source revolution.* Sebastopol, CA: O'Reilly & Aoosciates.

DiRomualdo, A., & Gurbaxani, V. (1998). Strategic intent for IT outsourcing. *Sloan Management Review*, *39*(4), 67–80.

Erber, G., & Sayed-Ahmed, A. (2005). Offshore outsourcing: A global shift in the present IT industry. *Inter Economics*, *40*(2), 100–112.

Finlay, P. N., & King, R. M. (1999). IT sourcing: A research framework. *International Journal of Technology Management*, *17*(1-2), 109–128. doi:10.1504/IJTM.1999.002703

Gupta, A., Seshasai, S., & Mukherji, S. (2007). Offshoring: The transition from economic drivers toward strategic global partnership and 24-hour knowledge factory. *Journal of Electronic Commerce in Organizations*, *5*(2), 1–23. doi:10.4018/jeco.2007040101

Jiang, B., Frazier, G. V., & Prater, E. L. (2006). Outsourcing effects on firms' operational performance: An empirical study. *International Journal of Operations & Production Management*, *26*(12), 1280–1300. doi:10.1108/01443570610710551

Kakabadse, A., & Kakabadse, N. (2005). Outsourcing: Current and future trends. *Thunderbird International Business Review*, *47*(2), 183–204. doi:10.1002/tie.20048

Khan, N., & Fitzgerald, G. (2004). Dimensions of offshore outsourcing business models. *Journal of Information Technology Cases and Applications*, *6*(3), 35–50.

King, N. (1994). The qualitative research interview. In Cassell, C., & Symon, G. (Eds.), *Qualitative methods in organizational research: A practical guide* (pp. 14–36). London, England: Sage.

King, W. R. (2008). The post-offshoring IS organization. *Information Resources Management Journal*, *21*(1), 77–87. doi:10.4018/irmj.2008010105

Liu, M., Wang, H. J., & Zhao, L. J. (2007). Achieving flexibility via service-centric community source: The case of Kuali. *Proceedings of the Americas Conference on Information Systems*, Colorado, USA.

Liu, M., Wheeler, B., & Zhao, J. L. (2008). On assessment of project success in community source development. *Proceeding of International Conference on Information Systems* Paris, France.

Liu, M., Wu, X., Zhao, J. L., & Zhu, L. (2010). Outsourcing of community source: Identifying motivations and benefits. *Journal of Global Information Management*, *18*(4), 36–52. doi:10.4018/jgim.2010100103

Liu, M., Zeng, D., & Zhao, L. J. (2008). A cooperative analysis framework for investment decisions in community source partnerships. *Proceedings of the Americas Conference on Information Systems*, Toronto, Canada.

Liu, M., & Zhao, L. J. (2007). Real options analysis of the community source approach: Why should institutions pay for open source? *Proceedings of the First China Summer Workshop on Information Management, Shanghai, China*.

McFarlan, F. W., & Nolan, R. L. (1995). How to manage an IT outsourcing alliance. *Sloan Management Review*, *36*(2), 9–23.

Perry, R., & Quirk, K. (2007). *IDC white paper: An evaluation of build versus buy for portal solutions*. ftp://ftp.software.ibm.com/software/bigplays/IDCBuyVsBuildPortal.pdf

Raymond, E. S. (1999). *The cathedral and the bazaar: Musings on Linux and open source by an accidental revolutionary*. Sebastapol, CA: O'Reilly & Associates.

Vixie, P. (1999). Software engineering. In Dibona, C., Ockman, S., & Stone, M. (Eds.), *Open sources: Voices from the open source revolution* (pp. 91–100). Sebastopol, CA: O'Reilly & Associates.

Wang, E. T. G., Barron, T., & Seidmann, A. (1997). Contracting structures for custom software development: The impacts of informational rents and uncertainty on internal development and outsourcing. *Management Science, 43*(12), 1726–1744. doi:10.1287/mnsc.43.12.1726

Wheeler, B. (2004). The open source parade. *EDUCAUSE Review, 39*(5), 68–69.

Yoshino, M. Y., & Rangan, U. S. (1995). *Strategic alliances – An entrepreneurial approach to globalization.* Boston, MA: Harvard Business Press.

Chapter 12

IS–Supported Managerial Control for China's Research Community:
An Agency Theory Perspective

Wen Tian
USTC-CityU Joint Advanced Research Center, China

Douglas Vogel
City University of Hong Kong, China

Jian Ma
City University of Hong Kong, China

Jibao Gu
University of Science and Technology of China, China

ABSTRACT

In the first decade of the 21st century, China's Research Community (CRC) is struggling to achieve better performance by increasing growth in knowledge quantity (e.g., publications), but has failed to generate sound growth in knowledge quality (e.g., citations). An innovative E-government project, Internet-based Science Information System (ISIS), was applied nationwide in 2003 with a variety of embedded incentives. The system has been well received and supports the National Natural Science Foundation of China (NSFC) to implement managerial control to cope with pressing demands relating to China's research productivity. This paper explores the impact of Information Systems (IS) from the perspective of agency theory based on CRC empirical results. Since the nationwide application of ISIS in 2003, CRC outcomes have markedly improved. The discussion and directions for future research examine implications of IS for E-government implementation and business environment building in developing countries.

DOI: 10.4018/978-1-61350-480-2.ch012

INTRODUCTION

Both success and failure of E-government cases in developing countries are reported in the literature (Heeks, 2002; Ndou, 2004). Among various factors addressed by scholars from different perspectives, mentioned in common is that the interests or goals of multiple stakeholders related to E-government projects should be given attention in keeping with results of E-government cases (Dada, 2006; Gil-Garcia et al., 2007; Heeks, 2002; Krishna & Walsham, 2005; Ndou, 2004). Consequently, one of the underpinning challenges to governance systems is to encapsulate the simultaneous demand for both stakeholder control and cooperation in organizational transformation brought about by E-government implementations (Tan et al., 2005). In certain areas such as healthcare, studies using empirical data from developed countries have shown that information systems (IS) can be leveraged as enablers to achieve goal alignment among different stakeholders when they are in a principal-agent (P-A) relationship (Wickramasinghe, 2000; Wickramasinghe & Silvers, 2003). Nevertheless, there is a deficiency in the literature addressing how IS could support governance endeavours when stakeholders are in principal-agent relationships in developing countries, particularly in China (Walsham et al., 2007; Walsham & Sahay, 2006).

As China's economy has enjoyed sustained and rapid growth since the mid 1990's, R&D funding has experienced annual growth, and government fiscal appropriations for Science & Technology (S&T) have increased steadily (Zhu & Gong, 2008). However, certain problems have drawn considerable global attention. For instance, concerns about misconduct (e.g., fraud and plagiarism) in the Chinese Research Community (CRC) have been reported by Nature (Cyranoski, 2006; Wang, 2006) and Science magazine (Xin, 2006); a significant contrast has appeared between the rapid increase in the number of scientific papers (e.g., the rank of China in Scientific Citation Index (SCI) counts) and the slow development

in terms of quality (e.g., the rank in Essential Science Indicators) and international influence of scientific papers (e.g., citations) (Jin, 2004). The pressing demands for China's research productivity have become one of the major driving factors for policy makers to look for innovative solutions so that China's basic research community can be reoriented toward high output, high quality, and high efficiency (Zhu & Gong, 2008). On such occasions, China's R&D related government agencies play the role of principals who are looking for a portfolio of agents capable of realizing the goal to perform high quality research.

As one of the main channels for funding basic research, the National Natural Science Foundation of China (NSFC) (see Appendix A) has placed continuous emphasis on innovation in E-government implementation to effectively and efficiently encourage sustained research and contribution to globally recognized literature. The Internet-based Science Information System (ISIS) [https://isis.nsfc.gov.cn], embedded nationwide in 2003, now annually manages qualified peer reviews, sharing of information and openness for critique, which in 2009 culminated in research funding distribution for over 97,755 grant submissions (NSFC, 2009c) with an annual increase rate of over 15%. Prior to ISIS, incentives in place did not produce desired results. It is with this application of ISIS that research quality improvement has attracted our attention.

Although there are many contributors to communities' research productivity, such as institutional, financial, manpower, technological and cultural factors (Bland & Ruffin Iv, 1992; Heinze et al., 2009), this paper focuses on the aspect of institutional factors introduced by a nationwide information system that contributes to the quality enhancement of research output. Taking an agency theory perspective in the context of ISIS's implementation for CRC, we explore how ISIS, as an E-government practice, influences goal alignment that gradually supports academic achievement and global recognition. Previous studies have focused mainly on the technical issues of ISIS (Ma et al., 1999; Tian et al., 2002). To fill this gap, research addressing managerial control mechanism embedded in ISIS in coping with agency opportunism problem will shed light on another facet of the system with both theoretical and practical implications. Thus, we generate the following research questions:

1. What is the nature of the principal-agency problem in NSFC's governance process?
2. How does ISIS function in addressing the agency problem and does it work?
3. What are the implications of ISIS for innovative E-government implementation in developing countries?

In order to answer our research questions, we review the literature to specify the meaning and context of concepts through the lens of agency theory. In the next section, we propose our conceptual framework and corresponding propositions. The research methodology is described in the fourth section, followed by the case overview. Details of data collection, data analysis and results are provided in the sixth section. The penultimate section discusses findings and suggests implications for research and practice, as well as giving directions for future research. Conclusions are drawn in the final section.

LITERATURE REVIEW

Agency theory provides a unique, realistic, and empirically testable perspective on problems of cooperative effort (Eisenhardt, 1989). Although this theory is rooted in the economic discipline, it has been extended to virtually all types of transactional exchanges that occur in a socio-economic system where information asymmetry, fears of opportunism, and bounded rationality exist (Milgrom & Roberts, 1992). Within the last decade, agency theory has gained attention

from researchers in IS or related fields, such as buyer–seller relationships (Ba & Pavlou, 2002; Pavlou et al., 2007; Singh & Sirdeshmukh, 2000), IT service purchaser-provider relationships (Fuloria & Zenios, 2001), IT outsourcing decisions (Gefen et al., 2008; Tiwana & Bush, 2007; Wu et al., 2004), supply-chain management (Bai & Yang, 2008; Feldmann & Muller, 2003; Qi & Zhang, 2008), and IS project management (Basu & Lederer, 2004; Choudhury & Sabherwal, 2003; Dologite et al., 2004; Kirsch, 1997; Mahaney & Lederer, 2003).

However, studies on the government-researcher relationship in the context of IS supported decision making and practice are scant, especially conducted through the lens of agency theory; while publications on the productivity of CRC, both domestic and overseas, are plentiful (Jonkers, 2008; Li, 2004; Meng et al., 2007; Wang, 2004; Zhao, 2003). This has been especially evident in recent years when China's rising economic power has brought about increasing investment in science and technology activities, but whether the investment has been taken advantage of is an important issue. Further, most of the currently published E-government strategies are based on successful experiences in developed countries, but which may not be directly applicable to developing countries (Chen et al., 2006). Therefore, this study is aimed at exploring the application of agency theory in an emerging E-government project in a developing country.

The outcome of CRC can be viewed as a consequence of the interaction between the investments of the Chinese government, the managerial control system embedded, and the researchers' desire to fulfil the requirements of the government of remaining within the budget. Therefore, in general, the government is treated as the principal, and the researchers are treated as the agents.

In the positivist stream of agency theory, a common approach is to identify a policy or behaviour in which stockholder and management interests diverge and then demonstrate that IS or outcome-based incentives solve the agency problem (Eisenhardt, 1989). In the case of CRC, we have adopted this approach. However, for a better understanding of the problem, we not only identify the agency opportunism problems and the relative managerial control mechanisms, but also introduce two contextual factors from IS design perspectives as a complement to agency theory, as recommended by Eisenhardt (1989). Therefore, three groups of concepts are elaborated upon: (1) agency opportunism which provides a theoretical explanation for examining the nature of the P-A problem in CRC involved in NSFC's governance, (2) control modes which shows what managerial control mechanisms are introduced, and (3) contextual factors which influence the P-A relationship apart from the control mechanisms.

Agency Opportunism

Agency theory is concerned with two types of principal and agent problems that can occur in P-A relationships: 1) agency problems when the desire or goals of the principal and agent conflict, and it is difficult or expensive for the principal to verify what the agent is actually doing; and 2) risk sharing problems where the principal and agent may prefer different actions because of the different risk preferences (Eisenhardt, 1989). The risk sharing problems are usually investigated through the economics or psychology paradigm, which is out of the core of the IS discipline. This paper mainly considers the first agency problem, referred to as agency opportunism, i.e., mis-alignment of goals when an agent (researcher) performs a task for the principal, and cost incurring for the principal to guard against sub-optimal behaviour (Wickramasinghe & Silvers, 2003). This problem contains two important aspects known as "moral hazard" and "adverse selection."

Moral hazard (hidden action) refers to the shirking of efforts by the agent, including actions or behaviours inconsistent with the goals of the principal (Tuttle et al., 1997); for example,

an employee working on a personal project on company time when it is difficult for managers to detect what he is actually doing.

Adverse selection (hidden information) concerns the agent's concealing of relevant information or misrepresenting his ability (Eisenhardt, 1989). In other words, the agent can retain private information that is hard for the principal to verify, for example, his background, interests, capabilities and similar characteristics that the principal cannot obtain without some cost (Basu & Lederer, 2004).

In the case of CRC, *moral hazard* bas been reported in terms of fraud, plagiarism, misrepresentation of the facts or problems of the complexity of projects in order to receive favour from the principal, recommending funding to one's "clan members" when there are outsiders who can perform better, etc. *Adverse selection* behaviours include forged publication records, education degrees and research experience, reluctance to fulfil academic obligations, etc. (Zhang, 2003).

Control Mechanism

The managerial control system prescribes ways for the organization to implement its strategies, and thus, to a large extent, determines its performance (Anthony & Govindarajan, 1998). There are four types of control modes: two formal modes (behaviour, outcome) and two informal modes (clan, self) (Kirsch, 1997). In the positive stream of agency theory, the two formal modes are adopted as two major governance mechanisms: 1) the outcome-based control mode that co-aligns the preferences of agents with those of the principal, and 2) the behaviour-based control mode (usually delivered by information processing mechanism) which informs the principal about what the agent is actually doing (Eisenhardt, 1989). The scope of this study is framed within these two mechanisms.

For outcome-based control, it is proposed that when the contracts between the P-A are outcome-based, the agent is more likely to behave in the interests of the principal, resulting in reduction of agency opportunism (Eisenhardt, 1989). Incentives based on measurable outcomes will increase goal alignment between the P-A, which is referred to as *incentive alignment* control.

Incentive alignment mechanism in the context of government-research community relationship is mainly the performance-based funding with output indicators, such as indicators derived from research activities (Kivistö, 2007). To cope with the CRC outcome, incentives are introduced through the selection of appropriate indicators to align the agent's goal so as to enhance the quality. To analyse IS supported *incentive alignment*, both positive and negative incentives are examined. They are: 1) reward for meeting the requirements of the afore-mentioned indicators, 2) punishment for misconduct that jeopardizes the endeavour to enhance peer-review quality and research quality.

For behaviour-based mechanism, it is proposed that when the principal has information to verify agent behaviour, the agency opportunism is low (Eisenhardt, 1989). Thus, *monitoring* is used as a feedback system to gather information of the agent behaviour and then provide feedback after verifying it, which will deters agents from pursuing risky investment strategies (Kirby & Davis, 1998). *Monitoring* mechanism in the context of government-research community relationship is primarily conducted in terms of the input-based funding and quality assessment procedure (Brennan & Shah, 2000; Jongbloed & Koelman, 2001). In NSFC grant allocation process, input-based funding, such as line-item budgeting or formula funding, has not been used; it is the grant proposal that mainly determines the funding result. Thus, input-based funding is not included in the measurement of monitoring.

There are two types of quality monitoring: *signaling* and *screening* (Brennan & Shah, 2000). *Signaling* refers to the researchers revealing of private information to the government about their capacity and willingness, while *screening* refers to those activities undertaken by the principal which are intended to separate good types of pro-

spective agents from bad types (Kivistö, 2008). In the case of NSFC, *signaling* is the initial step applicants make when they fill out the forms to provide necessary basic information which is relatively simple, while *screening* producers are more complicated. Therefore, *to* assess monitoring, screening procedures that help reduce the likelihood of moral hazard and adverse selection are extracted and then examined to see how helpful they are in screening the hidden action and hidden information.

By means of the above formal managerial control, agency opportunism can be curbed (Eisenhardt, 1989). That is why professional management at the micro-level is often required to provide for transparency of goals, rational resource allocation, reward, etc. (Biggart & Hamilton, 1992). However, when Eisenhardt (1989) encouraged scholars to put emphasis on IS to advance agency theory, the term in the 80's literature only referred to human systems for information processing without any IT artifacts involved. To address this gap, *control transparency* is adopted in our study, to assess the outcome of the control mechanisms mentioned above.

Control transparency refers to the amount and type of information available to interested parties (Finel & Lord, 1999). It relates to behaviour control that indicates the extent to which one makes available the information needed to provide some behavioural control of an exchange process (Kirsch, 1997; Nicolaou & McKnight, 2006). IS research suggests that a system design intervention - such as behaviour control transparency - can provide high quality current information for decision making to reduce opportunistic behaviour (Nicolaou & McKnight, 2006). Thus, the level of control transparency introduced by ISIS implementation is analysed in this case study.

Nevertheless, the implementation of an E-government project will introduce changes to organizations' management environment which, in turn, will influence the chance of agency opportunism. Hence, contextual factors are reviewed.

Contextual Factors

Control systems have to keep pace with the changing environment (Bruns, Jr., & McFarlan, 1987). Eisenhardt (1989) has identified five important independent variables as contextual factors that will influence P-A contracts. As the uncertainty of R&D outcomes and the scarcity of S&T resources are the inherent nature of S&T activities in any research community (Farina & Gibbons, 1981; Keefer, 1978), *the outcome uncertainty* and *goal conflict* are not included in our conceptual model. In addition, another contextual factor referred to as *the length of the agency relationship* is also not included, since it focuses on the time effect in P-A contracts. Instead, two salient contextual factors influencing the P-A relationship in CRC are considered: *outcome measurability* and *task programmability*.

Outcome measurability, which refers to the degree to which the outcome of the task by the agent can be measured (Eisenhardt, 1989), has generally been found to have a positive relationship with outcome control (Eisenhardt, 1989; Kirsch, 1996). The more measurable the outcome of the task, the more feasible the *incentive alignment* control of the agents.

Task programmability is defined as the degree to which appropriate behaviour by the agent in the task can be specified in advance (Eisenhardt, 1989). The more programmed the task, the more salient the agent's behaviour becomes. Both *outcome measurability* and *task programmability* in developing countries' E-government environment are considered to be low, reflected by the lack of transparency in the governance (Ciborra, 2005; Dada, 2006; Ndou, 2004), insufficient planning and complex legislative procedure (Basu, 2004), etc. The level of *outcome measurability* and *task programmability* of an E-government project is shaped by the interplay of the institutional and technological design, and evolves over time. Therefore, IS functionality, designed and developed in accordance with institutional environment changes during implementation and post adoption are included to represent the status of *outcome measurability* and *task programmability*.

Table 1 summarizes the key concepts of the agency opportunism, control modes and contextual factors in accordance with their relationships.

Ba et al. (2001) have proposed a framework to analyze whether the organizational processes embedded in IS are in line with the system's and organizational objectives, something which is critical to the success of the application of the system. It suggests that the mechanism incorporated in the IS will have impact on user behaviour and subsequently shape the outcomes. Drawing upon Ba's framework (2001), our conceptual research framework is suggested with propositions generated from agency theory.

RESEARCH FRAMEWORK

According to NSFC's E-governance arrangement, the ISIS system is used to collect researchers' background information, support researchers' competition for NSFC's grants and assist the panels' decision making procedure on grant allocation. Two types of CRC outcome are considered: grant allocation and research output. There are certainly more items of the CRC outcome, but these two outcomes are the core of the P-A relationship representing primary stakeholders in E-government project: 1) between NSFC (i.e., the principal) and experts in peer-review system (i.e., the agents), in that the agent makes decisions on the allocation of financial resources on behalf of the principal to attain the goal of *"selecting, supporting and managing innovative research projects"* (NSFC, 2009a); and 2) between NSFC (i.e., the principal) and researchers who are funded by it (i.e., the agents), in that the agents carry out research that meets the principal's strategy which is to execute the Medium to Long Term Plan (MLP, issued in January 2006) for the Development of Science and Technology of China (Linton, 2008).

Table 1. Summary of key concepts

Constructs	Concepts	Key characteristics
Agency opportunism	*Moral hazard*	Hidden action E.g., an agent *shirks* his duty claimed in the contract, or *distorts* information to manipulate decision making process.
	Adverse selection	Hidden information E.g., an agent *retains* his willingness or capability to perform the task difficult for the principal to verify.
Formal control	*Incentive alignment*	Outcome-based reward system E.g., rewards for enhancing the quality of publications are articulated; indicators for enhancing the peer-review quality are articulated.
	Monitoring	Behaviour-based feedback system E.g., policy, regulations and procedures in handling misconduct are articulated, while behaviour of controllees is observed so that punishment or preventive measures can be conducted.
Contextual factors conveyed by IS	*Outcome measurability*	the degree to which the outcome of the task by the agent can be measured
	Task programmability	the degree to which appropriate behaviour by the agent in the task can be specified in advance

According to the literature review, the system, embedded with appropriate features that promote corresponding control modes, reduces the likelihood of agency opportunism, which aligns the users' behaviour and then alters the outcomes: the results of grant allocation and the research output. Based on Eisenhardt (1989) and Ba et al. (2001), the conceptual framework and propositions are as follows:

The arrows in the figure do not necessarily indicate causal relationships. They represent the influences of one upon the other, as stated in the following propositions. First, understanding the two types of agent opportunism upon NSFC grant application in the context of the CRC is our focus for interpreting research question one. Second, determining which formal control mechanisms are used and how they are supported by the ISIS system in dealing with agency opportunism is the key to research question two.

In addition, the CRC today is so large that all information containing the goals, rules, and plans has to be disseminated through government websites, while procedures, processes, etc., are embedded in the routine of IS operations, emphasizing the role of IS in this context. Therefore, proposition 1 to 4 are made.

Proposition 1: Moral hazard is an agency problem in China's Research Community and it will affect grant allocation and research output.

Proposition 2: Adverse selection is an agency problem in China's Research Community and will affect grant allocation and research output.

Proposition 3: Appropriate incentive alignment mechanisms embedded in IS that are commonly used by the principal (NSFC) will compensate agents for achieving the goals related to the principal's announced performance and merit evaluation system, which will reduce the likelihood of moral hazard in the allocation of grants.

Proposition 4: Appropriate monitoring mechanisms embedded in IS that are commonly used by the principal (NSFC) will prevent agents from withholding private information, which will reduce the likelihood of adverse selection in the application and allocation of grants.

In order to reach the P-A relationship harmony, the performance and merit evaluation system (when incentive alignment takes place), and the process management level (when monitoring takes

Figure 1. Conceptual framework

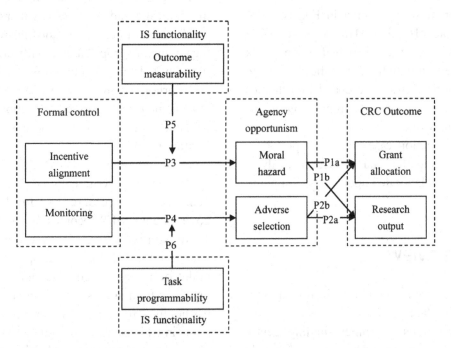

place) can be moderated by the implementation of the IS; consequently, the control mechanism can be further supported.

Proposition 5: IS provides better outcome measurability that will help achieve better incentive alignment control for China's Research Community.

Proposition 6: IS provides better task programmability that will help achieve better monitoring control for China's Research Community.

METHODOLOGY

Yin (1994) recommends case study as the preferred strategy when "the focus is on a contemporary phenomenon within some real-life context" (p. 1). The general methodology of this study is to rely on the propositions in the conceptual framework to organize the case analysis before and after ISIS's implementation.

For proposition 1 and 2, tactic pattern matching (Yin, 1994) is used to address internal validity. In order to evaluate the CRC outcome, the status of grant allocation and research output before and after the implementation of ISIS is compared. If dramatically different patterns are observed, the degree of agency opportunism is changed. Secondary data gathered from several databases is used to test the pattern change. Peer-review fairness is used to measure grant allocation quality and publication performance to measure research output.

For proposition 3 and 4, pattern comparison analysis is based on secondary data from the literature, media reports, press statements and public websites that reflect feedback from general researchers and agents. They are then triangulated with data from NSFC brochures, annual reports, yearbooks, and NSFC websites.

For proposition 5 and 6, IS functionality supporting outcome measurability and task programmability is analyzed through in-depth interviews with the team leader of the system's developers and discussions with two major system developers, which were conducted in English. The interview protocol (Yin, 1994) and sample coding of interview excerpt are provided in Appendix B. In conjunction with the data collection stage, transcribed interviews were coded into themes that map to key concepts and propositions in our research framework.

In the next section, case overview is provided. Details regarding the data collection, analysis and results are elaborated upon in the following section.

CASE OVERVIEW

The NSFC is a government organization directly affiliated with the State Council. It is the largest and most prestigious government funding agency for basic research in China. The pressing demands on NSFC come from several parties of strength.

First, as the research community from both the domestic and the overseas market expands quickly (Zweig et al., 2004), dealing with increasing volumes of proposals manually in constrained time limits becomes impossible. The efficiency and fairness of the peer-review system, R&D management and resource allocation are urgent demands from the agents who conduct the scientific research.

Second, as the number of grants authorized by the central government increases sharply, the management of resources to fulfil the central government's MLP becomes the most important administrative objective. A critical problem mentioned in MLP is that, despite the swelling ranks of research personnel and increasingly generous funding, research system performance has not lived up to expectations, i.e., quantitative gains in Chinese research productivity have not always been matched by qualitative gains (Cao et al., 2006). It is the effectiveness and corruption-free nature of the system that forms the major

Figure 2. NSFC Yearly Funding (2001 - 2007)

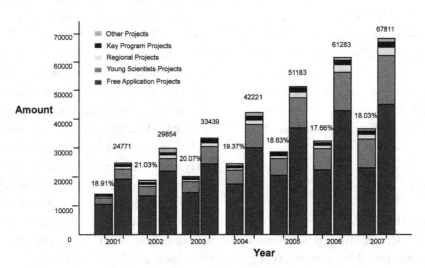

Note:

1.   In each year's result, the right bar represents the number of proposals with the sum number on top, while the left bar represents the number of accepted proposals with the average acceptance ratio on top (ratio is not to scale);

2.   Data is from Annual Report of NSFC and NSFC web site (http://nsfc.gov.cn).

concern of the principal's side. As reflected by MLP, "China's approach, unlike that of many western countries," establishes "a central role for the government in steering the quantity, quality and direction of R&D" (Linton, 2008).

Third, strength comes from the industry and the public, among whom are individuals and organizations keeping eyes on the output of the whole system. In many E-government projects, China's present priority lies mostly in establishment of an intranet in terms of networking government institutions and monitoring government work digitally rather than citizen-oriented front office work (Zhang, 2005). Such a situation could lead to limited ability of individuals to gain insight into governmental procedures and lack of governmental process transparency, both of which are critical components of an effective E-government programme (Qiu & Hachigian, 2005). NSFC's initiation of the ISIS project has been an attempt to cope with this external strength (NSFC, 2009b).

It is not an easy job to build such a nationwide information system to realize as many objectives as possible. However, ISIS has performed admirably under the circumstances. As highlighted by the team leader of the system developers: "We have tried our best to look for innovative applications of state of the art technology all the way through the project, including sustained innovation to deal with the increasing volume of tasks and disruptive innovation to deal with behaviours that might threaten the credibility of the system." Figure 2 illustrates NSFC funding from 2001-2007 with respective program proposal acceptance submission ratios. In general, we note the relatively increased equality and balance across all of the programs supported since the ISIS introduction in 2003 in conjunction with peer review in the presence of ever increasing submissions and funding. Competition remains keen since over four out of five proposals are unfunded.

In 2000, ISIS was introduced by NSFC and generalized to nation-wide use in 2003. A brief introduction of NSFC's implementation of ISIS is in Appendix A. ISIS is an end-to-end solution for researchers, universities, research institutes, and NSFC to manage and disseminate their research information (e.g., projects and research outputs). It has greatly simplified the administration processes for application, evaluation and management of NSFC projects. It avoids duplication of data entry and reduces the administrative workload and human errors. It also standardizes the processes and technologies for R&D project administration. Its core functions are:

- **Project application:** XML-based electronic document management and submission, decision-making support, and online review and analysis of application statistics;
- **Project management:** project risk control, analysis of project statistics and project progress, and completion reports; and
- **Dissemination of research outputs:** submission of research results, and search and publication of research results.

Any individual researcher can obtain general information on projects approved by the NSFC through ISIS, e.g., project history searches and duplication checks. NSFC program directors, research administrators in universities, and research institutions can use ISIS for managing and inquiring the progress of NSFC projects. A major advantage of ISIS is that it allows non-registered users to obtain project information for public supervision (Li, 2008). ISIS also accepts data exchanges from internet-based research information systems (IRIS) from participating universities and research institutions, thus providing opportunities for extended system applications, such as institutionally developed database-access-interface to ISIS, data-exchange software packages, and virtual research centres for international cooperation (He et al., 2007; Huang et al., 2003; Li, 2008; Xie et al., 2008).

With the application of ISIS, the R&D resource distribution process is becoming increasingly transparent. Scientists from all over the world, as well as any individuals who are interested in the CRC, are able to obtain information related to the grants, successful research projects and research outputs. Further, the mechanism of resource allocation goes toward an independent, rationally-driven, merit-based direction, while gradually separating government politics and academic achievement recognition. Growing participation and increased transaction memory of grant allocation have emerged, making it appropriate for us to study the agency problem in the context of E-government.

CASE ANALYSIS AND RESULTS

Identifying Agency Opportunism

The agency opportunism problem in CRC outcome (i.e., grant allocation and research output) can be identified through manifestations resulting from the P-A relationship between the government and the research community. This problem can happen at both micro and macro levels. Our study focuses on the former. In the grant allocation part, the unit of analysis is the individual level behaviour of the reviewers and researchers; in the research output part, the unit of analysis refers to the individual level behaviour of researchers.

The adverse selection problem is examined before researchers get funding, since it is pre-contractual opportunism that exploits informational asymmetries about future performance. The moral hazard problem, however, is examined as the peer-review for grant allocation finishes, and the research is conducted with visible output, since it is post-contractual opportunism that exploits informational asymmetries about current performance (Barney & Quchi, 1986). The operationalization of agency opportunism is derived from Kivistö's definition of behaviour and manifestations (Kivistö, 2008), as shown in Table 2.

We focus on manifestations in bold which are items that can be identified based on the source of data we obtained. Whether the review is slapdash or not, or whether the reviewer is reluctant to review or not, can only be evaluated by digging into the content of the peer-review system, which is difficult and sensitive. Therefore, to identify agency opportunism problems, the analysis consists of two parts: the quality of NSFC's peer-review fairness and the quality of NSFC funded publications. In grant allocation, if the reviewers' evaluation that leads to decision bias can be detected, then the moral hazard problem can be identified; regarding research output, if the quan-

Table 2. Manifestations of agency opportunism

Concepts	Behaviour	Manifestations	
		Grant allocation	**Research output**
Moral hazard	Shirking	Reviewers: Provide slapdash review	Researchers who get funded: **Publish large number of papers, regardless of the quality**
	Distortion of information	Reviewers: **Evaluation of proposals that leads to decision bias**	Researchers who get funded: **Plagiarism**
Adverse selection	Hiding the willingness	Reviewers: Reluctant to review	-
	Hiding the capacity	Applicants: **Forge publication records to add merit**	-

tity of papers outweighs the quality of papers, and misconducts (such as plagiarism and forging of publication records) are detected, the moral hazard and adverse selection problem can be identified. The before and after pattern analysis results are presented as follows.

The Quality Enhancement of Peer-Review Fairness

As mentioned previously, if agency opportunism existed, the fairness of peer-review decisions could be questioned. We now examine that issue in greater detail. From 2001 to 2005, the Chinese Academy of Sciences accepted 159 new academicians every other year. The number of newly elected academicians was 56 in 2001, 58 in 2003 and 45 in 2005 (http://www.cas.ac.cn). Based on the project history search function of ISIS, we found that among all the newly elected academicians, there were 29 (52%) in 2001, 15 (24%) in 2003, and 18 (40%) in 2005 that had been funded by NSFC. We tracked the record of the projects and counted the number of projects that they were in charge of from 1999 to 2007, and then we ran tests for 4 groups of data through Wilcoxon Test to see if there was a significant increase in the number of projects. The groups were:

- **Group 1:** 2 years' data before and after 2001 for newly elected Academicians in 2001
- **Group 2:** 2 years' data before and after 2003 for newly elected Academicians in 2003
- **Group 3:** 2 years' data before and after 2005 for newly elected Academicians in 2005
- **Group 4:** 4 years' data before and after 2003 for newly elected Academicians in 2003

In the results, we indicate whether the changes are significant before and after the raise in their status.

Table 3 shows that newly elected academicians in 2001 received 0.46 units more projects on average, and that this change is statistically significant. Newly elected academicians in 2003 and 2005 did not show this trend. Before 2001, scientists received more scientific resources after their academic status was raised. In 2003 and 2005, this situation no longer existed. This implies that fairness in the process of resource allocation was improved, especially after the application of ISIS in the year 2003. ISIS submission data from 2003 to 2006 also demonstrate that female researchers are making more submissions annually (from 24% to 31%). A similar situation exists for Associate Professors compared to Professors (from 13% to 16%, respectively), indicative of enhanced equality following the ISIS introduction.

The Quality Enhancement of Publications

Quality enhancement of publications is a long march. Table 4 indicates the number of papers with top 1% citations in SCI in the years 1996 and 2005, and top 0.1% citations in SCI in the years 2004-2005.

Compared with quantity contribution, China's research community's quality contribution is not a distinction. The desired quality is not matched with the increase of quantity. If we measure the percentage of best papers to total papers, China had only 0.36% in 1996 and 0.44% in 2005, while Japan had 0.60% to 0.64%, and other countries ranged from 0.86% to 1.44%, indicating that large numbers of China's publications went to B (or lower) journals.

To empirically examine enhancement of research quality conducted by agents sponsored by NSFC, three disciplines (Earth Science, Chemistry and Management Science) were randomly chosen from the six disciplines that NSFC supports: Physics, Chemistry, Maths, Earth Science, Biology and Management Science. Three or four top journals were selected (also randomly) in each discipline with high Impact Factor in Journal

Table 3. Mean statistics and Wilcoxon signed ranks test of different groups

Group	Sample	Time Section	Mean	Standard deviation	Z-value	Asymp.Sig (2-tailed)
1	29	[2000,2001]	0.3793	0.49380	-2.977*	0.003
		[2002,2003]	0.8621	0.63943		
2	15	[2002,2003]	0.6667	0.61721	-0.333	0.739
		[2004,2005]	0.6000	0.50709		
3	18	[2004,2005]	0.3333	0.59409	-0.632	0.527
		[2006,2007]	0.4444	0.51131		
4	15	[2000,2003]	1.2000	0.73679	0.000	1.000
		[2004,2007]	1.2000	0.77460		

Note: "*" represents significant under 95% confidence interval.

Table 4. Quality of the SCI articles by Chinese authors

	"Best Paper" (Top 1%)				"Hot Paper" (Top 0.1%)		Total Papers			
Year	1996	%	2005	%	2004-2005	%	1996	%	2005	%
USA	4415	64.4	4676	55.1	744	61.7	306270	33.9	420084	32.3
Britain	745	10.9	1133	13.3	182	15.1	81237	9.0	107783	8.3
Japan	428	6.2	603	7.1	102	8.5	70819	7.8	93748	7.2
Germany	614	9.0	1102	13.0	175	14.5	64537	7.1	95262	7.3
China	**53**	**0.8**	**340**	**4.0**	**55**	**4.6**	**14654**	**1.6**	**76931**	**5.9**
France	428	6.2	675	7.9	114	9.5	49787	5.5	65652	5.1

Note: data source is from Jin (2007).

Citation Reports database (JCR Science Edition 2006, ISI Web of Knowledge). The Nature and Science Magazine (from 2000-2007) were also selected to reflect the trend for all natural science research outcomes. Papers published by Chinese scientists from 1987-2007 were counted, within which papers sponsored by NSFC were calculated. If the percentage of the NSFC sponsored papers increased, we felt we had reason to believe that the NSFC was doing a better job than other foundations in China in the field of basic research funding. Since the total amount of funding did increase, a higher percentage illustrates the enhancement which, in part, eliminates other factors influencing paper quality.

Chinese scientists have started to gain global recognition by improved research quality, as noted through publications in top-tier journals. In Table 5, the number of papers published by CRC with acknowledgement to NSFC funding has grown in recent years in the selected top journals in Earth Science, Chemistry and Management Science. Taking into consideration that the total time span is 20 years (1987-2007), it is noteworthy that more than half of the papers have been published between 2004 and 2007. To further understand the influence of the fund, we checked the background of those researchers who published papers under the NSFC fund in Earth Science and Chemistry. It turned out that 64.3% of researchers in Earth Science and 65.7% researchers in Chemistry were funded by multiple NSFC grants. This implies that they are not only productive researchers, but also active ISIS contributors and users.

Table 5. NSFC funded papers in top journals

Journal	Impact Factor	Total	NSFC	Before	After
Area: Earth Science 1987-2007					
CLIM DYNAM	3.468	27	12	4(33.33%)	8(66.67%)
ACTA ASTRONOM	3.451	1	1	0(0.00%)	1(100.00%)
B AM METEOROL SOC	3.055	16	1	0(0.00%)	1(100.00%)
Area: Chemistry 1987-2007					
CHEM REV	26.054	25	13	4(30.77%)	9(69.23%)
ACCOUNTS CHEM RES	17.113	33	21	12(57.14%)	9(42.86%)
A NNU REV PHYS CHEM	11.25	3	2	0(0.00%)	2(100.00%)
Area: Management Science 1987-2007					
PROD OPER MANAG	2.516	12	3	1(33.33%)	2(66.67%)
J OPER MANAG	2.042	9	2	0(0.00%)	2(66.67%)
TRANSPORT RES B-METH	1.761	63	13	6(46.15%)	7(53.85%)
MANAGE SCI	1.687	31	0	0(-)	0(-)
Science 2000-2007	30.028	269	78	35(44.87%)	43(55.13%)
Nature 2000-2007	26.681	188	72	32(44.44%)	40(55.56%)

Note: Meanings of the columns are as below:

Total = total counts of papers with Chinese authors; NSFC = counts of papers sponsored by NSFC;

Before = paper was received before 2003 (Including 2003); After = paper was received after 2003.

Identifying IS-Supported Mechanism Improvement

Overall Mechanism Improvement Through IS Supported Credit Information Sharing

A historically held viewpoint is that the credit institution in China is very weak and has considerable discredit conducts which impede the development of the country (Wu, 2003b). Serious credit issues also appear in science and technology activities, such as plagiarizing others' intellectual property rights or abusing R&D funds (Wu, 2003b; Zhang, 2006).

The Credit Reporting System (CRS) has been developed to deal with moral hazard and adverse selection caused by information asymmetry between borrowers and lenders, as well as other similar relationships. The competition for NSFC's scientific funding among the researchers

can be treated as a research fund market where information asymmetry exists between the P-A relationship. The credit reporting relationship in our case is shown in Figure 3.

Economics literature has shown strong evidence of how the existence of CRS improves the functioning of the market by creating effective incentive and screening mechanisms, especially in less developed countries (Luoto et al., 2007). In our case, a senior executive member of one of China's top universities provided his feedback on ISIS, emphasizing that "the application of ISIS has gradually created a database which can be leveraged as a credit reporting system of CRC members who have applied for NSFC grant or served in peer-review process." Such facilitating of credit information flow as countermeasures to solve discredit conducts in CRC has been acclaimed by many Chinese scholars (Gu et al., 2008; Wu, 2003a; Zhang, 2003; Zhang, 2005, 2006).

Figure 3. Credit reporting relationship in NSFC

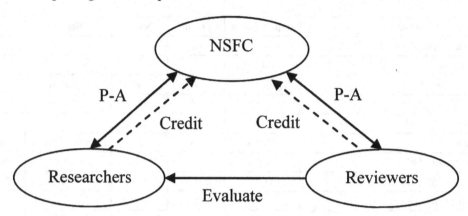

With the sharing of researchers' and reviewers' credit information, the principal is able to create incentive mechanisms, rewarding excellent, while deterring negligent, researcher behaviour; at the same time, the principal is able to create screening mechanisms that improve quality assessment of proposals. Thus, overall mechanism improvement can be achieved.

Higher Level of Control Transparency through Information Transparency and Richness

Predicted by agency theory, IS facilitated managerial control can enable a higher level of control transparency to provide high quality current information for decision making. As a result, goal alignment behaviour can be promoted and monitoring costs can be diminished as well. It is expected that, after the implementation of ISIS, studies using the ISIS generated data as data source for policy making will become more visible, thereby the principal can make better incentive decisions and the research community and the public can better participate in the monitoring process.

To analyze the level of control transparency, the number and content of articles using ISIS generated data as data source for policy making were gathered. With regard to the articles, we used the database provided by the China Journal Net database (known as CNKI), which is the largest and one of the most prestigious, academically comprehensive databases in China. Articles published between 2001 and 2009 were searched as reference.

Using the combination of keywords including "NSFC", "ISIS," "research management," "grant allocation" in CNKI database, we obtained hundreds of Chinese articles. After screening the abstract, content, references and journals, we ruled out those that were not related to our case in failing to address any aspects of NSFC or IS applications for NSFC. Thus, 22 Chinese articles emerged. The profile of these articles is indicated in Table 6.

It becomes evident from Table 6 that after 3 to 4 years of ISIS implementation, academia begins to benefit from the ISIS database, since it has become a rich source of community transaction memory and knowledge repositories for the CRC activities. Seven articles directly used ISIS generated data to analyze policy making regarding incentives to agents. Six articles mentioned the benefit of transparency brought about by ISIS. Three articles talked about the method of outcome measurability. Thus, it is demonstrated that information transparency and richness have been gained through the collection of behaviour data gener-

Table 6. Chinese studies related to the IS-supported NSFC governance

Topic	Year	Number of articles	Sum
Analyzing ISIS generated data for policy making	2007	1	7
	2008	5	
	2009	1	
Discussing information transparency benefit from ISIS	2006	2	6
	2008	3	
	2009	1	
Discussing method for outcome measurability	2007	2	3
	2008	1	
IS and peer-review	2006	1	2
	2007	1	
IS and screening	2007	1	2
	2008	1	
IS and organizational change	2002	1	1
ISIS system development	2002	1	1
Total			22

ated from the use of ISIS. As a result, the level of control transparency has been raised. This creates the incentive for external researchers to generate more explicit indicators for policy making on incentive alignment, while taking part in monitoring the agents' behaviour.

Stronger Disclosure of Misconduct through Information Exploitation

Disclosure of misconduct comes from several sources, such as the principal, the agents, the external research community and the public. For the principal, with the possession of a large volume of transaction memory, information technology can be applied to detect misconduct. According to our interview and discussion with the developers, technology-driven innovations are continuously brought to the improvement of ISIS to exploit abnormal behaviour. For example, the interview result shows that, "Business Intelligent (BI) methods have been applied into the Research & Information management stage, the R&D project management security, and the Social Network

formation stage, as means to detect suspicious behaviour that could have been resulted from misconduct." By means of incremental innovation, the system itself has exerted stronger ability to detect misconduct.

For the agents and the external research community, their expertise allows them to be able to verify peer performance. For the public, their localized knowledge plays a complementary role to discriminate certain types of misconduct (Gu et al., 2008). Open access to individual researchers and the public allows them to obtain general information from the ISIS database on projects approved by NSFC. Thus, the chance of disclosure of misconduct becomes larger. NSFC has pointed out that one of the goals of the ISIS project is to provide more government information disclosure to let the public take part in misconduct governance (NSFC, 2009b).

In addition, we observed the change of regulations and the report given by the Supervision Committee of NSFC. From the website of NSFC and the hyperlink to the Supervision Committee, there is no evidence that statistics on misconduct cases were published before 2003. There were only a few reported incidences in the Misconduct Report before 2003. Since 2004, the committee has started to publish annual reports on the details of misconduct cases. The punishment decision reports are even put online, providing detailed information of the misconducts. The number of publicized misconduct cases was 16, 20 and 10 in 2004, 2005 and 2006, respectively. Considering that the number of submissions and funded proposals nearly doubled during the period since the ISIS introduction in 2003, the trends are indeed encouraging.

In short, this section has analyzed the exemplifications of mechanism improvement. But we wonder how such improvement could be achieved from the perspective of system design intervention. Based on the literature review, IS innovations for outcome measurability and task programmability are considered.

IDENTIFYING IS FUNCTIONALITY INNOVATIONS

IS Supported Outcome Measurability to Enhance Mechanism Improvement

It is predicted by agency theory that the more explicitly these indicators are measured, the more quantitatively the goals of the principal are articulated, and the more effective the control becomes.

The implementation of ISIS affects the stage of grant allocation, the researchers' application, and tracing of the research output. It has, on the one hand, maintained the existing measurement that is used to assess the outcome; on the other hand, it has increased the opportunity of introducing innovative measurements of indicators. Thus, the level of outcome measurability can be different in the context of IS-supported grant allocation and research management. IS functionalities supporting the improvement of indicator explicitness and

appropriateness is evaluated so that the level of outcome measurability is differentiated.

According to the team leader of the system developers, there is tight collaboration between the developers and experts from the academy of mathematics and system science of the Chinese Academy of Science (CAS). The goal is to make better measurements of researcher merit. He raised one example of the innovations in the development of the *"creativity factor"* to measure the performance of a researcher.

In the past, a researcher's merit was mainly evaluated by how many SCI publications and how many total publications one had made, at the time quantity mattered a lot. The 'creativity factor' we co-developed with CAS experts is an integrated index in consideration of the number of publications, citations, author orders, impact factor of journals, etc. Weightings of these factors are calculated through their contributions so that different coefficients are given. Therefore, when the principal gets the result of one's creativity factor, it is explicitly that which the measurements are and how they are calculated. Meanwhile, the agent would be aware that he has to work not only on the quantity side, but also quality side to enhance his performance.

IS Supported Task Programmability to Enhance Mechanism Improvement

According to the description of the team leader of developers, the workflow of NSFC grant allocation is shown in Figure 4.

The darkest arrows represent the stages where business processes must be completed through ISIS. In general, grant application, peer-review process and follow-up research management are assisted by ISIS. Thus, task programmability in the ISIS project can be interpreted as the degree to which control mechanism is made routine

Figure 4. Workflow of NSFC grant allocation

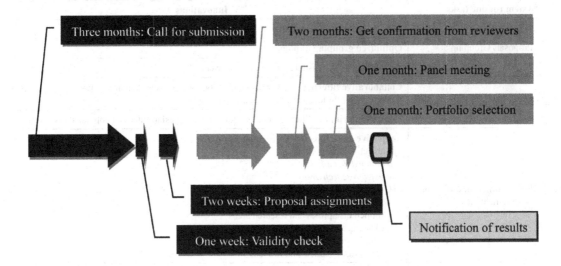

(Mahaney, 2000) in coping with these three stages of tasks. Table 7 summarizes the innovations for completing the tasks.

The information in Table 7 shows support for mechanism improvement in three ways.

First, the validity check task has become more routine because the system has greatly assisted "Researcher's Profile Collection and Verification." The collection, confirmation and verification of data become more simple, accurate and standardized, based on the subjective and objective integrated approach. ISIS helps the principal to import a large volume of proposals and finish validity checks in limited time. Therefore, the principal will find it much easier to collect and verify the information of an agent's behaviour so that the screening becomes more feasible.

Second, the reliability and fairness (Daniel, 2005; Norman, 1986) of peer-review process have been raised, based on optimized grouping of proposals and intelligent recommendations. This is because proposals can go to the more appropriate reviewers, and reviewers can get a group of similar proposals that are easier to make comparisons with. This mechanism reduces the likelihood of the blindness of proposal assignment or overwhelming manipulation of assignments, which enhances the fairness of peer-review.

Third, the periodical collection of research outputs provides the principal with more reliable information to reinforce the data verification in previous tasks. As collection is done by the system, more objective data will be gathered, leaving little space for opportunistic behaviour.

According to the developers, by means of certain design intervention, "data verification and management processes are made simple, accurate and standardized." As commented by the team leader of the developers, "Thanks to the help of ISIS, the staff member of NSFC has only increased slightly from 2001 to 2008, despite the rapid expanding of expert tank and proposal volume. It has largely improved the level of NSFC's personnel efficiency."

DISCUSSION

Returning to our research questions, we now place our results in the context of propositions made regarding ISIS impact and implications.

Table 7. The innovations for task programmability

System routine tasks	Innovations
Proposals Grouping	**Computer assisted assignment** of reviewers to proposals
	Optimized grouping of proposals
	Social Networks assisted grouping of reviewers
Intelligent Recommendation	**Collaborative filtering:** to filter information or patterns using techniques involving collaboration among multiple agents, viewpoints, and data sources
	Content filtering: to filter content or information. Bayesian filter is a popular statistical filter

Researcher's Profile Collection and Verification	**Subjective and Objective Integrated Approach**
	1) Subjective information:
	Self claimed disciplines
	Interested research areas
	2) Objective information
	Funded research projects
	    Applied disciplines and research areas
	    Abstracts and full text
	Published articles
	    Keywords, abstracts and full text

Research online for CRC and NSFC	**Periodical** collection of research outputs for Principal Investigators (PIs) Profile-based **Intelligent assignment** of research outputs to PIs **Collection and confirmation** of research outputs for finished projects **Cite@Write** function for project final reports and proposals

First, the nature of P-A problem in NSFC's governance is rooted in the information asymmetries in two types of P-A relationships, as shown in Figure 4. Due to the nature of the complexity of academic work, the quantity and quality outputs generated by the research community add difficulty in information verification and merit evaluation. Such information gap in CRC leads to moral hazard and adverse selection problems such as undesirable outcome in NSFC's grant allocation and CRC's research output. Our findings indicate that the level of agency opportunism was higher in terms of the imbalance of the quantitative and qualitative development of publications as well as peer-review bias before the implementation of ISIS. After IS supported control mechanism took place, quality enhancement of publications and peer-review was identified. Therefore, propositions 1 to 4 are illuminated in this context. Although the content and format of specific agency problems may vary in different cultures, the nature of the problem is similar.

Second, we have identified three ways that ISIS has facilitated the improvement of managerial mechanisms in coping with agency opportunism. The first is that the establishment of Credit Reporting System (CRS) in CRC was made attainable because of ISIS adoption and diffusion. In China, the institutional weakness of the management of CRC members' scientific credit has dramatically reduced the cost of discredit behaviour (Wu, 2003b; Zhang, 2006). The existence of CRS will create a strong and effective incentive as well as screening mechanism to align P-A relationships in the scientific fund market.

The second benefit of ISIS is that the level of control transparency has been enhanced because ISIS accumulates higher quality information for decision making. Although it takes time to leverage ISIS as a transaction memory and knowledge repository about CRC activities, NSFC could benefit from higher information quality in terms of more transparent and richer information in order to make better control decisions.

Third, multiple stakeholders have become more capable of disclosing scientific misconduct by exploiting ISIS information. Thus, negative incentives for misconduct are emphasized while monitoring costs are diminished.

Fourth, method innovations in the design of IS supported outcome measurability and task programmability are identified which have enhanced the mechanism improvement. Outcome measurability can be achieved through the improvement of indicator integration and explicitness. Task programmability can be achieved by standardizing task procedures and embedding the control mechanism into system operation routine. Thus, proposition 5 and 6 are supported. However, beyond proposition 6, the enhancement of task programmability has not only improved the monitoring, but also improved the peer-review quality, suggesting its role in promoting incentive alignment. Thus, the effect of task programmability is widened from that of previous literature (Eisenhardt, 1989).

In conjunction with IS supported mechanism improvement, these findings have enriched our understanding by adding empirical demonstrations in developing countries, that "the mutual influence between IS and social processes sometimes results in solutions that work relatively well" (Walsham et al., 2007).

Our findings offer the following implications for practice.

1. In E-governance environment when P-A relationship among stakeholders exists, IS can be leveraged as an efficient instrument to deal with the agency opportunism problem. Policy makers, system developers and users are encouraged to participate in organization's E-governance planning, so that the IS design can be aligned with intended control mechanism to handle the problem more effectively. However, it is noted that the benefit of IS takes time to occur. For instance, it takes ISIS 3 to 4 years to accumulate enough information for further exploitation. Therefore, the project investors have to be aware of the time lag and evolvement of the system performance.

2. E-government solutions cannot be achieved in one day, as incremental innovation is demanded in the post implementation stage to achieve a successful outcome. Many E-government projects have failed in developing countries because of the design-actuality gaps (Heeks, 2002). In our case, it shows that the project contexts, such as organization's project objective, financial capability, task volume, user diversity, technology maturity, external pressure and so on, are changing all the time. But from the interview results, we can see that the developers are very proactive in problem seeking and solution finding. Their incremental innovation has realized an effective IS supported control mechanism. As a developing country, China is making rapid economic development that has made these changes inevitable. Both the principal and the developer have to be aware of the design-actuality gaps, especially when incorporating continuous system improvement at the post implementation stage into project planning and financial arrangement.

3. Although IS do not have a simple deterministic impact on development (Walsham et al., 2007), the introduction of E-government can have a great impact on the country's institutional environment, thus leading to overall mechanism improvement. In this case, for example, the introduction of ISIS has made the establishment of Credit Reporting System (CRS) in CRC attainable. It has greatly influenced the institutional foundation of the scientific credit market, thus leading to overall mechanism improvement. In some developing countries the institutional arrangement taken for granted in developed countries might not exist, and

policy makers are encouraged to include scaling E-governance solutions into their vision as a disruptive innovation to tackle institutional problems.

4. The adoption and diffusion of IS to achieve information transparency has had a significant impact on the business environment in China. On the one hand, the traditional bureaucratic practice of "stratified distribution of knowledge" in China tends to cause information to be rarely disclosed to people outside of official hierarchies, creating disastrous information bottlenecks (Ding, 2002). In other words, individuals or business entities in China do not usually have much access to business databases, resulting in information asymmetric problems. On the other hand, it is argued that China's problems with accurate information are not unique to that nation (Chambers, 1994). In many developing countries, it is in the self-interest of lower-level officials to input data into governing control systems that are in line with uppers' expectations and to withhold information that might prove to be damaging (Chambers, 1994). In the Chinese context, though, these problems are accentuated by a culture in which this kind of deception has developed into somewhat of an art form (Kluver, 2005).

As ISIS, which shares free access to the public has successfully accumulated higher quality information for decision making, information asymmetry is reduced and the principal's monitoring cost is also lowered. In addition, as information about business transaction memory gets richer, new technology applications which help behaviour pattern analysis to further facilitate principal monitoring becomes more feasible. Further, carefully designed task programmability introduced by IS will help to collect work performance related data which could prevent agents from withholding information or conducting deceptions. Thus,

these kinds of Chinese E-government initiatives can be utilized as vehicles to support better the business environment.

5. IS supported outcome measurability and task programmability are useful tactics to enhance mechanism improvement. In knowledge-intensive industries, such as R&D management and health care management areas, the principal's monitoring cost is high because the knowledge worker agent's specialized skills and knowledge are hard to observe. In addition, the goal alignment can be low because the agents are empowered to make decisions that have far reaching consequences for the principal (Wickramasinghe & Silvers, 2003). In these particular areas, an increased level of outcome measurability and task programmability is expected to result in better P-A relationship control. Practitioners, such as Knowledge Management System (KMS), developers are encouraged to pay attention to these two aspects of system features to attain mechanism improvement.

To further investigate the proposed implications, future research could progress along the following trajectories.

1. IS practice has provided a great opportunity to advance agency theory. Study in the IT arena could enrich the meaning of existing variables, and extend the explanatory power of the theory. This study has only considered certain manifestations of agency opportunisms and control mechanisms; future research could examine other types, as well as their interactions.

2. IS research carried out in developing countries often emphasizes cultural perspectives but ignores important economic elements (Walsham & Sahay, 2006). This study has applied a well-established theory from

economic perspectives, and the qualitative results indicate that an important economic advancement of an IS-supported credit-reporting system has emerged to reposition the CRC into a more "market-oriented" place. We encourage researchers to pay attention to the economic impact of IS when conducting IS research in the context of developing countries.

3. Incremental innovations in IS projects dealing with institutional arrangement can shed light on the question of "how." From a design science perspective, research which illustrates incremental innovations in IS projects will disclose more relevant implications for practitioners. For example, since there are many other forms of IS supported peer review systems, comparisons can be made. For instance, 'Faculty of 1000' was created to help scientists cope with information overload and to return the focus of science to science rather than to journals (Daniel, 2005). The implications of other studies on IS supported peer review systems could be worthwhile for ISIS improvement.

Limitations of our research are evident in that China is only one country and has a unique history. There is no clear way to conclude that ISIS was the only contributor to the academic rise beyond helping manage the successful increase in funding support and proposal submissions. Although we are cautious about choosing indicators in the analysis, several pieces are still not included which might have alternative explanations; for example, the citation analysis to reflect publication quality, the time lags between R&D input and output, the increasing quantity of grants, or the enlarging body of research community. However, it is beyond the focus of this paper to list all the related factors to CRC outcomes.

Further, we have only interviewed a few key informants in our case study. More feedback from general researchers, principals and agents are expected in future studies. Due to the limited time

and financial constraints, we had to make trade-off decisions. But we have not stopped investigating the project update. Practical extensions of ISIS have embarked since 2009, such as Scholarmate. com which promote scientists' social networking activities in China and global research cooperation and exchange activities. More outcomes remain to be observed which can be used later to test our propositions.

CONCLUSION

In this paper we have sought to illustrate how an E-government project, ISIS, initiated by the National Science Foundation of China with embedded managerial control mechanisms consistent with China's goals of increased global recognition, can help rectify traditional agency opportunism problems and promote increased contributions for China's Research Community.

An IS supported scientific Credit Reporting System tackles the historical discredit issue which impedes the overall institutional environment of China's Research Community toward a healthy development. Aspects of reduced agency opportunism and demonstrated control mechanism improvement have been achieved along with the IS innovations in outcome measurability and task programmability that has led to an overall increase in quality proposals which, in turn, have resulted in increased global recognition, as witnessed through publications in top journals. Extensive peer review has been supported and quality control has been achieved. Global respect and credibility is evident, as is reduction in academic misconduct. In short, the incentive alignment and monitoring mechanisms embedded in ISIS have been successful.

Extensions are envisioned to further the E-governance endeavour, such as automating decision making effectiveness and efficiency in managing increasing global research cooperation demands and exchange activities.

REFERENCES

Anthony, R. N., & Govindarajan, V. (1998). *Management control systems* (9th ed.). Boston: Irwin/McGraw-Hill.

Ba, S., & Pavlou, P. A. (2002). Evidence of the effect of trust building technology in electronic markets: Price premiums and buyer behaviour. *Management Information Systems Quarterly, 26*(3), 243–268. doi:10.2307/4132332

Ba, S., Stallaert, J., & Whinston, A. B. (2001). Research commentary: Introducing a third dimension in information systems design: The case for incentive alignment. *Information Systems Research, 12*(3), 225–239. doi:10.1287/isre.12.3.225.9712

Bai, S. Z., & Yang, Y. (2008). *Study on the information sharing incentive and supervisory mechanism in supply chain.* Paper presented at the 2008 International Conference on Wireless Communications, Networking and Mobile Computing (WiCOM 2008).

Barney, J. B., & Quchi, W. G. (1986). *Organizational Economics* (3rd ed.). San Francisco, CA: Jossey-Bass Publishers.

Basu, S. (2004). E-government and developing countries: An overview. *International Review of Law Computers, 18*(1), 109–132. doi:10.1080/13600860410001674779

Basu, V., & Lederer, A. L. (2004). *An agency theory model of ERP implementation.* Paper presented at the SIGMIS04, Tucson, AZ.

Biggart, N. W., & Hamilton, G. G. (1992). *On the limits of a firm-based theory to explain business network: The Western bias of neoclassical economics.* Boston: Harvard Business School Press.

Bland, C. J., & Ruffin, M. T. Iv. (1992). Characteristics of a productive research environment: Literature review. *Academic Medicine, 67*(6), 385–397. doi:10.1097/00001888-199206000-00010

Brennan, J., & Shah, T. (2000). *Managing quality in higher education.* Buckingham: OECD, SRHE and Open University Press.

Bruns, W. J. Jr, & McFarlan, F. W. (1987). Information technology puts power in control systems. *Harvard Business Review, 65*(5), 89–94.

Cao, C., Suttmeier, R. P., & Simon, D. F. (2006). China's 15-year science and technology plan. *Physics Today, 59*(12), 38. doi:10.1063/1.2435680

Chambers, R. (1994). All power deceives. *IDS Bulletin, 25*(2), 14–26. doi:10.1111/j.1759-5436.1994.mp25002002.x

Chen, Y. N., Chen, H. M., Huang, W., & Ching, R. K. H. (2006). E-government strategies in developed and developing countries: An implementation framework and case study. *Journal of Global Information Management, 14*(1), 23–46.

Choudhury, V., & Sabherwal, R. (2003). Portfolios of control in outsourced software development projects. *Information Systems Research, 14*(3), 291–314. doi:10.1287/isre.14.3.291.16563

Ciborra, C. (2005). Interpreting e-government and development efficiency, transparency or governance at a distance? *Information Technology & People, 18*(3), 260–279. doi:10.1108/09593840510615879

Cyranoski, D. (2006). Named and shamed. *Nature, 441*(7092), 392–393. doi:10.1038/441392a

Dada, D. (2006). The failure of e-government in developing countries: A literature review. *The Electronic Journal on Information Systems in Developing Countries, 26*(7), 1–10.

Daniel, H.-D. (2005). Publications as a measure of scientific advancement and of scientists' productivity. *Learned Publishing, 18*, 143–148. doi:10.1087/0953151053584939

Ding, X. L. (2002). The challenges of managing a huge society under rapid transformation. In Wong, J., & Zheng, Y. (Eds.), *China's post-Jiang Leadership Succession: Problems and Perspectives* (pp. 189–213). Singapore: Singapore University Press. doi:10.1142/9789812706508_0008

Dologite, D. G., Mockler, R. J., Bai, Q., & Viszhanyo, P. F. (2004). IS change agents in practice in a US-Chinese joint venture. *Journal of Global Information Management*, *12*(4), 1–22.

Eisenhardt, K. M. (1989). Agency theory: An assessment and review. *Academy of Management Review*, *14*(1), 57–74. doi:10.2307/258191

Farina, C., & Gibbons, M. (1981). The concentration of research funds: The case of the Science Research Council. *R & D Management*, *11*(2), 63. doi:10.1111/j.1467-9310.1981.tb00451.x

Feldmann, M., & Muller, S. (2003). An incentive scheme for true information providing in Supply Chains. *Omega*, *31*(2), 63–73. doi:10.1016/S0305-0483(02)00096-8

Finel, B. I., & Lord, K. M. (1999). The surprising logic of transparency. *International Studies Quarterly*, *43*(2), 315–339. doi:10.1111/0020-8833.00122

Fuloria, P. C., & Zenios, S. A. (2001). Outcomes-adjusted reimbursement in a health-care delivery system. *Management Science*, *47*(6), 735–751. doi:10.1287/mnsc.47.6.735.9816

Gefen, D., Wyss, S., & Lichtenstein, Y. (2008). Business familiarity as risk mitigation in software development outsourcing contracts. *MIS Quarterly: Management Information Systems*, *32*(3), 531–551.

Gil-Garcia, J. R., Chengalur-Smith, I., & Duchessi, P. (2007). Collaborative e-government: impediments and benefits of information-sharing projects in the public sector. *European Journal of Information Systems*, *16*, 121–133. doi:10.1057/palgrave.ejis.3000673

Gu, J., Fan, R., & Liang, L. (2008). Research of economic analysis and countermeasure on the short-lighted behavior of scientific and technical personnel. [in Chinese]. *East China Economic Management*, *22*(1), 145–149.

He, X. X., Zhang, J., & Zhao, J. (2007). Based on the frontier of management to promote management excellency - the role of registered unit in the management of National Natural Science Fund. [in Chinese]. *Science Foundation in China*, *21*(5), 309–311.

Heeks, R. (2002). Information systems and developing countries: Failure, success, and local improvisations. *The Information Society*, *18*, 101–112. doi:10.1080/01972240290075039

Heinze, T., Shapira, P., Rogers, J. D., & Senker, J. M. (2009). Organizational and institutional influences on creativity in scientific research. *Research Policy*, *38*(4), 610–623. doi:10.1016/j.respol.2009.01.014

Huang, B. S., Meng, X., Zheng, Y. H., & Liang, W. P. (2003). Promoting the international cooperation in chemical sciences by rational use of science fund. *Science Foundation in China*, *17*(1), 44–46.

Jin, B. H. (2004). Considerations of high quantity of publications compared to low quantity of citations: Value orientation and quantitative index of R & D evaluation. *Science of Science and Management of S. & T.*, *3*, 9–11.

Jin, B. H. (2007). The advancing aircraft carrier of China's research output: SCI analysis in 2006. *Science Focus*, *2*(1), 20–44.

Jongbloed, B., & Koelman, J. (2001). Keeping up performances: an international survey of performance-based funding in higher education. *Journal of Higher Education Policy and Management, 23*(2), 127–145. doi:10.1080/13600800120088625

Jonkers, K. (2008). *An analytical framework to compare research systems applied to a diachronic analysis of the transformation of the Chinese research system.* Paper presented at the VI Globelics Conference.

Keefer, D. L. (1978). Allocation planning for R&D with uncertainty and multiple objectives. *IEEE Transactions on Engineering Management, 25*(1), 8.

Kirby, S. L., & Davis, M. A. (1998). A study of escalating commitment in principal-agent relationships: effects of monitoring and personal responsibility. *The Journal of Applied Psychology, 83*(2), 206–217. doi:10.1037/0021-9010.83.2.206

Kirsch, L. J. (1996). The management of complex tasks in organizations: Controlling the systems development process. *Organization Science, 7*(1), 1–21. doi:10.1287/orsc.7.1.1

Kirsch, L. J. (1997). Portfolios of control modes and IS project management. *Information Systems Research, 8*(3), 215–239. doi:10.1287/isre.8.3.215

Kivistö, J. (2007). *An agency theory as a framework for the government-university relationship.* Unpublished doctoral dissertation, University of Tampere, Tampere, Finland.

Kivistö, J. (2008). An assessment of agency theory as a framework for the government-university relationship. *Journal of Higher Education Policy and Management, 30*(4), 339–350. doi:10.1080/13600800802383018

Kluver, R. (2005). The architecture of control: a Chinese strategy for e-governance. *Journal of Public Policy, 25*(1), 75–97. doi:10.1017/S0143814X05000218

Krishna, S., & Walsham, G. (2005). Implementing public information systems in developing countries: Learning from a success story. *Information Technology for Development, 11*(2), 123–140. doi:10.1002/itdj.20007

Li, L. (2008). The study and suggestions for the internet-based science information system of NSFC. *Science Foundation in China, 22*(1), 52–54.

Li, Z. Z. (2004). Chinese science in transition: misconduct in science and analysis of the causes. *Science Research Management, 25*(3), 137–144.

Linton, K. C. (2008). *China's R&D Policy for the 21st Century: Government Direction of Innovation.* Retrieved from http://ssrn.com/abstract=1126651

Luoto, J., McIntosh, C., & Wydick, B. (2007). Credit information systems in less developed countries: A test with microfinance in Guatemala. *Economic Development and Cultural Change, 55*(2), 313–334. doi:10.1086/508714

Ma, J., Fan, Z. P., & Huang, L. H. (1999). A subjective and objective integrated approach to determine attribute weights. *European Journal of Operational Research, 112*, 397–404. doi:10.1016/S0377-2217(98)00141-6

Mahaney, R. C. (2000). *Information systems development project success and failure: An agency theory interpretation.* Unpublished doctoral dissertation, University of Kentucky, KY.

Mahaney, R. C., & Lederer, A. L. (2003). Information systems project management: an agency theory interpretation. *Journal of Systems and Software, 68*(1), 1. doi:10.1016/S0164-1212(02)00132-2

Meng, H., Zhou, L., & He, J. K. (2007). The co-integration analysis on NSFC input and S&T paper's output. *Studies in Science of Science, 25*(6), 1147–1155.

Milgrom, P., & Roberts, J. (1992). *Economics, organization and management.* Upper Saddle River, NJ: Prentice Hall.

Ndou, V. D. (2004). E-government for developing countries: Opportunities and challenges. *The Electronic Journal on Information Systems in Developing Countries, 18*(1), 1–24.

Nicolaou, A. I., & McKnight, D. H. (2006). Perceived information quality in data exchanges: Effects on risk, trust, and intention to use. *Information Systems Research, 17*(4), 332–351. doi:10.1287/isre.1060.0103

Norman, K. L. (1986). Importance of factors in the review of grant proposals. *The Journal of Applied Psychology, 71*(1), 156–162. doi:10.1037/0021-9010.71.1.156

NSFC. (2009a). *Guide to Programmes 2009.* NSFC, Department of Management Science.

NSFC. (2009b). *NSFC 2008: Annual report on government information disclosure.* Retrieved from http://www.nsfc.gov.cn/nsfc/cen/xxgk/baogao.html.

NSFC. (2009c). *The Result of the 2009 NSFC First Round Peer-Review for All Proposals.* Retrieved from http://www.nsfc.gov.cn/Portal0/InfoModule_375/27220.htm

Pavlou, P. A., Liang, H., & Xue, Y. (2007). Understanding and mitigating uncertainty in online exchange relationships: A principal-agent perspective. *MIS Quarterly: Management Information Systems, 31*(1), 105–135.

Qi, Y., & Zhang, Q. (2008). *Research on information sharing risk in supply chain management.* Paper presented at the 2008 International Conference on Wireless Communications, Networking and Mobile Computing (WiCOM 2008).

Qiu, J. L., & Hachigian, N. (2005). *A New Long March: E-Government in China.* Paris: OECD.

Singh, J., & Sirdeshmukh, D. (2000). Agency and trust mechanisms in consumer satisfaction and loyalty judgments. *Journal of the Academy of Marketing Science, 28*(1), 150–167. doi:10.1177/0092070300281014

Tan, C.-W., Pan, S. L., & Lim, E. T. K. (2005). Managing stakeholder interests in e-government implementation: Lessons learned from a Singapore e-government project. *Journal of Global Information Management, 13*(1), 31–53.

Tian, Q. J., Ma, J., & Liu, O. (2002). A hybrid knowledge and model system for R&D project selection. *Expert Systems with Applications, 23*(2), 265–271. doi:10.1016/S0957-4174(02)00046-5

Tiwana, A., & Bush, A. A. (2007). A comparison of transaction cost, agency, and knowledge-based theory predictors of IT outsourcing decisions: A U.S.-Japan cross-cultural field study. *Journal of Management Information Systems, 24*(1), 259–300. doi:10.2753/MIS0742-1222240108

Tuttle, B., Harrell, A., & Harrison, P. (1997). Moral hazard, ethical considerations, and the decision to implement an information systems. *Journal of Management Information Systems, 13*(4), 27–44.

Walsham, G., Robey, D., & Sahay, S. (2007). Foreword: Special issue on information systems in developing countries. *Management Information Systems Quarterly, 31*(2), 317–326.

Walsham, G., & Sahay, S. (2006). Research on information systems in developing countries: Current landscape and future prospects. *Information Technology for Development, 12*(1), 7–24. doi:10.1002/itdj.20020

Wang, C. H. (2004). Discussing on the manifestations, damages and characteristics of the phenomena of rent-seeking in scientific research field. *Science Research Management, 25*(3), 131–136.

Wang, Q. (2006). Misconduct: China needs university ethics courses. *Nature, 442*(7099), 132–132. doi:10.1038/442132b

Wickramasinghe, N. (2000). IS/IT as a tool to achieve goal alignment in the health care industry. *International Journal of Healthcare Technology and Management, 2*(1/4), 163–180. doi:10.1504/IJHTM.2000.001089

Wickramasinghe, N., & Silvers, J. B. (2003). IS/IT the prescription to enable medical group practices attain their goals. *Health Care Management Science, 6*(2), 75–86. doi:10.1023/A:1023376801767

Wu, D. J., Ding, M., & Hitt, L. M. (2004). Learning in ERP contracting: A principal-agent analysis. In *Proceedings of the Hawaii International Conference on System Sciences*.

Wu, X. (2003a). Credit, science credit and its institutional structure. *Studies in Science of Science, 21*, 65–70.

Wu, X. (2003b). *The Institutional Analysis on Scientific-credit Issues of China*. Unpublished master's thesis, Zhejiang University, Hangzhou, China.

Xie, Y. Q., Huang, Y., Li, D. N., Jiang, J., & Li, D. (2008). Analysis of National Natural Science Foundation of China management in Hainan Medical College. *Science Foundation in China, 22*(1), 57–58.

Xin, H. (2006). Scientific misconduct: Scandals shake Chinese science. *Science, 312*(5779), 1464–1466. doi:10.1126/science.312.5779.1464

Yin, R. K. (1994). *Case study research: Design and methods* (2nd ed.). Thousand Oaks, CA: Sage Publications.

Zhang, J. (2003). Overview and economic analysis on academic corrupt research. *Social Science of Beijing, 3*, 105–110.

Zhang, J. (2005). Good Governance Through E-Governance? Assessing China's E-Government Strategy. *Journal of E-Government, 2*(4), 39–71. doi:10.1300/J399v02n04_04

Zhang, M. (2005). Review and forecast of the system construction of scientific credit. *Scientific and Technology Progress and Policy, 11*, 20–22.

Zhang, M. (2006). Bad credit behaviors in scientific research projects and their countermeasures. *Scientific Management Research, 3*(24), 66–69.

Zhao, R. (2003). Transition in R&D management control system: Case study of a biotechnology research institute in China. *The Journal of High Technology Management Research, 14*(2), 213–229. doi:10.1016/S1047-8310(03)00022-1

Zhu, Z., & Gong, X. (2008). Basic research: Its impact on China's future. *Technology in Society, 30*(3-4), 293–298. doi:10.1016/j.techsoc.2008.05.001

Zweig, D., Chen, C., & Rosen, S. (2004). Globalization and transnational human capital: Overseas and returnee scholars to China. *The China Quarterly, 179*, 735–757. doi:10.1017/S0305741004000566

APPENDIX A

Brief introduction of NSFC's implementation of ISIS

There are seven scientific departments: four bureaus, one general office and three associated units in NSFC. The scientific departments are responsible for the selection and management of the projects, while the bureaus, general office and associated units are mainly responsible for policy making, administration and other related affairs.

Some senior staff at NSFC found it difficult to enter decision opinions in Chinese on computer terminals, thus assistant devices such as Chinese write pad should be provided. At the beginning, some staff thought that the use of the system would increase their workload. Thus, reward incentives should be designed and implemented so as to ensure the successful adoption of the system. The wide spread occurrence of SARS in China during April and June 2003 has encouraged the adoption and diffusion of ISIS. At the individual level, ISIS users are classified into the system administrator (the super user of the system), top manager, department manager, division manager, organizational user, principal investigator, external reviewer, and public user. Each user is assigned to a user group with access rights to certain system functions.

The ISIS decision support functions at the individual level provide supports for division managers and external reviewers who have to make individual decisions at the operational level.

At the group level, ISIS implements a group management component. Internal users like top managers, department managers, and division managers have permission to create their own decision-making groups. They can use the function to create a new group, maintain the group membership, coordinate the group work, and check the work progression. One manager may create different decision-making groups for different decision-making tasks.

At the organizational level, ISIS provides support to the three major phases of the R&D project selection process. (1) For the phase of task decomposition and group creation, ISIS supports the task decomposition (see Figure 1) in NSFC. Specific subsystems have been designed for decision groups on these tasks. Thus, the controller of a group can select and manage his/her group members. (2) For the phase of decision-making support and coordination, decision models and knowledge rules for proposals submission are implemented. Thus, the group controllers can check and coordinate their groups' work progression. Multiple views of the work progression are usually provided, including detailed lists, statistic tables and figures. (3) For information aggregation of final organizational decision results, the ISIS provides group controllers with ways to aggregate the information of members' decision results and the decision models are also developed to facilitate this process.

Thirteen, 37 and 52 (all) divisions in NSFC participated in the exercises of project selection through ISIS in 2001, 2002 and 2003, respectively. Results of the applications are very positive, especially as the success rate of electronic peer review has reached 96% in 2003. The main reasons include: (1) the system provides support at the individual, group, and organizational levels over the Internet. It therefore facilitates the coordination among groups of decision makers, and shortens the cycle time of project selection. (2) It provides decision makers with useful decision aids such as decision models and business logic rules with Web interfaces. (3) It helps to simplify the workflow and improve the work efficiency of NSFC.

APPENDIX B

Interview protocol

1. **Describe NSFC and ISIS project include organization history, organization structure, governance mechanism, project background, organizational goals and E-government objectives:** This serves to facilitate a complete understanding of the principal and the background of the E-government project.
2. **Identify how NSFC handle the P-A issue:** Open-ended questions are addressed to understand the P-A contracting process. Then, managerial procedures enabling incentive alignment and monitoring to occur are identified.
3. **Describe ISIS implementation process and design principles to meet the organization's managerial requirement:** The anticipated outcome from these inquiries is to obtain data to show how ISIS functions to incorporate institutional arrangement.
4. **Elucidate advantages and problems with the use of ISIS to facilitate outcome measurability and task programmability:** The anticipated outcome from these inquiries is to obtain data to show whether ISIS will influence contextual factors and how it functions to cope with them.

Sample coding of interview excerpt

Interview Comment	Code	Implication
Managerial: Misconduct disclosure	CM1	Control Mechanism 1 is articulated by the principal
IS: Innovative technology method	ISIV	IS Innovation in general is applied by developers
Managerial: Incentive indicators to evaluate the agent	CM1	Control Mechanism 1 is articulated by the principal
IS: Standardization of tasks	TP	Task programmability is realized by developers
Managerial: Data verification	CM2	Control Mechanism 2 is articulated by the principal
IS: Data mining method for misconduct *Managerial: Lower monitoring cost*	ISIV, CM2	IS Innovation in general help to attain Control Mechanism 2
IS: Proposal grouping method *Managerial: Better peer-review result*	ISIV, CM1	IS Innovation in general help to attain Control Mechanism 1

This work was previously published in International Journal of Global Information Management, Volume 18, Issue 4, edited by Felix B. Tan, pp. 53-81, copyright 2010 by IGI Publishing (an imprint of IGI Global).

Section 3
Government and Development Issues

Chapter 13

ICT Diffusion and Foreign Direct Investment:
A Comparative Study between Asia–Pacific and the Middle Eastern Economies

Farid Shirazi
Ryerson University, Canada

Dolores Añón Higón
University of Valencia, Spain

Roya Gholami
Aston Business School, UK

ABSTRACT

This chapter investigates the impact of inward and outward FDI on ICT diffusion in the Asia-Pacific and Middle East regions for the period 1996-2008. The results indicate that while inward FDI has generally had a positive and significant impact on ICT diffusion in Asia-Pacific economies, its impact on the Middle Eastern countries has been detrimental. In contrast, the results of this study also show that outward FDI has had, in general, the inverse effect, it has been in general positive and significant for the Middle East countries but insignificant for Asia-Pacific economies.

INTRODUCTION

In recent literature, Foreign Direct Investment (FDI) has been cited as a measure of globalization (Loungani & Razin 2001) and trade liberalization (Santos-Paulino & Thrilwall 2004) that can provide development opportunities to host economies (Soper et al., 2006). As developing countries become more open to international competition, organizations are increasingly forced to compete with multinational corporations (MNCs) in both domestic and foreign markets. In turn, FDI becomes an important component of the economic strategies of developing countries (UNCTAD, 2006a).

DOI: 10.4018/978-1-61350-480-2.ch013

In the last decade, several economies in the Asia-Pacific region have witnessed profound progress in Information and Communication Technology (ICT) and FDI inflows. Middle Eastern countries, on the other hand, have not been able to attract FDI inflows to the same extent despite their success in ICTs and an increase in the capacity of FDI inflows for the period 1996-2008. FDI flows in the Middle East and North Africa (MENA) accounted for less than 10% of total FDI investment in 2007 (Delasnerie, 2007).

Delasnerie (2007) argues that 60% of investors consider MENA a high-risk region due to political situation in its countries, poor infrastructure, corruption, fear of terrorism and insufficient intellectual resources in R&D and other knowledge-based services. Similarly, Krogtrup and Matar (2005) claim that Arab economies are less likely to possess the absorptive capacity necessary to benefit from FDI, due to the poor quality of the education system, the financial sector, and technological and institutional development.

World leaders noted "the importance of removing barriers to bridging the digital divide, particularly those that hinder the full achievement of the economic, social and cultural development of economies and the welfare of their people, in particular, in developing economies" (ITU Tunis Agenda, 2005, sec.10).

There are many attempts made to "quantify" and "qualify" the global digital divide with different conceptual frameworks, sets of variables and methodologies (see Wong 2002, Tipton 2002, Grubesic & Murray, 2002, Brown & Licker 2003, Oyelaran-Oyeyinka et al. 2003, Barroso & Martinez 2005, Dutton et al. 2004, Lu 2005, Yap et al. 2006, Cava- Ferreruela et al. 2006, La Rose et al., 2007, Zhao et al. 2007, Dwivedi & Lal 2007, Hitt & Tambe 2007, Picot & Wernick 2007, Robertson et al. 2007, Trkman et al. 2008, Howick & Whalley 2008).

Though there has been an intense interest among international institutions (e.g. UNCTAD, UNDP, UNESCO) in terms of the access to and benefits and impacts of ICT at regional and country levels, little attention has focused on the potential impact of factors like FDI inflows, trade openness, and government interventions in economic activities on the ICT diffusion in developing economies. Therefore, the aim of this study is to fill this gap by analyzing the extent to which FDI inflows, trade openness, government interventions in economic activities, and other socio-economic factors affect the disparities among developing economies, specifically in the Asia-Pacific and Middle East regions for the period 1996-2008.

For this study, our main research question is: Can FDI flows (inward and outward) provide insights into the differences in ICT diffusion between these two regions? To answer this question, we performed an econometric analysis for a balanced panel[1] of nine Islamic states[2] in the Middle East (Bahrain, Iran, Jordan, Kuwait, Oman, Qatar, Saudi Arabia, Syria, and the United Arab Emirates) and eight Asia-Pacific[3] economies (China, Hong Kong, Indonesia, Malaysia, the Philippines, Singapore, Thailand, and Vietnam). This paper is organized as follows: next section provides an overview of ICT diffusion and FDI flows in both regions while the following section reviews the literature on the determinants of ICT diffusion. Then the research approach is introduced, followed by regression results and findings. Finally a discussion and concluding comments based on these findings are presented.

BACKGROUND

ICT Diffusion in Asia-Pacific vs. the Middle East

Table 1 shows the level of digital access in Asia-Pacific and the Middle East. There are four levels of digital access as defined by the UN's ICT Task Force (2005), ranging from High Access to Low Access. While Hong Kong (China) and Singapore are considered to have high digital access, no

Middle East countries hold rank in this category. Four of the developed economies in the Middle East are categorized as having 'upper digital access' (Bahrain, Kuwait, Qatar and UAE), while in the Asia-Pacific region, only Malaysia is located in this category. The remaining (Asia-Pacific and Middle East) economies are categorized in the medium digital access category, while Syria is the only nation registered in the 'low digital access' category.

During the last decade, both Asia-Pacific and Middle East economies experienced an explosive ICT diffusion due to increased investment in this area. Among the Asia-Pacific economies, China was the second-fastest growing ICT economy after Vietnam due to its massive ICT investment within the period (see Table 1A in the Appendix). From 1996 to 2005, China's annual telecom investment grew from $91 billion USD to $210 billion USD, the result of which was telecom revenue of $584 billion USD in 2005, compared to $140 billion USD in 1996 (ITU, 2007a). In addition, the outward FDI from China reached historical high of $68 billion in year 2010 according to UNCTAD (2010) report. The ICT industry became not only the largest industry in China but also the fastest growing industry that attracted the most investors from across the globe.

As a result of this investment, its export of telecommunication equipment increased from $1.9 billion USD in 1996 to $24 billion USD in 2004 (ITU, 2007a). There were three main drivers behind the success of the Chinese IT industry: (a) the liberalization of investment and trade freedom, which encouraged FDI flow into the ICT sector in China; (b) the increased popularity of ICT products and services including PCs, mobile phones, and the Internet in both domestic and corporate environments; (c) the size of China's market and population provided an attractive environment for global ICT providers to invest heavily in ICT production in China (Tan 2004; Yin 2005).

Over the last decade, Hong Kong and Singapore have been able to fully liberalize their ICT markets and maximize an open ICT competition (Hiong 2006; ITU 2007a). In Hong Kong, there are no limits on foreign ownership of the ICT sector and FDI inflows are encouraged by the internal government. Hong Kong's openness towards its ICT market coupled with its clear market conditions makes the country an attractive place for ICT operators and investors. Singapore had the largest growing market for mobile cell phones during the last decade. In 2003, ICT export in Singapore accounted for 54.1% of ICT industry revenue, bypassing its domestic ICT revenue.

Table 1. Digital access level

	Digital Access Level			
	High Access	**Upper Access**	**Medium Access**	**Low Access**
Asia-Pacific	Hong Kong	Malaysia	China	
	Singapore		Indonesia	
			Philippines	
			Thailand	
			Vietnam	
Middle East		Bahrain	Iran	Syria
		Kuwait	Jordan	
		Qatar	Oman	
		UAE	Saudi Arabia	

Source: UN ICT Task Force, 2005.

Other Asia-Pacific economies including Malaysia, the Philippines, Thailand and Vietnam have shown tremendous expansion in their ICT core indicators (see Table 1A in Appendix). Vietnam is of particular interest for the study of ICT diffusion. Despite the fact that Vietnam has weak investment freedom, financial freedom, property rights, and freedom from corruption (Kane, 2007) the market reform provided a foundation to adopt an open-market economy. In 2003, approximately 570 companies engaged in ICT production and services in which 354 of them were foreign companies (Quynh, 2006). From 1996 to 2005, Vietnam was the fastest growing ICT nation in the Asia-Pacific region, particularly in the area of main telephone line and mobile cell phone usage. FDI inflows were a major drive for Vietnam's success in ICTs (Pipe 2003).

As for the Middle East region, in the late 1990s, governments invested heavily in ICTs, enabling renewal and expansion of ICT infrastructure through the development and implementation of new technologies. According to World Bank statistics (2006a), in 2000, the total telecom investment in the Middle East was 22.6% of total revenue. The result of this investment was an increase of more than 1.3 fold in telecom revenue in 2005, led by Qatar and Jordan (World Bank 2006b; ITU 2007b). ICTs play a crucial role in the oil production process, as modern petroleum technologies are information-intensive technologies. Highly advanced ICT technologies are used in this sector, which provide opportunities to improve economic performance at all stages of the oil supply chain from the production of crude oil to refinery and distribution. ICTs also provide possibilities for expanding proven crude oil reserves, improving the rate of crude oil extraction from existing wells, and providing means by which to discover new wells (UNCTAD, 2006b).

As shown in Table 2A (Appendix), the Middle East economies are divided into two main categories. The first category contains economies that rate high on the Human Development Index,

ICT infrastructure development index, and GDP per capita: Bahrain, Kuwait, Qatar, and the UAE. These economies were also able to attract more FDI inflows in their economies or had the best FDI outflow performance, whereas economies of Iran and Syria deviated from such practices. One of the main factors behind the successful ICT diffusion in these economies is the privatization of government-owned telecom sectors that occurred in the late 1990s in conjunction with social, economic, and political reform. In contrast, telecom sectors in less developed economies in this region remain under strict governmental control.

Kane (2007) states that Israel, Bahrain, and Jordan ranked highest in economic freedom among 17 Middle East and North African (MENA) economies; two other oil-producing economies, Oman and Kuwait, are among the top five. Syria, Iran, and Libya comprise a group of fairly disparate economies united by their lack of economic or political liberalism. From 1995 to 2005, the average telecommunications revenue as a percentage of the GDP increased by 1.3 fold with Qatar and Jordan seeing the greatest increase (World Bank, 2006a). While some of the most developed economies in the Middle East in terms of ICT reaped the benefits of a very high level of ICT standards in their economies, Syria had the lowest rate in this area.

Foreign Direct Investment Flows in Asia-Pacific vs. Middle East

FDI flows (inward and outward) from economies of this study picked up strongly, but at different speeds (UNCDAT, 2011) reflecting the strength of their economies. As shown in Table 2 during the period 1996-2008 the inward FDI stock (net increase in liabilities) in Asia-Pacific economies of this study has increased by 3.3 fold while the region's outward FDI stock (net increase in assets) for the same period has increased by 6.7 fold. These values constitute in part 43.1% and 51.6% of the total inward and outward FDI stocks accounted for developing countries in 2008. In other words the outward FDI stock for Asian-Pacific region bypassed the region's inward FDI stock. The Middle Eastern countries were also witness for increase of their FDI flows shares. During the above period, the total inward and outward FDI stocks in this region have increased by 8.5 fold and 9.6 fold respectively. These increase in FDI flows constitute for 6.3% and 4.4% of the total inward and outward FDI stocks accounted for developing economies in year 2008 according to UNCTAD (2011) report. This process has continued to a higher level of FDI flows in particular for economies such as Hong Kong and China. Outflow FDI for these economies for example rose by more than $10 billion each, reaching historical highs of $76 billion and $68 billion respectively (UNCTAD, 2010). Chinese companies continued to be on a buying spree, actively acquiring overseas assets in a wide range of countries and industries including telecommunications. Also, in regards to China's export success, Graham and Wada (2001) point out that most of the Chinese export expansion is the result of foreign-invested enterprises, and recommend that China seek the help of multinational corporations to acquire the necessary capital, and technological and managerial skills. In contrast, China's large market, strong infrastructure, low-cost labour and the government's open door economic policy are important factors for attracting foreign investors to China (Tan 2004).

The UNCTAD's 2006 World Investment Report divides economies into four main categories: a) Front-runners: economies with high FDI potential and performance (Bahrain, China, Hong Kong, Jordan, Qatar, Singapore, UAE); b) Above potential: economies with low FDI potential but strong FDI performance (Vietnam); c) Below potential: economies with high FDI potential but low FDI performance (Iran, Kuwait, Oman, Saudi Arabia); d) Under-performers: economies with both low FDI potential and performance (Indonesia, Syria). Only economies in the Middle East are categorized as having high FDI potential but low FDI performance. Similarly, Indonesia (in the Asia-Pacific region) and Syria (in the Middle East) are categorized as under-performing economies in terms of both FDI potential and performance.

The Middle Eastern economies that have liberalized their economy and created a positive environment for FDI inflows have not only attracted the most FDI into the region but have also been able to actively participate as FDI investors within and outside the region (particularly in North Africa). According to the LocoMonitor's recent report, the UAE was the largest contributor of FDI in the Islamic world in terms of the number of projects from 2002 to the beginning of 2007 (Chowdhry, 2007). The next section discusses determinants of ICT diffusion in developing world.

DETERMINANTS OF ICT DIFFUSION IN DEVELOPING ECONOMIES

Various studies have shown that ICT diffusion is correlated with the level of national wealth, resources for technology investments, human capital (Bontis, 2004), openness to trade (Addison and Heshmati 2003, Baliamoune-Lutz 2003, Gholami et al. 2006), the percentage of urban population (Forman et al. 2002, Chinn & Fairlie 2007), and degree of government intervention in business activities (ITU 1997; 1999, Hargittai 1999; Salmenkaita & Salo 2002).

It is important to note that, in the context of this research, the term ICT diffusion is adopted from the United Nations Conference on Trade and Development's Index of ICT Diffusion (UNCTAD, 2006c). This index measures the progress of ICT development in each country through two main components: 1) *Connectivity*, as measured by the number of Internet hosts, PCs, telephone mainlines and mobile telephone subscribers per capita. As such, it gives a measure of the level of infrastructure development; 2) *Access*, as mea-

Table 2. FDI flows in Asia-Pacific and Middle East

	Inward FDI Stock (million USD)				Outward FDI Stock (million USD)			
Asia Pacific (AP)	1996	2000	2006	2008	1996	2000	2006	2008
China	128,069	193,348	292,559	378,083	19,882	27,768	73,330	147,949
Hong Kong	237,992	455,469	742,416	816,184	99,710	388,380	677,153	762,038
Indonesia	26,871	25,060	54,534	72,227	6,496	6,940	1,042	2,802
Malaysia	36,028	52,747	53,710	73,601	8,891	15,878	36,127	66,926
Philippines	11,668	18,156	16,914	21,746	1,490	2,044	2,131	5,736
Singapore	89,494	110,570	241,570	326,790	39,675	56,755	160,668	207,130
Thailand	19,706	29,915	76,950	93,500	3,137	2,203	6,398	13,364
Vietnam	10,065	20,596	33,536	49,854	0	0	0	0
Total AP	561,889	907,863	1,514,195	1,833,994	181,277	501,968	958,854	1,207,954
Middle East (ME)								
Bahrain	4,452	5,906	11,191	14,741	1,349	1,752	6,051	9,340
Iran	2,308	2,597	17,682	20,967	137	572	1,171	1,853
Jordan	1,394	3,135	12,713	16,320	21	44	312	382
Kuwait	441	608	773	943	4,544	1,677	10,845	15,385
Oman	2,288	2,577	5,720	11,680	613	611	1,294	1,846
Qatar	781	1,912	10,655	17,769	14	74	1,076	12,265
Saudi Arabia	17,120	17,577	50,659	110,200	3,170	5,285	7,513	10,876
Syria	549	1,244	3,191	5,900	37	107	417	421
UAE	2,071	1,069	40,314	68,224	839	1,938	20,434	50,822
Total ME	31,403	36,625	152,898	266,744	10,724	12,061	49,113	103,189
World	3,869,965	7,445,637	13,833,355	15,294,653	4,093,273	7,962,170	15,617,992	15,987,901
Developing	984,174	1,731,604	3,330,619	4,251,668	384,109	857,354	1,723,325	2,342,750
Transitional	17,355	60,841	397,306	426,756	5,427	21,339	222,828	231,073
Developed	2,868,435	5,653,192	10,105,429	10,616,230	3,703,737	7,083,477	13,671,840	13,414,078

Source: UNCTAD (2011)

sured by the number of estimated Internet users, adult literacy rate, cost of a local call and GDP per capita (PPP US$).

Inward Foreign Direct Investment

ICT has influenced a global shift in the service industries, which are now relocating to developing economies. For example, MNCs providing business services are now large investors in Brazil, China, and India. Foreign investors, however,

are attracted to economies that have already established an ICT infrastructure. Consequently, poorer economies cannot attract ICT-intensive FDI because they do not have the ICT infrastructure, nor do they have sufficient private or public resources to develop such an infrastructure. Consequently, two groups of developing economies emerge: those that are attractive to ICT-intensive FDI and those that are not (Addison and Heshmati 2003). Baliamoune-Lutz (2003) argues that higher inward FDI contributes to ICT diffusion since inward

FDI usually allows recipient economies access to advanced technologies, managerial skills and a higher level of knowledge. Investing in new technologies requires the availability of capital, either from internal sources, or from external sources like FDI. For most developing economies, internal financial resources are not expected to be a significant source of capital. Rather, foreign financial resources are expected to play a significant role, as they can serve as a substitute for scarce domestic capital (Shih et al., 2008). In many developing economies, the level of ICT infrastructure is not yet high enough to attract FDI. Without external support, a developing country with a lower level of ICT infrastructure has less capability internally to finance high ICT capacity.

Therefore, these economies may need to bring in more FDI to promote their ICT industry (Gholami et al., 2006). Shih et al. (2008) found evidence that the level of foreign investment is significantly related to ICT diffusion. Although positive and significant for both groups, the effect of foreign investment was stronger in developing countries. This finding is consistent with the argument that developing countries have more to gain from inflows of knowledge associated with foreign investment. Examining the causal relationship between ICT and FDI, Gholami et al. (2006) found that a causal relationship exists between ICT and FDI in developed countries. In the context of developing countries, however, the ICT infrastructure was the main drive behind attracting foreign investments.

Outward Foreign Direct Investment

There has been a significant amount of FDI flow from developed countries where these countries invested in other developed countries (North-North investment) or developing countries (North-South investment). More recently, there has been the reverse process in terms of FDI — South-North investment flow, where some of the developing countries have gathered the know-how and the necessary capital to invest into developed countries (Bongalia, Goldstein & Mathews, 2006) as well as a significant amount of investments in ICTs in other developing and emerging economies (Alexander, 2010).

Some economies in the Middle East (e.g., Kuwait, Qatar and UAE) and Asia Pacific (e.g., China, Hong Kong, Singapore and Malaysia) are actively involved in investing in telecom sector in other countries. According to UNCTAD report (2008) the outward FDI from Asia-Pacific region is likely to grow in a higher speed and volume in future, as Asian firms are increasingly aspiring to become significant regional and global players in their respective industries in particular in the area of telecommunication. Some Middle Eastern telecommunication firms are the key players in ICT investment within and outside the region according to UNCTAD (2008), particularly telecom firms from countries such as Qatar, Bahrain and Kuwait played an active role in mobile communication investments in other Arab nations. According to UNCTAD (2008) one of the main barriers and challenges in outflow foreign direct investment is the limited domestic technological and engineering capabilities, as well as managerial and other expertise, preventing developing countries from undertaking infrastructure projects and related services. Thus technology transfer is among the most important potential contributions that TNC participation can make to host developing countries.

Dunning (1981) argues that there are five stages in a country's investment development path. These different stages reflect a country's net outward direct investment position in relation to the country's level of economic development.

Stage One

In the first stage, countries receive minimal FDI and these investments are generally resource based investments, taking advantage of the natural resources available in these countries. Hence, the focus is predominantly on the labour intensive manufacturing sectors (Dunning & Lundan, 2008). Typically, these are the least developed countries that have limited infrastructure and technology capabilities. In this context ICT does not play a significant role for countries in this stage (Alexander, 2010).

Stage Two

In the second stage, countries are still net recipients of FDI and focused on natural resources (Fonseca, 2007). The outward investment is still in its embryonic stages. There is huge increase in attention giving to institutions promoting secondary education, public health, transport and communication which potentially transforms a country to peruse an export-led development strategy (Dunning & Lundan, 2008). This would suggest that countries at this stage of development will consider ICT expansion as an important tool for internationalization (Alexander, 2010).

Stage Three

In third stage, countries begin to make investments abroad; however, these countries are still net recipients of FDI. There is an increased investment in education and research and development (Fonseca, 2007). Developing countries at this stage are approaching economic maturity; their income level and industrial structure are beginning to resemble those of a developed country and firms may begin to engage in efficiency-seeking or market seeking investments (Dunning & Lundan, 2008). It would imply that ICT plays a key role in these countries. As such, the strength of the relationship between ICT and FDI is expected to be of a greater magnitude for countries in the third stage of investment development path when compared with countries in the second stage of investment development path (Alexander, 2010).

Stage Four

In the fourth stage, the outward FDI is greater than inward FDI flows. The outward FDI is mainly in search for strategic assets in developed countries and also in search for new markets and cheaper labour (Vavilov, 2006). Countries at stage four are likely to be among the leading investors on R&D (Dunning & Lundan, 2008). ICT would play a significant role in brining about efficiencies in production systems and also facilitating innovation (Alexander, 2010).

Stage Five

In this stage, the outward and inward FDI cancel out each other as observed in industrialized nations (Dunning 1993; Duning and Narula 1998; Vavilov 2006). Firms become increasingly globalized (Sangder 2009) and therefore more dependent on ICT to facilitate business interactions (Alexander, 2010). Governments of the countries at stage five of investment development path play an important role in attracting FDI by continuously improving their location based advantages such as technological capabilities (Alexander, 2010).

GDP per Capita

The assumption that per capita income is a major determinant of ICT diffusion is fairly standard in both theoretical and empirical studies (Rogers, 1993, Hargittai 1999, Norris 2000, Kiiski and Pohjola 2002, Baliamoune-Lutz 2003, Dewan et al. 2005). Longitudinal data from developed countries indicate that adoption and diffusion of ICT are highly correlated with income. Countries with higher per capita income invest more in R&D, and hence are better equipped to discover and use advanced information technologies (Hardy 1980, Norton 1992, Shih et al., 2008). Beilock and Dimitrova (2003) examine the impact of gross national product (GNP), including the log and exponential forms, the level of civil liberties, infrastructure and regional variables on Internet use in a sample of 105 economies from a dataset published in 2000. They find that GNP is "by far" the most important determinant.

In this study, the average per capita income is proxied by the variable *GDP per capita*, as is common in country-level studies of this nature. Consistent with the evidence suggesting the existence of a digital divide, we would expect a positive relationship between GDP per capita and all the measures of ICT diffusion (see e.g., Dewan et al. 2005, Quibria et al. 2003).

Openness to Trade

Openness is measured as the ratio of the sum of exports and imports to GDP. Multinational corporations (MNCs) tend to bring with them business practices that rely more heavily on ICT, and thus are more likely to invest in ICT and bring knowledge of how to use it productively. Hence, MNCs require that their suppliers make similar ICT investments (Shih et al., 2008). The effective use of ICT requires a broad range of technical and managerial knowledge, much of which can be found beyond the borders of any country. Foreign trade facilitates the diffusion of such knowledge across borders as it "provides channels of communication that stimulate cross-border learning of production methods, product design, organizational methods, and market conditions" (Coe, Helpman & Hoffmaister, 1997).

FDI also has a positive impact on technical progress in the host country (Barrell & Pain, 1997). Baliamoune-Lutz (2003) shows that government trade policies influence lead to more rapid ICT diffusion in a country. Encouraging foreign investment by removing restrictions is likely to have a major impact on ICT diffusion. In Mexico and Brazil, the economic liberalization that led to investment by foreign MNCs stimulated ICT diffusion. These MNCs required suppliers to adopt IT, and created competitive pressure for domestic firms to invest in ICT (Dedrick et al,

2004). Recently, Shih et al. (2008) argue that the impacts would be more significant for developing economies, which are likely to be farther behind the global frontier in adopting ICT-enabled business practices and thus should benefit more from external sources of knowledge.

Human Capital

Shih et al. (2008) suggest that the presence of human capital with appropriate skills and access to sources of information on how to use technology is crucial for ICT diffusion in developing countries. Educated workers more readily adjust to the implementation of new technologies and an educated workforce reduces opposition to social changes associated with the adoption of new technologies (Robison et al., 2002).

Shin et al. (2008) also point out that the impact of human capital is expected to be greater in developing countries, which are still in the process of creating adequate levels of assets, than in developed countries. Once the levels of education, regarded as a complementary asset to ICT, reaches a certain level, the marginal impact of an extra level of education on ICT diffusion may be diminishing.

The empirical evidence on the impact of human capital on ICT diffusion is, however, inconclusive. Drawing on a sample of developing and OECD countries, Kiiski and Pohjola (2001) found that the tertiary education has a positive and statistically significant influence on ICT diffusion. Similarly, Shih et al. (2008) find that education has a significant effect on ICT diffusion in developing countries. In contrast, Hargittai (1999), Kiiski and Pohjola (2001) and more recently Shih et al. (2008) find that in the case of industrialized countries, education does not seem to influence ICT diffusion. Also, in a sample that included both developed and developing countries, Norris (2000) shows that education does not have a significant influence on ICT diffusion. Contrary to expectations, Baliamoune-Lutz (2003) did not find a positive association between ICT diffusion and education for a sample of developing countries.

Percentage of Urban Population

The size of the urban population, represented by the proportion of the population residing in urban areas, is expected to affect ICT diffusion. While the inclusion of this variable is motivated by prior studies (e.g., Forman et al. 2003, Chinn and Fairlie 2007), its effect on ICT diffusion is ambiguous. On the one hand, the larger the proportion of the urban population the higher the demand for information-intensive products and services (e.g. financial services) and therefore the stronger the demand for ICT. In addition, "urban density" theory predicts that ICT diffusion will be more extensive in urban areas, because such locations allow for the pooling of complementary assets, which lowers the cost of adoption of ICT. On the other hand, there are arguments in the literature that the more urban the population, the less pressing the need for ICT to compensate for distance-related communications (Forman et al. 2002, Chinn and Fairlie 2007[4]). According to UNCTAD's survey (2007) the growing negative externalities linked to "mega cities" might lead to a growing attractiveness of medium-sized towns for investment projects.

Dewan et al. (2005) find that while the size of the urban population has almost no impact on the number of Internet users, it does impact mainframe and PC penetration per capita. The negative coefficient on the size of the urban population is consistent with Chinn and Fairlie (2007) and supports the notion postulated by Forman et al. (2002) that IT is a substitute for other means of communication in urban environments.

Government's Intervention in Business Activities

The institutional legal environment in a country is also relevant to ICT diffusion because national policies can either enhance or hold back diffusion of a technology, depending on the approach to regulating mechanisms, privatization, and free competition outlined in such policies. These arguments suggest that economies with free competition in the telecom sector will have higher Internet connectivity than economies with monopolies in this sector of their respective countries (OECD, 1996, 1997a, 1997b, 1998, 1999; ITU 1997, 1999, Hargittai 1999).

Using a sample of 2001 data from 45 countries and focusing on variables of regulatory, regime characteristics and price regulation, Wallsten (2005) finds that the more formal and controlled a country's regulatory system, the fewer Internet users and hosts there are.

RESEARCH APPROACH

This research aims to determine the extent to which differences in inward and outward FDI impact ICT diffusion in Asia-Pacific and Middle Eastern economies for the period of 1996 to 2008. In other words, it aims to explain the dependent variable, *ICT*, using two main independent variables (*inward FDI* and *outward FDI*) and a set of control variables selected according to previous literature review. Kirkpatrick et al. (2006) point out that a large number of variables have been considered in the literature as possible determinants of ICT. We selected candidates for inclusion in our model based on their importance in past results as well as the availability of data.

This study does not investigate the causality linkage among variables, nor does it include other parameters that may impact the increase or decrease of FDI flows into host countries (e.g, paved roads, transportation infrastructure, socio-political climate, institutional democracy, economic freedom, and so on) as it is beyond the scope of this study.

Data Collection

Data for this analysis is drawn from various sources including the ITU, the World Bank, UNDP and UNCTAD. The ICT index, a composite index of network infrastructure and ICT machinery and equipment, was obtained from Orbicom-ITU (2005) and ITU (2007, 2009). This index includes main telephone lines per 100 inhabitants, number of waiting lines per mainlines, number of digital lines per mainlines, cell phones per 100 inhabitants, cable TV subscriptions per 100 households, internet hosts per 1000 inhabitants, number of secure servers per Internet hosts and International bandwidth (kb/s per inhabitant).

Data on FDI stocks and trade openness was obtained form UNCTAD Foreign Direct Investment database while data on GDP per capita and education was derived from the World Bank (2007) and UNDP (2006, 2007, 2009). To increase the emphasis on higher education and its impact on ICT diffusion, the formula used by ITU (2007) (education = (primary+ 2*secondary + 3* tertiary)/6) was applied. This research does not claim that the formal education as described above is the only source of acquiring knowledge in regards to ICTs. Other parameters, for instance, employees' training and/or trainings provided by various institutions (e.g., government sponsored centers) can help to increase citizens' knowledge in accessing and the use of ICT resources. However, these types of data are not available. Finally, we use the percentage of the urban population (World Bank, 2007; UNDP 2009) and the level of government intervention in business activities, provided by Heritage Foundation as control variables (Miles et al., 2006; Miller and Holems, 2009).

The Government Intervention Index is composed of government consumption as a percentage of the economy, government ownership of businesses and industries, the share of government revenues from state-owned enterprises, and government ownership of property and economic output produced by the government. A score of one indicates the least amount of intervention while a score of five indicates the highest intervention.

Research Model

To empirically assess the effect of inward and outward FDI on ICT diffusion, we use a panel of 17 Middle Eastern and Asia-Pacific countries over the period 1996-2008. The data has a pooled time-series cross section (TSCS) structure, allowing us to identify patterns across countries and over time. In the estimates presented below, we first capture the link between inward FDI and ICT in a model which can be expressed in the following log-linear form[5]:

$$\ln ICT_{it} = \alpha_0 + \alpha_1 \ln FDI\_in_{it} + \alpha_2 X_{it} + d_t + \varepsilon_{it} \tag{1}$$

where the subscripts refer to country i and year t. We assume that the index of ICT[6] depends on inward FDI (FDI_in), and a set of control variables which according to previous literature, may be related to ICT diffusion: trade openness (represented by $Open$), GDP per capita (GDP), education (Edu), an index of government intervention in economic activities (Gov) and the percentage of the urban population ($Urban$). Finally, d_t represents T-1 time dummy variables (for 1996-2008), which capture unobserved time-specific effects[7], and ε_{it} is the random error term in the equation, representing the net influence of all unmeasured factors. Additionally, we assume that ICT may also depend on outward FDI (FDI_out). Therefore, we also estimate the following model:

$$\ln ICT_{it} = \alpha_0 + \alpha_1 \ln FDI\_in_{it} + \alpha_2 \ln FDI\_out_{it} + \alpha_3 X_{it} + d_t + \varepsilon_{it} \tag{2}$$

In which outward FDI is represented by FDI_out. Before undertaking the econometric analysis of Model (1) and Model (2), it is important to check the degree of variation in the variables used for estimation and, in particular, the main variable of interest, ICT and its other two proxies, namely personal computer (PC) penetration rate and internet penetration rates, and inward and outward FDI. *Table 3* provides summary statistics for the whole sample as well as an indication of the country displaying, on average, the highest and lowest value of each variable of interest. More generally, *Table 3* shows that economies with the highest ICT diffusion (Singapore and Hong Kong) are also the best performers in terms of openness, outward FDI, educational attainment and urbanization. Summary statistics for the two regions are also presented. Major differences between the two regions are found for inward and outward FDI flows, with the Middle East ranking lowest, and also for urbanization, with Asia-Pacific ranking lowest. The significance of mean differences between the two regions was tested by means of the Student t-test. The difference in means between the two regions for all variables was statistically significant, except for the log of internet.

In addition, we analyze whether these disparities in terms of FDI across the two regions translate into different relationships between ICT

and inward FDI and additionally, between ICT and outward FDI. To do so, we first present in *Figure 1* the fitted value of the relationship between ICT and inward FDI for both regions (Middle East and Asia-Pacific), with their respective 95% confidence intervals. Different proxies of ICT (ICT network index, PC penetration rate, and internet penetration rate) are used to check the consistency of the results.

We observe that the relationship between ICT and inward FDI differs significantly across the two regions: while the relationship is positive for Asia-Pacific countries, it appears insignificant or even negative for Middle East countries. This relationship is consistent to any of the ICT proxies used. The difference between the two geographical regions, with the exception of the extreme values of the distribution, is statistically significant.

Additionally, we present in *Figure 2* the fitted value of the relationship between ICT and outward FDI for both regions (Middle East and Asia-Pacific), with their respective 95% confidence intervals. We observe also that the relationship between ICT and outward FDI differs, although to a lesser degree, across the two regions: while

Table 3. Summary statistics(1996-2008)

	Total				Asia_Pacific		Middle East	
	Mean	sd	Min*	Max*	Mean	Std. Dev.	Mean	Std. Dev.
ln(ICT)	4.089	(1.244)	Syria	Hong Kong	3.941	(1.379)	4.221**	(1.100)
ln(Internet)	1.512	(2.263)	Vietnam	Singapore	1.576	(2.360)	1.456	(2.183)
Ln(PC)	1.950	(1.275)	Vietnam	Singapore	1.778	(1.556)	2.103**	(0.941)
Ln(Edu)	4.635	(0.166)	Oman	Singapore	4.678	(0.104)	4.596***	(0.198)
ln(GDP_pc)	8.804	(1.182)	Vietnam	Qatar	8.623	(1.191)	8.965**	(1.155)
ln(Open)	4.702	(0.590)	Iran	Singapore	4.910	(0.703)	4.518***	(0.387)
ln(FDI_inward)	9.675	(1.787)	Kuwait	China	11.054	(1.219)	8.450***	(1.230)
ln(FDI_oward)[a]	8.102	(2.367)	Qatar	Hong Kong	9.807	(1.874)	6.776***	(1.794)
ln(Gov)	4.273	(0.411)	Kuwait	China	4.477	(0.060)	4.091***	(0.496)
ln(Urban)	4.162	(0.420)	Vietnam	Hong Kong	3.937	(0.485)	4.361***	(0.202)

Notes: Min* and Max* refer to the worst performing and best performing country for each variable, respectively. **, *** significantly different from the mean value of the Asia Pacific countries at the 5% and 1% respectively. [a] The outward FDI stock for Vietnam is zero for the period studied, therefore Vietnam is not included in the statistics presented for FDI_outward.

Figure 1. Inward FDI and ICT: Evidence for Middle East and Asia-Pacific countries

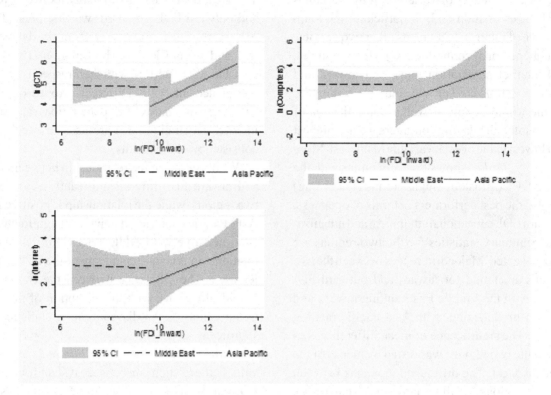

the relationship appears positive for both regions, the intercept and trend are apparently different between the two regions. This relationship is consistent to any of the ICT proxies used. The difference between the two geographical regions, with the exception of the extreme values of the distribution, is statistically significant.

In estimating the empirical Model (1) and Model (2), several issues need to be taken into account. First, multicollinearity is a problem linked to independent variables that are highly related to each other and may cause a wide shift in the parameter estimates due to small changes in the sample size, or the chosen variables. In detecting multicollinearity, a simple regression analysis of the models was conducted to examine the variance inflation factors (VIF). VIF measures the inflation in the variances of the parameter estimates as a result of the collinearities that exist among the independent variables[8]. The highest VIF score

was 5.17, in the model including all controls and the average VIF was 2.36, thus indicating low levels of multicollinearity (Myers, 1990). Further, the Shapiro-Francia test (Thode, 2002) was used to check for normality in each specification.

Second, despite its inferential advantages, TSCS data involves potentially more serious assumption violations of the ordinary least square (OLS) regression model than the non-panel design in terms of heteroskedasticity, autocorrelation, and contemporaneous correlation in the error term (Stimson, 1985). To address these potential violations, we follow Beck and Katz (1995) and estimate Model (1) using the panel-corrected standard error (PCSE) procedure which consists of presenting the OLS coefficients with standard errors corrected for the panel structure of the data[9]. The PCSE approach deals with the problems of heteroskedasticity and contemporaneous correlation. In addition, to address the problem of

Figure 2. Outward FDI and ICT: Evidence for Middle East and Asia-Pacific

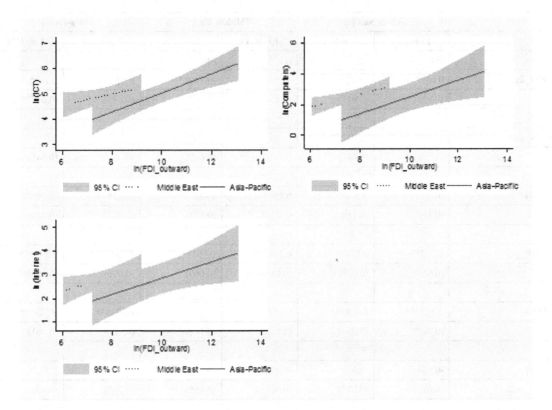

autocorrelation we use the corresponding PCSE option with the Prais and Winsten (1954) method for a first order autoregressive process, AR(1)[10].

EMPIRICAL RESULTS

Table 4 provides the results concerning the estimation of both models by the PCSE method. Column (1) and column (2) provide the results of estimating Model (1) and Model (2) respectively. The results presented in column (1) show that inward FDI has a significant positive effect on ICT diffusion, however, once we control for outward-FDI the positive effect is transformed in a negative one. What affects positively ICT diffusion appears to be outward FDI. On the other hand, trade openness, education, GDP per capita and urbanization all have a positive impact on ICT diffusion. These results are consistent with the

findings from previous studies in this area (Addison & Heshmati, 2003; Gholami *et al.*, 2006).

In columns (3) to (6), we show the results for the estimation of Model (1) and Model (2) for the Middle Eastern and Asia-Pacific economies, respectively. There are important differences with regards to the determinants of ICT diffusion in each of the regions. While openness[11], GDP per capita, and urbanization have a positive impact on ICT independently of the region, the role of FDI, education, and government intervention differs across regions. For Middle East countries, education and outward FDI have a positive influence on ICT whereas inward FDI shows a negative impact. In contrast, for Asia-Pacific economies both inward FDI and government intervention are positive and statistically significant determinants of ICT, but education has a negative role and outward FDI has an insignificant role.

Table 4. Determinants of ICT: 1996-2008

	Whole Sample		Middle East		Asia-Pacific	
	Ln(ICT)	Ln(ICT)	Ln(ICT)	Ln(ICT)	Ln(ICT)	Ln(ICT)
	(1)	(2)	(3)	(4)	(5)	(6)
Ln(FDI_in)	0.043*	-0.110**	-0.104**	-0.159***	0.141***	0.100*
	(0.024)	(0.047)	(0.052)	(0.057)	(0.039)	(0.057)
Ln(FDI_out)		0.148***		0.121***		0.053
		(0.041)		(0.040)		(0.045)
Ln(Edu)	0.441***	0.509***	0.456***	0.486***	-1.128*	-0.766**
	(0.170)	(0.159)	(0.128)	(0.131)	(0.577)	(0.388)
Ln(GDP_pc)	0.212**	0.124*	0.268***	0.152**	0.495***	0.334***
	(0.083)	(0.071)	(0.055)	(0.060)	(0.096)	(0.065)
Ln(Open)	0.365***	0.408***	0.359***	0.415***	0.333***	0.583***
	(0.090)	(0.092)	(0.115)	(0.112)	(0.100)	(0.101)
Ln(Gov)	-0.036	-0.027	-0.075	-0.056	1.622**	0.812*
	(0.052)	(0.047)	(0.065)	(0.065)	(0.725)	(0.436)
Ln(Urban)	1.112***	0.819***	1.145*	1.086*	0.645***	0.243**
	(0.269)	(0.198)	(0.625)	(0.589)	(0.139)	(0.117)
Time dummies	Yes	Yes	Yes	Yes	Yes	Yes
N	221	208	117	117	104	91 [a]
R²	0.637	0.686	0.603	0.638	0.904	0.947

Notes: The dependent variable is the ln(ICT). The estimation method used was PCSE with AR(1) process in the error term, following Beck and Katz (1995). Estimated (robust) standard errors are in parentheses. All estimations include non-reported year-specific dummy variables and a constant term. * indicates coefficient significant at 10% confidence interval, ** indicates 5% significance, and *** indicates 1% significance. [a] The outward FDI stock for Vitnam is zero for the period studied, therefore we lose those observations in the results provided.

In order to check the sensitivity of the results, we use two other proxies of ICT, namely PC penetration rate and Internet penetration rate[12]. The objective is to check whether the use of different proxies for the ICT composite alter the above findings. The results of estimating Model (1) and Model (2), using either of the two variables as dependent variables, are presented in Table 5 and Table 6, respectively. In accordance with the above results, we split the sample between the Middle Eastern and Asia-Pacific countries. The results show that GDP per capita plays a positive and significant role in explaining the progress in ICTs in these regions. With regards to openness, our results show that openness affects positively the diffusion of internet but it only has a significant

Table 5. The determinants of PC penetration: 1996-2008

	Middle East		Asia-Pacific	
	Ln(PC)	Ln(PC)	Ln(PC)	Ln(PC)
	(1)	(2)	(3)	(4)
Ln(FDI_in)	-0.014	-0.042	0.131**	0.042
	(0.035)	(0.033)	(0.067)	(0.086)
Ln(FDI_out)		0.067*		0.086
		(0.036)		(0.067)
Ln(Edu)	0.428*	0.460*	0.769	-0.950*
	(0.249)	(0.257)	(0.724)	(0.533)
Ln(GDP_pc)	0.380***	0.316***	0.237**	0.261***
	(0.056)	(0.059)	(0.120)	(0.094)
Ln(Open)	-0.006	0.006	0.656***	0.669***
	(0.188)	(0.187)	(0.147)	(0.153)
Ln(Gov)	0.022	0.035	-0.555	0.354
	(0.032)	(0.037)	(0.777)	(0.632)
Ln(Urban)	0.850**	0.846**	1.124***	1.060***
	(0.381)	(0.375)	(0.255)	(0.218)
Time dummies	Yes	Yes	Yes	Yes
N	117	117	104	91
R^2	0.625	0.646	0.796	0.840

Notes: The dependent variable is the log of PC penetration (lnPC). The estimation method used was PCSE with AR(1) process in the error term following Beck and Katz (1995). Estimated (robust) standard errors in parentheses. All estimations include non-reported, year-specific dummy variables and a constant term. * indicates coefficient significant at 10% confidence interval, ** indicates 5% significance, and *** indicates 1% significance.

positive effect for the diffusion of computers in Asia-Pacific countries. Educational attainment has a positive influence on technology diffusion in Middle East countries but negative in Asian Pacific countries. On the other hand, the rate of urbanization seems to positively affect the PC penetration for both regions but plays no significant role in Internet penetration.

Both, inward and outward FDI stocks, our two variables of interest, seem to play a different role depending on the region considered. Inward FDI exerts a negative influence on ICT and Internet penetration for Middle Eastern countries (Table 4 and Table 6), and non significant for PC penetration (Table 5). Perhaps there is a threshold level of foreign presence that most Middle Eastern countries have not yet reached, or, the level

Table 6. The determinants of Internet penetration: 1996-2008

	Middle East		Asia-Pacific	
	Ln(internet)	Ln(internet)	Ln(internet)	Ln(internet)
	(1)	(2)	(3)	(4)
Ln(FDI_in)	-0.267***	-0.277***	-0.078	0.189
	(0.083)	(0.082)	(0.097)	(0.140)
Ln(FDI_out)		0.023		-0.194*
		(0.075)		(0.105)
Ln(Edu)	3.260***	3.263***	-1.762	-2.465***
	(0.490)	(0.490)	(1.506)	(0.818)
Ln(GDP_pc)	0.343***	0.321***	1.379***	1.076***
	(0.085)	(0.094)	(0.274)	(0.201)
Ln(Open)	0.831**	0.841***	0.190	0.745***
	(0.330)	(0.326)	(0.327)	(0.235)
Ln(Gov)	-0.038	-0.033	-0.464	-2.806**
	(0.120)	(0.123)	(2.164)	(1.258)
Ln(Urban)	0.539	0.537	0.371	0.122
	(1.067)	(1.065)	(0.446)	(0.194)
Time dummies	Yes	Yes	Yes	Yes
N	117	117	104	91
R^2	0.855	0.855	0.841	0.939

Notes: Dependent variable is the log of Internet penetration (*ln internet*). The estimation method used was PCSE with AR(1) process in the error term following Beck and Katz (1995). Estimated (robust) standard errors are in parentheses. All estimations include non-reported year specific dummy variables and a constant term. * indicates coefficient significant at 10% confidence interval, ** indicates 5% significance, and *** indicates 1% significance.

of FDI investment in ICTs is overshadowed by the government investment in this area. In contrast inward FDI exerts either a positive (or insignificant) influence on ICT (or PC and Internet penetration) in Asian Pacific countries. In this respect, our findings support UNCTAD (2007) assumptions that "companies will increase their level of preference for South, East and South-East Asia [Asia-Pacific] as well as for the new EU-12 countries as investment locations; second, Western

Europe and North America will continue to be the leading destinations for investment flows; and, finally, other regions, notably Africa and West Asia [Middle East], will be given a lower degree of preference" (p.29).

On the other hand, while outward FDI has a positive and significant impact on ICT and PC penetration for Middle Eastern economies, the impact appears either insignificant for ICT and

PC penetration or negative for internet penetration in Asian Pacific countries.

DISCUSSION OF FINDINGS AND POLICY IMPLICATIONS

Currently, a number of economies in the Asia-Pacific region including China, Hong Kong, Singapore, Malaysia, the Philippines, and Thailand have emerged as global ICT suppliers and exporters. The main impetus for ICT diffusion in these economies during the period of 1996 to 2008 was rapid economic reform in the mid-1990s that affected the entire region. This reform imposed drastic changes in the form of new policies that encouraged FDI inflows into their economies, particularly in China and Vietnam. Tan (2004) points out that the current political environment favors the free flow of technology between China and other nations, and that the course of changes in China led to the loosening of previous restrictions on technology import and export behavior which has in turn resulted in abandonment of the ideological belief in nationalism and self-reliance among the Chinese as far as technology is concerned.

Tan and Leewongcharoen (2005) argue that Thailand's success in becoming one of the world leaders in IT hardware exports is related to the extent of FDI inflows into the country. FDI provided the necessary technology, capital, and skilled human resources to leverage Thailand's IT industry. These factors, combined with government policies for encouraging FDI inflows, were the main drivers behind this success. Despite the success of ICT diffusion in both regions during the period between 1996 and 2008, the digital divide between the Asia-Pacific and the Middle East regions increased during this period. In addition, the digital divide between the most ICT-developed economies and the least ICT-developed economies in each region increased by 5 fold in Asia-Pacific and 5.3 fold in the Middle East. In particular, the

digital divide increased in two main ICT areas: mobile cell phones and the Internet (see Table 1A in Appendix). Sciadas (2006) argues that empirical evidence widely supports and accepts that the diffusion and appropriate utilization of ICT not only presents enormous opportunities for economic and social development, but that its absence seriously threatens the existing and sizeable gaps between the 'haves' and 'have-nots'. In addition, the issue of digital divide occupies the area of overlap between economic, social and cultural matters, and it is rooted in the heart of the information society. Nwagwu (2006) points out the close statistical relationship between the diffusion of information technology, productivity and competitiveness for countries, regions, industries, and firms, as well as the importance of education in general, and technical education in particular for the design and productive use of the new technologies.

The empirical analysis of this study indicates openness to trade, education, and the existence of a competitive economic environment are an integral part of ensuring an atmosphere for attracting FDI and in turn ICT diffusion. The increase and expansion of the global marketplace, Internet economy, changing business ecosystems and pressure for accountability and transparency place a demand on the capability of governments to provide supporting infrastructures to increase FDI and the expansion of ICTs.

However, the delay in ICT diffusion in Middle Eastern economies indicates the low level of FDI inflows. The existing level of FDI inflows in the region is low in comparison to other regions and particularly in East Asia. The government intervention in business activities that dominates most of the resources as well as controls and directs use of scarce resources for government purposes deters investors from becoming more involved in business activities. For instance, in the area of ICT diffusion, economies that successfully partially privatized their telecom sector were able to increase their operations in both local and regional markets. On the other hand, those that imposed firm governmental control over their ICT diffusion unable to develop their ICT infrastructure at the same pace and hence increased the existing digital divide in the region increased.

Bahrain was able to establish itself as one of the most advanced economies in the Persian Gulf (Beach & Miles, 2006). It maintains a friendly business environment with an excellent banking and finance system, low regulation, and low barriers to foreign investment. Even though Iran should be a highly attractive market for a range of foreign firms given its extensive, underdeveloped natural resources, and young and increasingly educated population, it was unable to attract FDI inflows, due to the strict governmental control of economic resources, political instability and the current UN sanctions against Iran. These factors create an unfavorable environment for FDI inflows. The results of this study also indicate that the variables used ultimately have socio-economic implications. Their interdependence has a positive and negative impact on influencing the appeal of business attraction, as peace and regional stability are preconditions for attracting FDI investors.

In the context of the Middle East, the regional conflict that has been intensified during the last three decades (e.g., the Iran-Iraq war (1980-1988), the long lasting Arab-Israeli conflicts, the Persian Gulf War (1991), and the 2003 invention of Iraq by the coalition forces) has had a devastating social, political, and economic impact on many countries in the region and can be considered one of the major risk factors (UNCTAD 2007) for a low FDI inflow in the region.

The other reason for a low level of FDI inflows is related to an increased level of GDP, mainly caused by an increase in oil and natural gas prices in the global market, and as a consequence the increased production of oil and natural gas during this period. It is unexpected that internal financial resources would be a significant source of capital for most developing countries; however, for many Middle Eastern economies (in the context of this

paper), the income generated from oil and natural gas production has provided the necessary capital to be spent on infrastructure development projects including ICTs with limited or no need for capital from FDI investors.

Policy makers in the region as well as planners and decision makers in the private and public enterprises should consider FDI inflows from the lens of acquiring R&D and knowledge-based services in order to create an environment for new business opportunities, enhance productivity, improve information access, improve administration as well as product management quality control and facilitate collaboration with other companies (Kuwayama, Ueki, & Tsuji 2005). A process that started in late 2000 in the UAE with the construction of "Free Trade Zones", tax-free zones to all companies participating in the community, such as Dubai's "Internet City" (TBS Journal, 2001). This community is currently the host of major ICT-based international corporations. Bontis (2004) points out that research and development is a key parameter in the nation's future intellectual wealth, or so-called renewal capital. This significance comes from the direct relationship between the success of a country's future development and the effectiveness of its R&D sector.

CONCLUSION

For this study, our main research question was: Can FDI inflows and outflows provide insights into the differences in ICT diffusion between Middle East and Asia-Pacific? To answer this question, we performed an econometric analysis for a balanced panel of nine Islamic states in the Middle East and eight Asia-Pacific economies for the period 1996-2008. The results indicate that while inward FDI has generally had a positive and significant impact on ICT diffusion in Asia-Pacific economies, its impact on Middle Eastern countries has been detrimental. In contrast, the results of this study show that outward FDI had a positive and significant impact on ICT diffusion in Middle-Eastern economies, while its impact on Asia-Pacific economies has been statistically insignificant. Moreover, while dissimilarities exist between the economies included in this study in terms of their level of socio-economic and political development, GDP per capita and trade openness have had positive impacts on ICT diffusion in both regions.

We propose possible directions for future research. First, future research should investigate the causality linkage among these variables. This can enrich our understanding of how the degree of internationalization affects the diffusion of ICT in the long run. Similarly, future research may apply a more dynamic approach by linking the current ICT to prior ICT levels. Finally, a third direction for potential research is to conduct an extended analysis of the relationship between internationalization and ICT to other countries, and include in the analysis other parameters that may drive FDI inflows and outflows into host countries (e.g., paved roads, transportation infrastructure, socio-political climate, institutional democracy, and economic freedom).

REFERENCES

Addison, T., & Heshmati, A. (2003). *The new global determinants of FDI flows to developing economies: The impacts of ICT and democratization*. WIDER Discussion Papers.

Alexander, D. (2010). *The relationship between information and communication technologies and foreign direct investment at the different stages of investment development path*. MBA dissertation, University of Pretoria.

Baliamoune-Lutz, M. (2003). An analysis of the determinants and effects of ICT diffusion in developing economies. *Information Technology for Development, 10*(3), 151–169. doi:10.1002/itdj.1590100303

Barrell, R., & Pain, N. (1997). Foreign direct investment, technological change, and economic growth within Europe. *The Economic Journal, 107*(445), 1770–1786. doi:10.1111/1468-0297.00256

Barroso, J. L. G., & Martinez, J. P. (2005). The geography of the digital divide: broadband deployment in the Community of Madrid. *Univ Access Inf Soc, 3*, 264–271. doi:10.1007/s10209-004-0103-0

Beach, W., & Miles, A. M. (2006). *Explaining the factors of the index of economic freedom*. 2006 Index of Economic Freedom, The Heritage Foundation and Wall Street Journal.

Beck, N., & Katz, J. N. (1995). What to do (not to do) with time-series cross-section data. *The American Political Science Review, 89*, 634–647. doi:10.2307/2082979

Beilock, R., & Dimitrova, D. V. (2003). An exploratory model of inter-country Internet diffusion. *Telecommunications Policy, 27*(3-4), 237–252. doi:10.1016/S0308-5961(02)00100-3

Bongalia, F., Goldstein, A., & Mathews, J. (2006). Accelerated internationalization by emerging multinationals: The case of white goods sector. *Munich Personal RePEc Archive, 1485*, 1–37.

Bontis, N. (2004). National intellectual capital index: A United Nations initiative for Arab region. *Journal of Intellectual Studies, 5*(1), 13–39.

Brown, I., & Licker, P. (2003). Exploring differences in internet adoption and usage between historically advantaged and disadvantaged groups in South Africa. *Journal of Global Information Technology Management, 6*(4), 6–26.

Cava-Ferreruela, I., & Alabau-Muñoz, A. (2006). Broadband policy assessment: A cross-national empirical analysis. *Telecommunications Policy, 30*(8-9), 445–463. doi:10.1016/j.telpol.2005.12.002

Chinn, M. D., & Fairlie, R. W. (2007). The determinants of the global digital divide: A cross-country analysis of computer and internet penetration. *Oxford Economic Papers*, *59*, 16–44. doi:10.1093/oep/gpl024

Chowdhry, S. (2007). *Foreign direct investment (FDI) on the rise in OIC economies.* Retrieved from http://www.dinarstandard.com/current/OIC_FDI040907.htm

Coe, D. T., Helpman, E., & Hoffmaister, A. W. (1997). North-south R&D spillovers. *The Economic Journal*, *107*(440), 134–149. doi:10.1111/1468-0297.00146

Dedrick, J., Kraemer, K. L., Palacios, J. J., Tigre, P. B., & Botelho, A. J. J. (2001). Economic liberalization and the computer industry: Comparing outcomes in Brazil and Mexico. *World Development*, *29*(7). doi:10.1016/S0305-750X(01)00038-9

Delasnerie, S. (2007,October 3). View from Middle East & Africa. *fDimagazine.com*, (p. 69). Retrieved from http://www.fdimagazine.com/news/fullstory.php/aid/2179/View_from_the_Middle_East___Africa.hl

Dewan, S., Ganley, D., & Kraemer, K. L. (2005). Across the digital divide: A cross-country multi-technology analysis of the determinants of IT penetration. *Journal of the Association for Information Systems*, *6*(12), 409–432.

Dunning, J. H. (1981). *International production and the multinational enterprise.* London, UK: Allen and Unwin.

Dunning, J. H., & Lundan, S. (2008b). Institutions and the OLI paradigm of the multinational enterprise. *Asia Pacific Journal of Management*, *25*, 573–593. doi:10.1007/s10490-007-9074-z

Dutton, W., Gillett, S. E., McKnight, L., & Peltu, M. (2004). Bridging broadband Internet divides: Reconfiguring access to enhance communicative power. *Journal of Information Technology*, *19*, 28–38. doi:10.1057/palgrave.jit.2000007

Dwivedi, Y., & Lal, B. (2007). Socio-economic determinants of broadband adoption. *Industrial Management & Data Systems*, *107*(5), 654–671. doi:10.1108/02635570710750417

Forman, C., Goldfarb, A., & Greenstein, S. (2002). *Digital dispersion: An industrial and geographic census of commercial Internet use.* Cambridge, MA: National Bureau of Economic Research.

Forman, C., Goldfarb, A., & Greenstein, S. (2003). *How did location affect adoption of the commercial Internet? Global village, urban density and industry composition.* Working Paper No 9979 (NBER, Cambridge, MA).

Gholami, R., Lee, S. T., & Heshmati, A. (2006). The causal relationship between information and communication technology and foreign direct investment. *World Economy*, *29*(1), 43–62. doi:10.1111/j.1467-9701.2006.00757.x

Graham, M. E., & Wada, E. (2001). *Foreign direct investment in China: Effects on growth and economic performance.* Peterson Institute Working Paper Series WP01-3, Peterson Institute for International Economics.

Grubesic, T., & Murray, A. (2002). Constructing the divide: Spatial disparities in broadband access. *Regional Science*, *81*, 197–221.

Hardy, A. (1980). The role of the telephone in economic development. *Telecommunications Policy*, *4*, 278–286. doi:10.1016/0308-5961(80)90044-0

Hargittai, E. (1999). Weaving the Western Web: Explaining differences in Internet connectivity among OECD economies. *Telecommunications Policy*, *23*, 701–718. doi:10.1016/S0308-5961(99)00050-6

Hiong, G. (2006). *Digital review of Asia-Pacific 2005/2006: Singapore.* Retrieved from http://www.digital-review.org/.

Hitt, L., & Tambe, P. (2007). Broadband adoption and content consumption. *Information Economics and Policy, 19*(3-4), 362–378. doi:10.1016/j.infoecopol.2007.04.003

Howick, S., & Whalley, J. (2008). Understanding the drivers of broadband adoption: The case of rural and remote Scotland. *The Journal of the Operational Research Society.* doi:10.1057/palgrave.jors.2602486

ITU. (2003). *Declaration of principles.* Geneva: World Summit on the Information Society.

ITU. (2005). *Tunis agenda for the information society.* Tunis: World Summit on the Information Society.

ITU. (2007a). *World telecommunication/ ICT indicators.*

ITU. (2007b). *Measuring the information society 2007, ICT opportunity index and world telecommunication/ICT indicators.*

ITU. (2009). *Measuring the information society 2009, The ICT development index.* Geneva, Switzerland: ITU Publication.

Journal, T. B. S. (2001). *Dubai media city prepares for next phase, no. 5.* Retrieved from http://www.tbsjournal.com/Archives/Fall01/dubai.html

Kane, T. (2007). Economic freedom in five regions. In T. Kane, K. R. Holmes, & A. M. O'Grady (Eds.), *2007 index of economic freedom.* Washington, DC: The Heritage Foundation and Dow Jones & Company, Inc. Retrieved from www.heritage.org/index

Kiiski, C., & Matti Pohjola, M. (2002). Cross country diffusion of the Internet. *Information Economics and Policy, 14*, 297–310. doi:10.1016/S0167-6245(01)00071-3

Kirkpatrick, C., Parker, D., & Zhang, Y. (2006). Foreign direct investment in infrastructure in developing countries: does regulation make a difference? *Transnational Corporations, 15*(1), 143–171.

Kuwayama, M., Ueki, Y., & Tsuji, T. (2005). *Information Technology for development of small and medium-sized exporters in Latin America and East Asia.* Chile: United Nations Publication.

La Rose, R., Gregg, J. L., Strover, S., Straubhaar, J., & Carpenter, S. (2007). Closing the rural broadband gap: Promoting adoption of the internet in rural America. *Telecommunications Policy, 31*, 359–360. doi:10.1016/j.telpol.2007.04.004

Loungani, P., & Razin, A. (2001). How beneficial is foreign direct investment for developing economies? *Finance and Development Quarterly, 38*(2), 6–10.

Lu, M. (2005). Digital divide in developing economies. *Journal of Global Information Technology Management, 4*(3), 1–4.

Miles, M. A., Holmes, K. R., & Anastasia, M. A. (2006). *2006 index of economic freedom.* The Heritage Foundation and Wall Street Journal.

Miller, T., & Holems, K. R. (2009). *2009 index of economic freedom.* The Heritage Foundation and Wall Street Journal.

Myers, R. H. (1990). *Classical and modern regression application* (2nd ed.). CA: Duxbury press.

Norris, P. (2000). *The global divide: Information poverty and Internet access worldwide.* Retrieved from http://www.ksg.harvard.edu/people/.

Norton, S. (1992). Information costs, telecommunications, and the microeconomics of macroeconomic growth. *Economic Development and Cultural Change, 41*, 175–196. doi:10.1086/452002

Nwagwu, E. W. (2006). Integrating ICTs into the globalization of the poor developing countries. *Information Development, 22*(3), 167–179. doi:10.1177/0266666906069070

OECD. (1996). *Information infrastructure convergence and pricing: The Internet.* Paris.

OECD. (1997a). *Global information infrastructure-Global information society (GII-GIS): Policy requirements.*

OECD. (1997b). *Global information infrastructure-Global information society (GII-GIS): Policy recommendations for action.* Paris.

OECD. (1998). *Internet tra$c exchange: Developments and policy.* Paris, France: OECD. Retrieved from http://www.oecd.org/dsti/sti/it/cm/prod/tra$c.htm

OECD. (1999). *Building infrastructure capacity for electronic commerce: Leased line developments and pricing.* Paris, France: OECD.

Orbicom. (2005). *From the digital divide to digital opportunities: measuring infostates for development.* Montreal, Canada: Claude-Yves Charron.

Oyelaran-Oyeyinka, B., & Kaushalesh, L. (2003). The internet diffusion in Sub-Saharan Africa: A cross-country analysis. *Telecommunications Policy, 29,* 507–527. doi:10.1016/j.telpol.2005.05.002

Picot, A., & Wernick, C. (2007). The role of government in broadband access. *Telecommunications Policy, 31*(10-11), 660–674. doi:10.1016/j.telpol.2007.08.002

Pipe, G. R. (2003). *Vietnam promotes ICT to accelerate development. I-WAYS, The Journal of E-Government Policy and Regulation, 26(1).* Amsterdam, The Netherlands: ISO press.

Prais, S. J., & Winsten, C. B. (1954). Trend estimators and serial correlation. *Cowles Commission Discussion Paper,* 383.

Quibria, M. G., Ahmed, S. N., Tschang, T., & Reyes-Macasaquit, M. L. (2003). Digital divide: Determinants and policies with special reference to Asia. *Journal of Asian Economics, 13,* 811–825. doi:10.1016/S1049-0078(02)00186-0

Quynh, N. (2006). *Digital review of Asia-Pacific 2005/2006: Vietnam.* Retrieved from http://www.digital-review.org/

Robertson, A., Soopramanien, D., & Fildes, R. (2007). Segmental new-product diffusion of residential broadband services. *Telecommunications Policy, 31*(5), 265–275. doi:10.1016/j.telpol.2007.03.006

Robison, K. K., & Crenshaw, E. M. (2002). Post-industrial transformations and cyber-space: A cross-national analysis of Internet development. *Social Science Research, 31,* 334–363. doi:10.1016/S0049-089X(02)00004-2

Rogers, W. H. (1993). Regression standard errors in clustered samples. *Stata Technical Bulletin, 53,* 32–35.

Salmenkaita, J., & Ahti Salo, A. (2002). Rationales for government intervention in the commercialization of new technologies. *Technology Analysis and Strategic Management, 14*(2), 183–200. doi:10.1080/09537320220133857

Santos-Paulino, A., & Thirlwall, A. P. (2004). The impact of trade liberalisation on exports, imports and the balance of payments of developing economies. *The Economic Journal, 114*(493), 50–72. doi:10.1111/j.0013-0133.2004.00187.x

Sciadas, G. (2006). *Digital review of Asia-Pacific 2005/2006: Vietnam.* Retrieved from http://www.digital-review.org/

Shih, E., Kraemer, K. L., & Dedrick, J. (2008). IT diffusion in developing economies: Policy issues and recommendations. *Communications of the ACM, 51*(2), 43–48. doi:10.1145/1314215.1340913

Soper, D. S., Demirkan, H., Goul, M., & St. Louis, R. (2006). The impact of ICT expenditures on institutionalized democracy and foreign direct investment in developing economies. *Proceedings of the 39th Annual Hawaii International Conference on System Sciences* (HICSS'06).

Stimson, J. (1985). Regression in space and time: A statistical essay. *American Journal of Political Science, 29*(4), 914–947. doi:10.2307/2111187

Tan, F., & Leewongcharoen, K. (2005). Factors contributing to IT industry success in developing economies: The case of Thailand. *Information Technology for Development, 11*(2), 161–194. doi:10.1002/itdj.20009

Tan, Z. (2004). Evolution of China's telecommunications manufacturing industry: Competition, strategy and government policy. *Communication & Strategies, 53.*

The World Bank. (2006a). *Information and communications for development 2006: Global trends and policies.*

The World Bank. (2006b). *2006 information and communications for development: Global trends and policies.*

The World Bank. (2007). *World Bank development indicators (WDI)*. CD-ROM.

Thode, H. C. Jr. (2002). *Testing for normality*. New York, NY: Marcel Dekker. doi:10.1201/9780203910894

Tipton, F. (2002). Bridging the digital divide in Southeast Asia. *ASEAN Economic Bulletin, 19*(1), 83–99. doi:10.1355/AE19-1F

Trkman, P., Jerman-Blazic, B., & Turk, T. (2008). Factors of broadband development and the design of a strategic policy framework. *Telecommunications Policy, 32*(2), 101–115. doi:10.1016/j.telpol.2007.11.001

UN ICT Task Force. (2005). *Information and communication technology for peace - The role of ICT in preventing, responding to and recovering from conflict.* UN ICT Task Force Series 11.

UNCTAD. (2006a). *World investment report 2006: FDI from developing and transition economies: Implications for development.* UNCTAD Handbook of Statistics 2006-07. Retrieved from http://stats.unctad.org/fdi/ReportFolders/reportFolders.aspx?sCS_referer=&sCS_ChosenLang=en

UNCTAD. (2006b). *Information economy report 2006: The development perspective.*

UNCTAD. (2006c). *The digital divide report: ICT diffusion index 2005*. Retrieved from http://www.unctad.org/en/docs/iteipc20065_en.pdf

UNCTAD. (2007). *World investment prospects survey 2007–2009*. New York, NY: UN Publication.

UNCTAD. (2008). *World investment report: Transnational corporations and the infrastructure challenges.* New York, NY: UN Publication.

UNCTAD. (2010). *World investment report 2010.* Retrieved from http://www.unctad.org/en/docs/wir2010_en.pdf

UNCTAD. (2011). *Global investment trend monitor, No. 6. April 27.* Retrieved from http://www.unctad.org/en/docs/webdiaeia20114_en.pdf

UNDP. (2006). *Human development reports: Statistics*. Retrieved from http://hdr.undp.org/en/statistics/

UNDP. (2007). *Human development report 2007/2008*. Retrieved from http://hdr.undp.org/en/reports/global/hdr2007-2008/

UNDP. (2009). *Overcoming barriers: Human mobility and development.* Retrieved from http://hdr.undp.org/en/reports/global/hdr2009/

United Nations ICT Task Force. (2005). *Measuring ICT: The global status of ICT indicators: Partnership on measuring ICT for development.* New York, NY: United Nations ICT Task Force.

Vavilov, S. (2006). *Investment development path in petroleum exporters*. University of Paris.

Wallsten, S. (2005). Regulation and internet use in developing economies. *Economic Development and Cultural Change, 53*, 501–523. doi:10.1086/425376

Wong, P.-K. (2002). ICT production and diffusion in Asia: Digital dividends or digital divide? *Information Economics and Policy, 14*, 167–187. doi:10.1016/S0167-6245(01)00065-8

Yap, A., Das, J., Burbridge, J., & Cort, K. (2006). A composite-model for e-commerce diffusion: Integrating cultural and socio-economic dimensions to the dynamics of diffusion. *Journal of Global Information Management, 14*(3), 17–36. doi:10.4018/jgim.2006070102

Yin, X. (2005). China. In Kuwayama, M., Ueki, Y., & Tsuji, T. (Eds.), *Information Technology for development of small and medium-sized exporters in Latin America and East Asia*. Chile: United Nations Publication.

Zhao, H. X., Kim, S., Suh, T. W., & Du, J. J. (2007). Social institutional explanations of global internet diffusion: A cross-country analysis. *Journal of Global Information Management, 15*(2), 28–55. doi:10.4018/jgim.2007040102

ENDNOTES

[1]   In the context of this research, a 'balanced panel' refers to data set with equally long time series for all cross-sectional units.

[2]   Three other nations in the Middle East region were not included in this study: Israel, Lebanon and Iraq. Israel is not an Islamic state, there is missing data for Lebanon, and finally, no panel data was available for Iraq due to the country's long lasting international conflicts.

[3]   This study includes non-OECD economies from Asia-Pacific and the Middle East regions.

[4]   The marginal returns to the use of ICT are higher in remote locations lacking economies of densities.

[5]   A major advantage of a log-linear model is that the coefficients of the continuous variables (α) amount to elasticities. In other words, the coefficient for each continuous independent variable is the estimated percentage change in the dependent variable associated with a one percent increase in the independent variable, controlling for other factors in the model.

[6]   All continuous variables are transformed into natural logarithms.

[7]   The time dummies play an important role in the model by capturing the effects on ICT diffusion of the reduction in the global price of technology relative to the period mean.

[8]   For the independent variable x_j, VIF_j is determined by $VIF_j = 1/(1-R^2_j)$ where R^2_j is the multiple coefficient of determination for the regression model which relates the independent variable x_j to all other independent variables in the model.

[9]   Generalized Least Squares (GLS) yielded almost identical results. However, Beck and Katz (1995) argue that in many cases the feasible GLS method might underestimate the standard errors of the estimated coefficients. Further, the data matrix used in this study is precisely the sort for which Beck and Katz's specification is more efficient.

[10]   We use the statistical software package SATAv11 for estimation.

[11]   Trade openness in Asia-Pacific and Middle Eastern countries is ultimately the result of economic reforms in the majority of countries (except for Iran and Syria) during the period of 1996 to 2005.

[12]   The correlation between these two variables with the log of the ICT network index is of 0.92 and 0.86 respectively.

APPENDIX

Table 1A. ICT Core Indicators

Economy	Main Tel.			Cell Phones			Internet		
Asia-Pacific	1996	2005	2008	1996	2005	2008	1996	2005	2008
China	4.41	26.63	25.48	0.55	29.9	47.95	0.01	8.6	22.28
Hong Kong	53.63	53.94	58.72	21.16	123.47	165.85	4.66	50.08	67
Indonesia	2.11	5.73	13.36	0.28	21.06	61.84	0.06	7.18	7.92
Malaysia	17.81	16.79	15.89	7.18	75.17	102.59	0.85	42.37	55.8
Philippines	2.55	4	4.51	1.37	41.3	75.39	0.06	5.32	6.22
Singapore	42.58	42.39	40.24	16.7	100.76	138.15	8.17	51.84	73.02
Thailand	7.05	10.95	10.42	3.13	42.98	92.01	0.23	11.03	23.89
Vietnam	1.5	18.81	33.98	0.09	11.39	80.37	0	3.45	23.92
Middle East									
Bahrain	23.95	26.9	28.42	6.65	58.35	185.77	0.83	21.33	51.95
Iran	9.7	27.3	33.83	0.1	10.39	58.65	0.02	10.07	31.37
Jordan	7.78	11.1	8.46	0.54	28.93	86.6	0.05	11.22	26
Kuwait	20.68	19	18.53	7.97	88.57	99.59	0.79	26.05	34.26
Oman	8.83	10.32	9.84	0.67	51.94	115.58	0.2	11.1	20
Qatar	26.4	26.3	20.56	5.69	92.15	131.39	0.99	28.16	34.04
Saudi Arabia	9.36	15.5	16.27	0.99	54.12	142.85	0.03	6.62	30.8
Syria	8.2	15.25	17.12	0	15.49	33.24	0.02	5.78	16.8
UAE	29.64	27.49	33.63	7.78	100.86	208.65	0.39	31.08	65.15

Source: ITU World telecom Indicators, 2009

Table 2A. Demographic Data

Middle East	Population (2005) mil.	Life Expectancy (years) 2005	Adult Literacy 2005	GDP(PPP US$) 2005	HDI rank 2005	Inward FDI stock (mil US$) 2008	ICT rank (2007)
Bahrain	0.7	74.5	86.5%	17,773	41	14741	42
Iran	69.4	71.1	82.4%	2,781	94	20967	78
Jordan	5.5	72	91.1%	2,323	86	16320	76
Kuwait	2.7	77.5	93.3%	31,861	33	943	57
Oman	2.5	74.7	81.4%	9,584	58	11680	77
Qatar	0.8	74.1	89.0%	52,240	35	17769	44
Saudi Arabia	23.6	72.6	82.9%	13,399	61	110200	55
Syria	18.9	73.8	80.8%	1,382	108	5900	89
UAE	4.1	79.1	88.7%	28,612	39	68224	32
Asia-Pacific							
China	1313	72	90.9%	1,713	81	378083	73
Hong Kong	7.1	81.5	98.4%	25,592	21	816184	11
Indonesia	226.1	68.6	90.4%	1,302	107	72227	108
Malaysia	25.7	73	88.6%	5,142	63	73601	52
Philippines	84.6	70.3	92.6%	1,192	90	21746	91
Singapore	4.3	78.8	92.5%	26,893	25	326790	15
Thailand	63	68.6	92.6%	2,750	78	93500	63
Vietnam	85	72.4	90.3%	631	105	49854	92

Source: Population, Life Expectancy & Adult GDPP and HDI rank from UNDP-Human Development Report 2007; FDI flows are from UNCTAD-World Investment Report 2006; ICT rank is from ITU-Measuring the Information Society (2009).

Chapter 14

A Process Model for Successful E-Government Adoption in the Least Developed Countries:
A Case of Bangladesh

Ahmed Imran
The Australian National University, Australia

Shirley Gregor
The Australian National University, Australia

ABSTRACT

Least developed countries (LDCs), have been struggling to find a workable strategy to adopt information and communication technology (ICT) and e-government in their public sector organizations. Despite a number of high-level initiatives at national and international levels, the progress is still unsatisfactory in this area. Consequently, the countries are failing to keep pace in the global e-government race, further increasing the digital divide. This chapter reports on an exploratory study in a least developed country, Bangladesh, involving a series of focus groups and interviews with key stakeholders. A lack of knowledge and entrenched attitudes and mindsets are seen as the key underlying contributors to the lack of progress. The analysis of the relationships among the major barriers to progress led to a process model, which suggests a pathway for e-government adoption in an LDC such as Bangladesh. The chapter introduces important directions for the formulation of long-term strategies for the successful adoption of ICT in the public sector of LDCs and provides a basis for further theoretical development.

INTRODUCTION

Public Services in this era are increasingly experiencing a transformation in their work processes, modes of delivery and internal and external communications by virtue of Information and Communication Technology (ICT) applications. This

use of ICT is commonly termed *e-government*, the primary objective of which is "to improve the activities of public sector organizations" (Heeks, 2004, p.1). eGovernment has revolutionized traditional systems and added new dimensions to the functioning of modern government to provide the best service for its citizens and to improve its

DOI: 10.4018/978-1-61350-480-2.ch014

internal efficiency. eGovernment is also closely linked to the global transition to a knowledge-based society (KBS) that modern governments are aiming for. However, e-government in many least developed countries (LDCs) is in its infancy, with failure in the initial adoption and use of ICT in the public sectors hampering advances in e-government (Heeks & Bhatnagar, 2001). LDCs are distinguished by the Economic and Social Council of the United Nations in terms of three criteria: (1) low national income (gross national income per capita under US$905), (2) weak human assets (health, nutrition and education) and, (3) high economic vulnerability (instability of agricultural production and exports, inadequate diversification and a small economy) (UN OHRLLS, 2011). Forty eight countries are currently recognised as LDCs.

Our study investigates two research questions to advance work in this area:

First, what underlying issues are most critical for e-government adoption in an LDC and how are these issues interrelated? Second, how can e-government adoption be facilitated in this context?

The research has practical significance because it focuses on a particularly important sphere of e-government. The three spheres of e-government include; (i) improving the government processes, that is, G2G (government-to-government or e-Administration); (ii) connecting citizens, G2C (government-to-citizens or e-Services); and, (iii) building external interactions, G2B (government-to-business) (Backus, 2001; Heeks, 2004). The focus of this study is on the first sphere, the improvement of the internal government processes with ICT. Improvements in this sphere have the maximum potential to achieve greater benefit for LDCs. Being the largest user of an IT system, the public sector in an LDC can play the leading role in ICT diffusion throughout the country and can exert the greatest influence through its policies and regulations (e.g., Flamm, 1987; Nidumolu et al., 1996).

The original motivation for the study was provided by the lead author's involvement for a number of years in the public sector in Bangladesh and the problems encountered there. These experiences led to the belief that it was important to focus on G2G, at least as much as G2C, despite a common misconception that e-government is mainly about delivering government services over the Internet (Allan et al., 2006; Dawes, 2002a). This narrow vision of e-government does not take into account the "behind-the-scenes first-order changes" (Scholl, 2005, p.7) and the variety of government activities that occur within and between the government agencies. Thus, this focus fails to recognize the essential use of technologies other than the Internet. In many LDCs, universal access to the Internet and citizen-centric services may be far out of reach—for example, in Bangladesh the Internet penetration rate is 1.3%, (ITU, 2009)—yet there may still be scope for substantial improvements in the internal G2G processes, with significant outcomes in terms of improving productivity, efficiency and transparency. Evidence for the importance of G2G relative to G2C is provided in the Australian case, where only 38% of the population use online services to contact the government, despite 71% of household having Internet access. Yet Australia holds one of the leading positions in the world e-government rankings (8th) and the public sector efficiency index (3rd), with the bulk of the internal government processes being driven by ICT (AGIMO, 2009; ITU, 2009; Afonso et al., 2007; UN, 2010).

The study has further significance, because it contributes to theory in an important area. Walsham and Sahay (2006) noted the wide range of theories drawn upon in a survey of research on ICT in developing countries. These theories include structuration theory (as in Liu & Westrup, 2003; Walsham, 2002), actor-network theory (as in Braa et al., 2004; Stanforth, 2006) and institutional theory (as in Avgerou, 2002; Silva & Figueroa, 2002). The range of theories employed is not surprising, given the extent of the phenomena studied, ranging from the individual's usage of services, to cross-cultural issues, to the underlying technological issues. Furthermore, there are gaps in the theoretical underpinnings relating to ICT adoption in G2G in developing countries, because much of the theory has been developed in other contexts. Organizations in developing countries do not necessarily have management structures and the efficiency-driven rationality that is common in Western business environments. This divergence follows from the fundamental differences across cultures (Chen et al., 2006). In addition, ICT adoption in public sector organizations differs significantly from the case in private organizations (Bretschneider, 1990; Dawes et al., 2004; Kankanhalli & Kohli, 2009; Suomi & Kastu-Häikiö, 1998; Willcocks, 1994). Private organizations are mostly profit-driven and customer satisfaction is critical for their business success. Public sector organizations are mostly demand- or vote-driven and citizen satisfaction is sometimes ignored, especially in LDCs.

Furthermore, our frame of reference is different from many other studies of ICT in the government of developing countries that tend to focus on individual projects and initiatives and specific citizen service deliveries (G2C) (Indjikian & Siegel, 2005; Madon, 2004; Stanforth, 2006). We are taking a broader view of the public service and its critical interaction with ICT innovation against the historically-situated context of the Bangladesh public service. As Avgerou (2009, p.2) argues:

even if the technologies implemented in an IS project are already common elsewhere and widespread, the local experience of technology implementation and socio-organizational change constitutes an innovation for the organization concerned and may well constitute innovation for its socio-economic context.

Accordingly, the aim of the paper is to identify the issues surrounding ICT adoption in the public sector in an LDC, their relative importance and their interrelationships, from the perceptions and knowledge of the key stakeholders with a view to attaining a road map for bringing about change from the current to a future enhanced state. The empirical study was conducted in the context of a single LDC, Bangladesh, using a series of focus groups and interviews with the key stakeholders from diverse backgrounds to identify the root causes hindering successful adoption of ICT in the public sector. Some findings from this have already been put into action with considerable success in an applied research project funded by AusAID's public service linkage program (Imran et al., 2009; Gregor et al., 2010).

The remainder of the paper proceeds as follows. The next section outlines the context of the study in Bangladesh to provide a backdrop for the study's rationale and findings. The subsequent sections describe the conceptual background, the research method and the findings in respect to the main barriers identified as hindering e-government adoption. The paper then develops a model that indicates the interrelationships amongst the barriers and suggests a process by which e-government uptake might be facilitated. The paper concludes with some reflections on what the findings mean in terms of the prior literature.

STUDY CONTEXT: PUBLIC SECTOR IN BANGLADESH

It is important to provide a clear understanding of the research context before considering the relevant prior research, because of the importance of situating the research against the local context in developing countries (Avgerou & Walsham, 2000; Heeks, 2006a; Myers & Tan, 2002; Walsham & Sahay, 2006; Warschauer, 2003; Weisinger & Trauth, 2002; Westrup et al., 2003; Wilson, 2004). For example, Weisinger and Trauth (2002, p. 310) argue that multiple contexts of culture, tradition and socio-economic condition "are interwoven to produce a locally situated culture within the work environment". These unique circumstances of an LDC context play a predominant role in ICT adoption in those countries (Walsham et al., 2007; Yap et al., 2006). Thus, first, understanding and situating the study in the public sector environment of Bangladesh is extremely important.

Bangladesh, one of the 48 LDCs, is a small south Asian country and is an example of a typical LDC lagging in ICT adoption in the public sector. Bangladesh is densely populated with 162.2 million inhabitants (UNFPA, 2009). The country has a parliamentary form of government, with the President as the Head of State and the Prime Minister as the Head of the Government. The Prime Minister is assisted by a council of ministers, with the secretaries belonging to the civil service (BBS, 2009).

Bangladesh is also a country with "high-power distance" (Hofstede & McCrae 2004, p 62), where the typical hierarchical administrative culture inherited from the British colonial system has evolved into distinctive hierarchical attitudes with respect to the interactions with the common citizens, superior–subordinate relationships and the method of delivery of government services (Jamil, 2007; Siddiqui, 1996). Little change occurred in the culture and the regulation of the public service following British rule and then independence in 1971. Bureaucrats still feel comfortable working in high-power distant relationships rather than as equals (Jamil, 2007). The concepts of setting priorities and achieving cost-benefits are considered a foreign tradition; following established norms is more important than achieving results. Similarly, the principle that the consumers or subscribers have a right to good service is frequently ignored in the service sector of Bangladesh (Jamil, 2007). The issue of good governance is receiving increasing attention from the public as well as from developmental partners. The government is seen to be overly preoccupied with bureaucratic process. In this environment, creativity and innovation are not usually appreciated, and new ideas and new ways of doing

things are considered to be foreign values that foster competition and conflict, and may threaten the stability of the system. A government officer, therefore, is not expected to engage in finding new ways of solving societal problems; neither does he encourage subordinates to nurture innovative ideas (Jamil, 2007; Sein et al., 2008). Overall, the administration of Bangladesh is influenced by political instability and political influence and is seen as lacking in understanding, planning and initiative; all of which are further aggravated by a poor infrastructure and a lack of training and skill (Jamil, 2007).

Stage of ICT Development

Unfortunately, ICT in Bangladesh is still seen primarily in terms of a hardware and software industry and much of its wider implication for the national economy in terms of information processing is not generally understood (Imran, 2009). The lack of vision in this area is felt at all levels. Some want ICT to be implemented but do not know how to go about it, missing the "big picture" surrounding ICT innovation. As a result, many computerisation initiatives with a strong techno-centric focus have not changed the situation or improved efficiency and effectiveness significantly. A number of high level initiatives have been taken over the last decade, such as the formation of an "ICT Task Force" with the Prime Minister as the chair person, the formulation of *The National Information and Communication Technology Policy 2002* (MOSICT, 2002) and the Support for ICT Task Force (SICT) under the Ministry of Planning to realize various e-government projects. However, the results are not encouraging.

There is a lack of an integrated collaboration of work processes among the various government agencies in Bangladesh that hinders ICT-based networking and business procedures. Land administration provides one example. For any piece of land, the registration is handled by one department, it is surveyed by a second and maintained and taxed by a third agency. There is little collaboration amongst these agencies that are dealing with the same piece of land, whereas ideally, there could be a "one-stop shop". Inevitably, the lack of collaboration between the agencies results in different information from different agencies regarding the same piece of land.

Nevertheless, recently, there have been some encouraging signs. The government (2007–2008) successfully completed a massive national voter and ID project, which registered and issued ID cards for about 80 million people, one of the largest databases in the world. Although this project was aimed primarily at ensuring a correct voter list for the last election, it opened up a number of opportunities for multipurpose use. Furthermore, the newly-elected Government came into power in 2009 with a promise of 'Digital Government' in their election manifesto and showing a lot of enthusiasm in this area. However, it remains to be seen whether such promise remains as political rhetoric or it actually translates into reality. While, according to the UN *Global e-Government Readiness Report* (UN, 2005, 2008, 2010) Bangladesh had improved its ranking from 162 to 134 largely due to increased presence of government department websites, the core problem of overall public service efficiency and transparency remained at unsatisfactory level.

CONCEPTUAL BACKGROUND

Our study is about the adoption of information technology, albeit in a somewhat unusual context, and thus falls within the purview of theories relating to the adoption and diffusion of innovations. Given the nature of the study, we look first at the theory relating to the adoption of innovations and then specifically to that relating to organizations and ICT. We then consider the prior work that has investigated ICT adoption in the LDCs and some

specific work that relates to ICT adoption in the public sector organizations. Note that in reviewing the theory and prior work, we are particularly interested in the process of adoption and in the barriers that may hinder adoption in organizations in the LDCs. Frameworks, lessons and implications from some of those studies are discussed in the respective sections below.

Theories of Innovation and Technology Adoption

No single theory of innovation exists, although Rogers' (2003) classical model of diffusion has been extremely influential. The primary elements of Rogers' diffusion of innovation theory are: innovation, communication, time and the social system. With respect to innovation, five characteristics of an innovation are seen to explain its rate of adoption by individuals. These characteristics are relative advantage, compatibility, complexity, trialability and observability. The individual's decision process proceeds from knowledge of the innovation, through attitude formation, then on to adoption and implementation or alternatively to rejection. The first stage of the decision process entails acquiring knowledge of one or more types, the awareness that the innovation exists, the knowledge of how to use the innovation and the knowledge of how the innovation works. Interestingly, Rogers' theory recognizes the social system as a primary element within which the innovation is introduced and the importance of the opinion leaders within that social system in influencing the adoption decisions of the others. Rogers' theory incorporates the work of many other innovation diffusion researchers and draws on studies in many cultures. This theory has found application in non-western and developing countries for many social change innovations, owing to its base in cross-cultural research and its emphasis on recognizing the social context into which an innovation is introduced.

Despite the importance of Rogers' theory, it may not necessarily cover all innovations and all contexts. For example, it has been noted that it is not adequate for the diffusion of complex and networked technologies (Lyytinen & Damsgaard, 2001). Furthermore, the theory was originally developed for adoption by individuals and its coverage of adoption at the organizational level can be critiqued (Lundblad, 2003). When extended to organizational settings by Rogers, the innovation process was seen to involve several stages: Agenda-setting, Matching, Redefining/Restructuring, Clarifying and Routinizing. This extension of the theory is less well-grounded in practice and it is unclear whether the theory that applies to mature organizational settings in Western countries will apply to government organizations in the LDCs.

One further point should be noted. Adoption models, such as the Technology Acceptance Model (TAM) (Davis et al., 1989) and the Theory of Reasoned Action (TRA) (Fishbein & Ajzen, 1980), have been widely employed in explaining ICT adoption. These theories apply at the level of individual adoption and ignore organizational and social factors and the many constraints in the real world, such as, ability, time constraints, environmental limits and unconscious habits (Bagozzi et al., 1992; Legris et al., 2003). For these reasons, these theories are not seen as applicable in the context of our study.

Adoption of ICT in Organizations

The innovation and diffusion of ICT within organizations also encompasses organizational "change" or "transformation" of organizational structure and practices. There are numerous academic frameworks available in the management literature that seek to explain the issues related to organizational change. Amongst them a particular segment deals with change caused by technological innovation. Information technology (IT) is regarded as a major enabler of organizational change (Avgerou, 1996; Markus & Benjamin, 1997; Morton, 1991). Other work more focused on organizations includes Damanpour (1987), Lindquist and Mauriel (1989) and Marcus and Weber (1989).

An approach of particular interest is Scott's (2001) institutional theory, which is relevant in explaining ICT adoption in organizations through various institutional pressures (coercive, mimetic and normative). Institutional theory has been used in a number of studies in developing countries, discussing different problems using different dimensions of the theory. For example, Molla et al. (2006) examined the role of institutions in promoting and influencing the diffusion of eCommerce in Barbados and found it relevant for the macro level analysis of interventions and their impact in shaping the behaviours of firms and consumers. Silva and Figueroa (2002) examined ICTs in Chile and the role of institutional players, both private and government sector. Miscione (2007) studied a telemedicine project in Peru, comparing the public health system with traditional practices. Bada et al. (2004) investigated an ICT implementation in a Nigerian bank, and the importance of understanding local cultural factors.

Institutional theory, however, emerged from the organizational science literature and it may not be sufficient to resolve the complexity of the issues surrounding ICT adoption in the public sector and e-government in the context of a developing country. It is not clear whether the three institutional pressures mentioned by Scott (2001) and DiMaggio and Powell (1983) are sufficient in themselves or whether additional pressures specific to the context are needed to bring about institutional change in the public sector environment in LDCs. More study is thus needed to determine the most appropriate types of institutional interventions through investigation of practical experiences, rather than relying on theory alone (Montealegre, 1998).

Adoption of ICT in LDCs

The institutional studies above are only a small part of a stream of literature that is emerging on ICT in developing countries. The literature provides rich insights into contextual issues such as the culture, environment and typical characteristics of developing countries. Although, according to Molla et al. (2006, p. 2), "Much of the research on developing countries tends to be anecdotal and narrative, relying less on shared understanding of the underlying phenomena than intuitive and culture bound explanation and story telling".

Comprehensive reports on ICT and its impact in developing countries are produced by international bodies and organizations, including the UN (UN ICT Task Force), UNDP (2001) and the World Bank (2002, 2005). However, such work "strains for universality and global coverage and thereby weakens the theoretical focus. These works are written mainly for practitioners and are not theoretically explicit" (Wilson, 2004, p. 26).

Academic research into ICT adoption in the LDCs is diverse. In regard to the barriers and challenges to ICT adoption, the findings from the literature across countries are not consistent, which is not surprising given their different foci and the range of contexts. Some studies are techno-centric, with an emphasis on issues such as telecommunications infrastructure (Mbarika, 2002; Mursu et al., 2003; Nair & Kuppusamy, 2004; Silva & Figueroa, 2002; Wresch, 2003). Other studies focus on culture with prominence being given to conflict with the technologies developed for the Western cultures (Adam & Myers, 2003; Chau, 2008; Hagenaars, 2003; Hill et al., 1998; Shore & Venkatachalam, 1995; Straub et al., 2001). Some studies infer that the socio-economic or other unique conditions, such as the geopolitical situation of the LDCs, are the major hindrance to development (Korpela, 1996; Montealegre, 1999; Roche, 1996; Rodriguez & Wilson, 2000; Siau & Long, 2005). Furthermore, some studies have a narrow underlying viewpoint where the individual

factors, such as power, leadership commitment, political will or top management support, are seen as influential (Banerjee & Chau, 2004; Kandelin et al., 1998; Kelegai & Middleton, 2004; Kumar & Best, 2006; Raman & Yap, 1996; Sanford & Bhattacherjee, 2007). In total, a very wide range of factors has been found to hinder development at many levels, country and society, within organizations and at the personal level.

There have been some valuable attempts at developing high-level frameworks for ICT adoption in developing countries. Wilson (2004) highlights the massive structural changes at multiple levels and the complex dynamics that constitute the information revolution, based on his wide experience of working in a number of developing countries. He offers a high-level open architecture approach for an information revolution in a country covering a wide range of issues, including structures, institution, politics and policies for public-private, international-national relationships at the micro, meso and macro levels.

Despite the diversity of findings, there appears to be some consensus that the issues and barriers to ICT development in the LDCs are more human than technological and that close attention must be paid to the cultural context in each country. Straub et al. (2001) also insist that the best way to understand the cultural effect is to investigate each cases in their respective cultural contexts.

Adoption of ICT in the Public Sector of LDC

Our concern in this study is with public sector organizations in particular and how they adopt ICT to underpin e-government. Insights into the public sector organizations in general are gained from various public administration and organizational literature (Bretschneider, 1990; Jamil, 2007; Rainey, 1983; Rocheleau, 2006; Suomi & Kastu-Häikiö, 1998; Willcocks, 1994). These studies suggest that the institutional inertia formed by the tradition and culture of the public sector organization can be very deeply engraved and thus a major challenge for any change through innovation. In addition, the local norms, regulations and the availability of complementary factors are not always identical across the adopting units in a given social system, which may significantly affect their relative ability to adopt an innovation (Attewell, 1992; Brown, 1981).

Again, there is a range of literature in the LDC context and we will highlight those studies that have the most relevance for our purpose. First, we will look at some high-level frameworks that are pertinent, followed by some more focussed studies.

Chen et al. (2006) proposed a framework for e-government strategies that incorporates critical success factors (CSFs) for implementation and recognizes the differences between developed and developing countries. Their framework includes a large number of factors in three categories, infrastructure, cultural and society. In a similar vein, Grant and Chau (2005) provide a generic framework from a functional perspective, encompassing different application areas of e-government, such as e-participation and e-democracy.

A number of other studies and frameworks on e-government in developing countries also exist, but relate to more specific e-government projects or problem areas. Examples include the Heeks (2006a) lifecycle approach for managing e-government projects, Banerjee and Chau's (2004) evaluative framework for analysing "e-government convergence capability", Stanforth's (2006) study of a fiscal reform program in Sri Lanka, Sanford and Bhattacherjee's (2007) study of ICT implementation in a municipality in the Ukraine and Madon's (2003) framework for public service delivery in the FRIENDS G2C e-government initiative in southern India. Sharifi and Zarei (2004) highlight the importance of G2G in Iran in a unique case of a socio–political transition against a particular historical and religious context.

Implications of Prior Work

We conclude from our review of the prior literature that although this literature is extensive, it does not contain a well-developed theory that specifically deals with our research questions. The high-level frameworks, such as those of Chen et al. (2006) and the studies of the specific aspects of e-government in the LDCs, provide a backdrop for the current study. However, there is still a gap with respect to the question of what are the key factors that must be resolved first so that the process of bringing about change in the public sector as a whole in an LDC can be put firmly underway.

It appears that a greater understanding of the nature of the barriers and challenges for e-government adoption in the LDCs is required, particularly in respect to the actionable knowledge that can be used to bring about a much-needed change. In this situation, we require a mid-range theory, a theory that is developed through an inductive process and soundly grounded in empirical data (Keng & Yuan, 2006; Merton, 1968). The current study aims at providing the foundations for such a theory.

METHOD

The study adopts an approach for building a theory from data (Eisenhardt, 1989) with a contextual framework derived from the literature, but keeps an open mind for emerging themes during analysis.

It should be noted that the lead researcher in this study, a native Bangladeshi with experience of working in an LDC environment, has a background that allows insights into the culture, authority structures, decision-making and organizational behaviour in his country. This first-hand experience and familiarity with the systems under study enriches the analysis. Putnam (1993, p. 12) advocates:

marinating oneself in the minutiae of an institution — to experience its customs and practices, its successes and its failings. This immersion sharpens one's intuitions and provides innumerable clues about how the institution fits together and how it adapts to its environment.

Focus Groups

Focus groups were the primary method for data gathering for this study, using the "Nominal Group Technique" (NGT) that was developed as an organizational planning technique by Delbecq and Van de Ven in 1971. The NGT is "a consensus planning tool that helps to prioritise issues" in a more structured way than discussion alone, yet "takes the advantage of the synergy created by group participants" (Joppe, 2004, p. 1). The NGT is a variation of brainstorming, providing a more structured procedure that gives participants more opportunity to provide input on a particular issue (Sample, 1984). It is "most effective for generating information and for fact-finding concerning a problem", where "interacting group processes stimulate individuals to consider other dimensions of a problem and help synthesize and evaluate alternative solution possibilities" (Zogona et al., 1966, pp. 38–341 cited in Van de Ven & Delbecq, 1971, p. 209). The NGT method has been used in a wide range of studies, for example, to study

the characteristics of expert associate teachers (Boudreau, 2000).

The seven focus groups discussions (FGD) were conducted between August and December 2005, one in Canberra, Australia and the rest in Dhaka, Bangladesh. Table 1 shows details. Each had an approximate duration of 1.5 hours. The FGD in Canberra was carried out as a pilot test and also to obtain the perspective of the Bangladeshi-born people working in the public sector environment of a developed country. The groups in Bangladesh were composed of members of different bodies involved directly or indirectly with ICT adoption in the public sector of Bangladesh. The participants were chosen because of their ability to provide different perspectives on e-government in Bangladesh.

The focus groups discussed the broad question, "What are the barriers to ICT adoption in the public sector of an LDC, in the context of Bangladesh?" using the NGT procedures (Delbecq & Van De Ven, 1971; Dunham, 1998). A number of possible barriers emerged, which were grouped and regrouped on butcher paper in each FGD and discussed. In the final stage, the participants were asked to vote individually on which barriers they felt were the most important and to rank the top five (1–5) according to their relative importance. The aggregated value of the votes (reverse scaled) for the barriers from all the focus groups pro-

Table 1. Focus group details

Group	Location	Participants	Numbers
FGD 1	Canberra	Bangladeshi expatriate ICT professionals in the public sector in Australia	7
FGD 2	Dhaka	Software developers of a locally-based company	9
FGD 3	Dhaka	Academics from the Department of Computer Science, Management, Governance and Economics of a reputed university	10
FGD 4	Dhaka	ICT for Development experts from an international organization	3
FGD 5	Dhaka	Software, and ICT industry leaders and managers	9
FGD 6	Dhaka	ICT Journalists	5
FGD 7	Dhaka	Support to ICT Task Force officials, Government of Bangladesh and ICT experts	9
Total			52

vided an overall ranking of the perceived most important barriers to ICT adoption into the public sector in Bangladesh. Further analysis of the different themes from the discussions and qualitative interview data helped to identify relationships and the interdependence of the various factors.

Interviews

Eleven face-to-face interviews were conducted during the same period with high profile stakeholders in Bangladesh, including ICT professionals, chief executives of large organizations, senior academics and representatives from government bodies, donor agencies and independent think-tank organizations. A semi-structured questionnaire was used in the interviews, which included the question asked in the focus group discussions. The interview and FGD data were analysed using the

qualitative data analysis tool NVivo 7.0 and the grouping and categorization of the data followed the principles of "analytic progression" (Miles & Huberman, 1994, p. 91) and the "classification system" using the tree structure of coding by NVivo (Bazely, 2007, p. 101). The apparent connections and causal links between the constructs in the different categories were also noted. Such qualitative analysis is useful for producing flow diagrams or concept maps and a narrative description of the processes (Strauss & Corbin, 1990).

Some of the interviews and focus groups were conducted in Bengali, but most were in English. The first author speaks both Bengali and English fluently and has been responsible for the translations. Quotations are given in a form as close as possible to the original narrative, although, in some cases, this means the passage is not fully idiomatic English expression as spoken in Western cultures.

Figure 1. Major barriers by ranking scores

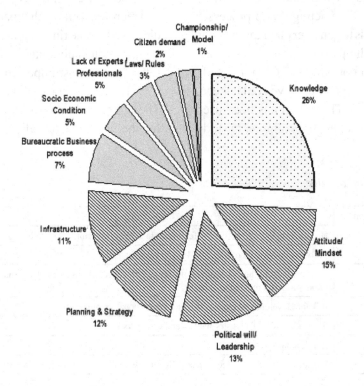

Other Procedures

Most importantly, the lead researcher's prior experience and connections in the Bangladesh Government and with important stakeholders served as a "gateway" for access to the public sector, which is usually very difficult to access and investigate. Access to the participants in the focus groups and interviews was made through different ICT and Government stakeholders who, in turn, helped to make contact with others. The participants were contacted by telephone or email to fix a time and place for a meeting. The authors' university provided the ethics approval. Standard ethical procedures were employed to obtain informed written consent and to ensure anonymity and confidentiality. Also, the permission of all the participants was sought to record the discussions. The FGD and interviews were recorded and transcribed with pseudonyms being used in place of real names.

RESULTS

When aggregated, the outcomes from the focus group sessions provided a ranked list of barriers to the adoption of ICT into the Bangladesh public sector. This ranking is important, because it resulted from an in-depth discussion during the group sessions and is not the result of superficial reflections. Figure 1 shows the analysis of ranking scores aggregated for each barrier. The barriers are clustered into three groups in terms of their relative importance. The first cluster contains one item, "Lack of Knowledge" (26%), which was by far the strongest identified barrier. The next cluster consisted of "attitude and mindset of decision makers" (2nd), "lack of political will and leadership" (3rd), "lack of planning and strategy" (4th) and "infrastructure" (5th). The third cluster was formed by a number of barriers, "bureaucratic business process" (6th), "lack of expertise

and professionals" (7th) and "socio-economic conditions" (8th). "Lack of laws and rules", "lack of citizen demand" and "lack of championship".

The results from the sessions taken individually are also interesting. The rankings were very similar across all the seven sessions, although they were conducted at different times and with different stakeholder groups. "Lack of knowledge" was ranked first in all sessions except session 1 and 4, where it was close to infrastructure or planning and strategy. The fact that similar views were fairly generally held across the diverse groups supports the view that these barriers are, in fact, real obstacles to progress.

A detailed description of each barrier follows, based on the in-depth analysis of the focus group and interview data.

Barrier 1: Lack of Knowledge

"Lack of knowledge" was overwhelmingly rated as the most important barrier with almost twice the score of the second barrier, "attitude and mindset". Lack of knowledge entails not only the lack of basic knowledge and education on ICT, but also implies a lack of perception and awareness amongst the leaders, stakeholders, government officials and the general public as a whole about ICT use and its implications. Figure 2 shows the sub-categories that comprised "Lack of Knowledge".

It was interesting to note that most of the discussions about other commonly-perceived barriers ultimately came back to this lack of awareness and correct knowledge about ICT and its use. The participants cited a number of examples, such as missing the great opportunity for Bangladesh to be connected with the rest of the world by submarine cable in the early nineties, on the pretext of a security risk. It was due to the lack of required knowledge and awareness that this milestone proposal was turned down. Now, more than a decade later, after spending a lot more than the initial expenditure it has finally been

Figure 2. Components of the knowledge barrier

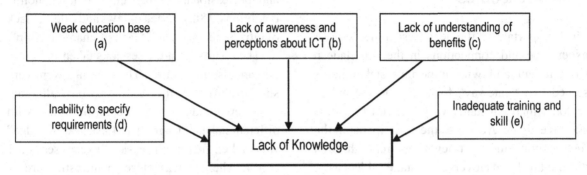

connected. By not connecting earlier, the country not only lost costly foreign currency, but also lost all those years that left the country far behind internationally (FGDs 2 and 3).

A weak and inadequate education system, especially in Information Systems (IS) and Information Technology (IT), is failing to produce the required suitable workforce for the country. Also, the public sector environment and the institutional resources are not promoting such knowledge, as one of the respondents said, "There's no knowledge base for supporting the e-government concept within the government" (FGD-2). Some justification for this claim was observed in another FGD, where a mid-level government official raised the question, "How can the e-government be applicable to the water resource ministry where there is no direct citizen dealing?" (FGD-1). Although, in some cases, clarification of such basic doubts consumed much of the FGD time, it also reflected similar perceptions held by many government officials that appeared in subsequent FGDs and interviews. Common misconceptions have emerged, largely because of the weak foundations of the education system where the curriculum is still very limited in ICT and, in most disciplines from where maximum public servants come from, ICT curriculum is totally absent. As such, the business value of ICT is not commonly understood, even by many tertiary-educated people in the country who still consider ICT to be a luxury. A respondent from

the ICT industry group (FGD-5) said, "The weakness in the basic education system is a major barrier — graduates are not capable of taking over the challenges of [the] IT industry". Many of the government officials who joined the public service after studying in public universities seldom come into contact with computers. Timidity and ignorance about computers remain a problem compounded by the existing work environments and the business process inherited by the legacy system. Those officials who have some interest usually do not get support from superiors who are mostly conservative and not "IT-savvy". The senior echelons of the ministry are usually not computer literate; thus, they do not have a general aptitude for developing computer-based business processes. There is some evidence of an existing fear that they might lose their grip over their staff. According to an expert from an international organization working in Bangladesh, most of the high officials in the public sector were educated 30–40 years ago, when ICT did not play any role, and it is their children and grandchildren who are now receiving ICT education.

There were many examples cited about the lack of awareness amongst the top-level government officials, where many of them either consider computers as a super typewriter or as a threat to their existing system. Another participant said:

Most of the government officers, who are known as pro-ICT, are actually pushing for hardware — more computers, to use them mainly as typewriters and not beyond that. Leaders are not serious because there is no public awareness. If you look at the newspapers, ICT occupies a very unimportant corner that mostly talks about hardware. Many people think it's kids play (FGD-4)

On the citizen side, language appeared to be a major obstacle. The computer's basic language, English, is not given much importance in the current education system. Software industry leaders in FGD-5 said they were losing many international markets because of the language barrier, allowing neighbouring countries (mainly India) to take the lead. On the other hand, many academics do not feel comfortable with computers because of their lack of skill and knowledge of ICT. A participant concluded by saying:

It is not necessarily a lack of willingness or a lack of sincerity. Many people are very sincere about ICT but it is more of an [a lack of] understanding from the management perspective of what [value] these tools can bring (FGD-4)

Barrier 2: Attitudes and Mindset

After knowledge, attitude and mindset were identified as the most important barrier by most of the groups. The typical mindset, especially amongst the decision-makers and government officials that are yet to be tuned into the modern ICT environment and work processes, is holding back ICT adoption in the public sectors. One respondent (FGD-4) termed mindset as an "operational arm" where another participant (FGD-2) commented "government officials do not want to be under any controlling system like computer software". The following sub categories together with the findings and quotations from the participants further explain the issues surrounding this category:

- **Motivation:** Motivation is completely missing among the major users and decision-makers where ICT coincides minimally with their personal benefit and agendas; rather, ICT poses a threat in some cases. One senior ICT Project Manager said:

People [Government servants] who are 45 plus, they don't like to use the system because they have a feeling, today or tomorrow, "I'll be transferred or promoted, I'll be retired, so what do I do with the new technology or a new system? It's not going to help me. My old system is good enough and I don't need to go into a new thing (FGD-6)

The lack of research showing the bottom line figures, the cost benefit analysis and the media's silence were also blamed for the failure of motivation in this regard, very closely linking this category to the knowledge category.

- **Resistance to change:** The politicians and the decision-makers were seen as not yet mentally ready to change the present status quo. The following quotes from the informants amply cover the aspect of resistance to change, "We have to recognize the fact that ICT adoption or e-government is not going to be easy. There will be resistance because it will eliminate inefficiency, it will eliminate jobs and above all it will eliminate corruption" (Interviewee-1).

A senior ICT advisor and academic commented in his interview, "To me it seems, although there will be resistance, because Government has a hangover, maybe because of centuries of working in colonial system, entire change will be difficult, but in some areas it may be possible" (Interviewee-3).

However, in his interview, another ex-senior executive from the Government mentioned while describing a successful implementation:

There was resistance from people, but I installed [software system] it. Even after installation there was a lot of hue and cry, its getting late, this and that is happening etc. Initially, everywhere there will be some problem, but gradually it becomes smooth (Interviewee-2)

My information going out means I'm nowhere. If I give that information on the web and if he gets the information easily, then that person will not come to me anymore. If he doesn't come to me then what value I've got sitting here, powerless! (Interviewee-10)

A veteran ICT specialist, while citing an example of an innovation in a government business process by a mid level officer, said:

The person who has spearheaded that one [software project] has been transferred. Because many people were unhappy; they thought the power they had may be curtailed, so very quickly he has been transferred to some other positions where he'll not be able to do it again. This is another case study where the system reacts against a person who tries to change it (Interviewee-3)

- **Fear of job loss:** This fear stemmed from the belief that if juniors became more efficient or if a system was too automated, then it would create unemployment for many. One of the government officials commented, "I think the staff resistance does not matter much. The main concern of the staff is whether he is secure or not. Along with getting a computer if he feels his job is at stake, only then he becomes worried, otherwise not" (Interviewee-10).

Barrier 3: Political Will and Leadership

The lack of political will and leadership scored third highest as a perceived barrier. According to a senior participant, for the last 10 years ICT was paid mainly "lip service only" and hardly any significant change occurred (FGD-7). Initiatives from the top level were believed to be crucial for ICT adoption, but many had concerns similar to this one, "senior levels, those who are on top, most of them are not interested in this matter" (FGD-1) and, "it seems no one is taking it very seriously. It's something like, other countries are doing; that is why we're also doing" (Interviewee-3).

Owing to the lack of political commitment, this sector is being ignored and given less priority in terms of financing and national strategy. As another respondent (FGD-7) said, "There are lots of interests amongst junior levels, but the drive has to come from the top. Many times a junior's initiative dies down owing to the lack of political will from the top". A senior ex-government officer and an ICT patron commented:

Because, they [the politicians in power] come here for five years. The first two years pass on past legacy and another one year for the budgetary discussion. Then, they are actually left with one effective year and in that one year they start preparing for the next election. So, whatever they are thinking they are thinking very haphazardly and they are thinking like, "OK World Bank is saying unless there is a transparency and e-governance, we'll not give you money", so suddenly, they become active. These are pointless, meaningless. (FGD-1)

Barrier 4: Lack of Planning and Strategy

The international organization ICT expert group rated this one as the major problem when they emphasized proper policy with achievable objectives in Bangladesh. Overall, this barrier scored fourth. The country does not have any master plan or road map for ICT, what has been achieved has occurred in a piecemeal or on an "ad hoc" basis. Comprehensive planning and strategy is missing on how to approach the whole issue. Although the Government has set up an ICT task force, a national committee and other initiatives, a systematic, coordinated approach towards ICT is absent. A long-term and a short-term vision to make the e-government a reality is not evident. A participant (FGD-4) commented that "ICT, unfortunately, is not seen as a strategic resource here". A senior ICT stakeholder expressed his frustration in this way:

In our country, there's IT policy, but there's no implementation, because the IT policy has never been translated into its action plan. We have policy but no strategy [agitated]... this is a fact, just go and see what sort of policy we have and what actions we've taken, there is no match (Interviewee-4)

Barrier 5: Infrastructure

Expatriate Bangladeshis identified infrastructure as the number one barrier together with attitude and mindset. Interestingly, infrastructure was not considered within the first four barriers by any of the groups in Bangladesh. This speculation amongst the expatriates may be driven mainly by their tendency to compare the highly-developed infrastructure they experience in a developed nation such as Australia and their long time detachment from the current realities in Bangladesh. Although some participants mentioned the legal infrastructure, knowledge infrastructure and business infrastructure, this category refers to the physical infrastructure: the network backbone capability, broadband, Internet connectivity, tele-density and so on. Many government offices do have PCs, but do not have network functioning and in many cases, officials do not get a PC. Consequently, a large group of people do not have access to a computer or the Internet. However, most of the participants believed that the infrastructure barrier could be overcome if other important barriers were removed and an adequate demand could be cre-

ated. One of the senior officials emphasized very strongly that resources actually are not a problem, rather "willingness" (to embrace the technology) is more important for adoption (FGD-7). A further justification of this statement was also found in the discussion on the submarine cable that took ten years for the government to understand and take a decision on, after it was first proposed in the early nineties.

Barrier 6: Existing Bureaucratic Processes

The legacy of existing business processes and the lack of accountability in the public sector were identified as a problem by at least 10 respondents (about 25%) and ranked 6th in terms of its severity. Discussions from various focus groups highlighted the existing bureaucratic processes that are incompatible with ICT-based smooth and transparent business processes. One ICT professional commented, "Whenever we want to do a good job, bureaucracy hinders that" (FGD-1). Another participant said, "It takes three days just to start the processing of a file in a government office" (FGD-7). There was seen to be no effective planning for restructuring and re-engineering the bureaucratic business processes inherited from the colonial era. Modern distributed information systems are often seen as a threat to the hierarchy of the organization. According to a prominent think-tank, "Bangladesh is becoming less and less competitive compared to its neighbours, owing to the poor quality of governance and bureaucracy in the civil service that is not properly trained to deal with contemporary challenges (Interviewee -5). Sub-components of this category are as follows:

- **Lack of Accountability and Transparency:** In the present system, there is no proper accountability and transparency in the way the government functions. According to an ex-bureaucrat, "the politician and the decision-makers are obviously not ready to share all the information as of today and subject themselves to scrutiny" (Interviewee-1). However, another senior executive and ex-government officer commented,

I don't think bureaucracy will be able to make or create a bottleneck in that sense. They create bottlenecks because they know that they are not accountable, so they don't bother. They think, my job is to finish the tenure, that is, in two years time (FGD-1)

The main problem in the government offices is that after someone launches an application, he can't trace out the file. Again if you can trace out, there are so many steps that you don't understand the status of that application. (Interviewee-10)

Barrier 7: Lack of Experts and Professionals

The lack of experts and professionals in the translating and implementing of ICT projects was identified as an important barrier by academics and government groups. The lack of qualified system analysts and experts to handle the computer-based information systems is quite evident. The lack of skilled personnel in system architecture and system planning, together with an experienced and skilled workforce in service support and troubleshooting, is posing problems in implementing and sustaining the use of ICT. Institutions are failing to produce proper ICT human resources in accordance with the industry requirements; consequently, many of the local software firms are importing people from neighbouring countries, whereas there are scores of educated jobless people underused in the country. Most of the respondents deliberately blamed the education system for this situation.

At the ICT industry level, the professional and qualified ICT consultants who can provide

comprehensive strategy and planning for a project are rare. Moreover, in many government organizations, the wrong people are put in the wrong places, thus resulting in the failure to provide expertise and correct direction. At the project implementation level, because of poor retention capability, experts are switched in the middle of a project. As one of the interviewees said, "when the project was in mid level, their [Vendor's] Oracle specialist was taken away by some other company on hire. Then our whole project was at stake" (Interviewee-10).

Barrier 8: Socio-Economic Condition

Another common problem, the socio-economic condition of the LDCs, was not ranked highly (8th) compared with other barriers - possibly because the poor socio-economic condition of Bangladesh is a known or given fact. That is, it is rather an effect than of a cause and ICT could be a means to improve this condition. As such, many participants did not consider it a major barrier. However, a few were skeptical, for example,

The LDC is a labour-intensive economy, and the outside world is a capital economy. Here, they employ thousands of people with hammers to break a building. In the West, they do it within a few minutes by dynamite. It (the former) may be more effective here. (FGD-3)

Policy-makers tend to prioritise the basic needs of the citizens. ICT is an unknown phenomenon amongst the majority of the population; therefore, it is often not high on the policy agenda. As one respondent commented, "That's why the lack of stress is coming here. The link has to be created with their benefits that need to transform into a political commitment" (FGD-2). Another respondent was rather optimistic in saying, "If the interest can be grown amongst them in whatever ways, they will merge themselves with it, and they will induce themselves with it" (FGD-7). The heavy penetration of the mobile phone at all levels is an example, although its fast adoption was also driven by other factors, such as its ease of use, with voice being used instead of typing and there being no requirement to use the English language.

Barrier 9: Laws and Rules

Legal infrastructure and administrative reform is a precondition for the correct implementation of ICT-oriented business processes. In one of the focus groups, one government official raised doubts about the feasibility of ICT use because it may contradict government rules and procedures including the Official Secret Act followed since the British period. An ICT project director commented, "If we want to go to the next step, we have to change the law and the Acts; and if these benefits are not given, citizens will not accept that" (FGD-7). Still, some viewed it as a demand-driven activity, such as, "Cyber law will be there when cyber activity will start" (FGD-4). However, it was agreed by most of the participants that without some digital certification authority, and the acceptance of online user registration by the court, ICT can not be put into full operation in day-to-day business processes. Regarding the progress of the cyber law, one official commented, "For the last two and half years, the file is going from down to top and top to down, but it's not yet done" (Interviewee-4).

Barrier 10: Citizen Demand

Some participants, especially academics from other disciplines, considered the lack of citizen demand as a major barrier, because the vast majority of Bangladesh people are struggling to fight for the basic needs of life such as food, shelter and medicine. On the other hand, the same participants were found to be not aware of the huge impact ICT could have on the overall economy. According to

these participants, the vast majority of voters are yet to be exposed to ICT and its benefits. Thus, the politicians do not consider ICT to be a strategic tool: "Unless pressure comes from the citizens, it's difficult to bring it to the forefront" (FGD-1).

But the participants acknowledged that people use ICT when they are given adequate exposure to the technology and when they see its benefits. "When the people will find value in it, they will equally adopt it in faster way... maybe these case studies or information is not flowing from one place to another" (FGD-2). When the last SSC (Secondary School Certificate) exam results were made available online for the first time in Bangladesh, many people were skeptical that anyone would use it, but in fact, there were long queues in front of the call centre and cyber cafes, wherever they were available across the country. Similarly, when the United States of America DV Lottery (Diversified Immigration Lottery) procedure was changed to an online application, villagers travelled to district headquarters to send their applications online. One ICT industry representative added that they had to import thousands of scanners overnight because of the huge demand and because a photo scan was required for visa applications. Participants from the international organizations also suggested, "If you create enough critical mass at the grass roots, you can have enough of demand for it. A demand-driven approach can be created" (FGD-4). In other discussions, the need to create human resources and education were identified as the key factors to redress this issue. Some participants also highlighted the lack of awareness and demand amongst the civil service as being crucial, because they usually spearhead the wider population demand overall.

Barrier 11: ICT Championship and Models

The lack of championship and models was ranked 10th in this study, because the respondents from the

ICT industry and international organizations were the only groups who had identified it as a barrier. They argued that creating champions amongst the government officials with proper recognition could show the path to the others. ICT champions among the government officials in many other developing countries have created significant impacts (Wilson, 2004), but such champions amongst the government officials in Bangladesh are rarely seen. The respondents highlighted the lack of role models and "demonstration through quantifiable measurement" in Bangladesh that could convince others and be able to create a significant awareness.

The lack of incentives and rewards was thought to be inhibiting the growth of championship in this area of Bangladesh. Some senior government officials have shown leadership by taking on the role of ICT champions. For example, the CEO of a government organisation in Bangladesh led a successful implementation of an ICT-based business process within a minimum time frame. His personal initiative, compulsion and commitment till the end of project execution provided a good example of ICT championship to others (FGD-7).

The need for and importance of strong championship at the top organisations was also emphasized in different interviews. People often cited the example of Malaysia, where the former Prime Minister, Mahatir Mohammad, played a significant role in the e-government initiatives. But skepticism was also expressed by one interviewee, who hinted at the influence of the political climate and the environment on the development of such a championship in Bangladesh, "There may sometimes be champions coming out who would like to introduce these kinds of things, but whether those champions will survive or not within the context, that is another issue" (Interviewee- 5).

In addition to the eleven major barriers discussed above, some participants also recognized other problems, such as, little focus on work process redesign and reengineering systems, the

lack of rewards and punishments, the ownership problems caused by the transfer of government officials, the lack of ICT support services, the absence of a dedicated organization to oversee e-government and vested interests in project creation and implementation.

A PROCESS MODEL OF E-GOVERNMENT ADOPTION

The barriers and their descriptions provide a rich picture of the ICT environment in the public sector of Bangladesh. Figure 3 shows a concept map that depicts the interrelationships between the barriers. The map was developed in an iterative process where there were cycles of consideration of the data and a search for causal influences with a discussion and reflection on the prior literature. There was also consultation with the stakeholders individually and in groups. However, it should be noted that no new barriers were added in this process: rather, a deeper understanding of their interrelationship was developed. The concept map can be regarded as a type of process model, because it shows the dynamics by which the outcomes and change follow a combination of conditions, actors and events (Newman & Robey, 1992; Langley, 1999).

The original eleven "barriers" were grouped into four main categories, "Barriers/Enablers", "People/Drivers", "Tools/Actions" and the "Status quo". First the "Status quo" represents both the contextual outcome of ICT adoption as well as existing environmental constraints on ICT based business processes. This status quo reflects two major issues: socio economic condition and the existing administrative culture and bureaucratic business process. While these issues can be viewed as barriers or context, they can also be viewed as an outcome of the innovation of ICT. For example, e-government adoption has the potential to change the socio-economic condition (Madon, 2003).

Figure 3. Process Model for eGovernment adoption in LDCs (Note: numbers represent factor rankings)

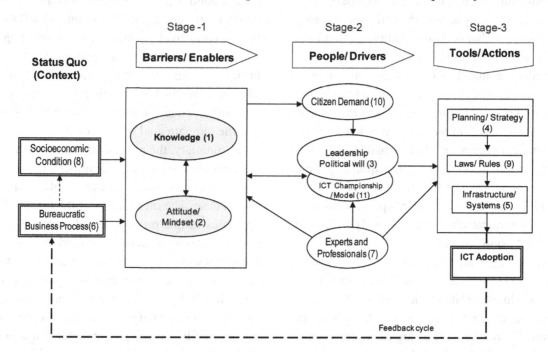

Similarly it will obviously change and streamline the bureaucratic business processes as well as the traditional administrative culture. All the elements of the socio economic condition including improvement of basic health, education, electricity, water, telecommunications networks, governance and the alleviation of poverty are in a way catalyst for new innovations (Musa et al., 2005).

The process model as a whole indicates the action points for resolving the problem of ICT adoption in the initial phases of e-government in an LDC. In the second category, each "barrier" can be viewed as an "enabler" if it is turned around and viewed as an opportunity for a limiting barrier to be overcome.

The steps are shown to give a complete form of a process model (Heeks, 2006b; Langley, 1999; Newman and Robey, 1992), first starting with knowledge deployment and creating a conducive environment for innovation in the public sector organisations and then followed by actual implementation. The intermediate step involves people

drivers who, being influenced by the actions of the first step would generate demand at the citizen level, educate the leaders and decision makers and, create enough of human resource experts and professionals to undertake the next step, which involves sustainable planning and strategies, changing laws and system implementation.

The detailed explanation and the justification of the model are as follows.

The socio–economic condition (8) is unique to the country of investigation, but is seen as influencing the other major barriers/enablers. A sound socio-economic condition (8) encourages knowledge-based activity and educational opportunity (1) that, in turn, influence the attitudes and mindsets (2) of the people and the decision-makers.

In the category of barriers/enabling conditions, the lack of knowledge (1) was found to be the most important barrier and to underlie the attitudes and mindsets of the people about ICT innovations. For example, the inaccurate perceptions of the novelty of ICT still persist on a large scale, with many officials thinking that ICT is akin to typewriting. The existing legacy bureaucratic processes (6) also discourage a change in the traditional mindsets and attitude towards technology.

Citizen demand in the human intermediary category of people/drivers is affected by the lack of knowledge and other barriers. As one respondent said, "Infrastructure problems will be automatically addressed if the demand is there, but to create a demand, training is required, people need to understand and see the benefit of the IT" (FGD-7). Furthermore, leadership and politics (3) will play a crucial role. Leadership and politics will require "vision", which is essential for formulating strategies. The participants did not categorize vision as a separate factor, possibly because it was seen as closely related to leadership. The question arises as to which comes first, knowledge or vision. Many argued that, logically, knowledge is a precondition of vision. By understanding the complexities and the long-term benefits of ICT, a strategic vision can be formulated and this can then be translated into leadership and political will. The ICT championship and IT models (11) also originate from knowledge and grow together with vision and leadership. The influence from leaders and champions can be bidirectional. It was found that while leaders and champions are influenced and motivated by correct knowledge, they often become the source of knowledge as well, because they can play an important role to initiate as well as implement knowledge deployment.

The experts and professionals in ICT (7) were found to be crucial as they can play an important role in different ways: (i) by education and transferring knowledge for production and progression; (ii) by creating awareness and championship amongst the leaders with the help of best practice and models; and, (iii) in the execution stage of planning, strategy, rule setting and building infrastructure and systems.

Finally, it is the leadership and championship that drive the necessary reforms of the existing law and administrative legislation (9) and the planning and strategizing for the future (4). All these factors collectively contribute to the building of adequate infrastructure and systems (5) that lead to improved business processes and ICT adoption. This outcome then in a cyclic process, improves the socio-economic conditions (8), which is usually the ultimate goal of any country.

In summary, the process model shows a possible pathway to break the deadlock of an apparently rigid organisation like the public sector in Bangladesh. The mechanism of the institutional apparatus in this process model was found important. Although there is no hard evidence of a linear relationship between the causal factors, these factors are tied in a mutually-influencing relationship. This cyclic process forms a dynamic structure, the result of which drives or inhibits the adoption of ICT in the public sector in a country such as Bangladesh. This process could be seen as an example of a slow institutional process that might lead to the "social change" from innovation, as conceptualized by Rogers.

DISCUSSION AND CONCLUSION

This study makes important contributions for resolving the challenges to ICT adoption in the public sector of LDCs. This study finds *lack of knowledge* as the underlying root cause of the problems, amongst the host of factors that have been identified as problematic in other studies, including the infrastructure, leadership and socioeconomic conditions. Previous studies sporadically mention different barriers and do not empirically measure the apparent weight of those problems nor do they prioritize the issues, which is important for subsequent planning. First, the ranking of barriers to ICT adoption prioritizes the more important barriers amongst the many perceived problem areas identified in other studies. Thus, this study indicates where effort should be directed to redress the underlying problems that lie at the heart of adopting ICT.

Second, the analysis of the relationships among the major barriers led to a process model that includes important barriers and enabling conditions — the drivers of change and change events — which suggest a pathway for the e-government adoption in an LDC such as Bangladesh. The model resolves the jigsaw puzzle consisting of many and scattered barriers to adoption.

It may be argued that some external influences such as donor and international pressures can initiate a process and innovation bypassing the first step of the process model. Such innovations, however, are found not likely to be successful and sustainable in the long run. For example, 80% of the government projects under "Support to ICT Task force" (SICT) in Bangladesh did not succeed (FGD-7; Interviewee-4). For successful and sustainable e-government innovation the internal institutional processes must be met. While it may be true that in many other circumstances, a technology-first approach can initiate change, in the entrenched public sector culture like Bangladesh, this may not hold true. Because the entrenched institutional forces and traditions of public sector organisations enact technology in ways that do not challenge the existing norms (Dawes, 2002b). This, perhaps, puts ICT or e-government innovation at risk and makes it more challenging than that of private sector organisations, where the traditions and institutions of public sectors impedes the very changes that are necessary to produce the real benefits from ICT.

Again, public sectors in LDCs consist of decision makers and powerful stakeholders with significant authority and power, who are often not interested in new technology. However, their motivation and understanding is crucial in order to embrace change. For example, senior bureaucrats and leaders off and on visit western countries and witness technological developments first hand, but are not necessarily interested in taking the initiative to apply these same technologies in

their context (Interviewee-5, FGD-6). So while there may be some overlap between step one and two, it may be argued that without focusing and concentrating on step one, the other two steps will not be accomplished effectively and successfully.

The study is congruent with the prior theoretical work, but also provides an advance on what was previously known. The findings support the arguments by Wilson (2004), Montealegre (1999) and others concerning the complexity of the environment and the interplay among the factors that drive the adoption process for ICT in an LDC. However, national culture, a factor mentioned in other literature, was not emphasized in this study, presumably because of the situation of the participants in one culture. However, some of the respondents did mention issues such as ICT's origination in the West, language problems and the need to reform the work culture prevailing in the country.

The identification of the knowledge factor as being of major importance was relatively hidden before this study. For example, although Rogers' innovation theory stresses the importance of the initial knowledge acquisition stage for individuals, his theory does not do the same for organizational adoption; perhaps, because that aspect of the theory was developed in more advanced technological societies (unlike Bangladesh) where a minimum level of knowledge, awareness and understanding is taken for granted. In the LDC context, this knowledge may be missing even at the senior decision-making levels in an organization. This missing element might not be readily acknowledged by many, because it would be seen as derogatory to people in a high position. Thus, understanding the nature of the knowledge barrier in the Western world is different from that in an LDC. Our point is that more attention should be paid to the knowledge problem. Rogers' theory also does not encompass other important factors related to knowledge in our model, including leadership and political will and planning and strategy.

Scott's institutional theory (2001) complements Rogers' theory because it suggests some distinct mechanisms (coercive, normative, mimetic pressures) to break the status quo and bring about change in organisations through innovations like ICT. Our findings are consistent with Scott's slow institutional process, which is likely to be more sustainable over a period of time. Institutional theory has many aspects, including economic, sociological and neo-institutional views. The aspect that is relevant to this study is that of the institutional mechanisms employed to bring about change within organisations. Institutional theory prescribes the institutional pillars or pressures and explains why mechanisms such as coercive and normative pressures are important for making expected changes within the organisation. However, while institutional theory talks about the mechanisms and processes of institutionalization, it does not identify the specific factors to be given priority in a particular context. It does not emphasize the knowledge development factor that has emerged as being of vital importance in the context of an LDC, where the knowledge of the ICT innovations is at a level far below that commonly found in Western institutions.

In other regards, our findings are congruent with institutional theory. Leadership and political will could be an important coercive pressure that could change institutional inertia through a strong mandate or an extensive knowledge deployment from the top. Clear leadership from the top is of increasing importance in organizations in a country such as Bangladesh where the ideas of hierarchy are entrenched and the power distance is high. We believe this would be a long-term institutional process rather than a quick fix that has to be accepted by the different stakeholders (Avgerou, 2000). With knowledge deployment and the right approach, a conducive environment can be created for innovation where it would be easy to mobilise political or other actors (Sein et al., 2008). We suggest that such a process, if acted upon directly, could avoid the large number of failures that have been associated with the techno-centric approaches.

The findings from an individual country case study such as that described here provide a basis for further studies to locate similarities in other countries and the opportunity to develop a more general theoretical model for all LDCs, subject, of course, to modifications to suit the local conditions. The causal model provides the foundations for a mid-range theory for e-government adoption in an LDC (Merton, 1968; Miles & Huberman, 1994).

Although the scope of this study did not permit investigating more than one LDC, the findings may be transferable to other situation with due regard to the context.

From a practitioner's perspective, the process model of a relatively complex phenomena can help educate agents to understand the dynamics of the adoption process and provide guidance on where and when to apply effort. We believe this model is a good starting point for breaking the deadlock of the e-government adoption problem. The model can be validated in action research in Bangladesh and other LDCs. In fact, we have used the model with considerable success already in an applied action research project in Bangladesh in 2008 (Imran et al., 2009), which was further enhanced and being implemented in 2010-todate.

To summarise, we have provided a process model to deal with the problem of e-government adoption in an LDC. This process model is more specific than the higher-level frameworks that have previously been advanced in that it describes in detail the key barriers to adoption. This model focuses attention on "lack of knowledge" as the underlying root cause of the problems, amongst the host of factors that have been identified as problematic in other studies, including the infrastructure and socio-economic conditions. Furthermore, the process model advanced provides a basis for the "ice breaking" actions that can lead to more successful e-government in the relatively rigid public sector world of an LDC.

ACKNOWLEDGMENT

The earlier version of this paper appeared as Imran, A., Gregor, S. (2010). Uncovering the hidden issues in e-Government Adoption in A Least Developed Country: The Case of Bangladesh. Journal of Global Information Management, 18(2): 30-56.

REFERENCES

Adam, M., & Myers, M. (2003). *Have you got anything to declare? Neo–colonialism, information systems, and the imposition of customs and duties in a third world country.* Paper presented at the International Federation of Information Processing, IFIP 9.4 and 8.2 Joint Conference on Organizational Information Systems in the Context of Globalization. Dordrecht, The Netherlands: Kluwer.

Afonso, A., Schuknecht, L., Tanzi, V., & Veldhuis, N. (2007). *Public sector efficiency: An International comparison.* The Fraser Institute. Retrieved September 7, 2011, from http://www.fraserinstitute.org/research-news/display.aspx?id=13328

AGIMO. (2009). *Interacting with government: Australians' use and satisfaction with e-government services.* Department of Finance and Administration, Australian Government Information Management Office, Commonwealth of Australia. Retrieved September 12, 2011, from http://www.finance.gov.au/publications/interacting-with-government-2009/ docs/ interacting-with-government-2009.pdf

Allan, J., Rambujan, N., Sood, S., Mbariak, V., Agrawal, R., & Saquib, Z. (2006). The e-government concept: A systematic review of research and practitioner literature. *IEEE, 4-9*(06).

Attewell, P. (1992). Technology diffusion and organizational learning: The case of business computing. *Organization Science, 3*(1), 1–19. doi:10.1287/orsc.3.1.1

Avgerou, C. (1996, August 16-19). Studying the socio–economic context of information systems. In *Proceedings of the Second Americas Conference on Information Systems.* Phoenix, Arizona.

Avgerou, C. (2000). IT and organizational change: An institutionalist perspective. *Information Technology & People, 13*(4), 234–262. doi:10.1108/09593840010359464

Avgerou, C. (2002). *Information systems and global diversity.* Oxford, UK: Oxford University Press.

Avgerou, C. (2009, May 6-28). *Discourses on innovation and development in information systems in developing countries' research.* Paper presented at the 10th International Conference of the IFIP 9.4 working group on Social Implications of Computers in Developing Countries Dubai, UAE.

Avgerou, C., & Walsham, G. (Eds.). (2000). *Information technology in context: Implementing systems in the developing world.* Brookfield, VT: Ashgate Publishing.

Backus, M. (2001). *E-Governance and developing countries – Introduction and examples* (Research Rep. No. 3). Retrieved July, 10, 2007, from http://www.ftpiicd.org/files/research/reports/report3.pdf

Bada, A. O., Aniebonam, M. C., & Owei, V. (2004). Institutional pressures as sources of improvisations: A case study from a developing country context. *Journal of Global Information Technology Management, 7*(3), 27–42.

Bagozzi, R. P., Davis, F. D., & Warshaw, P. R. (1992). Development and test of a theory of technological learning and usage. *Human Relations, 45*(7), 659–686. doi:10.1177/001872679204500702

Banerjee, P., & Chau, P. Y. K. (2004). An evaluative framework for analysing e–government convergence capability in developing countries. *Electronic Government, 1*(1), 29–48. doi:10.1504/EG.2004.004135

Bazely, P. (2007). *Qualitative data analysis with NVIVO*. London, UK: Sage.

BBS. (2009). Statistical pocket book of Bangladesh, 2009. *Bangladesh Bureau of Statistics: Planning division, Ministry of Planning, Government of the People's Republic of Bangladesh*. Retrieved September, 7, 2011, from http://www.bbs.gov.bd/WebTestApplication/userfiles/Image/SubjectMatterDataIndex/pk_book_09.pdf

Boudreau, P. (2000). L'expertise d'un enseignant associe. *McGill Journal of Education, 35*(1), 53–70.

Braa, J., Monteiro, E., & Sahay, S. (2004). Networks of action: Sustainable health information systems across developing countries. *Management Information Systems Quarterly, 28*(3), 337–362.

Bretschneider, S. (1990). Management Information Systems in public and private organizations: An empirical test. *Public Administration Review, 50*(5), 536–545. doi:10.2307/976784

Brown, L. (1981). *Innovation diffusion*. London, UK: Methuen.

Chau, P. Y. K. (2008). Cultural differences in diffusion, adoption and infusion of Web 2.0. *Journal of Global Information Management, 16*(1), i–ii.

Chen, Y. N., Chen, H. M., Huang, W., & Ching, R. K. H. (2006). E-government strategies in developed and developing countries: An implementation framework and case study. *Journal of Global Information Management, 14*(1), 23–46. doi:10.4018/jgim.2006010102

Damanpour, F. (1987). The adoption of technological, administrative and ancillary innovations: Impact of organizational factors. *Journal of Management, 13*, 675–688. doi:10.1177/014920638701300408

Davis, F. D., Bagozzi, R. P., & Warshaw, P. R. (1989). User acceptance of computer technology: A comparison of two theoretical models. *Management Science, 35*(8), 982–1003. doi:10.1287/mnsc.35.8.982

Dawes, S. (2002a). *The future of egovernment. An examination of New York City's e-government initiatives*. Retrieved August 18, 2011, from http://www.ctg.albany.edu/publications/reports/future_of_egov/future_of_egov.pdf

Dawes, S. (2002b). *Government and technology: User, not regulator. Book Review: Fountain. 2001. Building the virtual state: Information Technology and institutional change*. Washington, DC. Retrieved August 29, 2011, from http://jpart.oxfordjournals.org/cgi/reprint/12/4/627.pdf

Dawes, S., Pardo, T., Simon, S., Cresswell, A., LaVigne, M., Andersen, D., et al. (2004). *Making smart IT choices: Understanding value and risk in government IT investments*. Retrieved August 29, 2011, from http://www.ctg.albany.edu/publications/guides/smartit2

Delbecq, A. L., & Van de Ven, A. H. (1971). A group process model for problem identification and program planning. *The Journal of Applied Behavioral Science, 7*(4), 466. doi:10.1177/002188637100700404

DiMaggio, P. J., & Powell, W. W. (1983). The iron cage revisited: Institutional isomorphism and collective rationality in organizational fields. *American Sociological Review, 48*(2), 147–160. doi:10.2307/2095101

Dunham, R. B. (1998). *Nominal group technique: A users' guide*. School of Business, University of Wisconsin. Retrieved September 7, 2011, from http://www.peoplemix.com/documents/general/ngt.pdf

Eisenhardt, K. M. (1989). Building theories from case study research. *Academy of Management Review, 14*(4), 532–550.

Fishbein, M., & Ajzen, I. (1980). *Understanding attitudes and predicting social behaviour*. Eaglewood Cliffs, NJ: Prentice Hall.

Flamm, K. (1987). *Targeting the computer: Government support and international competition*. Washington, DC: Brookings Institution.

Grant, G., & Chau, D. (2005). Developing a generic framework for e-government. *Journal of Global Information Management, 13*(1), 1–30. doi:10.4018/jgim.2005010101

Gregor, S., Imran, A., & Turner, T. (2010). *Designing for a "'sweet spot' in an intervention in a least developed country: The case of e-government in Bangladesh*. Paper presented at the SIG GlobDev Third Annual Workshop, Saint Louis, USA.

Hagenaars, M. (2003). *Socio-cultural factors and ICT adoption*. Retrieved September 7, 2011, from http://www.iicd.org/articles/IICDnews.import2069

Heeks, R. (2004). *Basic definitions page*. eGovernment for Development project. IDPM, University of Manchester, UK. Retrieved June, 10, 2010, from http://www.egov4dev.org/success/definitions.shtml

Heeks, R. (2006a). *Implementing and managing e-government*. London, UK: Sage.

Heeks, R. (2006b, July 27-28). *Understanding and measuring egovernment: International benchmarking studies*. Paper presented at the UNDESA workshop on E-Participation and E-Government: Understanding the Present and Creating the Future, Budapest, Hungary

Heeks, R., & Bhatnagar, S. C. (Eds.). (2001). *Understanding success and failure in information age reform*. London, UK: Routledge.

Hill, C. E., Loch, K. D., Straub, D. W., & El-Sheshai, K. (1998). A qualitative assessment of Arab culture and information technology transfer. *Journal of Global Information Management, 6*(3), 29.

Hofstede, G., & McCrae, R. R. (2004). Personality and culture revisited: Linking traits and dimensions of culture. *Cross-Cultural Research, 38*(February), 52–88. doi:10.1177/1069397103259443

Imran, A., Gegor, S., & Turner, T. (2009, May 26-28). *eGovernment capacity building through knowledge transfer and best practice development in Bangladesh*. Paper presented at the 10th International Conference on Social Implications of Computers in Developing Countries, International Federation for Information Processing (IFIP), Dubai.

Indjikian, R., & Siegel, D. S. (2005). The impact of investment in IT on economic performance: Implications for developing countries. *World Development, 33*(5), 681. doi:10.1016/j.worlddev.2005.01.004

ITU. (2009). *Information society statistical profiles 2009 - Asia and the Pacific*. Market Information and Statistics Division within the Telecommunication Development Bureau. Retrieved from http://www.itu.int/ITU-D/ict/material/ISSP09-AP_final.pdf

Jamil, I. (2007). *Administrative culture in Bangladesh*. A H Development Publishing House.

Joppe, M. (2004). *The nominal group technique.* Retrieved from September 7, 2011, from http://www.uoguelph.ca/htm/MJResearch/ResearchProcess/841TheNominalGroupTechnique.htm

Kandelin, N. A., Lin, T. W., & Muntoro, K. R. (1998). A study of the attitudes of Indonesian managers toward key factors in Information System development and implementation. *Journal of Global Information Management, 6*(3), 17–26.

Kankanhalli, A., & Kohli, R. (2009, July 10-12). *Does public or private sector matter? An agenda for IS research in e-government.* Paper presented at the Pacific Asia Conference on Information Systems Hyderabad, India

Kelegai, L., & Middleton, M. (2004). Factors influencing information systems success in Papua New Guinea organisations: A case analysis. *Australasian Journal of Information Systems, 11*(2), 57–69.

Keng, S., & Yuan, L. (2006). Using social development lenses to understand e–government development. *Journal of Global Information Management, 14*(1), 47. doi:10.4018/jgim.2006010103

Korpela, M. (1996). Traditional culture or political economy? On the root causes of organizational obstacles of IT in developing countries. *Information Technology for Development, 7*(1), 29. doi:10.1080/02681102.1996.9627212

Kumar, R., & Best, M. L. (2006). Impact and sustainability of egovernment services in developing countries: Lessons learned from Tamil Nadu, India. *The Information Society, 22*, 1–12. doi:10.1080/01972240500388149

Langley, A. (1999). Strategies for theorizing from process data. *Academy of Management Review, 24*(4), 691–710.

Legris, P., Ingham, I., & Collerette, P. (2003). Why do people use information technology? A critical review of the technology acceptance model. *Information & Management, 40*(3), 191–204. doi:10.1016/S0378-7206(01)00143-4

Lindquist, K. M., & Mauriel, J. J. (Eds.). (1989). *Depth and breadth in innovation implementation: the case of school–based management.* New York, NY: Harper and Row.

Liu, W., & Westrup, C. (2003). *ICTs and organizational control across cultures: The case of a UK multinational operating in China.* Paper presented at the International Federation of Information Processing, IFIP 9.4 and 8.2 Joint Conference on Organizational Information Systems in the Context of Globalization Dordrecht, The Netherlands.

Lundblad, J. P. (2003). A review and critique of Rogers' diffusion of innovation theory as it applies to organizations. *Organization Development Journal, 21*(4), 50–64.

Lyytinen, K., & Damsgaard, J. (2001). What's wrong with the diffusion of innovation theory. *Diffusing Software Products and Process Innovations,* 173-190.

Madon, S. (2003). IT diffusion for public service delivery: Looking for plausible theoretical approaches. In Avgerou, C., & La Rovere, R. (Eds.), *Information systems and the economics of innovation* (pp. 71–88). Cheltenham, UK: Edward Elgar.

Madon, S. (2004). Evaluating the developmental impact of e–governance initiatives: An exploratory framework. *The Electronic Journal of Information Systems in Developing Countries, 20*(5), 1–13.

Marcus, A. A., & Weber, M. J. (Eds.). (1989). *Externally-induced innovation.* New York, NY: Harper and Row.

Markus, M. L., & Benjamin, R. I. (1997). The magic bullet theory in IT–enabled transformation. *Sloan Management Review, 38*(2), 55–68.

Mbarika, V. (2002). Re-thinking information and communications technology policy focus on internet versus teledensity diffusion for Africa's least developed countries. *The Electronic Journal of Information Systems in Developing Countries, 9*(1), 1–13.

Merton, R. K. (1968). *Social theory social structure.* New York, NY: Free Press.

Miles, M. B., & Huberman, A. M. (1994). *Qualitative data analysis.* Thousand Oaks, CA: Sage.

Miscione, G. (2007). Telemedicine in the upper Amazon: Interplay with local health care practices. *Management Information Systems Quarterly, 31*(2), 403–425.

Molla, A., Taylor, R., & Licker, P. (2006). E-commerce diffusion in small island countries: The influence of institutions in Barbados. *The Electronic Journal of Information Systems in Developing Countries, 28*(2), 1–15.

Montealegre, R. (1998). Waves of change in adopting the Internet: Lessons from four Latin American countries. *Information Technology & People, 11*(3), 235–260. doi:10.1108/09593849810228039

Montealegre, R. (1999). A temporal model of institutional interventions for Information Technology adoption in less developed countries. *Journal of Management Information Systems, 16*(1), 207–232.

Morton, M. S. (Ed.). (1991). *The corporation of the 1990s: Information Technology and organizational transformation.* New York, NY: Oxford University Press.

MOSICT. (2002). *National information and communication technology (ICT) policy.* Ministry of Science and Information & Communication Technology, Government of the People's Republic of Bangladesh.

Mursu, A., Lyytinen, K., Soriyan, H. A., & Korpela, M. (2003). Identifying software project risks in Nigeria: An international comparative study. *European Journal of Information Systems, 12*(3), 182–194. doi:10.1057/palgrave.ejis.3000462

Musa, P., Meso, P., & Mbarika, V. (2005, August 11-14). *Testing a modified TAM that accounts for realities of technology acceptance in Sub Saharan Africa.* Paper presented at the Americas Conference on Information Systems, Omaha, Nebraska.

Myers, M. D., & Tan, F. B. (2002). Beyond models of national culture in information systems research. *Journal of Global Information Management, 10*(1), 24–32. doi:10.4018/jgim.2002010103

Nair, M., & Kuppusamy, M. (2004). Trends of convergence and divergence in the information economy: Lessons for developing countries. *The Electronic Journal of Information Systems in Developing Countries, 18*(2), 1–32.

Newman, M., & Robey, D. (1992). A social process model of user-analyst relationships. *Management Information Systems Quarterly, 16*(2), 249–266. doi:10.2307/249578

Nidumolu, S. R., Goodman, S. E., Vogel, D. R., & Danowitz, A. K. (1996). Information technology for local administration support: The Governorates Project in Egypt. *Management Information Systems Quarterly, 20*(2), 197–224. doi:10.2307/249478

Putnam, R. D. (1993). *Making democracy work: Civic traditions in modern Italy.* Princeton, NJ: Princeton University Press.

Rainey, H. G. (1983). Public agencies and private firms: Incentive structures, goals and individual roles. *Administration & Society, 15*, 207–242. doi:10.1177/009539978301500203

Raman, K. S., & Yap, C. S. (1996). From a resource rich country to an information rich society: An evaluation on Information Technology policies in Malaysia. *Information Technology for Development, 7*(3), 109. doi:10.1080/02681102.1996.9525277

Roche, E. (1996). Closing the technology gap: Technological change in India computer industry: Global Information Technology and socio-economic development. *Journal of Global Information Management, 4*(3), 29.

Rocheleau, B. (2006). *Public management Information Systems*. Hershey, PA: IGI Global.

Rodriguez, F., & Wilson, E. J. (2000). *Are poor countries losing the information revolution?* (InfoDev Working Paper). Global Information and Communication Technologies Department.

Rogers, E. M. (2003). *Diffusion of innovations* (5th ed.). New York, NY: Free Press.

Sample, J. A. (1984). Nominal group technique: An alternative to brainstorming. *Journal of Extension, 22*(2).

Sanford, C., & Bhattacherjee, A. (2007). IT implementation in a developing country municipality: A sociocognitive analysis. *Journal of Global Information Management, 15*(3), 20. doi:10.4018/jgim.2007070102

Scholl, H. J. (2005). *Organizational transformation through egovernment: Myth or reality?* Berlin, Germany: Springer Verlag.

Scott, W. R. (2001). *Institutions and organizations*. Thousand Oaks, CA: Sage.

Sein, M., Ahmad, I., & Harindranath, G. (2008). *Sustaining ICT for development: The case of Grameenphone CIC*. Telektronikk.

Sharifi, H., & Zarei, B. (2004). An adaptive approach for implementing e–government in IR Iran. *Journal of Government Information, 30*(5-6), 600–619. doi:10.1016/j.jgi.2004.10.005

Shore, B., & Venkatachalam, A. R. (1995). The role of national culture in systems analysis and design. *Journal of Global Information Management, 3*(3), 5–14.

Siau, K., & Long, Y. (2005). Synthesizing e–government stage models – A meta-synthesis based on meta-ethnography approach. *Industrial Management & Data Systems, 105*(4), 443–458. doi:10.1108/02635570510592352

Siddiqui, K. (1996). *Towards good governance in Bangladesh: Fifty unpleasant essays*. Dhaka, Bangladesh: University Press Limited.

Silva, L., & Figueroa, E. B. (2002). Institutional intervention and the expansion of ICTs in Latin America: The case of Chile. *Information Technology & People, 15*(1), 8–25. doi:10.1108/09593840210421499

Stanforth, C. (2006). Using actor-network theory to analyze e-government implementation in developing countries. *Information Technologies and International Development, 3*(3), 35–60. doi:10.1162/itid.2007.3.3.35

Straub, D., Loch, K. D., & Hill, C. E. (2001). Transfer of information technology to the Arab world: A test of cultural influence modeling. *Journal of Global Information Management, 9*(4), 6–48. doi:10.4018/jgim.2001100101

Strauss, A., & Corbin, J. (1990). *Basics of qualitative research*. Newbury Park, CA: Sage.

Suomi, R., & Kastu–Häikiö, M. (1998). Cost–and service effective solutions for local administration – The Finnish case. *Total Quality Management, 9*, 335–346. doi:10.1080/0954412989180

UN. (2005). *Global egovernment readiness report 2005- From egovernment to e-inclusion*. Department of Economic and Social Affairs Division for Public Administration and Development Management United Nations.

UN. (2008). *The United Nations e-Government Survey 2008: From e-Government to Connected Governance. Department of Economic and Social Affairs Division for Public Administration and Development Management*. United Nations.

UN. (2010). *United nations global e-government survey: Leveraging e-government at a time of financial and economic crisis*. NY. Department of Economic and Social Affairs Division for Public Administration and Development Management, United Nations

UN OHRLLS. (2011). *List of LDCs; United Nations Office of the High Representative for the Least Developed Countries*. Retrieved September 5, 2011, from http://www.unohrlls.org/en/ldc/related/59/

UNDP. (2001). *Making new technologies work for human development*. New York, NY: UNDP.

UNFPA. (2009). *The state of world population 2009: Facing a changing world: Women, population and climate*. United Nations Population Fund. Retrieved December 9, 2010, from http://www.unfpa.org/swp/2009/en/pdf/EN_SOWP09.pdf

Van De Ven, A., & Delbecq, A. (1971). Nominal versus interacting group processes for committee decision-making effectiveness. *Academy of Management Journal, 14*(2), 203–212. doi:10.2307/255307

Walsham, G. (2002). Cross-cultural software production and use: A structurational analysis. *Management Information Systems Quarterly, 26*(2), 359–380. doi:10.2307/4132313

Walsham, G., Robey, D., & Sahay, S. (2007). Foreword: Special issue on information systems in developing countries. *Management Information Systems Quarterly, 31*(2), 317–326.

Walsham, G., & Sahay, S. (2006). Research on information systems in developing countries: Current landscape and future prospects. *Information Technology for Development, 12*(1), 7–24. doi:10.1002/itdj.20020

Warschauer, M. (2003). *Technology and social exclusion: Rethinking the digital divide*. London, UK: MIT Press.

Weisinger, J. Y., & Trauth, E. M. (2002). Situating culture in the global information sector. *Information Technology & People, 15*(4), 306–320. doi:10.1108/09593840210453106

Westrup, C., Liu, W., El Sayed, H., & Al Jaghoub, S. (Eds.). (2003). *Taking culture seriously: ICTs, cultures and development*. Hampshire, UK: Ashgate.

Willcocks, L. (1994). Managing information systems in UK public administration: Issues and prospects. *Public Administration, 72*(Spring).

Wilson, E. J. (2004). *The information revolution and developing countries*. London, UK: MIT Press.

World Bank. (2002). *The egovernment handbook for developing countries*. Center for Democracy and Technology.

World Bank. (2005). *Lessons learned from seventeen*. InfoDev Projects Gamos Limited, Big World.

Wresch, W. (2003). Initial e–commerce efforts in nine least developed countries: A review of national infrastructure, business approaches, and product selection. *Journal of Global Information Management, 11*(1), 67–78. doi:10.4018/jgim.2003040105

Yap, A., Das, J., Burbridge, J., & Cort, K. (2006). A composite-model for e-commerce diffusion: Integrating cultural and socio-economic dimensions to the dynamics of diffusion. *Journal of Global Information Management, 14*(3), 17–38. doi:10.4018/jgim.2006070102

Zogona, S., Willis, J. E., & MacKinnon, W. J. (1966). Group effectiveness in creative problem–solving tasks: An examination of relevant variables. *The Journal of Psychology, 62,* 111–137.

Chapter 15
What Shapes Global Diffusion of E-Government:
Comparing the Influence of National Governance Institutions

Bijan Azad
American University of Beirut, Lebanon

Samer Faraj
McGill University, Canada

Jie Mein Goh
University of Maryland, USA

Tony Feghali
American University of Beirut, Lebanon

ABSTRACT

Prior research has established the existence of a differential between industrialized and other countries for e-Government diffusion. It attempts to explain this divide by identifying economic and technical variables. At the same time, the role of national governance institutions in e-Government diffusion has been relatively under-theorized and under-studied. The authors posit that, the existing national governance institutions shape the diffusion and assimilation of e-Government in any country via associated institutions in three key sectors: government, private sector and non-governmental organizations. This paper develops and tests a preliminary model of e-Government diffusion using the governance institutional climate as represented via democratic practices, transparency of private sector corporate governance, corruption perception, and the free press. The results indicate that the level of development of national governance institutions can explain the level of e-Government diffusion over and above economic and technical variables. The authors' research contributes to the literature by providing initial evidence that the existing national governance institutions influence and shape e-Gov diffusion and assimilation beyond the adoption stage.

DOI: 10.4018/978-1-61350-480-2.ch015

INTRODUCTION

Electronic government[1] (e-Gov) researchers have found that in some countries the focal systems are being assimilated at a much slower rate than their initial adoption would have implied (Mayer-Schönberger & Lazer, 2007). These countries have yet to embed the underlying systems in their government operations. As a result, these nations fail to realize the promised benefits of e-Gov, i.e., more effective and efficient delivery of services and information to citizens and firms. Indeed, there appears to be an e-Gov differential developing among nations akin to the so called digital divide. This has led some researchers to draw theoretical and empirical parallels between the two (Helbig, Gil-García, & Ferro, 2009). Initially researchers had attributed this e-Gov divide to differences in IT and network infrastructure as well as a variety of national characteristics (e.g., Norris, 2001). However, more recently some authors have persuasively argued that there is more to this e-Gov divide than technology infrastructure maturity and national culture (e.g., Heeks & Bailur, 2007). A growing stream of research points to the potentially significant role of *institutional norms* and *practices* in most countries in shaping e-Gov assimilation (e.g., Fountain, 2001).

Indeed there is theoretical precedence for this perspective. Hernando de Soto (2000), in his book *The Mystery of Capital: Why Capitalism Triumphs in the West and Fails Everywhere Else,* eloquently argues that a supportive national climate of *governance institutions* has been an essential but invisible feature of the growth of capitalism in the Western world. His point is that the developing world has often struggled unsuccessfully with transitioning to mature capitalism precisely because there is inadequate attention to the formation and functioning of the focal national governance institutions in these countries (e.g., Stinchcombe, 1997). Focusing on the IT sector within national economies, King and colleagues (King, Gurbaxani, Kraemer, McFarlan, Raman, &

Yap, 1994) in their pioneering research have highlighted the critical role that national governance institutions play in the diffusion of IT innovations. These national institutions of governance are often embedded within individual public or private organizations and also operate at the societal levels (March & Olsen, 1989). They are referred to as the governance ethos or simply institutions of governance in a country (North, 1990). The institutions of governance are defined broadly as "rules of the game" (World Bank, 2002), that both enable and constrain the actions of social actors as they engage in high stake efforts to carry out e-Gov programs. As a result, some researchers have asserted that diffusion and assimilation of e-Gov, as a public sector IT undertaking, is influenced by the national governance institutions in a given country (e.g., Gronlund & Horan, 2004). Therefore, to produce a deeper understanding of e-Gov divide it is essential to study the potential influence of these institutions on the diffusion of e-Gov at the national level across countries.

This paper aims to do so and is inspired by both de Soto's and King et al.'s works: specifically we want to explore the role of the *national governance institutions* in e-Gov assimilation across countries globally beyond the influence of technology infrastructure maturity and national economic factors. We propose an exploratory framework which casts diffusion as the level of development of e-Gov. We then link the status of e-Gov development to variables which we propose can represent national governance institutions within any country. This framework is inspired by the work of institutional economists (e.g., North, 1990; Campos & Nugent, 1999; de Soto, 2000), and scholars of IT and institutions (e.g., King et al., 1994; Wilson, 2004).

We use data from published sources to explore relationship between the national governance institutional climate and the level of e-Gov development via six (6) preliminary hypotheses. The data sources include: UN Global e-Readiness Report for e-Gov development; the POLITY-IV database on democratic development; Transparency International data on Corruption Perception Index; the World Bank data on Corporate Governance; Reporters without Borders data on press freedom; the World Bank World Development database for Gross Domestic Product; and World Economic Forum data on Internet penetration.

The contributions of this research are threefold. First, it proposes that the national diffusion of e-Gov is shaped by the climate of national governance institutions. Second, it suggests a mechanism for this shaping process via a parsimonious model that represents the national governance institutions. Third, it provides a comparative analysis of diffusion of e-Gov across 60 countries for which complete data were available.

THEORETICAL FOUNDATIONS

E-Government Divide: Nascent Theorizing on Diffusion Differences Across Countries

As experience with the adoption and implementation e-Gov has accumulated, researchers have provided insight into what appears to work in practice as well as what does not work (e.g., Dwivedi, Weerakkody, & Williams, 2009). The focal unit of analysis of the latter research is often a government agency or a few agencies at a time—i.e., one or more organizations engaging in e-Gov. These studies have provided us with much needed insight and valuable knowledge of making e-Gov systems workable and routinzing their use (e.g., Chen, Chen, Huang, & Ching, 2006; Azad & Faraj, 2009). However, their results are of limited generalizability since their applicability to other organizations and locales often cannot be extended beyond the original country context.

As a result, there has been a rising interest in analyzing the diffusion of e-Gov at the country and national level. Two streams of research have emerged which examine the diffusion of e-Gov

at the country level. One stream has focused on analyzing the level of diffusion of Internet and e-Gov at the national level in individual countries. The work of Layne and Lee (2001) has served as an initial theme for some of this stream of research among scholars. These studies of e-Gov have often utilized a general innovation diffusion (Rogers, 1995) frame of reference to analyze and assess the stage of e-Gov diffusion within a single country and its correlates (e.g., Ke & Wei, 2004). A second stream of research has focused on cross-country analysis of e-Gov diffusion and its relationship to a large number of underlying factors. The objective of this stream of research has been to uncover basic "drivers" of e-Gov adoption (e.g., West, 2005; Moon, Welch, & Wong, 2005; Singh, Das, & Joseph, 2007; Katchanovski & La Porte, 2009). The results are promising and they point to insights on adoption factors including broadband availability and economic development, among others. These research streams acknowledge the divide, i.e., though countries have embraced digital government some are failing to take advantage of it beyond the simple adoption stage (West, 2007). However, most researchers puzzle over the behavior and rarely theorize beyond pointing to the presence of bewildering array of factors that are purported to be behind the divide.

Shaping E-Government: Importance of National Governance Institutions

Since its inception, e-Gov has been employed to improve both the internal operations of government and also service delivery to citizens and businesses. The adoption of these electronically-enabled improvements is ultimately tied to notions of effective governance (Ciborra, 2005). Governance is defined as the transparency and accountability "traditions and institutions by which authority in a country is exercised" (Kaufmann, Kraay, & Zoido-Lobatón, 1999). That is, better *governance* is associated with enhanced transparency and accountability in selecting, monitoring, and replacing governments. The improved governance is also related to the national capacity to successfully formulate and enforce sound market policies, as well as the respect of the state, citizens and businesses for these institutional mechanisms which govern social, political and economic exchange (Keohane & Nye, 2000). More specifically, the focal *institutions* of governance are "the humanly devised constraints that structure human interaction. They are made up of formal constraints (e.g., rules, laws, constitutions), informal constraints (e.g., norms of behavior, conventions, self-imposed codes of conduct), and their enforcement characteristics. Together they define the incentive structure of societies" (North, 1994, p. 360). This notion of governance institutions has risen to the top of the policy and research agenda in the 1990s and 2000s within the international development community (Kaufman et al., 1999; Campos & Nugent, 1999; Ahrens & Meurers, 2002). In particular, the recent challenges of implementing sustainable reforms in developing and transition countries have proved insurmountable and are largely attributed to weak or non-existent "rules of the game" systems—i.e., quintessential institutions of national governance (Rodrik, 2003).

The manifest rationales and justifications for undertaking e-Gov are often tied to improvements in the governance environments, especially accountability and transparency (Dunleavy, Margetts, Bastow, & Tinkler, 2006). However, we recognize that, the diffusion of e-Gov is also expected to be shaped in the first place, at least in part, by the existing national governance institutional climate (Kamarck & Nye, 2002). To be sure, the extant research has linked the adoption of e-Gov systems to a variety of information and communication technologies (ICT) factors including Internet use, web site characteristics, types of services, and penetration (e.g., West, 2005). However, considerations to do with influence of the national governance institutions may be more *fundamental.* That is, they provide the appropriate

incentives and rules (Wilson, 2004) within which the focal e-Gov programs and projects are likely to be chosen, designed and implemented. This link between pre-existing institutions of governance and e-Gov diffusion may appear obvious, however, it has rarely been empirically analyzed in a theory-driven manner. Building up this theoretical scaffolding will be the starting point for the framework suggested next.

Mechanism of Influence on E-Government: Role of National Governance Institutions

Indeed, if national governance institutions do shape e-Gov adoption as is posited above, then what can the expected mechanism of this influence be? Wilson (2004) has proposed a framework for analyzing the influence of governance institutions at the national level on ICT and Internet policy making. This framework includes the following elements: the government, the private sector, and the non-governmental organizations (NGOs). These three sectors are posited to collectively have a significant hand in formation and execution of national Internet/ICT decisions—albeit the degree of influence is subject to specific constellations of institutional forces in these sectors within a specific country context. According to Wilson, it is to a large degree, the governance norms and practices of these sectors that tend to shape key actor participation and in turn influences the processes and outcomes of ICT and Internet diffusion at the national level.

Wilson trains his analytical lens on the national governance institutions which are of great interest to this paper. His goal is to analyze the commensurate institution-based micro-interactions. Thus, he adopts a qualitative approach to empirically analyze the latter. In that context, his focus is on the key individual as well as organizational actors who appear to have made a difference in specific decisions to shape the ICT and Internet policy making and implementation. He chooses three countries (i.e., Brazil, China and Ghana) to illustrate key aspects of his framework and how specific actions of focal individuals and organizations were critical to key outcomes. For example, he shows how specific changes in the telecommunication regulations of Brazil to facilitate and lower the cost of Internet diffusion were spearheaded by a group of individuals whose social network and material interests happened to coincide.

For us, his framework can serve as a point of departure for the following five reasons. First, the similarities between ICT/Internet and e-Gov both technologically and institutionally are far greater than the potential differences. Thus, it makes theoretical sense to presume that Wilson's skeletal framework (the sectors approach) is a good base to start with in analyzing e-Gov diffusion using a governance institutions lens. Second, he convincingly argues for a theoretical framework which attaches enormous significance to the national governance institutional realm—a phenomena that may have been largely under-theorized by the e-Gov diffusion researchers so far. That is, according to Wilson, fundamentally it is these institutions that shape the decisions and outcomes of the ICT and Internet diffusion in the focal countries. Third, his framework is parsimonious, consisting of the abovementioned sectors, and thus very manageable. This simplicity is a major plus in the on-going search for potentially bewildering array of factors that influence diffusion of e-Gov (e.g., Moon et al., 2005; Boyer-Wright & Kotterman, 2008). Fourth, he makes a convincing empirical case via a rich analysis of the data on three countries that indeed, the institutions of national governance make a difference to ICT and Internet diffusion. Finally, his choice of qualitative method is both a boon and an opening for others to extend his research. The case study method used by Wilson is a boon because it provides rich details and an in-depth perspective on the institutional influences. However, precisely because of its qualitative limitations, it offers an opportunity to expand it to a large number of countries, al-

beit with significant adaptation of the theoretical framework to fit the latter type of analysis. This is the task of the next section.

E-Government Diffusion: A Preliminary Model Incorporating the Role of Institutions

The objective of this section is to propose a preliminary framework that helps us explore the potential role of national governance institutions in shaping the diffusion of e-Gov systems across countries. In particular, we are interested in the influence of the institutions above and beyond the traditionally advanced explanations for the e-Gov divide. With this objective in mind, we start with the economic development and technology infrastructure maturity categories which we refer to as the Control Model. We then add the elements from the above sectors (government, private and NGO) highlighting what the mechanisms of institutional shaping of e-Gov systems diffusion are expected to be. This model proposes indicators of the underlying institutions to do with each of these sectors in a manner that can be more or less representative of the national governance institutional climate. We summarize the model via a set of six (6) hypotheses that reflect the shaping of the e-Gov diffusion within a country by variables that represent the focal national governance institutions.

The Control Model: National Economic Development and Technology Infrastructure Maturity: Gross Domestic Product and Internet Penetration.

Our starting point is the traditional economic growth argument. That is, we want to explore the role of traditional variables, namely, the influence of the levels economic development and technology infrastructure maturity in shaping e-Gov diffusion. In other words, in a more economically developed society by definition there is a greater tendency to employ systems to deliver e-Gov services. Similarly, the greater the ICT infrastructure maturity, the greater the penetration of e-Gov is expected to be. Indeed, the work of Norris (2001) has substantiated that, higher levels of economic development are correlated with more sophisticated ICT and Internet environments worldwide. The propositions by Singh et al. (2007) and West (2007) go a step further. They posit that, as countries develop, ICT becomes more embedded within the organizational structures especially in the decision making processes. Furthermore, these structures are expected to be positively associated in the shaping of ICT and Internet decision making. Consistent with prior research we represent the level of economic development as Gross Domestic Product and Technology Infrastructure Maturity as Internet Penetration (Norris, 2001). Thus, we summarize the above traditional reasoning in the form of following two hypotheses:

H1: The higher the level of the Gross Domestic Product the greater the level of e-Gov diffusion

H2: The higher the level of the Internet Penetration the greater the level of e-Gov diffusion

Government Governance Institutions: Democratic Practices.

A premise of most e-Gov practice and research has been that e-Gov is about greater efficiency and accountability in the public sector (Fountain, 2007). Especially enhancing accountability by making government information and services more responsive to the needs of citizens and businesses has dominated the e-Gov project agendas (Dunleavy et al., 2006). Indeed, Dunleavy et al. (2006) in their assessment of e-Gov initiatives underscore the significance of accountability: "Underpinning everything that governments do is a dynamic of political *accountability*, that is both an obligation to explain major decisions convinc-

ingly before they are undertaken and a duty to answer in public for actual performance" (*italics* added, p. 85). In practice, accountability is exercised and experienced as a matter of degree. More specifically, this can be expected to be largely a reflection of the institutional climate of *democratic* governance. That is, we expect governments that exhibit greater degree of democratic orientation to be more favorably disposed to advancing a higher level of accountability and also a more advanced e-Gov development agenda, all else being equal. In our adaptation of the public sector national governance institutions, we posit that government accountability can be represented as the degree to which *democratic practices* are prevalent within a national environment. We further suggest that these practices influence the diffusion of e-Gov positively. Subsequently, we summarize our reasoning as the following hypothesis:

H3: The more prevalent the national Democratic Practices the greater the level of e-Gov diffusion.

Private Sector Governance Institutions: Transparency of Corporate Governance and Corruption Perception

Based on the theoretical interests of this paper, we link the private sector national governance institutions to e-Gov development as follows: (a) the transparency of private sector corporate governance; and (b) the transparency of exchange between public and private sectors. First, the traditional governance literature often provides two separate interpretations of institutions: a private one via Williamson's (1999) sense of corporate governance as rules regulating private orderings; and another one via public institutions of governance (Wilson, 1989) which are frequently limited to ethos of professional and independent public sector management. However, consistent with Ahrens and Meurers (2002), our formulation of national governance institutions integrates these private and public notions holistically within a unified governance institutional environment—whereby they tend to reinforce each other and have a symbiotic relationship (e.g., Evans & Rauch, 1999). Stated differently, the norms and practices of private corporate governance and public sector governance, namely, their institutions can be thought of as mutually interdependent social structures. That is, as market governance institutions develop and mature they tend to be emulated in the public sector and vice verse. As such, more mature market governance regimes are often associated with a more enlightened public sector governance structure affording more transparency to government operations. As a whole, we expect that the deployment of tools and methods to enhance transparency of public sector to be shaped by greater maturity in the private sector corporate governance structures. Thus, we further hypothesize that, e-Gov is more likely to be adopted and diffused, when there is a perception of greater transparency in private sector corporate governance institutions.

Second, we focus on the transparency of private-public sector interactions, or more precisely its opposite, corrupt practices. That is, we are interested in the openness of the dealings between public and private sectors. This general approach postulates that the existing transparency ethos within the public sector facilitates the move toward electronic government. More specifically, within the context of national governance institutional climate, we adopt the perspective that transparency refers to the manner and openness of strategic interactions "between policymakers, private businesses, and intermediate organizations" as a whole (Ahrens, 2006, p. 8). Thus, we posit that, the transparency of dealings between public and private sectors, i.e., (inverse of) *corruption perception*, can represent a dimension of the private sector governance institutions within the focal national environment.

Combining these two elements, we suggest that the greater the *transparency* of *corporate governance* and the lower the *corruption perception*, the more the positive influence on the development and assimilation of e-Gov systems. Subsequently, we summarize our reasoning as the following two hypotheses:

H4: The higher the perception of Transparency of Corporate Governance the greater level of e-Gov diffusion.
H5: The lower the Corruption Perception the greater level of e-Gov diffusion.

NGO Sector Governance Institutions: Free Press

Generally speaking, a more democratic polity goes hand in hand with a freer press as a key governance institution (Islam, 2006). That is, traditionally we expect to observe freer flow of information in more open societies. For example, Martin and Feldman (1998) have argued that that a free press is essential to the dissemination of information. Also, there is a growing body of evidence that relates the national governance institution of a free press as an underpinning of the "watchdog" role onto the public sector workings and a check against the abuse of governmental authority (e.g., Besley, Burgess, & Prat, 2002; Djankov, McLiesh, Nenova, & Shleifer, 2003).

That is, governments have strong incentives for maintaining secrecy regarding their operations, transactions, and decision-making. As a result, some societies have attempted to constrain this tendency in behavior by limiting the government's ability to curb those who might bring about greater openness. That is, there are institutions and practices which protect either free speech or a free press or both via freedom of access to public information (Stiglitz, 2008). Thus, the role of the media in enhancing transparency and openness of public records can be regarded as a fundamental "rules-of-the-game" and is often considered a key national governance institution. However, for such institutions to work effectively the focal artifactual means of access to government information play a large part in guaranteeing access to it—be they hand-copying, fax, photocopy, digital reproduction, etc. Indeed, one of the touted ideals of e-Gov has been to enhance accessibility and availability of this governmental information via online digital means. Based on this formulation, we hypothesize that, the presence and maturity of a free press is associated with greater levels of development of e-Gov. Subsequently, we summarize our reasoning as the following hypothesis:

H6: The greater the degree of Press Freedom the greater the level of e-Gov diffusion

Having described the hypothesized relationships between national governance institutional climate and e-Gov development and assimilation, they are summarized in Figure 1 graphically.

METHODOLOGY

Data Sources

We start with our dependent variable, i.e., e-Gov Development. We drew on the United Nations (UN) Global E-Government Report 2005 (United Nations, 2005a, 2005b) to derive this dependent variable. This data source has been used extensively in earlier e-Gov studies (e.g., Siau & Long, 2006; Srivastava & Teo, 2008). The UN Global E-Government Readiness Survey 2005 provides an assessment of how the governments that are members of UN, use the ICTs to provide access to government functions. This assessment constructs the patterns of diffusion of e-Gov and also includes items that the government may not taken advantage of. The data set contains quantitative measures rather than perceptual measures. These are counts of features and services as provided through the government websites. This is a positive charac-

Figure 1. Proposed model of e-government development across countries: Influence of national governance institutions

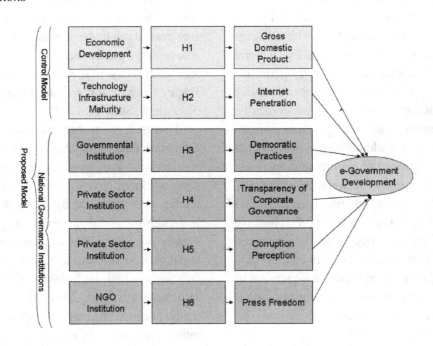

teristic of the underlying data since it relies less on the perception of the e-Gov websites design.

Proceeding to our independent variables we relied on the following sources. First, we used the measure "democracy of a government" to capture Democratic Practices which is derived from the POLITY IV project dataset (Marshall & Jaggers, 2007). The main purpose of the POLITY IV project has been to establish quantitative measures that characterize democratic features among different governments in the world for political scientists and governance researchers. The "democracy of a government" codes the general authority traits that characterize a polity. It is widely used in studying regime authority related factors (e.g., Treier & Jackman, 2008).

Second, we utilized the World Bank's Ethics Index (2004) of Transparency of Corporate Governance and the Corruption Perceptions index from Transparency International[2]. The World Bank has been compiling an index of Transparency of Corporate Governance in its governance database. The Transparency International also constructs and maintains a composite index which measures the degree to which corruption is perceived to exist in a country in the public-private interactions and transactions. This index is drawn from corruption-related information and surveys and then reviewed by an advisory committee drawn from experts on corruption, econometrics and statistics. Both indices have been used by economists in a variety of national governance assessments (e.g., Ahrens & Meurers, 2002; Campos & Nugent, 1999) and by e-Gov researchers (e.g., Srivastava & Teo, 2008).

Third, we utilized data from Reporters Without Borders (2006) to obtain measures for media Press Freedom. Reporters Without Borders is an association, which examines the level of Press Freedom, to protect the rights of reporters or journalists around the world. This data is used widely within governance research for a variety purposes including foreign direct investment oversight and anti-corruption campaigns (e.g., Li, 2005; Islam, 2008).

Fourth, for the level of Economic Development we obtained data from the World Bank World Development database indicators on Gross Domestic Product (GDP) per capita purchasing parity. The data from this database are widely used in economic development research (Campos & Nugent, 1999). For Technology Infrastructure Maturity we used Internet Penetration data from INSEAD's Global Information Technology Report (World Economic Forum, 2006). This measure is among most widely used data to capture the information technology infrastructure maturity within a country (e.g., West, 2007).

Operational Definition of Variables

Here we present the operational definition of select variables (e-Gov Development, Democratic Practices, and Internet Penetration) that require further explanation—the rest of variables have a one-to-one correspondence with the construct being measured in the secondary data set.

E-Gov Development

This variable was used to measure the digital government development in our exploratory model. It is a measure of e-Gov sophistication which signifies the level of e-Gov penetration in year 2005 based on a five stage model of e-government readiness according to the development and sophistication of the digital government functions. Similar operationalization has been employed in prior research on e-Gov (e.g., Siau & Long, 2006). The measure is similar (though not identical) to the Layne and Lee (2001) model of e-Gov stages. The UN evaluation of e-Gov websites was conducted during July-August 2005. In particular, it contains five stages which are described as follows:

- **Emerging:** The government provides a website containing a basic web page linking to other ministries or departments.

- **Enhanced:** The government provides sources of current and archived information on public agencies. Users can perform basic search of pages and e-Gov website with a help feature. However, it only permits unidirectional interaction.

- **Interactive:** The government provides a website which allows downloadable forms, some online services and contains multimedia information. Contact information is provided for government representatives and made available through different online means. There are frequent updates on sites.

- **Transactional:** The government website allows bi-directional interaction. Public services are payable through some form of electronic transaction involving the use of credit bank or debit card. Links to financial institutions are also made available on the website.

- **Networked:** A networked e-Gov website allows the integration of government to government (G2G), government to citizen (G2C), and citizen to government (C2G) interactions. In addition to two-way interactions, it allows citizens to participate in policy making and give feedback through online means.

- **Democratic Practices:** The Democratic Practices variable is based on the measures from POLITY IV database (Marshall & Waggers, 2007). Specifically we employed the POLITY2 index. This index is meant to signify political regime types from a scale of -10 (fully autocratic) to +10 (fully democratic). The index combines data on five factors that capture the institutional differences between democracies and autocracies: (1) the competitiveness of the process for selecting country's chief exec-

utive; (2) the openness of this process; (3) the extent to which institutional constraints limit a chief executive's decision-making authority; (4) the competitiveness of political participation within a country; and (5) the degree to which binding rules govern political participation within it.

- **Internet Penetration:** As a proxy for the level of *technology infrastructure maturity* in the country we adopted a readily available data for a large number of countries, i.e., Internet Penetration. This is measured using the number of Internet users per 1000 people (World Economic Forum, 2006).

Method of Analysis

The analysis was largely driven by data availability from separate data sources. As a result, there were some missing values for some countries. We decided the best way to reduce the impact of this and not introduce inconsistencies was to remove the cases (countries) with missing values for any one variable. As a result, we ended up with 60 complete country observations in the entire data set. All the independent variables were lagged by at least 1 year. The descriptive statistics for the key variables are shown in Table 1.

Pairwise correlations as shown in Table 1 indicates that Gross Domestic Product and Internet Penetration were highly correlated ($r = 0.74$) with the Level of e-Gov Development. Further analysis indicates that the variance inflation factors were 7.6 and 6.2 for Gross Domestic Product and Internet Penetration respectively. This suggests that multi-collinearity was not a major issue and will not affect the interpretation of coefficients (Maddala, 1992).

In order to test the Control and the Proposed Models, we employed hierarchical regression method using STATA 9.0 software which helped us perform the hypotheses testing. In order to ascertain the robustness of the results, the assumptions of regression were examined. Towards this end, variable distributions were inspected to find out if they did or did not conform to normality requirements. As a result, if they did not, they were log-transformed.

Our analysis proceeds in two steps. We first test the economic factors model consisting of the Gross Domestic Product and Internet Penetration levels. In the second step, and in accordance with hierarchical regression tenets, the national institutions of governance variables are introduced to assess their impact *over and above* those of the control model. The hierarchical regression approach thus allows a comparative evaluation of two models each including a set of theoretically related variables. If the second model achieves a significantly higher R^2, then it is surmised that the inclusion of latter variables have a unique and significant impact over and above the variables in the first model.

FINDINGS

Results

We tested hypotheses H1 through H6 using the entire data set in two phases (two models). Details of the results are presented in Table 2. Below, we present an overview of the results both for the Control Model and the Proposed Model.

Control Model (H1 and H2 alone)

The results of Control Model support traditional reasoning and confirm prior studies by generating an R^2 of 0.7—the second column of Table 2. That is, Gross Domestic Product is related to Level of e-Gov Development and thus H1 of the control model was supported (β=0.050, p<0.097). In addition, we found further support for H2 of the control model which suggested that Internet Penetration positively influences the Level of e-Gov Development as well (β=0.086, p<0.003).

Table 1. Variable definitions, descriptive statistics and correlations

	Variable	Mean	Std Dev	Min	Max	Dep Var	H1	H2	H3	H4	H5	H6
	DEPENDENT VARIABLE: LEVEL OF E-GOVERNMENT DEVELOPMENT	0.51	0.25	0	1	1						
H1	GROSS DOMESTIC PRODUCT	8.11	1.51	4.73	10.58	0.74	1					
H2	INTERNET PENETRATION	4.51	1.62	0.077	6.41	0.74	0.87	1				
H3	DEMOCRATIC PRACTICES	6.52	4.63	-7	10	0.5	0.47	0.39	1			
H4	TRANSPARENCY OF CORPORATE GOVERNANCE	3.78	0.46	2.42	4.56	0.62	0.63	0.52	0.19	1		
H5	CORRUPTION PERCEPTION	1.93	0.16	1.65	2.3	-0.34	-0.35	-0.32	-0.31	-0.26	1	
H6	PRESS FREEDOM	2.57	1.14	0.41	4.54	-0.55	-0.62	-0.55	-0.73	-0.44	0.42	1

All variables were log-transformed except for DEMOCRATIC PRACTICES

Table 2. Regression results

Hypothesis No.—Variable	Control Model: Traditional Approach	Proposed Model Expected Effect: Governance Institutional Climate: (+) Positive or (-) Negative	Proposed Model Results: National Governance Institutions	Support for Hypotheses: Proposed Model
H1—GROSS DOMESTIC PRODUCT	0.050	+	0.019	Not supported
	(0.097)‡		(0.496)	
H2—INTERNET PENETRATION	0.086	+	0.095	Supported
	(0.003)**		(0.000)**	
H3—DEMOCRATIC PRACTICES		+	0.009	Supported (weak)
			(0.055) ‡	
H4—TRANSPARENCY OF CORPORATE GOVERNANCE		+	0.164	Supported
			(0.000)**	
H5—CORRUPTION PERCEPTION		-	-0.309	Supported
			(0.004)**	
H6—PRESS FREEDOM		+	0.065	Supported
			(0.003)**	
Constant	-0.281		-0.331	
	0.0440		(-1.322)	
Observations	60		60	
R-squared	0.70		0.80	
F for $\Delta R2$	N/A		6.53**	

‡ significant at 10% level; * significant at 5% level; ** significant at 1% level; p-value in parentheses

All variables were log-transformed except for DEMOCRATIC PRACTICES

Proposed Model (H1 through H6 combined)

In the Proposed Model, we present the regression results based on and including all the hypothesized relationships, i.e., both the traditional argument and the national governance institutions together. In other words, above and beyond GDP/Internet Penetration, the Proposed Model includes the four additional variables which we have theorized above: Democratic Practices; Transparency of Corporate Governance; Corruption Perception; and Press Freedom. The difference in predictive power between the control model and our proposed model was significant (ΔR^2=0.1, F for ΔR^2=6.53,

p<0.000). In other words, these results provided initial and preliminary support for the hypothesis that national governance institutions as captured via the four (4) variables in our model are important and contribute to e-Gov development beyond the influence of the Gross Domestic Product and the Internet Penetration.

Overall, we found strong support for most of the hypothesized influences of national governance institutional climate on e-Gov development (details are presented in fourth column, Table 2). First, we found strong support for H2 within the proposed model, which suggested that Internet Penetration positively influences the Level of e-Gov Development (β=0.095, p<0.000). Second,

H4 regarding the Transparency of Corporate Governance ($\beta=0.164$, $p<0.000$), H5 regarding the Corruption Perception ($\beta=-0.309$, $p<0.01$) and H6 regarding the Press Freedom ($\beta=0.065$, $p<0.01$) were all strongly supported. In other words, the greater Transparency of Corporate Governance, and the lower Corruption Perception both positively contribute to the further development and diffusion of e-Gov systems. In addition, the watchdog function of the press over public sector authority and functioning, i.e., Press Freedom, was also found to positively influence the e-Gov development. Third, we found marginal support for H3 regarding Democratic Practices ($\beta=0.009$, $p<0.1$). That is, the influence of Democratic Practices on e-Gov diffusion though positive was the weakest of the all factors making up the national governance institutional climate in the proposed model. Fourth, within the proposed model, however, the Gross Domestic Product appeared not to be related to Level of e-Gov Development and thus H1 of the proposed model was not supported. That is, within the proposed model there appears to be little influence if any of economic development on e-Gov system diffusion and assimilation. In the discussion section, next, we have more to say regarding this counter-intuitive but interesting result.

Discussion

Using a theoretical framework which focuses on the significance of *national governance institutions* and the associated country indicators from several global databases, we have provided a fresh perspective on why the institutions of governance can shape e-Gov development and serve as an explanation for the e-Gov divide. Our analysis provides evidence that that indeed a more sophisticated national governance institutional climate positively influences the development and diffusion of e-Gov. In particular, the analysis shows that national governance institutions represented via *democratic practices, transparency of corporate* *governance, corruption perception,* and *press freedom* tend to positively shape the adoption and diffusion of e-Gov all else being equal. Overall, we surmise that our formulation of the national governance institutions—as forces which can facilitate or impede e-Gov development—is a credible framework for analyzing and explaining the e-Gov diffusion differentials among countries.

The e-Gov researchers have established that indeed there is a sort of a "divide" in the e-Gov diffusion worldwide (e.g., Helbig et al., 2009). That is, e-Gov as an ensemble of techniques, processes and systems to improve government has become popular among governments both in the industrialized as well as developing and transition economies (Mayer-Schönberger & Lazer, 2007). However, the examination of e-Gov penetration beyond the initial adoption in different countries reveals a pattern whereby the diffusion among the industrialized countries appears to be greater than within the developing and transition nations. This difference among the two groups of industrial and developing/transition economies has been the subject of research and more specifically a search for the key drivers that can explain the difference (e.g., Boyer-Wright & Kotterman, 2008). Indeed, the prior research has revealed useful factors including the level of economic development, availability of broadband, cost of telecommunications, and web site design which may explain some of the differences in diffusion rates (e.g., West, 2007). However, persistence of this difference over the last decade begs for additional theoretical insights to augment our knowledge of existing factors and drivers (e.g., Heeks & Bailur, 2007).

Our research, following in the foot steps of institutional economics researchers (North, 1990, 1994) has pointed to the national governance institutions as important shaping mechanisms that can largely enable or constrain e-Gov diffusion. In particular, we started with Wilson's (2003, 2004) qualitative framework and analysis (of Brazil, China, and Ghana) on the importance of national governance institutional forces (private sector, public sector, and NGO sector) for the diffusion of ICT and Internet. We then adapted this framework in order to analyze the influence of these governance institutions on e-Gov development across a wide range of countries globally via our four alternative hypotheses and quantitative indicators. Thus, one contribution of our research is that it complements the prior inventory of drivers and factors of e-Gov development by adding the distinctive role of governance institutional climate as a fundamental consideration. That is, we provide preliminary evidence that the national governance institutions play a significant role in the diffusion of e-Gov systems.

The extant research has documented that e-Gov adoption is on the rise while at the same time pointing to the difficulties some countries face in moving beyond providing simple web site and information catalogue functions (e.g., Coursey & Norris, 2008; Helbig et al., 2009). Indeed, the importance of institutional factors as opposed to traditional metrics to explain these difficulties is alluded to in related research (e.g., Tolbert, Mossberger, & McNeal, 2008). Our research affirms this difficulty but then goes a step further. We extend the three country qualitative study and framework of Wilson (2004) by adapting it to address e-Gov diffusion quantitatively and globally. We demonstrated the distinctive influences of national governance institutional climate captured as *democratic practices (H3), transparency of corporate governance (H4)* and *corruption perception (H5)*, and the *free press (H6)*.

A major thrust of e-Gov in practice and the associated research has been the push towards enhanced accountability via the use of focal systems "simplifying the institutional structure of government, making it more transparent" (Dunleavy, 2007, p. 422). Our research, though preliminary and exploratory, proposes to reverse this causality. That is, in fact such institutions may be pre-requisites for diffusion of e-Gov beyond the simple adoption of systems. It is indeed the theoretical argument of our research that national governance institutions provide the underpinning on which foundational e-Gov systems are designed, developed and implemented. A corollary of this is that, e-Gov systems may be adopted initially. However, if the national governance institutions within the three sectors (private sector, NGO and government) are lacking and/or underdeveloped, then further development and assimilation of e-Gov systems may be negatively influenced or slowed down or simply stalled. Our data in all the national governance institutional categories, namely, *democratic practices, transparency of corporate governance* and *corruption perception*, and *press freedom* support such reasoning. Thus, our research has extended the literature by providing initial evidence of the supportive role of national governance institutions in shaping the development of e-Gov beyond the simple adoption stage.

However, a counter-intuitive and unexpected result of our proposed model was the lack of a significant relation between economic level of development and e-Gov development and diffusion (i.e., *H1* in the Proposed Model). The methodological and secondary data problems notwithstanding, there appears to be an interesting mechanism at work. Although, it is only a conjecture at this stage, we surmise that, our Proposed Model is not equipped to account for the so called symbolic adoption and institutional decoupling phenomena (Meyer & Rowan, 1977). That is, although countries are adopting e-Gov systems in large numbers (e.g., Boyer-Wright & Kotterman, 2008), some are not moving beyond the adoption stage. In particular, only those that have fairly developed national governance institutions may be able to actually exploit the underlying technologies in their operations to move towards the assimilation stages of e-Gov. This phenomenon is referred to as "symbolic adoption" or "decoupling"[3] by neo-institutionalist researchers (Meyer & Rowan, 1977; Di Maggio & Powell, 1983; Strang & Meyer, 1993). More recent empirical analysis (Meyer, Boli, Thomas,

& Ramirez, 1997; Lee & Strang, 2006; Dobbin, Simmons, & Garrett, 2007) has shed additional light on this institutional decoupling mechanism. These researchers argue that, there is a trend in global adoption of "rational systems" (e.g., be they organizational forms or policies, or computer systems) among nation states. However, underpinning this trend there can be a pattern of action at work whereby the adoption of "rational systems" is decoupled from the follow-up implementation and assimilation. In essence, some polities in the "world society of nation-states" appear to enact the so called rational systems only symbolically as a means to gain legitimation but often do not follow up (or are not able to) the adoption stage with actual implementation (Dobbin et al., 2007). Some e-Gov researchers have gone so far as characterizing this trend as "Potemkin e-Villages" in reference to the Soviet officials who in the Stalin era used to create fake settings to impress visiting government dignitaries (Katchanovski & La Porte, 2005). It is beyond the scope of our research to ascertain whether such mechanisms are at work within the focal countries, a point that can be taken up in future research.

Our research has limitations which can be addressed through further research. First, the data elements that are missing can be worked on and rectified to increase the pool of countries in the sample—i.e., to increase the number of nations in the sample. This will require rectification and reconstruction of missing data from diverse sources—a challenge we could not attempt to meet in this paper due to resource constraints. Second, as additional data are obtained, we can increase the pool of countries to between 100-150 cases, then Structural Equation Modeling can be utilized to perform confirmatory analysis and produce more robust test of our Proposed Model. Third, more recently researchers have documented and theoretically accounted for the potential of e-Gov systems to initiate changes in the underlying institutional environment, e.g., reducing corruption (Kim, Kim, & Lee, 2009). On the surface, this type of theoretical argument and evidence runs counter to the logic of the model specification we have presented—e.g., low perception of corruption as a pre-requisite. However, by expanding the framework into a structurational model (Wilson, 2004[4]; Jones & Karsten, 2008) changes in underlying institutional environment can be allowed to vary. Furthermore, they can be represented as co-evolutionary phenomena in parallel to the technology-based changes introduced via e-Gov systems. As a result, national governance institutional climate becomes a set of dynamic social norms and patterns rather than static givens which then can be also be allowed to co-evolve with the technology. This is a noteworthy research direction to explore which can build on the current paper's theoretical and empirical horizon.

CONCLUSION

We started by highlighting the parallels between the digital divide and e-Gov diffusion differentials among nations globally and the scarcity of theoretical explanations. That is, researchers have found that globally e-Gov is being assimilated at a much slower rate subsequent to its adoption. In some countries the take-up of e-Gov is deep, while in a large number of other countries neither much information nor many services have been placed online. In other words, the latter group of nations appear not to be taking advantage of the features of the underlying IT and Internet to assimilate e-Gov in their operations. Indeed, although authors and policy makers have touted the transformational potential of these e-Gov features, most citizens and businesses have yet to reap these benefits in practice—at least within the latter countries. Our research aimed to offer a preliminary theoretical explanation for this divide. We are inspired by the works of de Soto (2000) on importance of governance institutions for economic development among the developing nations and King et al. (1994) on the central role of governance institutions in enabling or constraining IT innovation at the national level. Our exploratory study empirically underscores that institutions of national governance matter in the assimilation of e-Gov systems beyond the adoption phase.

In particular, our research establishes that the growth and nurturing of e-Gov systems are not just due to maturity of technology infrastructure maturity and economic development. They are helped by a whole array of national governance institutions that are public, private and tertiary sector (non-government). These institutions serve as stable social structures that can support (or impede) e-Gov assimilation often in an invisible manner via *democratic practices, transparency of corporate governance, corruption perception, and press freedom.*

Our research has management implications in regards to designing and implementing e-Gov systems. The extant research has often trained its analytical lens on many social and political factors that have proved challenging during the design and implementation of e-Gov (Chen et al., 2006). The search for the key social and political factors and the subsequent advice how to guide e-Gov projects has left many practitioners faced with a bewildering array of factors which in practice prove intractable (e.g., Moon et al., 2005). In our research, we have provided a glimpse of the significance of national governance institutions and their influence on the e-Gov system diffusion. As a result, the practitioners may be well advised to heed these findings as a harbinger of the type of issues that may surface when they move beyond the simple adoption of e-Gov systems into their diffusion and assimilation phases. Indeed, sometimes not undertaking e-Gov projects whose logic presume the existence of national governance institutions, which are weak or lacking in the focal national context, may be a wise choice. Instead, working on the institutional strengthening front or at least working on implementing e-Gov systems and institutional changes simultaneously may prove much more fruitful in the long run.

We conclude by suggesting the following. First, viewing the e-Gov assimilation in countries as being shaped by national governance institutional climate is a sensible way to analyze the reality of e-Gov divide. Second, representing the underlying national governance institutions via *democratic practices, transparency of corporate governance* and *corruption perception* as well as *press freedom* provided evidence that hypothesized relationship of influence over e-Gov development does exist. Our analysis of the influence of the level of *economic development* over e-Gov diffusion was inconclusive. We advanced a conjecture that symbolic adoption and decoupling are potential explanations for this result, although actual behavior needs to be explored empirically in future research. Third, the national governance institutional climate can provide either a context which impedes or facilitates e-Gov development and thus, should be taken into account in practical project design and implementation. That is, ignoring the influence of national governance institutional climate can be deleterious to the e-Gov receptivity in the long-run leading to failed or stalled projects.

REFERENCES

Ahrens, J. (2006). Governance in the process of economic transformation. *Private University of Applied Sciences Goettingen*. Retrieved from www.oecd.org/dataoecd/52/20/37791185.pdf

Ahrens, J., & Meurers, M. (2002). How governance affects the quality of policy reform and economic performance: new evidence for economies in transition. *Journal for Institutional Innovation, Development and Transition, 6,* 35–56.

Andersen, K. V., & Henriksen, H. Z. (2006). E-government maturity models: extension of the Layne and Lee model. *Government Information Quarterly, 23*(2), 236–248. doi:10.1016/j.giq.2005.11.008

Azad, B., & Faraj, S. (2009). E-Government institutionalizing practices of a land registration mapping system. *Government Information Quarterly*, *26*(1), 5–14. doi:10.1016/j.giq.2008.08.005

Besley, T., Burgess, R., & Prat, A. (2002). Mass media and political accountability. In Islam, R. S. (Ed.), *The right to tell: the role of mass media in economic development* (pp. 45–60). Washington, DC: World Bank Publications.

Boyer-Wright, K. M., & Kotterman, J. E. (2008). *High-level factors affecting global availability of online government services*. Paper presented at Proceedings of the 41st Annual Hawaii International Conference on System Sciences.

Campos, N. F., & Nugent, J. B. (1999). Development performance and the institutions of governance: evidence from East Asia and Latin America. *World Development*, *27*(3), 439–452. doi:10.1016/S0305-750X(98)00149-1

Chen, Y. N., Chen, H. M., Huang, W., & Ching, R. K. H. (2006). E-government strategies in developed and developing countries: an implementation framework and case study. *Journal of Global Information Management*, *14*(1), 23–46.

Ciborra, C. (2005). Interpreting e-government and development: efficiency, transparency or governance at a distance. *Information Technology & People*, *18*(3), 260–279. doi:10.1108/09593840510615879

Coursey, D., & Norris, D. F. (2008). Models of e-government: are they correct? An empirical assessment. *Public Administration Review*, *68*(3), 523–536. doi:10.1111/j.1540-6210.2008.00888.x

de Soto, H. (2000). *The mystery of capital: Why capitalism triumphs in the west and fails everywhere else*. New York: Basic Books.

DiMaggio, P. J., & Powell, W. W. (1983). The iron cage revisited: institutional isomorphism and collective rationality in organizational fields. *American Sociological Review*, *48*(2), 147–160. doi:10.2307/2095101

Djankov, S., McLiesh, C., Nenova, T., & Shleifer, A. (2003). Who owns the media? *The Journal of Law & Economics*, *46*(2), 341–381. doi:10.1086/377116

Dobbin, F., Simmons, B., & Garrett, G. (2007). The global diffusion of public policies: social construction, coercion, competition, or learning? *Annual Review of Sociology*, *33*(1), 449–472. doi:10.1146/annurev.soc.33.090106.142507

Dunleavy, P. (2007). Governance and state organization in the digital era. In Mansell, R., Avgerou, C., & Quah, D. (Eds.), *The Oxford Handbook of Information and Communication Technologies* (pp. 440–426). New York: Oxford University Press.

Dunleavy, P., Margetts, H., Bastow, S., & Tinkler, J. (2006). *Digital Era Governance: IT Corporations, the State, and E-government*. New York: Oxford University Press.

Dwivedi, Y. K., Weerakkody, V., & Williams, M. D. (2009). From implementation to adoption: Challenges to successful e-government diffusion. *Government Information Quarterly*, *26*(1), 3–4. doi:10.1016/j.giq.2008.09.001

Evans, P., & Rauch, J. E. (1999). Bureaucracy and growth: a cross-national analysis of the effects of "Weberian" state structures on economic growth. *American Sociological Review*, *64*(5), 748–765. doi:10.2307/2657374

Fountain, J. E. (2001). *Building the virtual state: Information technology and institutional change*. Washington, D.C: Brookings Institution Press.

Fountain, J. E. (2007). Challenges to organizational change: multi-level integrated information structures (MIIS). In Mayer-Schönberger, V., & Lazer, D. (Eds.), *Governance and Information Technology: From Electronic Government to Information Government*. Cambridge, MA: MIT Press.

Gronlund, A., & Horan, T. (2004). Introducing E-government: History, definitions, and issues. *Communications of the Association for Information Systems, 15*(39).

Heeks, R., & Bailur, S. (2007). Analyzing E-government research: perspectives, philosophies, theories, methods, and practice. *Government Information Quarterly, 24*(2), 243–265. doi:10.1016/j.giq.2006.06.005

Helbig, N., Ramón Gil-García, J., & Ferro, E. (2009). Understanding the complexity of electronic government: implications from the digital divide literature. *Government Information Quarterly, 26*(1), 89–97. doi:10.1016/j.giq.2008.05.004

Islam, R. (2006). Does more transparency go along with better governance? *Economics and Politics, 18*(2), 121–167. doi:10.1111/j.1468-0343.2006.00166.x

Islam, R. (Ed.). (2008). *Information and public choice: From media markets to policy making*. Washington, DC: World Bank.

Jones, M. R., & Karsten, H. (2008). Giddens' structuration theory and information systems research. *Management Information Systems Quarterly, 32*(1), 127–157.

Kamarck, E. C., & Nye, J. S. (Eds.). (2002). *Governance.com: Democracy in the information age*. Washington, DC: Brookings Institution Press.

Katchanovski, I., & La Porte, T. (2005). Cyberdemocracy or Potemkin e-villages? Electronic governments in OECD and post-communist countries. *International Journal of Public Administration, 28*(7), 665–681. doi:10.1081/PAD-200064228

Katchanovski, I., & La Porte, T. (2009). Democracy, Colonial Legacy, and the Openness of Cabinet-Level Websites in Developing Countries. *Journal of Comparative Policy Analysis: Research and Practice*, http://www.informaworld.com/ smpp/title~db=all~content=t713672306~t ab =issueslist~branches=11 - v11*11*(2), 213-232.

Kaufmann, D., Kraay, A., & Zoido-Lobatón, P. (1999). *Governance matters* (World Bank Policy Research Working Paper No. 2196). Washington, DC: World Bank.

Ke, W., & Wei, K. K. (2004). Successful e-government in Singapore. *Communications of the ACM, 47*(6), 95–99. doi:10.1145/990680.990687

Keohane, R. O., & Nye, J. S. (2000). Introduction. In Nye, J. S., & Donahue, J. D. (Eds.), *Governance in a globalizing world*. Washington, DC: Brookings Institution Press.

Kim, S., Kim, H. J., & Lee, H. (2009). An institutional analysis of an e-government system for anti-corruption: The case of OPEN. *Government Information Quarterly, 26*(1), 42–50. doi:10.1016/j.giq.2008.09.002

King, J. L., Gurbaxani, V., Kraemer, K. L., McFarlan, F. W., Raman, K. S., & Yap, C. S. (1994). Institutional factors in information technology innovation. *Information Systems Research, 5*(2), 139–169. doi:10.1287/isre.5.2.139

Layne, K., & Lee, J. (2001). Developing fully functional e-government: A four stage model. *Government Information Quarterly, 18*(2), 122–136. doi:10.1016/S0740-624X(01)00066-1

Lee, C. K., & Strang, D. (2006). The International diffusion of public-sector downsizing: Network emulation and theory-driven learning. *International Organization, 60*(4), 883–909. doi:10.1017/S0020818306060292

Li, S. (2005). Why a poor governance environment does not deter foreign direct investment: The Case of China and its implications for investment protection. *Business Horizons, 4*(4), 297–302. doi:10.1016/j.bushor.2004.06.002

Maddala, G. S. (1992). *Introduction to Econometrics*. New York: Prentice Hall.

March, J. G., & Olsen, J. P. (1989). *Rediscovering institutions: The organizational basis of politics*. New York: Free Press.

Marshall, M. G., & Jaggers, K. (2007). Polity IV project: Political regime characteristics and transitions 1800-2007. *Center for Systemic Peace*. Retrieved from www.systemicpeace.org/polity

Martin, R., & Feldman, E. (1998). Why is Access Important in Developing Countries. In Transparency International (Ed.), *TI working paper: Access-to-information in Developing Countries*. Berlin, Germany.

Mayer-Schönberger, V., & Lazer, D. (2007). *Governance and Information Technology: From Electronic Government to Information Government*. Cambridge, MA: MIT Press.

Meyer, J. W., Boli, J., Thomas, G. M., & Ramirez, F. O. (1997). World society and the nation-state. *American Journal of Sociology, 103*(1), 144–181. doi:10.1086/231174

Meyer, J. W., & Rowan, B. (1977). Institutionalized organizations: Formal structure as myth and ceremony. *American Journal of Sociology, 83*(2), 340–363. doi:10.1086/226550

Moon, M. J., Welch, E. W., & Wong, W. (2005). *What drives global e-governance? An exploratory study at a macro level*. Paper presented at Proceedings of the 38th Annual Hawaii International Conference on System Sciences (HICSS'05).

Norris, P. (2001). *Digital divide? Civic engagement, information poverty and the internet worldwide*. New York: Cambridge University Press.

North, D. C. (1990). *Institutions, institutional change and economic performance*. New York: Cambridge University Press.

North, D. C. (1994). Economic performance through time. *The American Economic Review, 84*(3), 359–368.

Reporters Without Borders. (2006). *Press freedom index 2006*. Retrieved from http://www.rsf.org/

Rodrik, D. (2003). Institutions, integration, and geography: In search of the deep determinants of economic growth. In Rodrik, D. (Ed.), *Search of prosperity: analytic country studies on growth*. Princeton, NJ: Princeton University Press.

Rogers, E. M. (1995). *Diffusion of innovations* (4th ed.). New York: Free Press.

Siau, K., & Long, L. (2006). Using social development lenses to understand e-government development. *Journal of Global Information Management, 14*(1), 47–62.

Singh, H., Das, A., & Joseph, D. (2007). Country-level determinants of e-government maturity. *Communications of the Association for Information Systems, 20*, 632–648.

Srivastava, S. C., & Teo, T. (2008). The relationship between e-government and national competitiveness: The moderating influence of environmental factors. *Communications of the Association for Information Systems, 23*(59).

Stiglitz, J. (2008). Fostering an independent media with a diversity of views. In Islam, R. (Ed.), *Information and public choice: From media markets to policy making* (pp. 139–152). Washington, DC: World Bank.

Stinchcombe, A. L. (1997). On the virtues of the old institutionalism. *Annual Review of Sociology, 23*(1), 1–18. doi:10.1146/annurev.soc.23.1.1

Strang, D., & Meyer, J. W. (1993). Institutional conditions for diffusion. *Theory and Society, 22*(4), 487–511. doi:10.1007/BF00993595

Tolbert, C. J., Mossberger, K., & McNeal, R. (2008). Institutions, policy innovation, and e-government in the American states. *Public Administration Review, 68*(3), 549–563. doi:10.1111/j.1540-6210.2008.00890.x

Treier, S., & Jackman, S. (2008). Democracy as a latent variable. *American Journal of Political Science, 52*(1), 201–217.

United Nations. (2005a). *The digital divide report: ICT diffusion index.* Paper presented at the United Nations Conference on Trade and Development. New York: UN.

United Nations. (2005b). *UN global e-government readiness report: From e-government to e-inclusion.* New York: UN.

West, D. M. (2005). *Digital government: Technology and public sector performance.* Princeton, NJ: Princeton University Press.

West, D. M. (2007). Global perspectives on e-government. In Mayer-Schonberger, V., & Lazer, D. (Eds.), *Governance and information technology: From electronic government to information government* (pp. 17–32). Cambridge, Massachusetts: MIT Press.

Williamson, O. E. (1999). Public and private bureaucracies: A transaction cost economics perspective. *Journal of Law Economics and Organization, 15*(1), 306–342. doi:10.1093/jleo/15.1.306

Wilson, E. J., III. (2003). Forms and dynamics of leadership for a knowledge society: The quad. *Center for International Development and Conflict Management, University of Maryland, College Park.* Retrieved from http://www.cidcm.umd.edu/leadership/quad2.pdf

Wilson, E. J. III. (2004). *The information revolution and developing countries.* Cambridge, MA: MIT Press.

Wilson, J. Q. (1989). *Bureaucracy: What government agencies do and why they do it.* New York: Basic Books.

World Bank. (2002). *Building institutions for markets: World Bank Development Report.* Washington, DC.

World Bank. (2004). *World Bank corporate corruption and ethics indices.* Retrieved from http://www.worldbank.org/wbi/governance/pdf/ETHICS.xls

World Economic Forum. (2006) *The global information technology report 2006-2007.* Retrieved from http://www.weforum.org/pdf/gitr/contents2007.pdf

Yoon, J., & Chae, M. (2009). Varying criticality of key success factors of national e-Strategy along the status of economic development of nations. *Government Information Quarterly, 26*(1), 25–34. doi:10.1016/j.giq.2008.08.006

ENDNOTES

[1]   We define e-Government broadly as the delivery of public sector information and services via the Internet and other digital means consistent with Andersen and Henriksen (2006).

[2]   Transparency International was founded in 1993 and has been monitoring and spearheading a global fight against corruption.

3   Via "*decoupling*…program implementation is…*ceremonialized*…organization can mobilize support from a broader range of… constituents. Thus, decoupling enables organizations…to be *similar*…but may show much diversity in actual practice." (Meyer & Rowan, 1977, p. 357, *italics* added)

4   Wilson's (2004) notion of "strategic restructuring" can serve an entrée into the adaptation of the structurational framework for exploring such institutional changes over time.

This work was previously published in International Journal of Global Information Management, Volume 18, Issue 2, edited by Felix B. Tan, pp. 85-104 copyright 2010 by IGI Publishing (an imprint of IGI Global).

Compilation of References

Abdalla, I., & Al-Homoud, M. (2001). Exploring the Implicit Leadership Theory in Arabian Gulf States. *Applied Psychology: An International Review*, *50*(4), 506–531. doi:10.1111/1464-0597.00071

Abdul-Gader, A. H. (1999). *Managing Computer-Based Information Systems in Developing Countries: A Cultural Perspective*. Hershey, PA: IGI-Global.

Abdul-Gader, A. H., & Al-Angari, K. (1995). *Information Technology Assimilation in the Government Sector: An Empirical Study*. King Abdul-Aziz City for Science and Technology.

Adam, M., & Myers, M. (2003). *Have you got anything to declare? Neo–colonialism, information systems, and the imposition of customs and duties in a third world country.* Paper presented at the International Federation of Information Processing, IFIP 9.4 and 8.2 Joint Conference on Organizational Information Systems in the Context of Globalization. Dordrecht, The Netherlands: Kluwer.

Adams, D. A., Nelson, R. R., & Todd, P. A. (1992). Perceived usefulness, ease of use, and usage of information technology: a replication. *Management Information Systems Quarterly*, *16*(2), 227–247. doi:10.2307/249577

Addison, T., & Heshmati, A. (2003). *The new global determinants of FDI flows to developing economies: The impacts of ICT and democratization*. WIDER Discussion Papers.

Afonso, A., Schuknecht, L., Tanzi, V., & Veldhuis, N. (2007). *Public sector efficiency: An International comparison*. The Fraser Institute. Retrieved September 7, 2011, from http://www.fraserinstitute.org/research-news/display.aspx?id=13328

Agarwal, R., & Prasad, J. (1998). A conceptual and operational definition of personal innovativeness in the domain of information technology. *Information Systems Research*, *9*(2), 204–215. doi:10.1287/isre.9.2.204

Agarwal, R. (2000). Individual acceptance of information technologies. In Zmud, R. W. (Ed.), *Framing the domains of IT management: Projecting the future...through the past*. Cincinnati, OH: Pinnaflex Educational Resources.

Agbonlahor, R. O. (2006). Motivation for use of information technology by university faculty: A developing country perspective. *Information Development*, *22*(4), 263–277. doi:10.1177/0266666906072955

Agerfalk, P. J., & Fitzgerald, B. (2008). Outsourcing to an unknown workforce: Exploring opensourcing as a global sourcing strategy. *Management Information Systems Quarterly*, *32*(2), 385–409.

AGIMO. (2009). *Interacting with government: Australians' use and satisfaction with e-government services*. Department of Finance and Administration, Australian Government Information Management Office, Commonwealth of Australia. Retrieved September 12, 2011, from http://www.finance.gov.au/publications/interacting-with-government-2009/ docs/ interacting-with-government-2009.pdf

Ahrens, J., & Meurers, M. (2002). How governance affects the quality of policy reform and economic performance: new evidence for economies in transition. *Journal for Institutional Innovation. Development and Transition*, *6*, 35–56.

Ahrens, J. (2006). Governance in the process of economic transformation. *Private University of Applied Sciences Goettingen*. Retrieved from www.oecd.org/dataoecd/52/20/37791185.pdf

Ahuja, M. K., & Thatcher, J. B. (2005). Moving beyond intentions and toward the theory of trying: Effects of work environment and gender on post-adoption information technology use. *Management Information Systems Quarterly*, *29*(3), 427–459.

Ahuja, M. K., & Carley, K. M. (1999). Network structure in virtual organizations. *Organization Science*, *10*(6), 741–757. doi:10.1287/orsc.10.6.741

Ajzen, I. (1991). The Theory of Planned Behavior. *Organizational Behavior and Human Decision Processes*, *50*(2), 179–211. doi:10.1016/0749-5978(91)90020-T

Ajzen, I., & Fishbein, M. (1980). *Understanding Attitudes and Predicting Social Behavior*. Englewood Cliffs, NJ: Prentice-Hall.

Ajzen, I. (1991). The theory of planned behavior. *Organizational Behavior and Human Decision Processes*, *50*(2), 179–211. doi:10.1016/0749-5978(91)90020-T

Ajzen, I. (2002). Perceived behavioral control, self-efficacy, locus of control, and the theory of planned behavior. *Journal of Applied Social Psychology*, *32*(4), 665–683. doi:10.1111/j.1559-1816.2002.tb00236.x

Akhter, S. H. (2003). Digital divide and purchase intention: Why demographic psychology matters. *Journal of Economic Psychology*, *24*(3), 321–327. doi:10.1016/S0167-4870(02)00171-X

Akmanligil, M., & Palvia, P. (2004). Strategies for Global Information Systems Development. *Information & Management*, *42*(1), 45–59.

Alanazi, F., & Rodrigues, A. (2003). Power Bases and Attribution in Three Cultures. *The Journal of Social Psychology*, *143*(3), 375–395. doi:10.1080/00224540309598451

Albers, M. J., & Kim, L. (2000). *User Web browsing characteristics using palm handheld for Information Retrieval*. Paper presented at the Proceedings of IEEE Professional Communication Society International Professional Communication Conference and Proceedings of the 18th Annual ACM International Conference on Computer Documentation: Technology and Teamwork, Cambridge, UK.

Alexander, D. (2010). *The relationship between information and communication technologies and foreign direct investment at the different stages of investment development path*. MBA dissertation, University of Pretoria.

Al-Farsy, F. (1996). *Modernity and Tradition: The Saudi Equation*. London: Panarc International Ltd.

Al-Gahtani, S. S. (2004). Computer Technology Acceptance Success Factors in Saudi Arabia: An Exploratory Study. *Journal of Global Information Technology Management*, *7*(1), 5–29.

Al-Gahtani, S. S., Hubona, G. S., & Wang, J. (2007). Information Technology (IT) in Saudi Arabia: Culture and the Acceptance and Use of IT. *Information & Management*, *44*(8), 681–691. doi:10.1016/j.im.2007.09.002

Al-Gahtani, S. S., & Shih, H.-P. (2009). The Influence of Organizational Communication Openness on the Post-Adoption of Computers: An Empirical Study in Saudi Arabia. *Journal of Global Information Management*, *17*(3), 20–41.

Ali, A. J. (1990). Management Theory in a Transitional Society: The Arab's Experience. *International Studies of Management and Organization*, *20*(3), 7–35.

Allan, J., Rambujan, N., Sood, S., Mbariak, V., Agrawal, R., & Saquib, Z. (2006). The e-government concept: A systematic review of research and practitioner literature. *IEEE, 4-9*(06).

Al-Qirim, N. (2006). Personas of E-Commerce adoption in small businesses in New Zealand. *Journal of Electronic Commerce in Organizations*, *4*(3), 18–45.

Al-Rafee, S., & Cronan, T. P. (2006). Digital piracy: Factors that influence attitude towards behavior. *Journal of Business Ethics*, *63*(3), 237–259. doi:10.1007/s10551-005-1902-9

Ambos, B., & Reitsperger, W. (2002). Governing Knowledge Processes in MNCs: The Case of German R&D Units Abroad. In *Proceedings of the 28th European International Business Academy Meeting*, Athens, Greece.

Amoako-Gyampah, K., & Salam, A. F. (2004). An extension of the technology acceptance model in an ERP implementation environment. *Information & Management*, *41*(6), 731–745. doi:10.1016/j.im.2003.08.010

Anandarajan, M., Igbaria, M., & Uzoamaka, P. A. (2002). IT acceptance in a less-developed country: A motivational factor perspective. *International Journal of Information Management, 22*(1), 47–65. doi:10.1016/S0268-4012(01)00040-8

Andersen, K. V., & Henriksen, H. Z. (2006). E-government maturity models: extension of the Layne and Lee model. *Government Information Quarterly, 23*(2), 236–248. doi:10.1016/j.giq.2005.11.008

Anderson, J., & Gerbing, D. (1988). Structural equation modeling in practice: A review and recommended two-step approach. *Psychological Bulletin, 103*(3), 411–423. doi:10.1037/0033-2909.103.3.411

Andrés, A. R. (2006). Software piracy and income inequality. *Applied Economics Letters, 13*(2), 101–105. doi:10.1080/13504850500390374

Ang, S., & Slaughter, S. A. (2001). Work outcomes and job design for contract versus permanent information systems professionals on software development teams. *Management Information Systems Quarterly, 25*(3), 321–350. doi:10.2307/3250920

Anthes, G. (2001). Think globally, act locally. *Computerworld, 35*(22), 36–37.

Anthony, R. N., & Govindarajan, V. (1998). *Management control systems* (9th ed.). Boston: Irwin/McGraw-Hill.

Armstrong, J., & Overton, T. (1977). Estimating Non-Response Bias in Mail Surveys. *JMR, Journal of Marketing Research, 14*(8), 396–402. doi:10.2307/3150783

Armstrong, C. P., & Sambamurthy, V. (1999). Information technology assimilation in firms: The influence of senior leadership and IT infrastructures. *Information Systems Research, 10*(4), 304–327. doi:10.1287/isre.10.4.304

Aronson, Z. H., Reilly, R. R., & Lynn, G. S. (2006). The impact of leader personality on new product development teamwork and performance: The moderating role of uncertainty. *Journal of Engineering and Technology Management, 23*(3), 221–247. doi:10.1016/j.jengtecman.2006.06.003

Aspray, W., Mayadas, F., & Vardi, M. Y. (2006). *Globalization and offshoring of Software: A report of the ACM Job Migration Task Force.*

Atiyyah, H. S. (1989). Determinants of Computer System Effectiveness in Saudi Arabian Public Organizations. *International Studies of Management and Organization, 19*(2), 85–103.

Atkinson, M. A., & Kydd, C. (1997). Individual characteristics associated with World Wide Web use: An empirical study of playfulness and motivation. *The Data Base for Advances in Information Systems, 28*(2), 53–62.

Attewell, P. (1992). Technology diffusion and organizational learning: The case of business computing. *Organization Science, 3*(1), 1–19. doi:10.1287/orsc.3.1.1

Au, K. Y. (1999). Intra-cultural variation: Evidence and implications for international business. *Journal of International Business Studies, 30*(4), 799–812. doi:10.1057/palgrave.jibs.8490840

August, T., & Tunca, T. I. (2008). Let the pirates patch? An economic analysis of software security patch restrictions. *Information Systems Research, 19*(1), 48–72. doi:10.1287/isre.1070.0142

Avgerou, C. (2000). IT and organizational change: An institutionalist perspective. *Information Technology & People, 13*(4), 234–262. doi:10.1108/09593840010359464

Avgerou, C. (2002). *Information systems and global diversity.* Oxford, UK: Oxford University Press.

Avgerou, C., & Walsham, G. (Eds.). (2000). *Information technology in context: Implementing systems in the developing world.* Brookfield, VT: Ashgate Publishing.

Avgerou, C. (1996, August 16-19). Studying the socio–economic context of information systems. In *Proceedings of the Second Americas Conference on Information Systems.* Phoenix, Arizona.

Avgerou, C. (2009, May 6-28). *Discourses on innovation and development in information systems in developing countries' research.* Paper presented at the 10th International Conference of the IFIP 9.4 working group on Social Implications of Computers in Developing Countries Dubai, UAE.

Avolio, B. J., & Dodge, G. E. (2001). E-leadership: Implications for Theory, Research, and Practice. *The Leadership Quarterly, 11*(4), 615–668. doi:10.1016/S1048-9843(00)00062-X

Ayoun, B. M., & Moreo, P. J. (2008). The influence of the cultural dimension of uncertainty avoidance on business strategy development: A cross-national study of hotel managers. *International Journal of Hospitality Management, 27*(1), 65–75. doi:10.1016/j.ijhm.2007.07.008

Azad, B., & Faraj, S. (2009). E-Government institutionalizing practices of a land registration mapping system. *Government Information Quarterly, 26*(1), 5–14. doi:10.1016/j.giq.2008.08.005

Ba, S., & Pavlou, P. A. (2002). Evidence of the effect of trust building technology in electronic markets: Price premiums and buyer behaviour. *Management Information Systems Quarterly, 26*(3), 243–268. doi:10.2307/4132332

Ba, S., Stallaert, J., & Whinston, A. B. (2001). Research commentary: Introducing a third dimension in information systems design: The case for incentive alignment. *Information Systems Research, 12*(3), 225–239. doi:10.1287/isre.12.3.225.9712

Baba, M. L., Gluesing, J., Ratner, H., & Wagner, K. H. (2004). The contexts of knowing: natural history of a globally distributed team. *Journal of Organizational Behavior, 25*(5), 547–587. doi:10.1002/job.259

Backus, M. (2001). *E-Governance and developing countries – Introduction and examples* (Research Rep. No. 3). Retrieved July, 10, 2007, from http://www.ftpiicd.org/files/research/reports/report3.pdf

Bada, A. O., Aniebonam, M. C., & Owei, V. (2004). Institutional pressures as sources of improvisations: A case study from a developing country context. *Journal of Global Information Technology Management, 7*(3), 27–42.

Bae, S. H., & Choi, J. P. (2006). A model of piracy. *Information Economics and Policy, 18*(3), 303–320. doi:10.1016/j.infoecopol.2006.02.002

Baets, W. (1992). Aligning information systems with business strategy. *The Journal of Strategic Information Systems, 1*(4), 205–213. doi:10.1016/0963-8687(92)90036-V

Bagchi, K., Kirs, P., & Cerveny, R. (2006). Global software piracy: Can economic factors alone explain the trend? *Communications of the ACM, 49*(6), 70–75. doi:10.1145/1132469.1132470

Bagozzi, R. P., Davis, F. D., & Warshaw, P. R. (1992). Development and test of a theory of technological learning and usage. *Human Relations, 45*(7), 659–686. doi:10.1177/001872679204500702

Bai, S. Z., & Yang, Y. (2008). *Study on the information sharing incentive and supervisory mechanism in supply chain.* Paper presented at the 2008 International Conference on Wireless Communications, Networking and Mobile Computing (WiCOM 2008).

Bajaj, A., & Nidumolu, S. R. (1998). A feedback model to understand information system usage. *Information & Management, 33*(4), 213–224. doi:10.1016/S0378-7206(98)00026-3

Baker, E. W., Al-Gahtani, S. S., & Hubona, G. S. (2007). The Effects of Gender and Age on New Technology Implementation in a Developing Country: Testing the Theory of Planned Behavior. *Information Technology & People, 20*(4), 352–375. doi:10.1108/09593840710839798

Baliamoune-Lutz, M. (2003). An analysis of the determinants and effects of ICT diffusion in developing economies. *Information Technology for Development, 10*(3), 151–169. doi:10.1002/itdj.1590100303

Ball, L., & Harris, R. (1982). SMIS Membership Analysis. *Management Information Systems Quarterly, 6*(1), 19–38. doi:10.2307/248752

Ballantine, J., Levy, M., & Powel, P. (1998). Evaluating information systems in small and medium-sized enterprises: issues and evidence. *European Journal of Information Systems, 7*(4), 241–251. doi:10.1057/palgrave.ejis.3000307

Bancroft, N. H. (1998). *Implementing SAP R/3: How to introduce a large system into a large organization.* London: Manning/Prentice Hall.

Bandura, A. (2001). Social cognitive theory: An agentive perspective. *Annual Review of Psychology, 52*(1), 1–26. doi:10.1146/annurev.psych.52.1.1

Banerjee, D., Khalid, A. M., & Sturm, J.-E. (2005). Socio-economic development and software piracy: An empirical assessment. *Applied Economics, 37*(18), 2091–2097. doi:10.1080/00036840500293276

Banerjee, P., & Chau, P. Y. K. (2004). An evaluative framework for analysing e–government convergence capability in developing countries. *Electronic Government*, *1*(1), 29–48. doi:10.1504/EG.2004.004135

Bankole, F. O., Bankole, O. O., & Brown, I. (2011). Mobile banking adoption in Nigeria. *The Electronic Journal of Information Systems in Developing Countries*, *47*(2), 1–23.

Barney, J. (1996). *Gaining and Sustaining Competitive Advantage*. Reading, MA: Addison Wesley.

Barney, J. B., & Quchi, W. G. (1986). *Organizational Economics* (3rd ed.). San Francisco, CA: Jossey-Bass Publishers.

Baron, R. M., & Kenny, D. A. (1986). The moderator-mediator variable distinction in social psychological research: conceptual, strategic, and statistical considerations. *Journal of Personality and Social Psychology*, *51*(1), 173–182.

Barrell, R., & Pain, N. (1997). Foreign direct investment, technological change, and economic growth within Europe. *The Economic Journal*, *107*(445), 1770–1786. doi:10.1111/1468-0297.00256

Barroso, J. L. G., & Martinez, J. P. (2005). The geography of the digital divide: broadband deployment in the Community of Madrid. *Univ Access Inf Soc*, *3*, 264–271. doi:10.1007/s10209-004-0103-0

Bartlett, C., & Ghoshal, S. (1989). *Managing Across Borders: The Transnational Solution*. Boston: Harvard Business School Press.

Baskerville, R. F. (2003). Hofstede Never Studied Culture. *Accounting, Organizations and Society*, *28*(1), 1–14. doi:10.1016/S0361-3682(01)00048-4

Basu, S. (2004). E-government and developing countries: An overview. *International Review of Law Computers*, *18*(1), 109–132. doi:10.1080/13600860410001674779

Basu, V., & Lederer, A. L. (2004). *An agency theory model of ERP implementation*. Paper presented at the SIGMIS04, Tucson, AZ.

Bazely, P. (2007). *Qualitative data analysis with NVIVO*. London, UK: Sage.

BBS. (2009). Statistical pocket book of Bangladesh, 2009. *Bangladesh Bureau of Statistics: Planning division, Ministry of Planning, Government of the People's Republic of Bangladesh*. Retrieved September, 7, 2011, from http://www.bbs.gov.bd/WebTestApplication/userfiles/Image/SubjectMatterDataIndex/pk_book_09.pdf

Beach, W., & Miles, A. M. (2006). *Explaining the factors of the index of economic freedom*. 2006 Index of Economic Freedom, The Heritage Foundation and Wall Street Journal.

Beard, J. W., & Sumner, M. (2004). Seeking strategic advantage in the post-net era: viewing ERP systems from the resource-based perspective. *Strategic Information Systems*, *13*(2), 129–150. doi:10.1016/j.jsis.2004.02.003

Beck, N., & Katz, J. N. (1995). What to do (not to do) with time-series cross-section data. *The American Political Science Review*, *89*, 634–647. doi:10.2307/2082979

Beilock, R., & Dimitrova, D. V. (2003). An exploratory model of inter-country Internet diffusion. *Telecommunications Policy*, *27*(3-4), 237–252. doi:10.1016/S0308-5961(02)00100-3

Bem, D. J. (1972). *Self-perception theory*. New York: Academic Press.

Benbasat, I., & Barki, H. (2007). Quo Vadis, TAM? *Journal of AIS*, *8*(4), 211–218.

Benbasat, I., & Zmud, R. W. (1999). Empirical research in Information Systems: the practice of relevance. *Management Information Systems Quarterly*, *23*(1), 3–16. doi:10.2307/249403

Berry, J. W. (1989). Imposed etics-emics-derived etics: The operationalization of a compelling idea. *International Journal of Psychology*, *24*, 721–735.

Besley, T., Burgess, R., & Prat, A. (2002). Mass media and political accountability. In Islam, R. S. (Ed.), *The right to tell: the role of mass media in economic development* (pp. 45–60). Washington, DC: World Bank Publications.

Beverakis, G., Dick, G. N., & Cecez-Kecmanovic, D. (2009). Taking information systems business process outsourcing offshore: The conflict of competition and risk. *Journal of Global Information Management*, *17*(1), 32–48. doi:10.4018/jgim.2009010102

Bhagat, R. G., Kedia, B. L., Harveston, P. D., & Triandis, H. C. (2002). Cultural variations in the cross-border transfer of organizational knowledge: An integrative framework. *Academy of Management Review, 27*(2), 204–239.

Bharadwaj, A. S. (2000). A resource-based perspective on information technology capability and firm performance: An empirical investigation. *MIS Quarterly, 24*(1), 169–197. doi:10.2307/3250983

Bhattacherjee, A. (2001). Understanding information systems continuance: An expectation- confirmation model. *Management Information Systems Quarterly, 25*(3), 351–370. doi:10.2307/3250921

Bhuian, S., Abdulmuhmin, A., & Kim, D. (2001). Business Education and its Influence on Attitudes to Business, Consumerism and Government Saudi Arabia. *Journal of Education for Business, 76*(4), 226–230. doi:10.1080/08832320109601315

Biddle, B. J., & Thomas, E. J. (1966). *Role theory: Concepts and research.* New York, NY: John Wiley and Sons.

Biggart, N. W., & Hamilton, G. G. (1992). *On the limits of a firm-based theory to explain business network: The Western bias of neoclassical economics.* Boston: Harvard Business School Press.

Bingi, P., Godla, J. K., & Sharma, M. K. (1999). Critical issues affecting an ERP implementation. *Information Systems Management, 16*(3), 7–14. doi:10.1201/1078/43197.16.3.19990601/31310.2

Bjerke, B., & Al-Meer, A. (1993). Culture's Consequences: Management in Saudi Arabia. *Leadership and Organization Development Journal, 14*(2), 30–35. doi:10.1108/01437739310032700

Bland, C. J., & Ruffin, M. T. Iv. (1992). Characteristics of a productive research environment: Literature review. *Academic Medicine, 67*(6), 385–397. doi:10.1097/00001888-199206000-00010

Bollen, K. A. (1989). *Structural equations with latent variables.* New York: John Wiley & Sons.

Bolloju, N., & Turban, E. (2007). Organizational assimilation of web services technology: A research framework. *Journal of Organizational Computing and Electronic Commerce, 17*(1), 29–52.

Bongalia, F., Goldstein, A., & Mathews, J. (2006). Accelerated internationalization by emerging multinationals: The case of white goods sector. *Munich Personal RePEc Archive, 1485,* 1–37.

Bontis, N. (2004). National intellectual capital index: A United Nations initiative for Arab region. *Journal of Intellectual Studies, 5*(1), 13–39.

Borg, I., Groenen, P. J. F., Jehn, K. A., Bilsky, W., & Schwartz, S. H. (2011). Embedding the organization culture profile into Schwartz's theory of universals in values. *Journal of Personnel Psychology, 10*(1), 1–12. doi:10.1027/1866-5888/a000028

Boudreau, P. (2000). L'expertise d'un enseignant associe. *McGill Journal of Education, 35*(1), 53–70.

Bowman, B., Davis, G., & Wetherbe, J. (1983). Three Stage Model of MIS Planning. *Information & Management, 6*(1), 11–25. doi:10.1016/0378-7206(83)90016-2

Boyer-Wright, K. M., & Kotterman, J. E. (2008). *High-level factors affecting global availability of online government services.* Paper presented at Proceedings of the 41st Annual Hawaii International Conference on System Sciences.

Braa, J., Monteiro, E., & Sahay, S. (2004). Networks of action: Sustainable health information systems across developing countries. *Management Information Systems Quarterly, 28*(3), 337–362.

Bradford, M., & Florin, J. (2003). Examining the role of innovation diffusion factors on the implementation success of enterprise resource planning systems. *International Journal of Accounting Information Systems, 4*(3), 205–225. doi:10.1016/S1467-0895(03)00026-5

Brancheau, J. C., Janz, B. D., & Wetherbe, J. C. (1996). Key Issues in Information Systems Management: 1994-1995 SIM Delphi Results. *Management Information Systems Quarterly,* 225–242. doi:10.2307/249479

Brennan, J., & Shah, T. (2000). *Managing quality in higher education.* Buckingham: OECD, SRHE and Open University Press.

Bretschneider, S. (1990). Management Information Systems in public and private organizations: An empirical test. *Public Administration Review, 50*(5), 536–545. doi:10.2307/976784

Brislin, R. (1986). The Wording and Translation of Research Instruments. In Lonner, W., & Berry, J. (Eds.), *Field Methods in Cross-Cultural Research* (pp. 137–164). Beverly Hills, CA: Sage Publications.

Brown, S. A., Massey, A. P., Montoya-Weiss, M. M., & Burkman, J. R. (2002). Do I really have to? User acceptance of mandated technology. *European Journal of Information Systems*, *11*(4), 283–295. doi:10.1057/palgrave.ejis.3000438

Brown, I. T. J. (2002). Individual and technological factors affecting perceived ease of use of Web-based learning technologies in a developing country. *The Electronic Journal on Information Systems in Developing Countries*, *9*(5), 1–15.

Brown, I., & Licker, P. (2003). Exploring differences in internet adoption and usage between historically advantaged and disadvantaged groups in South Africa. *Journal of Global Information Technology Management*, *6*(4), 6–26.

Brown, L. (1981). *Innovation diffusion*. London, UK: Methuen.

Brunelle, J. P. (2001). The impact of community service on adolescent volunteers' empathy, social responsibility, and concern for others. Unpublished doctoral dissertation, Virginia Common Wealth University, Virginia.

Bruner, G. C., & Kumar, A. (2005). Explaining consumer acceptance of handheld Internet devices. *Journal of Business Research*, *58*(5), 553–558. doi:10.1016/j.jbusres.2003.08.002

Bruns, W. J. Jr, & McFarlan, F. W. (1987). Information technology puts power in control systems. *Harvard Business Review*, *65*(5), 89–94.

BSA. (2003). *Eighth Annual BSA Global Software Piracy Study* (available online at: http://global.bsa.org/global-study/2003_GSPS.pdf).

BSA. (2008). *Fifth Annual BSA and IDC Global Software Piracy Study* (available online at: http://global.bsa.org/idcglobalstudy2007/).

Buonanno, G., Faverio, P., Pigni, F., Ravarini, A., Sciuto, D., & Tagliavini, M. (2005). Factors affecting ERP system adoption. *Journal of Enterprise Information Management*, *18*(4), 384–426. doi:10.1108/17410390510609572

Burn, J., Davison, R., & Jordan, E. (1997). The Information Society - A Cultural Fallacy. *The Journal of Failures and Lessons Learned in IT Management*, *1*(4), 219–232.

Burn, J., Saxena, K., Ma, L., & Cheung, H. (1993). Critical Issues of IS Management in Hong Kong: A Cultural Comparison. *Journal of Global Information Management*, *1*(4), 28–37.

Byrne, B. M., Shavelson, R. J., & Muthen, B. (1989). Testing for the equivalence of factor covariance and mean structures: the issue of partial measurement invariance. *Psychological Bulletin*, *105*(3), 456–466. doi:10.1037/0033-2909.105.3.456

Calantone, R. J., Cavusgil, S. T., & Zhao, Y. (2002). Learning orientation, firm innovation capability, and firm performance. *Industrial Marketing Management*, *31*(6), 515–524. doi:10.1016/S0019-8501(01)00203-6

Campbell, D. T., & Fiske, D. W. (1959). Convergent and Discriminant Validation by the Multitrait-Multimethod Matrix. *Psychological Bulletin*, *56*(2), 81–105. doi:10.1037/h0046016

Campos, N. F., & Nugent, J. B. (1999). Development performance and the institutions of governance: evidence from East Asia and Latin America. *World Development*, *27*(3), 439–452. doi:10.1016/S0305-750X(98)00149-1

Cao, C., Suttmeier, R. P., & Simon, D. F. (2006). China's 15-year science and technology plan. *Physics Today*, *59*(12), 38. doi:10.1063/1.2435680

Caprara, G., Vecchione, M., & Schwartz, S. H. (2009). Mediational role of values in linking personality traits to political orientation. *Asian Journal of Social Psychology*, *12*, 82–94. doi:10.1111/j.1467-839X.2009.01274.x

Carbonell, P., & Rodriguez, A. I. (2006). Designing teams for speedy product development: The moderating effect of technological complexity. *Journal of Business Research*, *59*(2), 225–232. doi:10.1016/j.jbusres.2005.08.002

Carlson, J. R., & Zmud, R. (1999). Channel expansion theory and the experiential nature of media richness perceptions. *Academy of Management Journal*, *42*(2), 153–170. doi:10.2307/257090

Caudle, S. L., Gorr, W. L., & Newcomer, K. E. (1991). Key Information System Management Issues for the Public Sector. *Management Information Systems Quarterly*, 171–188. doi:10.2307/249378

Cava-Ferreruela, I., & Alabau-Muñoz, A. (2006). Broadband policy assessment: A cross-national empirical analysis. *Telecommunications Policy*, 30(8-9), 445–463. doi:10.1016/j.telpol.2005.12.002

Chae, M. H., Kim, J. W., Kim, H. Y., & Ryu, H. S. (2002). Information quality for mobile data services: A theoretical model with empirical validation. *Electronic Markets*, 12(1), 38–46. doi:10.1080/101967802753433254

Chai, L., & Pavlou, P. (2004). From 'Ancient' to 'Modern': A Cross-Cultural Investigation of Electronic Commerce Adoption in Greece and the United States. *Journal of Enterprise Information Management*, 17(6), 416–423. doi:10.1108/17410390410566706

Chambers, R. (1994). All power deceives. *IDS Bulletin*, 25(2), 14–26. doi:10.1111/j.1759-5436.1994.mp25002002.x

Chan, Y., & Huff, S. (1992). Strategy: An Information Systems Research Perspective. *The Journal of Strategic Information Systems*, 1(4), 191–204. doi:10.1016/0963-8687(92)90035-U

Chan, R. (1999). Knowledge management for implementing ERP in SMEs. In *Proceedings of the 3rd Annual SAP Asia Pacific*. Singapore: Institute of Higher Learning Forum.

Chang, S. I., Chang, H. C., & Chang, I. C. (2008). The key factors of innovation ability and industry's high value in small and medium-sized enterprises: Evidence from information services industry. *Taiwan Business Performance Journal*, 1(2), 175–201.

Chang, S. I., & Gable, G. G. (2002). A comparative analysis of major ERP lifecycle implementation, management and support issues in Queensland Government. *Journal of Global Information Management*, 10(3), 36–54.

Chatterjee, D., Grewal, R., & Sambamurthy, V. (2002). Shaping up for e-commerce: Institutional enablers of the organizational assimilation of web technologies. *MIS Quarterly*, 26(2), 65–89. doi:10.2307/4132321

Chau, P. Y. K. (2008). Cultural Differences in Diffusion, Adoption, and Infusion of Web 2.0. *Journal of Global Information Management*, 16(1), I–III.

Chau, P. Y. K., Cole, M., Massey, A. P., Montoya-Weiss, M., & O'Keefe, R. M. (2002). Cultural differences in the online behavior of consumers. *Communications of the ACM*, 45(10), 138–143. doi:10.1145/570907.570911

Chau, P. Y. K., & Tam, K. Y. (1997). Factors affecting the adoption of open systems: An exploratory study. *MIS Quarterly*, 21(1), 1–24. doi:10.2307/249740

Chau, P. Y. K. (2008). Cultural differences in diffusion, adoption and infusion of Web 2.0. *Journal of Global Information Management*, 16(1), i–ii.

Chen, G., Wu, R., & Guo, X. (2007). Key Issues in Information Systems Management in China. *Journal of Enterprise Information Management*, 20(2), 198–208. doi:10.1108/17410390710725779

Chen, C. C., Chen, X., & Meindl, J. R. (1998). How can cooperation be fostered? The cultural effects of individualism-collectivism. *Academy of Management Review*, 23(2), 285–304.

Chen, J. C., & Williams, B. C. (1998). The impact of electronic data interchange on SMEs: Summary of eight British case studies. *Journal of Small Business Management*, 36(4), 68–72.

Chen, R. S., Sun, C. M., Helms, M. M., & Jin, W. J. (2008). Role negotiation and interaction: An exploratory case study of the impact of management consultants on ERP system implementation in SMEs in Taiwan. *Information Systems Management*, 25(2), 159–173. doi:10.1080/10580530801941371

Chen, G., Wu, R., & Guo, X. (2007). Key issues in information systems management in China. *Journal of Enterprise Information Management*, 20(2), 198–208. doi:10.1108/17410390710725779

Chen, Y. N., Chen, H. M., Huang, W., & Ching, R. K. H. (2006). E-government strategies in developed and developing countries: an implementation framework and case study. *Journal of Global Information Management*, 14(1), 23–46.

Cheung, H. (2006). *Windows Vista Ultimate for $3.50* (available online at: http:// www.tgdaily.com/ content/ view/30080/118/).

Chin, W. W., Marcolin, B. L., & Newsted, P. R. (2003). Partial Least Squares Latent Variables Modeling Approach for Measuring Interaction Effects: Results from a Monte Carlo Simulation Study and an Electronic-Mail Emotion/ Adoption Study. *Information Systems Research, 14*(2), 189–217. doi:10.1287/isre.14.2.189.16018

Chin, W. W. (1998). Issues and opinion on structural equation modeling. *Management Information Systems Quarterly, 22*(1), 7–16.

Chin, W. W., & Gopal, A. (1995). Adoption intention in GSS: Relative importance of beliefs. *The Data Base for Advances in Information Systems, 26*(2&3), 42–64.

Chin, W. (1998). The partial least squares approach for structural equation modeling. In Marcoulides, G. (Ed.), *Modern Methods for Business Research*. Hillsdale, NJ: Lawrence Erlbaum Associates.

Chin, W. W. (2001). *PLS-Graph User's Guide, Version 3.0*. Unpublished manuscript.

Chin, W. W., Marcolin, B. L., & Newsted, P. R. (1996). *A Partial Least Squares Latent Variables Modelling Approach for Measuring Interaction Effects: Results from a Monte-Carlo Simulation Study and Voice Mail Emotion/Adoption Study*. Paper presented at the International Conference on Information Systems, Cleveland, OH.

China Enterprise IT application development report. (2003). China National Information Center. Retrieved from http://www.cirm.net.cn

Chinn, M. D., & Fairlie, R. W. (2007). The determinants of the global digital divide: A cross-country analysis of computer and internet penetration. *Oxford Economic Papers, 59*, 16–44. doi:10.1093/oep/gpl024

Chiu, H.-C., Hsieh, Y.-C., & Wang, M.-C. (2008). How to encourage customers to use legal software. *Journal of Business Ethics, 80*(3), 583–595. doi:10.1007/s10551-007-9456-7

Chiu, F. Y., Wu, H. C., & Ho, T. F. (2001). *The strategy framework of ERP implementation for small and medium enterprise in Taiwan- A case study of manufacturing industry*. Paper presented at the National Computer Symposium, Taipei, Taiwan.

Choi, M., Lee, I. S., Choi, H., & Kim, J. W. (2003). *A cross-cultural study on the post-adoption behavior of mobile internet users*. Paper presented at the Proceedings of the Digit 2003 (Pre-ICIS), Seattle, WA.

Choudhury, V., & Sabherwal, R. (2003). Portfolios of control in outsourced software development projects. *Information Systems Research, 14*(3), 291–314. doi:10.1287/isre.14.3.291.16563

Chowdhry, S. (2007). *Foreign direct investment (FDI) on the rise in OIC economies*. Retrieved from http://www.dinarstandard.com/current/OIC_FDI040907.htm

Churchill, G. A. (1979). A paradigm for developing better measures of marketing constructs. *Journal of Marketing, 16*(1), 64–73. doi:10.2307/3150876

Chwelos, P., Benbasat, I., & Dexter, A. S. (2001). A Dexter research report: empirical test of an EDI adoption model. *Information Systems Research, 12*(3), 304–321. doi:10.1287/isre.12.3.304.9708

Ciborra, C. (2005). Interpreting e-government and development: efficiency, transparency or governance at a distance. *Information Technology & People, 18*(3), 260–279. doi:10.1108/09593840510615879

Clemons, E., & Row, M. (1991). Information Technology at Rosenbluth Travel: Competitive Advantage in a Rapidly Growing Global Service Company. *Journal of Management Information Systems, 8*(2), 53–79.

Clemons, E. K., & Kimbrough, S. O. (1986). Information Systems, Telecommunications, and Their Effects on Industrial Organization. In *Proceedings of the 7th International Conference on Information Systems* (pp. 99-108).

Clemons, E. K., & Row, M. (1987, December 1-9). Structural Differences among Firms: A Potential Source of Competitive Advantage in the Application of Information Technology. In *Proceedings of the 8th International Conference on Information Systems*.

CNNIC report-Top News. (2005). *Chinese Information System Engineering, 16*.

Coe, D. T., Helpman, E., & Hoffmaister, A. W. (1997). North-south R&D spillovers. *The Economic Journal*, *107*(440), 134–149. doi:10.1111/1468-0297.00146

Cohen, J. (1988). *Statistical power analysis for the behavioral sciences*. Hillsdale, NJ: Erlbaum.

Compeau, D. R., & Higgins, C. A. (1995). Application of Social Cognitive Theory to Training for Computer Skills. *Information Systems Research*, *6*(2), 118–143. doi:10.1287/isre.6.2.118

Compeau, D. R., Higgins, C. A., & Huff, S. L. (1999). Social Cognitive Theory and Individual Reactions to Computing Technology: A Longitudinal Study. *Management Information Systems Quarterly*, *23*(2), 145–158. doi:10.2307/249749

Compeau, D. R., & Higgins, C. A. (1995). Computer self-efficacy: Development of a measure and initial test. *Management Information Systems Quarterly*, *19*(2), 189–211. doi:10.2307/249688

Cooper, R. B. (2000). Information technology development creativity: A case study of attempted radical change. *Management Information Systems Quarterly*, *24*(2), 245–276. doi:10.2307/3250938

Cooper, R. B., & Zmud, R. W. (1990). Information technology implementation research: A technological diffusion approach. *Management Science*, *36*(2), 123–139. doi:10.1287/mnsc.36.2.123

Coursey, D., & Norris, D. F. (2008). Models of e-government: are they correct? An empirical assessment. *Public Administration Review*, *68*(3), 523–536. doi:10.1111/j.1540-6210.2008.00888.x

Cragg, P. B., & King, M. (1993). Small-Firm computing: Motivators and inhibitors. *Management Information Systems Quarterly*, *17*(1), 47–60. doi:10.2307/249509

Cravens, D. (1988). Gaining strategic marketing advantage. *Business Horizons*, *31*(5), 44–54. doi:10.1016/0007-6813(88)90054-7

Cronan, T. P., & Al-Rafee, S. (2008). Factors that influence the intention to pirate software and media. *Journal of Business Ethics*, *78*(4), 527–545. doi:10.1007/s10551-007-9366-8

Cronan, T. P., Foltz, C. B., & Jones, T. W. (2006). Piracy, computer crime, and IS misuse at the University. *Communications of the ACM*, *49*(6), 85–90. doi:10.1145/1132469.1132472

Cui, L., Zhang, C., Zhang, C., & Huang, L. (2008). Exploring IT adoption process in Shanghai firms: An empirical study. *Journal of Global Information Management*, *16*(2), 1–17.

Cyranoski, D. (2006). Named and shamed. *Nature*, *441*(7092), 392–393. doi:10.1038/441392a

Dabholkar, P. A. (1996). Consumer evaluations of new technology-based self-service options: An investigation of alternative models. *International Journal of Research in Marketing*, *13*(1), 29–51. doi:10.1016/0167-8116(95)00027-5

Dada, D. (2006). The failure of e-government in developing countries: A literature review. *The Electronic Journal on Information Systems in Developing Countries*, *26*(7), 1–10.

Dadfar, A., Norberg, R., Helander, E., Schuster, S., & Zufferey, A. (2003). *Intercultural Aspects of Doing Business with Saudi Arabia*. Linkoping, Sweden: Linkoping University.

Dadfar, H. (1990). *Industrial Buying Behavior in the Middle East*. Linkoping, Sweden: Linkoping University.

Daft, R. L., & Lengel, R. H. (1986). Organizational information requirements, media richness and structural design. *Management Science*, *32*(5), 554–571. doi:10.1287/mnsc.32.5.554

Damanpour, F. (1992). Organization size and innovation. *Organization Studies*, *13*(3), 375–402. doi:10.1177/017084069201300304

Damanpour, F. (1987). The adoption of technological, administrative and ancillary innovations: Impact of organizational factors. *Journal of Management*, *13*, 675–688. doi:10.1177/014920638701300408

Daniel, H.-D. (2005). Publications as a measure of scientific advancement and of scientists' productivity. *Learned Publishing*, *18*, 143–148. doi:10.1087/0953151053584939

Das, S., Zahra, S., & Warkentin, M. (1991). Integrating the Content and Process of Strategic MIS Planning with Competitive Strategy. *Decision Sciences, 22*(1), 953–984.

Dasgupta, S., Agarwal, D., Ioannidis, A., & Gopalakrishnan, S. (1999). Determinants of information technology adoption: An extension of existing models to firms in a developing country. *Journal of Global Information Management, 7*(3), 30–40.

Davenport, T. H. (1998). Putting the enterprise into the enterprise system. *Harvard Business Review, 76*(4), 121–131.

Davenport, T. H. (2000). *Mission critical: Realizing the promise of enterprise systems.* Boston: Harvard Business School Press.

Davenport, T. H., & Markus, M. L. (1999). Rigor vs. relevance revisited: response to Benbasat and Zmud. *Management Information Systems Quarterly, 23*(1), 19–23. doi:10.2307/249405

Davis, F. D. (1989). Perceived Usefulness, Perceived Ease of Use, and User Acceptance of Information Technology. *Management Information Systems Quarterly, 13*(3), 319–340. doi:10.2307/249008

Davis, F. D., Bagozzi, R. P., & Warshaw, P. R. (1989). User Acceptance of Computer Technology: a Comparison of Two Theoretical Models. *Management Science, 35*(8), 982–1003. doi:10.1287/mnsc.35.8.982

Davis, F. D. (1989). Perceived Usefulness, Perceived Ease of Use, and User Acceptance of Information Technology. *Management Information Systems Quarterly, 13*(3), 319–340. doi:10.2307/249008

Davis, F. D., Bagozzi, R. P., & Warshaw, P. R. (1989). User acceptance of computer technology a comparison of two theoretical models. *Management Science, 35*(8), 982–1003. doi:10.1287/mnsc.35.8.982

Davis, F. D., Bagozzi, R. P., & Warshaw, P. R. (1992). Extrinsic and intrinsic motivation to use computers in the workplace. *Journal of Applied Social Psychology, 22*(14), 1111–1132. doi:10.1111/j.1559-1816.1992.tb00945.x

Davis, J. H., & Ruhe, J. A. (2003). Perceptions of country corruption: Antecedents and outcomes. *Journal of Business Ethics, 43*(4), 275–288. doi:10.1023/A:1023038901080

Davis, F. D. (1989). Perceived usefulness, perceived ease of use, and user acceptance of information technology. *Management Information Systems Quarterly, 13*(3), 319–340. doi:10.2307/249008

Davis, G. B., Ein-Dor, P., King, W. R., & Torkzadeh, R. (2006). IT offshoring: History, prospects and challenges. *Journal of the Association for Information Systems, 7*(11), 770–795.

Davis, F. D., Bagozzi, R. P., & Warshaw, P. R. (1989). User acceptance of computer technology: A comparison of two theoretical models. *Management Science, 35*(8), 982–1003. doi:10.1287/mnsc.35.8.982

Davison, R. M., & Martinsons, M. G. (2003). Cultural Issues and IT Management: Past and Present. *IEEE Transactions on Engineering Management, 50*(1), 3–7. doi:10.1109/TEM.2003.808249

Davison, R., & Jordan, E. (1998). Group Support Systems: Barriers to Adoption in a Cross-Cultural Setting. *Journal of Global Information Technology Management, 1*(2), 37–50.

Dawes, S. (2002a). *The future of egovernment. An examination of New York City's e-government initiatives.* Retrieved August 18, 2011, from http://www.ctg.albany.edu/publications/reports/future_of_egov/future_of_egov.pdf

Dawes, S. (2002b). *Government and technology: User, not regulator. Book Review: Fountain. 2001. Building the virtual state: Information Technology and institutional change.* Washington, DC. Retrieved August 29, 2011, from http://jpart.oxfordjournals.org/cgi/reprint/12/4/627.pdf

Dawes, S., Pardo, T., Simon, S., Cresswell, A., LaVigne, M., Andersen, D., et al. (2004). *Making smart IT choices: Understanding value and risk in government IT investments.* Retrieved August 29, 2011, from http://www.ctg.albany.edu/publications/guides/smartit2

Day-Hookoomsing, P. N. (2000). Leadership training for improved quality in a post-colonial, multicultural society. *The International Journal of Sociology and Social Policy, 20*(8), 23–32. doi:10.1108/01443330010789025

De Guinea, A., Kelley, H., & Hunter, M. G. (2005). Information systems effectiveness in small businesses: Extending a Singaporean model in Canada. *Journal of Global Information Management, 13*(3), 55–79.

De Mooij, M. (2004). *Consumer behavior and culture: Consequences for global marketing and advertising.* Thousand Oaks, CA: Sage Publications.

De Mooij, M., & Hofstede, G. (2002). Convergence and divergence in consumer behavior: implications for international retailing. *Journal of Retailing, 78*(1), 61–69. doi:10.1016/S0022-4359(01)00067-7

de Soto, H. (2000). *The mystery of capital: Why capitalism triumphs in the west and fails everywhere else.* New York: Basic Books.

Deans, P. C., Karwan, K., Goslar, M., Ricks, D., & Toyne, B. (1991). Identification of Key International Information Systems Issues in U.S-based Multinational Corporations. *Journal of Management Information Systems, 7*(4), 27–50.

Deans, P. C., Karwan, K. R., Goslar, M. D., Ricks, D. A., & Toyne, B. (1991). Identification of key international information systems issues in U.S.-based multinational corporations. *Journal of Management Information Systems, 7*(4), 27–50.

Dedrick, J., Kraemer, K. L., Palacios, J. J., Tigre, P. B., & Botelho, A. J. J. (2001). Economic liberalization and the computer industry: Comparing outcomes in Brazil and Mexico. *World Development, 29*(7). doi:10.1016/S0305-750X(01)00038-9

Dekleva, S., & Zupancic, J. (1996). Key Issues in Information Systems Management: a Delphi Study in Slovenia. *Information & Management, 31*, 1–11. doi:10.1016/S0378-7206(96)01066-X

Del Galdo, E., & Nielsen, J. (Eds.). (1996). *International User Interfaces.* New York: John Wiley and Sons.

Delasnerie, S. (2007,October 3). View from Middle East & Africa. *fDimagazine.com,* (p. 69).

Delbecq, A. L., & Van de Ven, A. H. (1971). A group process model for problem identification and program planning. *The Journal of Applied Behavioral Science, 7*(4), 466. doi:10.1177/002188637100700404

Delone, W. H. (1988). Determinants of success for computer usage in small business. *Management Information Systems Quarterly, 12*(1), 51–61. doi:10.2307/248803

Depken, C. A., & Simmons, L. C. (2004). Social construct and the propensity for software piracy. *Applied Economics Letters, 11*(2), 97–100. doi:10.1080/1350485042000200187

Desai, M., Fukuda-Parr, S., Johansson, C., & Sagasti, F. (2002). Measuring the technology achievement of nations and the capacity to participate in the network age. *Journal of Human Development, 3*(1), 95–122. doi:10.1080/14649880120105399

DeSanctis, G., & Monge, P. (1999). Introduction to the special issue: Communication processes for virtual organizations. *Organization Science, 10*(6), 693–703. doi:10.1287/orsc.10.6.693

Deutsch, M., & Harold, G. B. (1955). A study of normative and informational social influences upon individual judgment. *Journal of Abnormal and Social Psychology, 51*(3), 629–636. doi:10.1037/h0046408

Dewan, S., Ganley, D., & Kraemer, K. L. (2010). Complementarities in the diffusion of personal computers and the Internet: Implications for the global digital divide. *Information Systems Research, 21*(4), 925–940. doi:10.1287/isre.1080.0219

Dewan, S., & Riggins, F. J. (2005). The digital divide: Current and future research directions. *Journal of the Association for Information Systems, 6*(12), 298–337.

Dewan, S., Ganley, D., & Kraemer, K. L. (2005). Across the digital divide: A cross-country multi-technology analysis of the determinants of IT penetration. *Journal of the Association for Information Systems, 6*(12), 409–432.

Diamantopoulos, A., & Winklhofer, H. (2001). Index Construction with Formative Indicators: An Alternative to Scale Development. *JMR, Journal of Marketing Research, 38*(2), 269–277. doi:10.1509/jmkr.38.2.269.18845

Dibbern, J., Goles, T., Hirschheim, R., & Jayatilaka, B. (2004). Information systems outsourcing: A survey and analysis of the literature. *The Data Base for Advances in Information Systems, 35*(4), 6–97.

DiBona, C., Ockman, S., & Stone, M. (1999). *Open sources: Voices from the open source revolution.* Sebastopol, CA: O'Reilly & Aoosciates.

Dickson, G. W., Leitheiser, R. L., Wetherbe, J. C., & Nechis, M. (1984). Key information Systems Issues for the 1980's. *Management Information Systems Quarterly*, *8*(3), 135–159. doi:10.2307/248662

DiMaggio, P. J., & Powell, W. W. (1983). The iron cage revisited: institutional isomorphism and collective rationality in organizational fields. *American Sociological Review*, *48*(2), 147–160. doi:10.2307/2095101

Ding, X. L. (2002). The challenges of managing a huge society under rapid transformation. In Wong, J., & Zheng, Y. (Eds.), *China's post-Jiang Leadership Succession: Problems and Perspectives* (pp. 189–213). Singapore: Singapore University Press. doi:10.1142/9789812706508_0008

DiRienzo, C. E., Das, J., Cort, K. T., & Burbridge, J. (2007). Corruption and the role of information. *Journal of International Business Studies*, *38*(2), 320–332. doi:10.1057/palgrave.jibs.8400262

DiRomualdo, A., & Gurbaxani, V. (1998). Strategic intent for IT outsourcing. *Sloan Management Review*, *39*(4), 67–80.

Djankov, S., McLiesh, C., Nenova, T., & Shleifer, A. (2003). Who owns the media? *The Journal of Law & Economics*, *46*(2), 341–381. doi:10.1086/377116

Dobbin, F., Simmons, B., & Garrett, G. (2007). The global diffusion of public policies: social construction, coercion, competition, or learning? *Annual Review of Sociology*, *33*(1), 449–472. doi:10.1146/annurev.soc.33.090106.142507

Dodds, W. B., & Monroe, K. B. (1985). The effect of brand and price information on subjective product evaluations. *Advances in Consumer Research. Association for Consumer Research (U. S.)*, *12*(1), 85–90.

Doherty, N. F., Marples, C. G., & Suhaimi, A. (1999). The Relative Success of Alternative Approaches to Strategic Information Systems Planning: An Empirical Analysis. *The Journal of Strategic Information Systems*, *8*(3), 263–283. doi:10.1016/S0963-8687(99)00024-4

Doll, W. J., Deng, Z., Raghunathan, T. S., Gholamreza, T., & Xia, W. (2004). The meaning and measurement of user satisfaction: A multi-group invariance analysis of the end-user computing satisfaction instrument. *Journal of Management Information Systems*, *21*(1), 227–262.

Doll, W. J., Hendrickson, A., & Deng, X. (1998). Using Davis' perceived usefulness and ease of use instrument for decision making: A confirmatory and multigroup invariance analysis. *Decision Sciences*, *29*(4), 839–870. doi:10.1111/j.1540-5915.1998.tb00879.x

Dologite, D. G., Mockler, R. J., Bai, Q., & Viszhanyo, P. F. (2004). IS change agents in practice in a US-Chinese joint venture. *Journal of Global Information Management*, *12*(4), 1–22.

Donthu, N., & Yoo, B. (1998). Cultural influences on service quality expectations. *Journal of Service Research*, *1*(2), 178–186. doi:10.1177/109467059800100207

Douglas, D. E., Cronan, T. P., & Behel, J. D. (2007). Equity perceptions as a deterrent to software piracy behavior. *Information & Management*, *44*(5), 503–512. doi:10.1016/j.im.2007.05.002

Doz, Y., Bartlett, C., & Prahalad, C. (1981). Global Competitive Pressures and Host Country Demands: Managing Tensions in MNCs. *California Management Review*, *23*(3), 63–74.

Doz, Y., & Prahalad, C. K. (1984). Patterns of Strategic Control within MNCs. *Journal of International Business Studies*, *15*(2), 55–72. doi:10.1057/palgrave.jibs.8490482

Doz, Y. (1976). *National Policies and Multinational Management*. Unpublished doctoral dissertation, Harvard Business School, Boston.

Drury, P. (2011). *Kenya's Pasha centres: Development ground for digital villages*. Cisco Internet Business Solutions Group. Retrieved from http://www.cisco.com/web/about/ac79/docs/case/Kenya-Pasha-Centres_Engagement_Overview_IBSG.pdf

Dunham, R. B. (1998). *Nominal group technique: A users' guide*. School of Business, University of Wisconsin. Retrieved September 7, 2011, from http://www.peoplemix.com/documents/general/ngt.pdf

Dunleavy, P., Margetts, H., Bastow, S., & Tinkler, J. (2006). *Digital Era Governance: IT Corporations, the State, and E-government*. New York: Oxford University Press.

Dunleavy, P. (2007). Governance and state organization in the digital era. In Mansell, R., Avgerou, C., & Quah, D. (Eds.), *The Oxford Handbook of Information and Communication Technologies* (pp. 440–426). New York: Oxford University Press.

Dunning, J. H. (1981). *International production and the multinational enterprise*. London, UK: Allen and Unwin.

Dunning, J. H., & Lundan, S. (2008b). Institutions and the OLI paradigm of the multinational enterprise. *Asia Pacific Journal of Management*, *25*, 573–593. doi:10.1007/s10490-007-9074-z

Dutton, W., Gillett, S. E., McKnight, L., & Peltu, M. (2004). Bridging broadband Internet divides: Reconfiguring access to enhance communicative power. *Journal of Information Technology*, *19*, 28–38. doi:10.1057/palgrave.jit.2000007

Dwivedi, Y., & Lal, B. (2007). Socio-economic determinants of broadband adoption. *Industrial Management & Data Systems*, *107*(5), 654–671. doi:10.1108/02635570710750417

Dwivedi, Y. K., Weerakkody, V., & Williams, M. D. (2009). From implementation to adoption: Challenges to successful e-government diffusion. *Government Information Quarterly*, *26*(1), 3–4. doi:10.1016/j.giq.2008.09.001

Dwivedi, Y., Williams, M. D., Lal, B., & Schwarz, A. (2008). Profiling adoption, acceptance and diffusion research in the information systems discipline. *Proceedings of the European Conference on Information Systems*, Galway, Ireland.

Edwards, J. (2001). Multidimensional constructs in organizational behavioral research: An integrative analytical framework. *Organizational Research Methods*, *4*(2), 144–192. doi:10.1177/109442810142004

Egri, C. P., & Ralston, D. A. (2004). Generation cohorts and personal values: A comparison of China and the United States. *Organization Science*, *15*(2), 210–220. doi:10.1287/orsc.1030.0048

Ein-Dor, P., Segev, E., & Orgad, M. (1992). The Effect of National Culture on IS: Implications for International Information Systems. *Journal of Global Information Management*, *1*(1), 33–44.

Eisenhardt, K. M. (1989). Agency theory: An assessment and review. *Academy of Management Review*, *14*(1), 57–74. doi:10.2307/258191

Eisenhardt, K. M. (1989). Building theories from case study research. *Academy of Management Review*, *14*(4), 532–550.

Elbeltagi, I., McBride, N., & Hardaker, G. (2005). Evaluating the factors affecting DSS usage by senior managers in local authorities in Egypt. *Journal of Global Information Management*, *13*(2), 42–65. doi:10.4018/jgim.2005040103

Erber, G., & Sayed-Ahmed, A. (2005). Offshore outsourcing: A global shift in the present IT industry. *Inter Economics*, *40*(2), 100–112.

Erez, M., & Gati, E. (2004). A dynamic, multi-level model of culture: From the micro level of the individual to the macro level of a global culture. *Applied Psychology: An International Review*, *53*(4), 583–598. doi:10.1111/j.1464-0597.2004.00190.x

Ettlie, J. E., & Pavlou, P. A. (2006). Technology-based new product development partnerships. *Decision Sciences*, *37*(2), 117–147. doi:10.1111/j.1540-5915.2006.00119.x

Evans, P., & Rauch, J. E. (1999). Bureaucracy and growth: a cross-national analysis of the effects of "Weberian" state structures on economic growth. *American Sociological Review*, *64*(5), 748–765. doi:10.2307/2657374

Evanschitzky, H., & Wunderlich, M. (2006). An examination of moderator effects in the four-stage loyalty model. *Journal of Service Research*, *8*(4), 330–345. doi:10.1177/1094670506286325

Evaristo, R. (2003). Cross-cultural research in IS. *Journal of Global Information Management*, *11*(4), i–iii.

Evers, V., & Day, D. (1997). *The role of culture in interface acceptance.* Paper presented at the Proceedings of the IFIPTC13 International Conference on Human-Computer Interaction, London, UK.

Fang, T. (2003). A critique of Hofstede's fifth national culture dimension. *International Journal of Cross Cultural Management*, *3*(3), 347–368. doi:10.1177/1470595803003003006

Farina, C., & Gibbons, M. (1981). The concentration of research funds: The case of the Science Research Council. *R & D Management*, *11*(2), 63. doi:10.1111/j.1467-9310.1981.tb00451.x

Feeny, D., & Ives, B. (1990). In Search of Sustainability: Reaping Long-Term Advantage from Investments in Information Technology. *Journal of Management Information Systems*, *7*(1), 27–46.

Feldmann, M., & Muller, S. (2003). An incentive scheme for true information providing in Supply Chains. *Omega*, *31*(2), 63–73. doi:10.1016/S0305-0483(02)00096-8

Fichman, R. G. (2000). The diffusion and assimilation of information technology innovations. In R. W. Zmud (Ed.), *Framing the domains of IT management: Projecting the future through the past* (pp.105-127). Cincinnati, OH: Pinnaflex Publishing.

Finel, B. I., & Lord, K. M. (1999). The surprising logic of transparency. *International Studies Quarterly*, *43*(2), 315–339. doi:10.1111/0020-8833.00122

Fink, D. (1998). Guidelines for the successful adoption of information technology in small and medium enterprises. *International Journal of Information Management*, *18*(4), 243–253. doi:10.1016/S0268-4012(98)00013-9

Finlay, P. N., & King, R. M. (1999). IT sourcing: A research framework. *International Journal of Technology Management*, *17*(1-2), 109–128. doi:10.1504/IJTM.1999.002703

Fishbein, M., & Ajzen, I. (1975). *Belief, attitude, intention, and behavior: An introduction to theory and research*. Reading, PA: Addison-Wesley.

Fishbein, M., & Ajzen, I. (1980). *Understanding attitudes and predicting social behaviour*. Eaglewood Cliffs, NJ: Prentice Hall.

Fishbein, M., Triandis, H. C., Kanfer, F. H., Becker, M., Middlestadt, S. E., & Eichler, A. (2001). Factors influencing behavior and behavior change. In Baum, A., Revenson, T. A., & Singer, J. E. (Eds.), *Handbook of health psychology*. Mahwah, NJ: Lawrence Erlbaum.

Flamm, K. (1987). *Targeting the computer: Government support and international competition*. Washington, DC: Brookings Institution.

Flynn, B. B., Sakakibara, S., Schroeder, R. G., Bates, K. A., & Flynn, E. J. (1990). Empirical research methods in operations management. *Journal of Operations Management*, *11*(4), 339–366. doi:10.1016/S0272-6963(97)90004-8

Ford, D. P., Connelly, C. E., & Meister, D. B. (2003). Information Systems Research and Hofstede's Culture's Consequences: An Uneasy and Incomplete Partnership. *IEEE Transactions on Engineering Management*, *50*(1), 8–25. doi:10.1109/TEM.2002.808265

Forman, C., Goldfarb, A., & Greenstein, S. (2002). *Digital dispersion: An industrial and geographic census of commercial Internet use*. Cambridge, MA: National Bureau of Economic Research.

Forman, C., Goldfarb, A., & Greenstein, S. (2003). *How did location affect adoption of the commercial Internet? Global village, urban density and industry composition*. Working Paper No 9979 (NBER, Cambridge, MA).

Fornell, C., & Larcker, D. F. (1981). Evaluating structural equations with unobservable variables and measurement error. *JMR, Journal of Marketing Research*, *18*(1), 39–50. doi:10.2307/3151312

Foster, W., Goodman, S., Osiakwan, E., & Bernstein, A. (2004). Global diffusion of the internet IV: The internet in Ghana. *Communications of the Association for Information Systems*, *13*(38), 654–670.

Fountain, J. E. (2001). *Building the virtual state: Information technology and institutional change*. Washington, D.C: Brookings Institution Press.

Fountain, J. E. (2007). Challenges to organizational change: multi-level integrated information structures (MIIS). In Mayer-Schönberger, V., & Lazer, D. (Eds.), *Governance and Information Technology: From Electronic Government to Information Government*. Cambridge, MA: MIT Press.

Frank, L., Sundqvist, S., Puumalainen, K., & Taalikka, S. (2001). *Cross-cultural comparison of innovators: Empirical evidence from wireless services in Finland, Germany and Greece*. Paper presented at the ANZMAC Conference, Auckland, New Zealand.

Frenzel, C. W. (1991). *Information Technology Management*. Boston: Boyd & Fraser.

Friedman, T. L. (2005). *The World Is Flat: A Brief History of the Twenty-first Century*. New York: Farrar, Straus and Giroux.

Fuloria, P. C., & Zenios, S. A. (2001). Outcomes-adjusted reimbursement in a health-care delivery system. *Management Science, 47*(6), 735–751. doi:10.1287/mnsc.47.6.735.9816

Furuholt, B., Kristiansen, S., & Wahid, F. (2005). Information dissemination in a developing society: Internet café users in Indonesia. *The Electronic Journal of Information Systems in Developing Countries, 22*(3), 1–16.

Fusilier, M., & Durlabhji, S. (2005). An exploration of student internet use in India: The technology acceptance model and the theory of planned behavior. *Campus-Wide Information Systems, 22*(4), 233–246. doi:10.1108/10650740510617539

Gabel, M., & Bruner, H. (2003). *Global Inc.: An Atlas of the Global Corporation: A Visual Exploration of The History, Scale, Scope, and Impacts of Multinational Corporations*. New York: New Press.

Gable, G. G. (1991). Consultant engagement for first time computerization: a pro-action client role in small business. *Information & Management, 20*(2), 83–93. doi:10.1016/0378-7206(91)90046-5

Gable, G. G., & Raman, K. S. (1992). Government initiatives for IT adoption in small businesses: experiences of the Singapore Small Enterprise Computerization Programme. *International Information Systems, 1*(1), 68–93.

Gales, L. (2008). The role of culture in technology management research: National character and cultural distance frameworks. *Journal of Engineering and Technology Management, 25*(1-2), 3–22. doi:10.1016/j.jengtecman.2008.01.001

Gallivan, M. S. (2005). Information Technology and Culture: Identifying Fragmentary and Holistic Perspectives of Culture. *Information and Organization, 15*(4), 295–338. doi:10.1016/j.infoandorg.2005.02.005

Gan, L. L., & Koh, H. C. (2006). An empirical study of software piracy among tertiary institutions in Singapore. *Information & Management, 43*(5), 640–649. doi:10.1016/j.im.2006.03.005

Gartner Group and Dataquest. (1999). *ERP software publishers: service strategies and capabilities*. New York: Gartner Group and Dataquest.

Garvin, D. A. (1993). Building a learning organization. *Harvard Business Review, 71*(4), 78–91.

Gefen, D., & Straub, D. (2005). A Practical Guide to Factorial Validity Using PLS-Graph: Tutorial and Annotated Example. *Communications of AIS, 16*(1), 91–109.

Gefen, D., & Straub, D. (2005). A practical guide to factorial validity using PLS-Graph: Tutorial and annotated example. *Communications of the Association for Information Systems, 16*, 91–109.

Gefen, D., & Straub, D. W. (1975). Gender differences in the perception and use of e-mail: An extension to the technology acceptance model. *Management Information Systems Quarterly, 21*(3), 389–400.

Gefen, D., Wyss, S., & Lichtenstein, Y. (2008). Business familiarity as risk mitigation in software development outsourcing contracts. *MIS Quarterly: Management Information Systems, 32*(3), 531–551.

Getz, K. A., & Volkema, R. J. (2001). Culture, perceived corruption, and economics: A model of predictors and outcomes. *Business & Society, 40*(1), 7–30. doi:10.1177/000765030104000103

Gholami, R., Lee, S. T., & Heshmati, A. (2006). The causal relationship between information and communication technology and foreign direct investment. *World Economy, 29*(1), 43–62. doi:10.1111/j.1467-9701.2006.00757.x

Gil-Garcia, J. R., Chengalur-Smith, I., & Duchessi, P. (2007). Collaborative e-government: impediments and benefits of information-sharing projects in the public sector. *European Journal of Information Systems, 16*, 121–133. doi:10.1057/palgrave.ejis.3000673

Goles, T., Jayatilaka, B., George, B., Parsons, L., Chambers, V., Taylor, D., & Brune, R. (2008). Softlifting: Exploring determinants of attitude. *Journal of Business Ethics, 77*(4), 481–499. doi:10.1007/s10551-007-9361-0

Goode, S., & Cruise, S. (2006). What motivates software crackers? *Journal of Business Ethics, 65*(2), 173–201. doi:10.1007/s10551-005-4709-9

Goodhue, D. L., & Thompson, R. L. (1995). Task-Technology Fit and Individual Performance. *Management Information Systems Quarterly, 19*(2), 213–226. doi:10.2307/249689

Goodman, S. E., & Green, J. D. (1992). Computing in the Middle East. *Communications of the ACM, 35*(8), 21–25. doi:10.1145/135226.135236

Gopal, R. D., & Sanders, G. L. (2000). Global software piracy: You can't get blood out of a turnip. *Communications of the ACM, 43*(9), 82–89. doi:10.1145/348941.349002

Gottschalk, P. (1999a). Implementation predictors of strategic information systems plans. *Information & Management, 36*(2), 77–91. doi:10.1016/S0378-7206(99)00008-7

Gottschalk, P. (1999b). Strategic information systems planning: The IT strategy implementation matrix. *European Journal of Information Systems, 8*(2), 107–118. doi:10.1057/palgrave.ejis.3000324

Gottschalk, P. (1999c). Implementation of formal plans: The case of information technology strategy. *Long Range Planning, 32*(3), 362–372. doi:10.1016/S0024-6301(99)00040-0

Graen, G. B. (2006). In the eye of the beholder: Cross-cultural lesson in leadership from project GLOBE: A response viewed from the third culture bonding (TCB) model of cross-cultural leadership. *The Academy of Management Perspectives, 20*(4), 95–101.

Graham, M. E., & Wada, E. (2001). *Foreign direct investment in China: Effects on growth and economic performance.* Peterson Institute Working Paper Series WP01-3, Peterson Institute for International Economics.

Grandon, E. E., & Pearson, J. M. (2004). Electronic commerce adoption: An empirical study of small and medium US businesses. *Information & Management, 42*(1), 197–216.

Grant, G., & Chau, D. (2005). Developing a generic framework for e-government. *Journal of Global Information Management, 13*(1), 1–30. doi:10.4018/jgim.2005010101

Gregor, S., Imran, A., & Turner, T. (2010). *Designing for a "'sweet spot' in an intervention in a least developed country: The case of e-government in Bangladesh.* Paper presented at the SIG GlobDev Third Annual Workshop, Saint Louis, USA.

Gronlund, A., & Horan, T. (2004). Introducing E-government: History, definitions, and issues. *Communications of the Association for Information Systems, 15*(39).

Grover, V., & Segars, A. H. (2005). An Empirical Evaluation of Stages of Strategic Information Systems Planning: Patterns of Process Design and Effectiveness. *Information & Management, 42*(5), 761–779. doi:10.1016/j.im.2004.08.002

Grover, V., & Goslar, M. D. (1993). The initiation, adoption, and implementation of telecommunications technologies in U.S. organizations. *Journal of Management Information Systems, 10*(1), 141–163.

Grover, V. (1993). An empirically derived model for the adoption of customer-based interorganizational systems. *Decision Sciences, 24*(3), 603–640. doi:10.1111/j.1540-5915.1993.tb01295.x

Grubesic, T., & Murray, A. (2002). Constructing the divide: Spatial disparities in broadband access. *Regional Science, 81*, 197–221.

Gu, J., Fan, R., & Liang, L. (2008). Research of economic analysis and countermeasure on the short-lighted behavior of scientific and technical personnel. [in Chinese]. *East China Economic Management, 22*(1), 145–149.

Gudykunst, W. B., & Ting-Toomey, S. (1988). *Culture and interpersonal communication.* Newbury Park, CA: Sage Publications.

Guo, X., & Chen, G. (2005). Internet diffusion in Chinese companies. *Communications of the ACM, 48*(4), 54–58. doi:10.1145/1053291.1053318

Gupta, A. (1987). SBU strategies, corporate-SBU relations, and SBU effectiveness in strategy implementation. *Academy of Management Journal, 30*(3), 477–500. doi:10.2307/256010

Gupta, B., Dasgupta, S., & Gupta, A. (2008). Adoption of ICT in a government organization in a developing country: An empirical study. *The Journal of Strategic Information Systems, 17*(2), 140–154. doi:10.1016/j.jsis.2007.12.004

Gupta, A., Seshasai, S., & Mukherji, S. (2007). Offshoring: The transition from economic drivers toward strategic global partnership and 24-hour knowledge factory. *Journal of Electronic Commerce in Organizations, 5*(2), 1–23. doi:10.4018/jeco.2007040101

Hagenaars, M. (2003). *Socio-cultural factors and ICT adoption.* Retrieved September 7, 2011, from http://www.iicd.org/articles/IICDnews.import2069

Hair, J. F., Anderson, R. E., Tatham, R. L., & Black, W. C. (2006). *Multivariate data analysis.* Upper Saddle River, NJ: Prentice-Hall.

Hall, E. T. (1976). *Beyond culture.* Garden City, NY: Anchor Books, Doubleday and Company.

Hall, E. T., & Hall, M. R. (1990). *Understanding cultural differences.* Yarmouth, ME: Intercultural Press.

Halman, L., Inglehart, R., Díez-Medrano, J., Luijkx, R., Moreno, A., & Basáñez, M. (2007). *Changing Values and Beliefs in 85 Countries.* Boston (MA): Brill.

Hanke, M., & Teo, T. (2003). Meeting the Challenges in Globalizing Electronic Commerce at United Airlines. *Journal of Information Technology Cases and Applications, 5*(4), 21.

Hardy, A. (1980). The role of the telephone in economic development. *Telecommunications Policy, 4,* 278–286. doi:10.1016/0308-5961(80)90044-0

Hargittai, E. (1999). Weaving the Western Web: Explaining differences in Internet connectivity among OECD economies. *Telecommunications Policy, 23,* 701–718. doi:10.1016/S0308-5961(99)00050-6

Harris, R., & Davison, R. (1999). Anxiety and Involvement: Cultural Dimensions of Attitudes Toward Computers in Developing Societies. *Journal of Global Information Management, 7*(1), 26–38.

Harrison, W. L., & Farn, C. K. (1990). A Comparison of Information Management Issues in the United States of America and the Republic of China. *Information & Management, 18*(4), 177–188. doi:10.1016/0378-7206(90)90038-J

Hart, P., & Saunders, C. S. (1998). Emerging electronic partnerships: Antecedents and dimensions of EDI use from the supplier's perspective. *Journal of Management Information Systems, 14*(4), 87–111.

Hartog, C., & Herbert, M. (1986). 1985 Opinion Survey of MIS Managers: Key Issues. *Management Information Systems Quarterly, 10*(4), 351–361. doi:10.2307/249189

Hartwick, J., & Barki, H. (1994). Explaining the Role of User Participation in Information System Use. *Management Science, 40*(4), 440–465. doi:10.1287/mnsc.40.4.440

Hasan, H., & Ditsa, G. (1999). The impact of culture on the adoption of IT: An interpretive study. *Journal of Global Information Management, 7*(1), 5–15.

Hashmi, M. (1988). *National Culture and Management Practices: United States and Saudi Arabia Contrasted.* Paper presented at the Proceedings of the Annual Eastern Michigan University Conference on Languages for Business and the Professions, Ann Arbor, MI.

HCI-Lab. (2004). *4th Worldwide mobile data study report.* South Korea: Yonsei University.

He, X. X., Zhang, J., & Zhao, J. (2007). Based on the frontier of management to promote management excellency - the role of registered unit in the management of National Natural Science Fund. [in Chinese]. *Science Foundation in China, 21*(5), 309–311.

Heeks, R., & Kanashiro, L. (2009). Telecentres in mountain regions: A Peruvian case study of the impact of information and communication technologies on remoteness and exclusion. *Journal of Mountain Science, 6*(4), 320–330. doi:10.1007/s11629-009-1070-y

Heeks, R. (2002). Information systems and developing countries: Failure, success, and local improvisations. *The Information Society, 18,* 101–112. doi:10.1080/01972240290075039

Heeks, R. (2006a). *Implementing and managing e-government.* London, UK: Sage.

Heeks, R., & Bhatnagar, S. C. (Eds.). (2001). *Understanding success and failure in information age reform.* London, UK: Routledge.

Heeks, R., & Bailur, S. (2007). Analyzing E-government research: perspectives, philosophies, theories, methods, and practice. *Government Information Quarterly, 24*(2), 243–265. doi:10.1016/j.giq.2006.06.005

Heeks, R. (2004). *Basic definitions page.* eGovernment for Development project. IDPM, University of Manchester, UK. Retrieved June, 10, 2010, from http://www.egov4dev.org/success/definitions.shtml

Heeks, R. (2006b, July 27-28). *Understanding and measuring egovernment: International benchmarking studies*. Paper presented at the UNDESA workshop on E-Participation and E-Government: Understanding the Present and Creating the Future, Budapest, Hungary

Heinze, T., Shapira, P., Rogers, J. D., & Senker, J. M. (2009). Organizational and institutional influences on creativity in scientific research. *Research Policy*, *38*(4), 610–623. doi:10.1016/j.respol.2009.01.014

Helbig, N., Ramón Gil-García, J., & Ferro, E. (2009). Understanding the complexity of electronic government: implications from the digital divide literature. *Government Information Quarterly*, *26*(1), 89–97. doi:10.1016/j.giq.2008.05.004

Hempel, P. S., & Kwong, Y. K. (2001). B2B e-Commerce in emerging economies: i-metal.com's non-ferrous metals exchange in China. *The Journal of Strategic Information Systems*, *10*(4), 335–355. doi:10.1016/S0963-8687(01)00058-0

Henderson, J. C., Rockart, J. F., & Sifonis, J. G. (1987). Integrating Management Support Systems into Strategic Information Systems Planning. *Journal of Management Information Systems*, *4*(1), 5–24.

Henderson, J. C., & Venkatraman, N. (1993). Strategic Alignment: Leveraging Information Technology for Transforming Organizations. *IBM Systems Journal*, *32*(1), 4–16. doi:10.1147/sj.382.0472

Hill, C. E., Loch, K. D., Straub, D. W., & El-Sheshai, K. (1998). A qualitative assessment of Arab culture and information technology transfer. *Journal of Global Information Management*, *6*(3), 29.

Hill, C. E., Straub, D. W., Loch, K. D., Cotterman, W. W., & El-Sheshai, K. (1994). *The Impact of Arab Culture on the Diffusion of Information Technology: A Culture-Centered Model*. Paper presented at the The Impact of Informatics on Society: Key Issues for Developing Countries, IFIP 9.4, Havana, Cuba.

Hiller, M. (2003). The role of cultural context in multilingual website usability. *Electronic Commerce Research and Applications*, *2*(1), 2–14. doi:10.1016/S1567-4223(03)00005-X

Hiong, G. (2006). *Digital review of Asia-Pacific 2005/2006: Singapore*. Retrieved from http://www.digital-review.org/.

Hitt, L., & Tambe, P. (2007). Broadband adoption and content consumption. *Information Economics and Policy*, *19*(3-4), 362–378. doi:10.1016/j.infoecopol.2007.04.003

Ho, K. K. W., Yoo, B., Yu, S., & Tam, K. Y. (2007). The Effect of Culture and Product Categories on the Level of Use of Buy-It-Now (BIN) Auctions by Sellers. *Journal of Global Information Management*, *15*(4), 1–19.

Hoffman, R. (1988). The general management of foreign subsidiaries in the USA: An exploratory study. *Management Information Review*, *28*(2), 41–55.

Hofstede, G. (2001). *Culture's Consequences: Comparing Values, Behaviors, Institutions, and Organizations across Nations* (2nd ed.). Thousand Oaks, CA: Sage Publications.

Hofstede, G., & Bond, M. H. (1988). The Confucius Connection: From Cultural Roots to Economic Growth. *Organizational Dynamics*, *16*(4), 5–21. doi:10.1016/0090-2616(88)90009-5

Hofstede, G. (1980). *Culture's Consequences: International Differences in Work-Related Values*. Beverly Hills, CA: Sage Publications.

Hofstede, G. (1984). *Culture's Consequences: International Differences in Work-Related Values*. Beverly Hills, CA: Sage Publications.

Hofstede, G. (1991). *Cultures and Organizations: Software of the Mind*. London: McGraw-Hill.

Hofstede, G. (2001). *Culture's Consequences* (2nd ed.). Thousand Oaks, CA: Sage Publications.

Hofstede, G. (2006). What did GLOBE really measure? Researchers' minds versus respondents' minds. *Journal of International Business Studies*, *37*(6), 882–896. doi:10.1057/palgrave.jibs.8400233

Hofstede, G., & Bond, M. (1988). The Confucian Connection: From Cultural Roots to Economic Growth. *Organizational Dynamics*, *16*(4), 4–21. doi:10.1016/0090-2616(88)90009-5

Hofstede, G. (1984). *Culture's Consequences: International Differences in Work-Related Values*. Beverly Hills, CA: SAGE Publications.

Hofstede, G. (2001). *Culture's Consequences: Comparing Values, Behaviors, Institutions, and Organizations Across Nations*. Thousand Oaks, CA: Sage Publications.

Hofstede, G. (1980). *Culture's consequences: International differences in work-related values*. Beverly Hills, CA: Sage Publications.

Hofstede, G. (1997). *Cultures and organizations: Software of the mind*. New York: McGraw-Hill.

Hofstede, G. (1984). *Culture's consequences: International differences in work related values*. Thousand Oaks, CA: Sage Publications.

Hofstede, G. (2001). *Culture's consequences*. Thousand Oaks, CA: Sage Publications.

Hofstede, G., & McCrae, R. R. (2004). Personality and culture revisited: Linking traits and dimensions of culture. *Cross-Cultural Research*, *38*(February), 52–88. doi:10.1177/1069397103259443

Hofstede, G., & Hofstede, G. J. (2005). *Cultures and Organizations: Software of the Mind* (2nd ed.). New York (NY): McGraw-Hill.

Holland, C., & Light, B. (1999). A critical success factors model for ERP implementation. *IEEE Software*, *16*(3), 30–36. doi:10.1109/52.765784

Hollifield, C. A., & Donnermeyer, J. F. (2003). Creating demand: influencing information technology diffusion in rural communities. *Government Information Quarterly*, *20*(2), 135–150. doi:10.1016/S0740-624X(03)00035-2

Holsapple, C. W., & Sena, M. P. (2005). ERP plans and decision-support benefits. *Decision Support Systems*, *38*(4), 575–590. doi:10.1016/j.dss.2003.07.001

Hong, S. E., Kim, J. K., & Lee, H. S. (2008). Antecedents of use-continuance in information systems: Toward an integrative view. *Journal of Computer Information Systems*, *48*(3), 61–73.

Hong, S. J., & Tam, K. Y. (2006). Understanding the adoption of multipurpose information appliances: The case of mobile data services. *Information Systems Research*, *17*(2), 162–179. doi:10.1287/isre.1060.0088

Hong, W., Thong, J. Y. L., Wong, W. M., & Tam, K. Y. (2001). Determinants of user acceptance of digital libraries: An empirical examination of individual differences and system characteristics. *Journal of Management Information Systems*, *18*(3), 97–124.

Houben, G., Lenie, K., & Vanhoof, K. (1999). A knowledge-based SWOT-Analysis System as an Instrument for Strategic Planning in Small and Medium Sized Enterprises. *Decision Support Systems*, *26*(2), 125–135. doi:10.1016/S0167-9236(99)00024-X

House, R. J., Hanges, P. J., Javidan, M., Dorfman, P. W., & Gupta, V. (2004). *Culture, Leadership, and Organizations: The GLOBE Study of 62 Societies*. Thousand Oaks, CA: Sage Publications.

House, R. J., Hanges, P. J., Javidan, M., Dorfman, P. W., & Gupta, V. (2004). *Culture, Leadership, and Organizations: The GLOBE Study of 62 Societies*. Thousand Oaks (CA): Sage Publications.

Howick, S., & Whalley, J. (2008). (in press). Understanding the drivers of broadband adoption: The case of rural and remote Scotland. *The Journal of the Operational Research Society*. doi:10.1057/palgrave.jors.2602486

Hsu, M. H., Yen, C. H., Chiu, C. M., & Chang, C. M. (2006). A longitudinal investigation of continued online shopping behavior: An extension of the theory of planned behavior. *International Journal of Human-Computer Studies*, *64*(9), 889–904. doi:10.1016/j.ijhcs.2006.04.004

Hsu, M., & Chiu, C. (2004). Internet self-efficacy and electronic service acceptance. *Decision Support Systems*, *38*(3), 369–381. doi:10.1016/j.dss.2003.08.001

Huang, L., Lu, M., & Wong, B. K. (2003). The Impact of Power Distance on Email Acceptance: Evidence from the PRC. *Journal of Computer Information Systems*, *44*(1), 93–101.

Huang, B. S., Meng, X., Zheng, Y. H., & Liang, W. P. (2003). Promoting the international cooperation in chemical sciences by rational use of science fund. *Science Foundation in China*, *17*(1), 44–46.

Hubona, G. S., Truex, D. P., Wang, J., & Straub, D. W. (2006). Cultural and Globalization Issues Impacting the Organizational Use of Information Technology. In Galletta, D. F., & Zhang, P. (Eds.), *Human-Computer Interaction and Management Information Systems - Applications*. Armonk, NY: M. E. Sharpe.

Huh, Y. E., & Kim, S. H. (2008). Do early adopters upgrade early? Role of post-adoption behavior in the purchase of next-generation products. *Journal of Business Research, 61*(1), 40–46. doi:10.1016/j.jbusres.2006.05.007

Hult, G. T. M., Ketchen, D. J. Jr, & Nichols, E. L. Jr. (2003). Organizational learning as a strategic resource in supply management. *Journal of Operations Management, 21*(5), 541–556. doi:10.1016/j.jom.2003.02.001

Hung, S. Y., Ku, C. Y., & Chang, C. M. (2003). Critical factors of WAP services adoption: An empirical study. *Electronic Commerce Research and Applications, 2*(1), 42–60. doi:10.1016/S1567-4223(03)00008-5

Hung, S. Y., & Liang, T. P. (2001). Cross-Cultural Applicability of group decision support systems. *International Journal of Management Theory and Practices, 2*(1), 1–12.

Hunton, J. E., Lippincott, B., & Reck, J. L. (2003). Enterprise resource planning systems: comparing firm performance of adopters and nonadopters. *International Journal of Accounting Information Systems, 4*(3), 165–184. doi:10.1016/S1467-0895(03)00008-3

Husted, B. W. (2000). The impact of national culture on software piracy. *Journal of Business Ethics, 26*(3), 197–211. doi:10.1023/A:1006250203828

Hwang, Y. J. (2004). *An empirical examination of individual-level cultural orientation as an antecedent to ERP systems adoption*. Paper presented at the The Pre-ICIS Workshop on Cross-Cultural Research in Information Systems, Washington D.C.

Iacovou, C. L., Benbasat, I., & Dexter, A. S. (1995). Electronic Data interchange and small organizations: adoption and impact of technology. *Management Information Systems Quarterly, 19*(4), 465–485. doi:10.2307/249629

Idris, A. M. (2007). Cultural Barriers to Improved Organizational Performance in Saudi Arabia. *SAM Advanced Management Journal, 72*(2), 36–53.

Ifinedo, P. (2006). Acceptance and continuance intention of web-based learning technologies (WLT) use among university students in a Baltic country. *The Electronic Journal of Information Systems in Developing Countries, 23*(6), 1–20.

Igbaria, M. (1992). An examination of microcomputer usage in Taiwan. *Information & Management, 22*(10), 19–28. doi:10.1016/0378-7206(92)90003-X

Igbaria, M., & Iivari, J. (1995). The Effects of Self-efficacy on Computer Usage. *OMEGA International Journal of Management Science, 23*(6), 587–605. doi:10.1016/0305-0483(95)00035-6

Igbaria, M., & Zviran, M. (1991). End-User Effectiveness: A Cross-Cultural Examination. *Omega, 19*(5), 369–379. doi:10.1016/0305-0483(91)90055-X

Igbaria, M., & Zviran, M. (1996). Comparison of End-User Computing Characteristics in the US, Israel, and Taiwan. *Information & Management, 30*(1), 1–13. doi:10.1016/0378-7206(95)00044-5

Imran, A., Gegor, S., & Turner, T. (2009, May 26-28). *eGovernment capacity building through knowledge transfer and best practice development in Bangladesh*. Paper presented at the 10th International Conference on Social Implications of Computers in Developing Countries, International Federation for Information Processing (IFIP), Dubai.

Indjikian, R., & Siegel, D. S. (2005). The impact of investment in IT on economic performance: Implications for developing countries. *World Development, 33*(5), 681. doi:10.1016/j.worlddev.2005.01.004

International-Telecommunication-Union. (2002). *ITU Internet Reports 2002: Internet for a Mobile Generation*.

Introna, L. D. (2007). Singular justice and software piracy. *Business Ethics. European Review (Chichester, England), 16*(3), 264–277.

Isaacs, S. (2007). *Survey of ICT and education in Mauritius. Survey of ICT and Education in Africa (volume 2): 53 country reports*. Washington, DC: infoDev / World Bank. Retrieved from http://www.infodev.org/en/Publication.354.html

Iskander, B. Y., Kurokawa, S., & LeBlanc, L. J. (2001). Adoption of EDI: The role of buyer-supplier relationships. *IEEE Transactions on Engineering Management, 48*(4), 505–517. doi:10.1109/17.969427

Islam, R. (2006). Does more transparency go along with better governance? *Economics and Politics, 18*(2), 121–167. doi:10.1111/j.1468-0343.2006.00166.x

Islam, R. (Ed.). (2008). *Information and public choice: From media markets to policy making*. Washington, DC: World Bank.

Ito, M., & Nakakoji, K. (1996). Impact of Culture on User Interface Design. In Galdo, E. M., & Nielsen, J. (Eds.), *International User Interfaces* (pp. 41–73). New York: Wiley.

ITU. (2003). *Declaration of principles*. Geneva: World Summit on the Information Society.

ITU. (2005). *Tunis agenda for the information society*. Tunis: World Summit on the Information Society.

ITU. (2009). *Measuring the information society 2009, The ICT development index*. Geneva, Switzerland: ITU Publication.

ITU. (2007a). *World telecommunication/ICT indicators*.

ITU. (2007b). *Measuring the information society 2007, ICT opportunity index and world telecommunication/ICT indicators*.

ITU. (2009). *Information society statistical profiles 2009 - Asia and the Pacific*. Market Information and Statistics Division within the Telecommunication Development Bureau. Retrieved from http://www.itu.int/ITU-D/ict/material/ISSP09-AP_final.pdf

Ives, B., & Learmonth, G. (1984). The Information System as a Competitive Weapon. *Communications of the ACM, 27*(12), 1193–1201. doi:10.1145/2135.2137

Jamil, I. (2007). *Administrative culture in Bangladesh*. A H Development Publishing House.

Jarvenpaa, S. L., Lang, K. R., Takeda, Y., & Tuunainen, V. K. (2003). Mobile commerce at crossroads. *Communications of the ACM, 46*(12), 41–44. doi:10.1145/953460.953485

Jarvenpaa, S., Knoll, & Leidner, D. (1998). Is anybody out there? Antecedents of trust in global virtual teams. *Journal of Management Information Systems, 14*(4), 29–64.

Jarvis, C., MacKenzie, S., & Podsakoff, P. (2003). A Critical Review of Construct Indicators and Measurement Model Misspecification in Marketing and Consumer Research. *The Journal of Consumer Research, 30*(2), 199–218. doi:10.1086/376806

Jasperson, J., Carter, P. E., & Zmud, R. W. (2005). A comprehensive conceptualization of post-adoption behaviors associated with information technology enabled work systems. *Management Information Systems Quarterly, 29*(3), 525–557.

Jiacheng, W., Lu, L., & Francescob, C. A. (2009). A cognitive model of intra-organizational knowledge-sharing motivations in the view of cross-culture. *International Journal of Information Management*, 10–1016.

Jiang, B., Frazier, G. V., & Prater, E. L. (2006). Outsourcing effects on firms' operational performance: An empirical study. *International Journal of Operations & Production Management, 26*(12), 1280–1300. doi:10.1108/01443570610710551

Jin, B. H. (2004). Considerations of high quantity of publications compared to low quantity of citations: Value orientation and quantitative index of R & D evaluation. *Science of Science and Management of S.& T., 3*, 9–11.

Jin, B. H. (2007). The advancing aircraft carrier of China's research output: SCI analysis in 2006. *Science Focus, 2*(1), 20–44.

Jones, M. R., & Karsten, H. (2008). Giddens' structuration theory and information systems research. *Management Information Systems Quarterly, 32*(1), 127–157.

Jongbloed, B., & Koelman, J. (2001). Keeping up performances: an international survey of performance-based funding in higher education. *Journal of Higher Education Policy and Management, 23*(2), 127–145. doi:10.1080/13600800120088625

Jonkers, K. (2008). *An analytical framework to compare research systems applied to a diachronic analysis of the transformation of the Chinese research system*. Paper presented at the VI Globelics Conference.

Joppe, M. (2004). *The nominal group technique*. Retrieved from September 7, 2011, from http://www.uoguelph.ca/htm/MJResearch/ResearchProcess/841TheNominalGroupTechnique.htm

Journal, T. B. S. (2001). *Dubai media city prepares for next phase, no. 5*. Retrieved from http://www.tbsjournal.com/Archives/Fall01/dubai.html

Kaba, B., Diallo, A., Plaisent, M., Bernard, P., & N'Da, K. (2006). Explaining the factors influencing cellular phones use in Guinea. *The Electronic Journal of Information Systems in Developing Countries*, *28*(6), 1–7.

Kabasakal, H., & Bodur, M. (2002). Arabic Cluster: a Bridge between East and West. *Journal of World Business*, *37*, 40–54. doi:10.1016/S1090-9516(01)00073-6

Kagan, A., Lau, K., & Nusgart, K. R. (1990). Information system usage within small business firms. *Entrepreneurship: Theory and Practice*, *14*(3), 25–37.

Kahn, R. L., Wolfe, D. M., Quinn, R. P., Snoek, J. D., & Rosenthal, R. A. (1964). *Organizational stress: Studies in role conflict and ambiguity*. New York, NY: Wiley.

Kahneman, D. (2003). A perspective on judgment and choice: Mapping bounded rationality. *The American Psychologist*, *58*(9), 696–720. doi:10.1037/0003-066X.58.9.697

Kakabadse, A., & Kakabadse, N. (2005). Outsourcing: Current and future trends. *Thunderbird International Business Review*, *47*(2), 183–204. doi:10.1002/tie.20048

Kamarck, E. C., & Nye, J. S. (Eds.). (2002). *Governance.com: Democracy in the information age*. Washington, DC: Brookings Institution Press.

Kandelin, N. A., Lin, T. W., & Muntoro, K. R. (1998). A study of the attitudes of Indonesian managers toward key factors in Information System development and implementation. *Journal of Global Information Management*, *6*(3), 17–26.

Kane, T. (2007). Economic freedom in five regions. In T. Kane, K. R. Holmes, & A. M. O'Grady (Eds.), *2007 index of economic freedom*. Washington, DC: The Heritage Foundation and Dow Jones & Company, Inc. Retrieved from www.heritage.org/index

Kankanhalli, A., Tan, B. C. Y., Wei, K. K., & Holmes, M. (2004). Cross-Cultural Differences and Information Systems Developer Values. *Decision Support Systems*, *38*(2), 183–195. doi:10.1016/S0167-9236(03)00101-5

Kankanhalli, A., & Kohli, R. (2009, July 10-12). *Does public or private sector matter? An agenda for IS research in e-government*. Paper presented at the Pacific Asia Conference on Information Systems Hyderabad, India

Karahanna, E., Evaristo, J. R., & Srite, M. (2005). Levels of Culture and Individual Behavior: An Integrative Perspective. *Journal of Global Information Management*, *13*(2), 1–20.

Karahanna, E., Straub, D. W., & Chervany, N. L. (1999). Information technology adoption across time: A cross-sectional comparison of pre-adoption and post-adoption beliefs. *Management Information Systems Quarterly*, *23*(2), 183–213. doi:10.2307/249751

Katchanovski, I., & La Porte, T. (2005). Cyberdemocracy or Potemkin e-villages? Electronic governments in OECD and post-communist countries. *International Journal of Public Administration*, *28*(7), 665–681. doi:10.1081/PAD-200064228

Katchanovski, I., & La Porte, T. (2009). Democracy, Colonial Legacy, and the Openness of Cabinet-Level Websites in Developing Countries. *Journal of Comparative Policy Analysis: Research and Practice*, http://www.informaworld.com/smpp/title~db=all~content=t713672306~tab =issueslist~branches=11 - v11*11*(2), 213-232.

Kaufmann, D., Kraay, A., & Zoido-Lobatón, P. (1999). *Governance matters* (World Bank Policy Research Working Paper No. 2196). Washington, DC: World Bank.

Ke, W., & Wei, K. K. (2004). Successful e-government in Singapore. *Communications of the ACM*, *47*(6), 95–99. doi:10.1145/990680.990687

Keefer, D. L. (1978). Allocation planning for R&D with uncertainty and multiple objectives. *IEEE Transactions on Engineering Management*, *25*(1), 8.

Kelegai, L., & Middleton, M. (2004). Factors influencing information systems success in Papua New Guinea organisations: A case analysis. *Australasian Journal of Information Systems*, *11*(2), 57–69.

Kelman, H. C. (1958). Compliance, Identification and Internalization: Three Processes of Attitude Change. *The Journal of Conflict Resolution, 2*(1), 51–60. doi:10.1177/002200275800200106

Keng, S., & Yuan, L. (2006). Using social development lenses to understand e–government development. *Journal of Global Information Management, 14*(1), 47. doi:10.4018/jgim.2006010103

Keohane, R. O., & Nye, J. S. (2000). Introduction. In Nye, J. S., & Donahue, J. D. (Eds.), *Governance in a globalizing world*. Washington, DC: Brookings Institution Press.

Khalifa, M., & Liu, V. (2001). Satisfaction with Internet-based services: a longitudinal study. *Journal of Global Information Management, 10*(3), 1–14.

Khalifa, M., & Liu, V. (2003). Determinant of satisfaction at different adoption stage of internet-based services. *Journal of the Association for Information Systems, 4*(5), 206–232.

Khan, N., & Fitzgerald, G. (2004). Dimensions of offshore outsourcing business models. *Journal of Information Technology Cases and Applications, 6*(3), 35–50.

Kiiski, C., & Matti Pohjola, M. (2002). Cross country diffusion of the Internet. *Information Economics and Policy, 14*, 297–310. doi:10.1016/S0167-6245(01)00071-3

Kim, B., Choi, M., & Han, I. (2009). User behaviors toward mobile data services: The role of perceived fee and prior experience. *Expert Systems with Applications, 36*(4), 8528–8536. doi:10.1016/j.eswa.2008.10.063

Kim, B., & Han, I. (2009). The role of trust belief and its antecedents in a community-driven knowledge environment. *Journal of the American Society for Information Science and Technology, 60*(5), 1012–1026. doi:10.1002/asi.21041

Kim, H. Y., Lee, I. S., & Kim, J. W. (2008). Maintaining continuers vs. converting discontinuers: Relative importance of post-adoption factors for mobile data services. *International Journal of Mobile Communications, 6*(1), 108–132. doi:10.1504/IJMC.2008.016007

Kim, S. S., & Malhotra, N. K. (2005). A longitudinal model of continued IS use: An integrative view of four mechanisms underlying post adoption phenomena. *Management Science, 51*(5), 741–755. doi:10.1287/mnsc.1040.0326

Kim, S., Kim, H. J., & Lee, H. (2009). An institutional analysis of an e-government system for anti-corruption: The case of OPEN. *Government Information Quarterly, 26*(1), 42–50. doi:10.1016/j.giq.2008.09.002

King, W. (1978). Strategic planning for MIS. *Management Information Systems Quarterly, 2*(1), 27–37. doi:10.2307/249104

King, W. (1988). How effective is your IS planning? *Long Range Planning, 21*(5), 103–112. doi:10.1016/0024-6301(88)90111-2

King, W. R., & He, J. (2006). A meta-analysis of the technology acceptance model. *Information & Management, 43*(6), 740–755. doi:10.1016/j.im.2006.05.003

King, W. R. (2008). The post-offshoring IS organization. *Information Resources Management Journal, 21*(1), 77–87. doi:10.4018/irmj.2008010105

King, J. L., Gurbaxani, V., Kraemer, K. L., McFarlan, F. W., Raman, K. S., & Yap, C. S. (1994). Institutional factors in information technology innovation. *Information Systems Research, 5*(2), 139–169. doi:10.1287/isre.5.2.139

King, N. (1994). The qualitative research interview. In Cassell, C., & Symon, G. (Eds.), *Qualitative methods in organizational research: A practical guide* (pp. 14–36). London, England: Sage.

Kini, R. B., Ramakrishna, H. V., & Vijayaraman, V. (2004). Shaping moral intensity regarding software piracy: A comparison between Thailand and U.S. students. *Journal of Business Ethics, 49*(1), 91–104. doi:10.1023/B:BUSI.0000013863.82522.98

Kirby, S. L., & Davis, M. A. (1998). A study of escalating commitment in principal-agent relationships: effects of monitoring and personal responsibility. *The Journal of Applied Psychology, 83*(2), 206–217. doi:10.1037/0021-9010.83.2.206

Kirkman, G. S. (2002). Executive summary. In G. S. Kirkman, P. K. Cornelius, J. D. Sachs, & K. Schwab (Eds.), *The global information technology report 2001-2002: Readiness for the networked world* (pp. xiii-xvi). Retrieved January 28, 2008, from http:// www.cid.harvard. edu/ archive/ cr/pdf/ gitrr2002_execsumm.pdf

Kirkpatrick, C., Parker, D., & Zhang, Y. (2006). Foreign direct investment in infrastructure in developing countries: does regulation make a difference? *Transnational Corporations*, *15*(1), 143–171.

Kirsch, L. J. (1996). The management of complex tasks in organizations: Controlling the systems development process. *Organization Science*, *7*(1), 1–21. doi:10.1287/orsc.7.1.1

Kirsch, L. J. (1997). Portfolios of control modes and IS project management. *Information Systems Research*, *8*(3), 215–239. doi:10.1287/isre.8.3.215

Kishore, R., Lee, J., & McLean, E. M. (2001). The role of personal innovativeness and self-efficacy in information technology acceptance: An extension of TAM with notions of risk. *Proceedings of the International Conference on Information Systems*. New Orleans, USA.

Kivistö, J. (2008). An assessment of agency theory as a framework for the government-university relationship. *Journal of Higher Education Policy and Management*, *30*(4), 339–350. doi:10.1080/13600800802383018

Kivistö, J. (2007). *An agency theory as a framework for the government-university relationship*. Unpublished doctoral dissertation, University of Tampere, Tampere, Finland.

Klaus, H., Rosemann, M., & Gable, G. G. (2000). What is ERP? *Information Systems Frontiers*, *2*(2), 141–162. doi:10.1023/A:1026543906354

Kluckhohn, F. R., & Strodtbeck, F. L. (1961). *Variations in value orientations*. Evanston, IL: Row, Peterson.

Kluver, R. (2005). The architecture of control: a Chinese strategy for e-governance. *Journal of Public Policy*, *25*(1), 75–97. doi:10.1017/S0143814X05000218

Kock, N., Parente, R., & Verville, J. (2008). Can Hofstede's Model Explain National Differences in Perceived Information Overload? A Look at Data from the US and New Zealand. *IEEE Transactions on Professional Communication*, *51*(1), 33–49. doi:10.1109/TPC.2007.2000047

Korpela, M. (1996). Traditional culture or political economy? On the root causes of organizational obstacles of IT in developing countries. *Information Technology for Development*, *7*(1), 29. doi:10.1080/02681102.1996.9627212

Kreitner, R., & Kinicki, A. (2001). *Organizational Behavior* (5th ed.). New York: McGraw-Hill.

Krishna, S., & Walsham, G. (2005). Implementing public information systems in developing countries: Learning from a success story. *Information Technology for Development*, *11*(2), 123–140. doi:10.1002/itdj.20007

Kuan, K. K. Y., & Chau, P. Y. K. (2001). A perception-based model for EDI adoption in small business using a technology-organization-environment framework. *Information & Management*, *38*(8), 507–521. doi:10.1016/S0378-7206(01)00073-8

Kumar, V., Maheshwari, B., & Kumar, U. (2003). An investigation of critical management issues in ERP implementation: evidence from Canadian organizations. *Technovation*, *23*(10), 793–807. doi:10.1016/S0166-4972(02)00015-9

Kumar, R., & Best, M. L. (2006). Impact and sustainability of egovernment services in developing countries: Lessons learned from Tamil Nadu, India. *The Information Society*, *22*, 1–12. doi:10.1080/01972240500388149

Kundu, S. K., Niederman, F., & Boggs, D. J. (2003). The relevance of global management Information Systems to international business. *Journal of Global Information Management*, *11*(1), 1–4.

Kuo, F., & Hsu, M. (2001). Development and validation of ethical computer self-efficacy measure: The case of softlifting. *Journal of Business Ethics*, *32*(4), 299–315. doi:10.1023/A:1010715504824

Kurth, M. (1993). The limits and limitations of transaction log analysis. *Library Hi Tech*, *11*(2), 98–104. doi:10.1108/eb047888

Kuwayama, M., Ueki, Y., & Tsuji, T. (2005). *Information Technology for development of small and medium-sized exporters in Latin America and East Asia*. Chile: United Nations Publication.

Kwahk, K. Y., & Lee, J. N. (2008). The role of readiness for change in ERP implementation: theoretical bases and empirical validation. *Information & Management*, *45*(7), 474–481. doi:10.1016/j.im.2008.07.002

Kwon, T. H., & Zmud, R. W. (1997). Unifying the fragmented models information systems implementation. In R.J. Boland & R.A. Hirschheim (Eds.), *Critical issues in information systems research* (pp. 227-251). Chichester, England: John Wiley and Sons Ltd.

Kwong, K. K., Yau, O. H. M., Lee, J. S. Y., Sin, L. Y. M., & Tse, A. C. B. (2003). The effects of attitudinal and demographic factors on intention to buy pirated CDs: The case of Chinese consumers. *Journal of Business Ethics*, *47*(3), 223–235. doi:10.1023/A:1026269003472

La Rose, R., Gregg, J. L., Strover, S., Straubhaar, J., & Carpenter, S. (2007). Closing the rural broadband gap: Promoting adoption of the internet in rural America. *Telecommunications Policy*, *31*, 359–360. doi:10.1016/j.telpol.2007.04.004

Lai, V. S., & Li, H. (2005). Technology acceptance model for internet banking: An invariance analysis. *Information & Management*, *42*(2), 373–386. doi:10.1016/j.im.2004.01.007

Lam, J. C. Y., & Lee, M. K. O. (2006). Digital inclusiveness: Longitudinal study of internet adoption by older adults. *Journal of Management Information Systems*, *22*(4), 177–206. doi:10.2753/MIS0742-1222220407

Lang, J. R., Calantone, R. J., & Gudmundson, D. (1997). Small firm information seeking as a response to environmental threats and opportunities. *Journal of Small Business Management*, *35*(1), 11–23.

Langley, A. (1999). Strategies for theorizing from process data. *Academy of Management Review*, *24*(4), 691–710.

Laroche, M., Bergeron, J., & Goutaland, C. (2003). How intangibility affects perceived risk: The moderating role of knowledge and involvement. *Journal of Services Marketing*, *17*(2), 122–140. doi:10.1108/08876040310467907

Laughlin, S. P. (1999). An ERP game plan. *The Journal of Business Strategy*, *20*(1), 32–37. doi:10.1108/eb039981

Layne, K., & Lee, J. (2001). Developing fully functional e-government: A four stage model. *Government Information Quarterly*, *18*(2), 122–136. doi:10.1016/S0740-624X(01)00066-1

Lederer, A., & Sethi, V. (1996). Key Prescriptions for Strategic Information Systems Planning. *Journal of Management Information Systems*, *13*(1), 35–62.

Lee, I., Choi, B., Kim, J., & Hong, S. J. (2007). Culture-technology fit: Effects of cultural characteristics on the post-adoption beliefs of mobile internet users. *International Journal of Electronic Commerce*, *11*(4), 11–51. doi:10.2753/JEC1086-4415110401

Lee, I., Kim, J. S., & Kim, J. W. (2005). Use contexts for the mobile data: A longitudinal study monitoring actual use of mobile data services. *International Journal of Human-Computer Interaction*, *18*(3), 269–292. doi:10.1207/s15327590ijhc1803_2

Lee, M. K. O., Cheung, C. M. K., & Chen, Z. (2005). Acceptance of Internet-based learning medium: the role of extrinsic and intrinsic motivation. *Information & Management*, *42*(8), 1095–1104. doi:10.1016/j.im.2003.10.007

Lee, Y. S., Kim, J. W., Lee, I. S., & Kim, H. Y. (2002). A cross-cultural study on the value structure of mobile internet usage: Comparison between Korea and Japan. *Journal of Electronic Commerce Research*, *3*(4), 227–239.

Lee, J. (2004). Discriminant analysis of technology adoption behavior: A case of internet technologies in small businesses. *Journal of Computer Information Systems*, *44*(4), 57–66.

Lee, A. S. (1999). Rigor and relevance in MIS research: beyond the approach of positivism alone. *Management Information Systems Quarterly*, *23*(1), 29–34. doi:10.2307/249407

Lee, C. K., & Strang, D. (2006). The International diffusion of public-sector downsizing: Network emulation and theory-driven learning. *International Organization*, *60*(4), 883–909. doi:10.1017/S0020818306060292

Lee, I. S., Choi, B., Kim, J., Hong, S., & Tam, K. (2004). *Cross-cultural comparison for cultural aspects of mobile internet: Focusing on Korea and Hong Kong.* Paper presented at the Proceedings of the 2004 Americas Conference on Information Systems (AMCIS), New York.

Lee-Ross, D. (2005). Perceived job characteristics and internal work motivation: An exploratory cross-cultural analysis of the motivational antecedents of hotel workers in Mauritius and Australia. *Journal of Management Development*, *24*(3), 253–266. doi:10.1108/02621710510584062

Lees, J. D. (1987). Successful development of small business information systems. *Journal of Systems Management*, 38(8), 32–39.

Lefever, S., Dal, M., & Matthíasdóttir, Á. (2006). Online data collection in academic research: Advantages and limitations. *British Journal of Educational Technology*, 38(4), 574–582. doi:10.1111/j.1467-8535.2006.00638.x

Legris, P., Ingham, I., & Collerette, P. (2003). Why do people use information technology? A critical review of the technology acceptance model. *Information & Management*, 40(3), 191–204. doi:10.1016/S0378-7206(01)00143-4

Leidner, D. E., & Kayworth, T. (2006). A Review of Culture in Information Systems Research: Toward a Theory of Information Technology Culture Conflict. *Management Information Systems Quarterly*, 30(2), 357–399.

Lenartowicz, T., & Roth, K. (2001). Does subculture within a country matter? A cross-cultural study of motivational domains and business performance in Brazil. *Journal of International Business Studies*, 32(2), 305–325. doi:10.1057/palgrave.jibs.8490954

Leung, K., Au, A., Huang, X., Kurman, J., Niit, T., & Nii, K. (2007). Social axioms and values: A cross-cultural examination. *European Journal of Personality*, 21, 91–111. doi:10.1002/per.615

Levy, M., & Powell, P. (2000). Information systems strategy for small-medium-sized enterprises: an organizational perspective. *The Journal of Strategic Information Systems*, 9(1), 63–84. doi:10.1016/S0963-8687(00)00028-7

Levy, M., Powell, P., & Galliers, R. (1999). Assessing information systems strategy development frameworks in SMEs. *Information & Management*, 36(5), 247–261. doi:10.1016/S0378-7206(99)00020-8

Lewis, W., Agarwal, R., & Sambamurthy, V. (2003). Sources of Influence on Beliefs about Information Technology Use: An Empirical Study of Knowledge Workers. *Management Information Systems Quarterly*, 27(4), 657–678.

Li, C. (1999). ERP packages: what's next? *Information Systems Management*, 16(3), 31–35. doi:10.1201/1078/43197.16.3.19990601/31313.5

Li, L. (2008). The study and suggestions for the internet-based science information system of NSFC. *Science Foundation in China*, 22(1), 52–54.

Li, Z. Z. (2004). Chinese science in transition: misconduct in science and analysis of the causes. *Science Research Management*, 25(3), 137–144.

Li, S. (2005). Why a poor governance environment does not deter foreign direct investment: The Case of China and its implications for investment protection. *Business Horizons*, 4(4), 297–302. doi:10.1016/j.bushor.2004.06.002

Liang, H., Saraf, N., Hu, Q., & Xue, Y. (2007). Assimilation of enterprise systems: The effect of institutional pressures and the mediating role of top management. *MIS Quarterly*, 31(1), 59–87.

Lim, J. (2004). The Role of Power Distance and Explanation Facility in Online Bargaining Utilizing Software Agents. *Journal of Global Information Management*, 12(2), 27–43.

Lin, F., Fofanah, S. S., & Liang, D. (2011). Assessing citizen adoption of e-Government initiatives in Gambia: A validation of the technology acceptance model in information systems success. *Government Information Quarterly*, 28(2), 271–279. doi:10.1016/j.giq.2010.09.004

Lindquist, K. M., & Mauriel, J. J. (Eds.). (1989). *Depth and breadth in innovation implementation: the case of school-based management*. New York, NY: Harper and Row.

Linton, K. C. (2008). *China's R&D Policy for the 21st Century: Government Direction of Innovation*. Retrieved from http://ssrn.com/abstract=1126651

Lippert, S. K., & Volkmar, J. A. (2007). Cultural effects on technology performance and utilization. *Journal of Global Information Management*, 15(2), 56–90. doi:10.4018/jgim.2007040103

Little, T. D. (1997). Mean and covariance structures (MACS) analyses of cross-cultural data: Practical and theoretical issues. *Multivariate Behavioral Research*, 32(1), 53–76. doi:10.1207/s15327906mbr3201_3

Liu, B. S. C., Olivier, F., & Subharshan, D. (2001). The relationships between culture and behavioral intentions toward services. *Journal of Service Research*, 4(2), 118–129. doi:10.1177/109467050142004

Liu, M., Wu, X., Zhao, J. L., & Zhu, L. (2010). Outsourcing of community source: Identifying motivations and benefits. *Journal of Global Information Management, 18*(4), 36–52. doi:10.4018/jgim.2010100103

Liu, M., & Zhao, L. J. (2007). Real options analysis of the community source approach: Why should institutions pay for open source? *Proceedings of the First China Summer Workshop on Information Management, Shanghai, China.*

Liu, M., Wang, H. J., & Zhao, L. J. (2007). Achieving flexibility via service-centric community source: The case of Kuali. *Proceedings of the Americas Conference on Information Systems,* Colorado, USA.

Liu, M., Wheeler, B., & Zhao, J. L. (2008). On assessment of project success in community source development. *Proceeding of International Conference on Information Systems* Paris, France.

Liu, M., Zeng, D., & Zhao, L. J. (2008). A cooperative analysis framework for investment decisions in community source partnerships. *Proceedings of the Americas Conference on Information Systems,* Toronto, Canada.

Liu, W., & Westrup, C. (2003). *ICTs and organizational control across cultures: The case of a UK multinational operating in China.* Paper presented at the International Federation of Information Processing, IFIP 9.4 and 8.2 Joint Conference on Organizational Information Systems in the Context of Globalization Dordrecht, The Netherlands.

Loch, K. D., Straub, D. W., & Kamel, S. (2003). Diffusing the Internet in the Arab World: The Role of Social Norms and Technological Culturation. *IEEE Transactions on Engineering Management, 50*(1), 45–63. doi:10.1109/TEM.2002.808257

Locke, S. (2005). Farmer adoption of ICT in New Zealand. *Proceedings of the Global Management & Information Technology Research Conference,* New York, USA.

Looney, R. (2004). Saudization and Sound Economic Reforms: Are the Two Compatible? *Strategic Insights, 3*(2).

Loungani, P., & Razin, A. (2001). How beneficial is foreign direct investment for developing economies? *Finance and Development Quarterly, 38*(2), 6–10.

Lu, J., Yu, C. S., Liu, C., & Yao, J. E. (2003). Technology acceptance model for wireless Internet. *Internet Research: Electronic Networking Applications and Policy, 13*(3), 206–222. doi:10.1108/10662240310478222

Lu, M. (2005). Digital divide in developing economies. *Journal of Global Information Technology Management, 4*(3), 1–4.

Luftman & Kempaiah. (2008). Key Issues For IT Executives 2007. *MIS Quarterly Executive, 7*(2), 99–112.

Luftman, Kempaiah, & Nash. (2006). Key Issues For IT Executives 2005. *MIS Quarterly Executive, 5*(2), 81-99.

Lundblad, J. P. (2003). A review and critique of Rogers' diffusion of innovation theory as it applies to organizations. *Organization Development Journal, 21*(4), 50–64.

Lundgren, L. (1998). The Technical Communicator's Role in Bridging the Gap between Arab and American Business Environments. *Journal of Technical Writing and Communication, 28*(4), 335–343. doi:10.2190/U8AH-MQWD-F9L7-QAFA

Luoto, J., McIntosh, C., & Wydick, B. (2007). Credit information systems in less developed countries: A test with microfinance in Guatemala. *Economic Development and Cultural Change, 55*(2), 313–334. doi:10.1086/508714

Luthans, F. (2002). *Organizational Behavior* (9th ed.). New York: McGraw-Hill.

Lyytinen, K., & Damsgaard, J. (2001). What's wrong with the diffusion of innovation theory. *Diffusing Software Products and Process Innovations,* 173-190.

Ma, J., Fan, Z. P., & Huang, L. H. (1999). A subjective and objective integrated approach to determine attribute weights. *European Journal of Operational Research, 112,* 397–404. doi:10.1016/S0377-2217(98)00141-6

Mabert, V. A., Soni, A., & Venkataramanan, M. A. (2003). Enterprise resource planning: Managing the implementation process. *European Journal of Operational Research, 146*(2), 302–314. doi:10.1016/S0377-2217(02)00551-9

MacDuffie, J. P. (2008). HRM and distributed work: Managing people across distances. In Walsh, J. P., & Brief, A. P. (Eds.), *The Academy of Management Annals* (*Vol. 1,* pp. 549–616). New York, NY: Lawrence Erlbaum.

MacKay, N., Oarent, M., & Gemino, A. (2004). A model of electronic commerce adoption by small voluntary organizations. *European Journal of Information Systems, 13*(2), 147–159. doi:10.1057/palgrave.ejis.3000491

Maddala, G. S. (1992). *Introduction to Econometrics.* New York: Prentice Hall.

Madon, S. (2005). Governance lessons from the experience of telecentres in Kerala. *European Journal of Information Systems, 14*(4), 401–416. doi:10.1057/palgrave.ejis.3000576

Madon, S. (2004). Evaluating the developmental impact of e–governance initiatives: An exploratory framework. *The Electronic Journal of Information Systems in Developing Countries, 20*(5), 1–13.

Madon, S. (2003). IT diffusion for public service delivery: Looking for plausible theoretical approaches. In Avgerou, C., & La Rovere, R. (Eds.), *Information systems and the economics of innovation* (pp. 71–88). Cheltenham, UK: Edward Elgar.

Mahaney, R. C. (2000). *Information systems development project success and failure: An agency theory interpretation.* Unpublished doctoral dissertation, University of Kentucky, KY.

Maheswaran, D., & Shavitt, S. (2000). Issues and New Directions in Global Consumer Psychology. *Journal of Consumer Psychology, 9*(2), 59–66. doi:10.1207/S15327663JCP0902_1

Maitland, C. (1999). Global diffusion of interactive networks: The impact of culture. *AI & Society, 13*(4), 341–356. doi:10.1007/BF01205982

Makino, S., Isobe, T., & Chan, C. (2004). Does Country Matter? *Strategic Management Journal, 25*(10), 1027–1043. doi:10.1002/smj.412

Malek, M., & Al-Shoaibi, A. (1998). *Information Technology in the Developing Countries: Problems and Prospects.* Paper presented at the Conference on Administrative Sciences: New Horizons and Roles in Development, King Fahd University of Petroleum and Minerals, Dhahran, Saudi Arabia.

Mao, E., & Palvia, P. C. (2006). Testing an Extended Model of IT Acceptance in the Chinese Cultural Context. *The Data Base for Advances in Information Systems, 37*(2/3), 20–32.

March, J. G., & Olsen, J. P. (1989). *Rediscovering institutions: The organizational basis of politics.* New York: Free Press.

Marcus, A., & Emilie, W. G. (2000). Crosscurrents: Cultural dimensions and global web user-interface design. *Interaction, 7*(4), 32–46. doi:10.1145/345190.345238

Marcus, A. A., & Weber, M. J. (Eds.). (1989). *Externally-induced innovation.* New York, NY: Harper and Row.

Markus, M. L., Axline, S., Petrie, D., & Tanis, C. (2000). Learning from adopters' experiences with ERP: problems encountered and success achieved. *Journal of Information Technology, 15*(4), 245–265. doi:10.1080/02683960010008944

Markus, M. L., & Benjamin, R. I. (1997). The magic bullet theory in IT–enabled transformation. *Sloan Management Review, 38*(2), 55–68.

Marron, D. B., & Steel, D. G. (2000). Which countries protect intellectual property? The case of software piracy. *Economic Enquiry, 38*(2), 159–174. doi:10.1093/ei/38.2.159

Marsh, H. W., Balla, J. R., & McDonald, R. P. (1988). Goodness-of-fit indexes in confirmatory factor analysis: The effect of sample size. *Psychological Bulletin, 103*(3), 391–410. doi:10.1037/0033-2909.103.3.391

Marshall, M. G., & Jaggers, K. (2007). Polity IV project: Political regime characteristics and transitions 1800-2007. *Center for Systemic Peace.* Retrieved from www.systemicpeace.org/polity

Martin, R., & Feldman, E. (1998). Why is Access Important in Developing Countries. In Transparency International (Ed.), *TI working paper: Access-to-information in Developing Countries.* Berlin, Germany.

Martins, L. L., Gilson, L. L., & Maynard, M. T. (2004). Virtual teams: What do we know and where do we go from here? *Journal of Management, 30*(6), 805–835. doi:10.1016/j.jm.2004.05.002

Martinsons, M. G. (2005). Transforming China. *Communications of the ACM, 48*(4), 44–48. doi:10.1145/1053291.1053316

Martinsons, M. G., & Davison, R. M. (2007). Culture's Consequences for IT Application and Business Process Change: A Research Agenda. *International Journal of Internet and Enterprise Management, 5*(20), 158–177. doi:10.1504/IJIEM.2007.014087

Martinsons, M. G. (2004). ERP in China: One package, two profiles. *Communications of the ACM, 47*(7), 65–68. doi:10.1145/1005817.1005823

Mascarenhas, B. (1989). Domains of state-owned, privately held, and publicly traded companies in international competition. *Administrative Science Quarterly, 34*(4), 582–597. doi:10.2307/2393568

Mata, F. J., Fuerst, W. L., & Barney, J. B. (1995). Information Technology and Sustained Competitive Advantage: A Resource-based Analysis. *Management Information Systems Quarterly, 19*(4), 487–505. doi:10.2307/249630

Mathieson, K. (1991). Predicting User Intentions: Comparing the Technology Acceptance Model with the Theory of Planned Behavior. *Information Systems Research, 2*(3), 173–191. doi:10.1287/isre.2.3.173

Mathieson, K. (1991). Predicting User Intentions: Comparing the Technology Acceptance Model with the Theory of Planned Behavior. *Information Systems Research, 2*(3), 173–191. doi:10.1287/isre.2.3.173

Mauritius ICT Indicators Portal. (2011). *International indices.* Mauritius National Computer Board. Retrieved from http://www.gov.mu/ portal/sites/indicators/ International_Indices. html

Mauritius National Computer Board. (2006). *Universal ICT education programme.* Retrieved from http://www.gov.mu/portal/sites/uieptest/about.html

Mauritius National Computer Board. (2007). *Republic of Mauritius national ICT strategic plan 2007-2011.* Retrieved from http://www. gov.mu/ portal/goc/ telecomit/ files/ NICTSP.pdf

Mawhinney, C., & Lederer, C. (1990). A study of personal computer utilization by managers. *Information & Management, 18*(5), 243–253. doi:10.1016/0378-7206(90)90026-E

Mayer-Schönberger, V., & Lazer, D. (2007). *Governance and Information Technology: From Electronic Government to Information Government.* Cambridge, MA: MIT Press.

Mbarika, V., & Byrd, T. A. (2009). An exploratory study of strategies to improve Africa's Least developed economies' telecommunications infrastructure: The stakeholders speak. *IEEE Transactions on Engineering Management, 56*(2), 312–328. doi:10.1109/TEM.2009.2013826

Mbarika, V. (2002). Re-thinking information and communications technology policy focus on internet versus teledensity diffusion for Africa's least developed countries. *The Electronic Journal of Information Systems in Developing Countries, 9*(1), 1–13.

McCoy, S., Everard, A., & Jones, B. M. (2005). An examination of the technology acceptance model in Uruguay and the US: A focus on culture. *Journal of Global Information Management, 8*(2), 27–45.

McCoy, S., Galletta, D. F., & King, W. R. (2005). Integrating national culture into IS research: The need for current individual level measure. *Communications of the Association for Information Systems, 15*(12), 211–224.

McDougall, S., Curry, M., & Bruijn, O. (1998). Understanding What Makes Icons Effective: How Subjective Rating Can Inform Design. In Hanson, J. (Ed.), *Contemporary Ergonomics* (pp. 285–289). London: Taylor & Francis.

McFarlan, F. W. (1984). Information technology changes the way you compete. *Harvard Business Review, 62*(3), 98–103.

McFarlan, F. W., & Rockart, J. (2004). China and information technology: An interview with Warren McFarlan from the Harvard Business School. *MIS Quarterly Executive, 3*(2), 83–88.

McFarlan, F. W., & Nolan, R. L. (1995). How to manage an IT outsourcing alliance. *Sloan Management Review, 36*(2), 9–23.

McFarlan, F. W, Chen, G., & Lane, D. (2005). Information technology at COSCO. *Harvard Business Review Case,* Product No.: 9-305-080.

McFarlin, D. B., & Sweeney, P. D. (1992). Distributive justice and procedural justice as predictors of satisfaction with personal and organizational outcomes. *Academy of Management Journal, 35,* 626–637. doi:10.2307/256489

Mejias, R. J., Shepherd, M. M., Vogel, D. R., & Lazaneo, L. (1996). Consensus and perceived satisfaction levels: a cross-cultural comparison of GSS and non-GSS outcomes within and between the United States and Mexico. *Journal of Management Information Systems, 13*(3), 137–161.

Melone, N. P. (1990). A Theoretical Assessment of the User-Satisfaction Construct in Information Systems Research. *Management Science, 36*(1), 76–91. doi:10.1287/mnsc.36.1.76

Melville, N., & Ramirez, R. (2008). Information technology innovation diffusion: An information requirements paradigm. *Information Systems Journal, 18*(3), 247–273. doi:10.1111/j.1365-2575.2007.00260.x

Meng, H., Zhou, L., & He, J. K. (2007). The co-integration analysis on NSFC input and S&T paper's output. *Studies in Science of Science, 25*(6), 1147–1155.

Mentzas, G. (1997). Implementing an IS strategy - A team approach. *Long Range Planning, 30*(1), 84–95. doi:10.1016/S0024-6301(96)00099-4

Merton, R. K. (1968). *Social theory social structure.* New York, NY: Free Press.

Meso, P., Musa, P., & Mbarika, V. (2005). Towards a model of consumer use of mobile information and communication technology in LDCs: The case of Sub-Saharan Africa. *Information Systems Journal, 15*(2), 119–146. doi:10.1111/j.1365-2575.2005.00190.x

Meyer, A. D., & Goes, J. B. (1988). Organizational assimilation of innovations: A multilevel contextual analysis. *Academy of Management Journal, 31*(4), 897–923. doi:10.2307/256344

Meyer, J. W., Boli, J., Thomas, G. M., & Ramirez, F. O. (1997). World society and the nation-state. *American Journal of Sociology, 103*(1), 144–181. doi:10.1086/231174

Meyer, J. W., & Rowan, B. (1977). Institutionalized organizations: Formal structure as myth and ceremony. *American Journal of Sociology, 83*(2), 340–363. doi:10.1086/226550

MIC (Market Intelligence & Consulting Institute). (2008). *Current and tendency of information technology investment in Taiwan large company during 2007-2009.* Taiwan: MIC.

Milberg, S., Smith, H. J., & Burke, S. (2000). Information Privacy: Corporate Management and National Regulation. *Organization Science, 11*(1), 35–57. doi:10.1287/orsc.11.1.35.12567

Miles, M. B., & Huberman, A. M. (1994). *Qualitative data analysis.* Thousand Oaks, CA: Sage.

Miles, M. A., Holmes, K. R., & Anastasia, M. A. (2006). *2006 index of economic freedom.* The Heritage Foundation and Wall Street Journal.

Milgrom, P., & Roberts, J. (1992). *Economics, organization and management.* Upper Saddle River, NJ: Prentice Hall.

Miller, T., & Holems, K. R. (2009). *2009 index of economic freedom.* The Heritage Foundation and Wall Street Journal.

Min, Q., Li, Y., & Ji, S. (2009, June 27-28). *The effects of individual-level culture on mobile commerce adoption: An empirical study.* Paper presented at the 2009 Eighth International Conference on Mobile Business, Liaoning, China.

Ministry of Commerce of the People's Republic of China (MoC). (2007). *Statistics Data of Imports and Exports.* Retrieved Nov. 11, 2007, from http://zhs.mofcom.gov.cn/tongji.shtml

Miscione, G. (2007). Telemedicine in the upper Amazon: Interplay with local health care practices. *Management Information Systems Quarterly, 31*(2), 403–425.

Mohdzain, M. B., & Ward, J. M. (2007). A Study of Subsidiaries' Views of Information Systems Strategic Planning in Multinational Organizations. *The Journal of Strategic Information Systems, 16*(4), 324–352. doi:10.1016/j.jsis.2007.02.003

Molla, A., & Licker, P. S. (2005a). E-commerce adoption in development countries: A model and instrument. *Information & Management, 42*(6), 877–899. doi:10.1016/j.im.2004.09.002

Molla, A., & Licker, P. S. (2005b). Perceived e-readiness factors in e-commerce adoption: An empirical investigation in a developing country. *International Journal of Electronic Commerce*, *10*(1), 83–110.

Molla, A., Taylor, R., & Licker, P. (2006). E-commerce diffusion in small island countries: The influence of institutions in Barbados. *The Electronic Journal of Information Systems in Developing Countries*, *28*(2), 1–15.

Montealegre, R. (1998). Waves of change in adopting the Internet: Lessons from four Latin American countries. *Information Technology & People*, *11*(3), 235–260. doi:10.1108/09593849810228039

Montealegre, R. (1999). A temporal model of institutional interventions for Information Technology adoption in less developed countries. *Journal of Management Information Systems*, *16*(1), 207–232.

Moon, M. J., Welch, E. W., & Wong, W. (2005). *What drives global e-governance? An exploratory study at a macro level*. Paper presented at Proceedings of the 38th Annual Hawaii International Conference on System Sciences (HICSS'05).

Moore, G. C., & Benbasat, I. (1991). Development of an instrument to measure the perceptions of adopting an information technology innovation. *Information Systems Research*, *2*(3), 192–222. doi:10.1287/isre.2.3.192

Moore, G., & Benbasat, I. (1993). *An empirical examination of a model of the factors affecting utilization of information technology by end-users (Tech. Rep.)*. Vancouver, Canada: University of British Columbia, Faculty of Commerce.

Moore, G. C., & Benbasat, I. (1991). Development of an instrument to measure perceptions of adopting an information technology innovation. *Information Systems Research*, *2*(3), 192–222. doi:10.1287/isre.2.3.192

Moores, T. T., & Gregory, F. H. (2000). Cultural Problems in Applying SSM for IS Development. *Journal of Global Information Management*, *8*(1), 14–19.

Moores, T. T. (2003). The effect of national culture and economic wealth on global software piracy rates. *Communications of the ACM*, *46*(9), 207–215. doi:10.1145/903893.903939

Moores, T. T. (2008). An analysis of the impact of economic wealth and national culture on the rise and fall of software piracy rates. *Journal of Business Ethics*, *81*(1), 39–51. doi:10.1007/s10551-007-9479-0

Moores, T. T., & Chang, J. C. J. (2006). Ethical decision making in software piracy: Initial development and test of a four-component model. *MIS Quarterly*, *30*(1), 167–180.

Moores, T. T., & Dhaliwal, J. (2004). A reversed context analysis of software piracy issues in Singapore. *Information & Management*, *41*(8), 1037–1042. doi:10.1016/j.im.2003.10.005

Morgan, D. L. (1997). *Focus Group as Qualitative Research*. Thousand Oaks, CA: Sage Publications.

Morris, M. W., Leung, K., Ames, D., & Lickel, B. (1999). Views from Inside and Outside: Integrating Emic and Etic Insights about Culture and Justice Judgment. *Academy of Management Review*, *24*(4), 781–796. doi:10.2307/259354

Morris, M. G., & Dillon, A. (1997). How user perceptions influence software use. *IEEE Software*, *14*(4), 58–65. doi:10.1109/52.595956

Morton, M. S. (Ed.). (1991). *The corporation of the 1990s: Information Technology and organizational transformation*. New York, NY: Oxford University Press.

MOSICT. (2002). *National information and communication technology (ICT) policy*. Ministry of Science and Information & Communication Technology, Government of the People's Republic of Bangladesh.

Moynihan, T. (1990). What Chief Executives and Senior Managers Want from Their IT Departments. *Management Information Systems Quarterly*, *14*(1), 15–25. doi:10.2307/249303

Munter, M. (1993). Cross-cultural communication for managers. *Business Horizons*, *36*(3), 69–78. doi:10.1016/S0007-6813(05)80152-1

Mursu, A., Lyytinen, K., Soriyan, H. A., & Korpela, M. (2003). Identifying software project risks in Nigeria: An international comparative study. *European Journal of Information Systems*, *12*(3), 182–194. doi:10.1057/palgrave.ejis.3000462

Musa, P., Meso, P., & Mbarika, V. (2005, August 11-14). *Testing a modified TAM that accounts for realities of technology acceptance in Sub Saharan Africa*. Paper presented at the Americas Conference on Information Systems, Omaha, Nebraska.

Myers, M., & Tan, F. (2002). Beyond Models of National Culture in Information Systems Research. *Journal of Global Information Management, 10*(1), 24–32.

Myers, R. H. (1990). *Classical and modern regression application* (2nd ed.). CA: Duxbury press.

Nachum, G. (2003). *Owning the mobile content customer*. Retrieved June 11, 2005 from http://www.totaltele.com/interviews/display.asp?InterviewID=254

Nah, F. F., Tan, X., & Teh, S. H. (2004). An Empirical Investigation on End-Users' Acceptence of Enterprise Systems. *Information Resources Management Journal, 17*(3), 32–53.

Nair, M., & Kuppusamy, M. (2004). Trends of convergence and divergence in the information economy: Lessons for developing countries. *The Electronic Journal of Information Systems in Developing Countries, 18*(2), 1–32.

Nakakoji, K. (1996). Beyond language translation: Crossing the cultural divide. *IEEE Software, 13*(6), 42–46. doi:10.1109/52.542293

Ndou, V. D. (2004). E-government for developing countries: Opportunities and challenges. *The Electronic Journal on Information Systems in Developing Countries, 18*(1), 1–24.

Newkirk, H., Lederer, A., & Srinivasan, C. (2003). Strategic Information Systems Planning: Too Little or Too Much. *The Journal of Strategic Information Systems, 12*(3), 201–228. doi:10.1016/j.jsis.2003.09.001

Newman, M., & Robey, D. (1992). A social process model of user-analyst relationships. *Management Information Systems Quarterly, 16*(2), 249–266. doi:10.2307/249578

Ng, S. I., Lee, J. A., & Soutar, G. N. (2007). Are Hofstede's and Schwartz's value frameworks congruent? *International Marketing Review, 24*(2), 164–180. doi:10.1108/02651330710741802

Nicolaou, A. I., & McKnight, D. H. (2006). Perceived information quality in data exchanges: Effects on risk, trust, and intention to use. *Information Systems Research, 17*(4), 332–351. doi:10.1287/isre.1060.0103

Nidumolu, S., & Subramani, M. (2003). The Matrix of Control: Combining Process and Structure Approaches to Managing Software Development. *Journal of Management Information Systems, 20*(3), 159–196.

Nidumolu, S. R., Goodman, S. E., Vogel, D. R., & Danowitz, A. K. (1996). Information technology for local administration support: The Governorates Project in Egypt. *Management Information Systems Quarterly, 20*(2), 197–224. doi:10.2307/249478

Niederman, F., Brancheau, J. C., & Wetherbe, J. C. (1991). Information Systems Management Issues for the 1990s. *Management Information Systems Quarterly, 15*(4), 475–502. doi:10.2307/249452

Niedleman, L. D. (1979). Computer usage by small and medium sized European firms: an empirical study. *Information & Management, 2*(2), 67–77. doi:10.1016/0378-7206(79)90008-9

Nisbett, R. E., & Wilson, T. D. (1977). Telling more than we can know: Verbal reports on mental processes. *Psychological Review, 84*(3), 231–259. doi:10.1037/0033-295X.84.3.231

Norman, K. L. (1986). Importance of factors in the review of grant proposals. *The Journal of Applied Psychology, 71*(1), 156–162. doi:10.1037/0021-9010.71.1.156

Norris, P. (2001). *Digital divide? Civic engagement, information poverty and the internet worldwide*. New York: Cambridge University Press.

Norris, P. (2000). *The global divide: Information poverty and Internet access worldwide*. Retrieved from http://www.ksg.harvard.edu/people/.

North, D. C. (1990). *Institutions, institutional change and economic performance*. New York: Cambridge University Press.

North, D. C. (1994). Economic performance through time. *The American Economic Review, 84*(3), 359–368.

Norton, S. (1992). Information costs, telecommunications, and the microeconomics of macroeconomic growth. *Economic Development and Cultural Change, 41*, 175–196. doi:10.1086/452002

NSFC. (2009a). *Guide to Programmes 2009.* NSFC, Department of Management Science.

NSFC. (2009b). *NSFC 2008: Annual report on government information disclosure.* Retrieved from http://www.nsfc.gov.cn/nsfc/cen/xxgk/baogao.html.

NSFC. (2009c). *The Result of the 2009 NSFC First Round Peer-Review for All Proposals.* Retrieved from http://www.nsfc.gov.cn/Portal0/InfoModule_375/27220.htm

Nunnally, J. C. (1978). *Psychometric theory* (2nd ed.). New York, NY: McGraw-Hill Book Company.

Nunnaly, J. (1978). *Psychometric theory.* New York: McGraw-Hill.

Nwagwu, E. W. (2006). Integrating ICTs into the globalization of the poor developing countries. *Information Development, 22*(3), 167–179. doi:10.1177/0266666906069070

Nysveen, H., Pedersen, P. E., & Thorbjnsen, H. (2005). Intentions to use mobile services: Antecedents and cross-service comparisons. *Journal of the Academy of Marketing Science, 33*(3), 330–346. doi:10.1177/0092070305276149

OECD. (1999). *Building infrastructure capacity for electronic commerce: Leased line developments and pricing.* Paris, France: OECD.

OECD. (1996). *Information infrastructure convergence and pricing: The Internet.* Paris.

OECD. (1997b). *Global information infrastructure-Global information society (GII-GIS): Policy recommendations for action.* Paris.

OECD. (1998). *Internet tra$c exchange: Developments and policy.* Paris, France: OECD. Retrieved from http://www.oecd.org/dsti/sti/it/cm/prod/tra$c.htm

Oh, H. (1999). Service quality, customer satisfaction, and customer value: A holistic perspective. *International Journal of Hospitality Management, 18*(1), 67–82. doi:10.1016/S0278-4319(98)00047-4

Olikowski, W. J., & Iacono, C. S. (2001). Desperately seeking the 'IT' in IT research – a call to theorizing the IT artifact. *Information Systems Research, 12*(2), 121–134. doi:10.1287/isre.12.2.121.9700

Oliver, R. L. (1980). A cognitive model of the antecedents and consequences of satisfaction decisions. *JMR, Journal of Marketing Research, 17*(3), 460–469. doi:10.2307/3150499

Orbicom. (2005). *From the digital divide to digital opportunities: measuring infostates for development.* Montreal, Canada: Claude-Yves Charron.

Organization for Economic Cooperation and Development (OECD). (2008). *Information Technology Outlook.*

Ouellette, J. A., & Wood, W. (1998). Habit and intention in everyday life: The multiple processes by which past behavior predicts future behavior. *Psychological Bulletin, 124*(1), 54–74. doi:10.1037/0033-2909.124.1.54

Oyelaran-Oyeyinka, B., & Kaushalesh, L. (2003). The internet diffusion in Sub-Saharan Africa: A cross-country analysis. *Telecommunications Policy, 29*, 507–527. doi:10.1016/j.telpol.2005.05.002

Pai, J. C., Lee, G. G., Tseng, W. G., & Chang, Y. L. (2007). Organizational, technological and environmental factors affecting the implementation of ERP systems: Multiple–case study in Taiwan. *Electronic Commerce Studies, 5*(2), 175–196.

Palmer, M., Alghofaily, I., & Alminir, S. (1984). The Behavioral Correlates of Rentier Economies: A Case Study of Saudi Arabia. In Stookey, R. (Ed.), *The Arabian Peninsula: Zone of Ferment.* Stanford, CA: Hoover Institution Press.

Palvia, P. (1998). Research issues in global information technology management. *Information Resources Management Journal, 11*(2), 27–36.

Palvia, P., Palvia, S., & Whitworth, J. (2002). Global Information Technology: A Meta Analysis of Key Issues. *Information & Management, 39*(5), 403–414. doi:10.1016/S0378-7206(01)00106-9

Palvia, P., Means, D. B., & Jackson, W. M. (1994). Determinants of computing in very small businesses. *Information & Management, 27*(3), 161–174. doi:10.1016/0378-7206(94)90044-2

Pan, M. J., & Jang, W. Y. (2008). Determinants of the adoption of enterprise resource planning within the technology organization environment framework: Taiwan's communications industry. *Journal of Computer Information Systems, 48*(3), 94–102.

Park, H. (2003). Determinants of corruption: A cross-national analysis. *Multinational Business Review, 11*(2), 29–48.

Parthasarathy, M., & Bhattacherjee, A. (1998). Understanding post-adoption behavior in the context of online services. *Information Systems Research, 9*(4), 362–379. doi:10.1287/isre.9.4.362

Pavlou, P., & Chai, L. (2002). What Drives Electronic Culture Across Cultures – A Cross-Cultural Empirical Investigation. *Journal of Electronic Commerce Research, 3*(4), 240–253.

Pavlou, P., & Fygenson, M. (2006). Understanding and Predicting Electronic Commerce Adoption: An Extension of the Theory of Planned Behavior. *Management Information Systems Quarterly, 30*(1), 115–143.

Pavlou, P. A., Liang, H., & Xue, Y. (2007). Understanding and mitigating uncertainty in online exchange relationships: A principal-agent perspective. *MIS Quarterly, 31*(1), 105–136.

Pavlou, P. A., Liang, H., & Xue, Y. (2007). Understanding and mitigating uncertainty in online exchange relationships: A principal-agent perspective. *MIS Quarterly: Management Information Systems, 31*(1), 105–135.

Peace, A. G., Galletta, D. F., & Thong, J. Y. L. (2003). Software piracy in the workplace: A model and empirical test. *Journal of Management Information Systems, 20*(1), 153–177.

Pedersen, P. E., & Ling, R. R. (2003). *Modifying adoption research for mobile Internet service adoption: Cross-disciplinary interactions.* Paper presented at the Proceedings of the 36th Hawaii International Conference on System Sciences (HICSS-36), Big Island, Hawaii.

Pedersen, P. E., Methlie, L. B., & Thorbjonsen, H. (2002). *Understanding mobile commerce end-user adoption: A triangulation perspective and suggestions for an exploratory service evaluation framework.* Paper presented at the Proceedings of the 36th Hawaii International Conference on System Sciences (HICSS-36), Big Island, Hawaii.

Peng, T. K., Peterson, M. F., & Shyi, Y.-P. (1991). Quantitative Methods in Cross-National Management Research: Trends and Equivalence. *Journal of Organizational Behavior, 12*(2), 87–107. doi:10.1002/job.4030120203

Peng, M. W. (2003). Institutional transitions and strategic choices. *Academy of Management Review, 28*(2), 275–286.

Peng, M. W., & Heath, P. S. (1996). The growth of the firm in planned economies in transition: Institutions, organizations, and strategic choice. *Academy of Management Review, 21*(2), 492–528. doi:10.2307/258670

Peng, M. W., Tan, J., & Tong, T. (2004). Ownership types and strategic groups in an emerging economy. *Journal of Management Studies, 41*(7), 1105–1109. doi:10.1111/j.1467-6486.2004.00468.x

Peppard, J. (1999). Information management in the global enterprise: An organizing framework. *European Journal of Information Systems, 8*(2), 77–94. doi:10.1057/palgrave.ejis.3000321

Pereira, R. (1998). Cross-cultural influences on global electronic commerce. In *Proceedings of the AIS*, Baltimore, MD (pp. 318-320).

Perry, R., & Quirk, K. (2007). *IDC white paper: An evaluation of build versus buy for portal solutions.* ftp://ftp.software.ibm.com/software/bigplays/IDCBuyVs-BuildPortal.pdf

Peterson, M. F., & Castro, S. L. (2006). Measurement Metrics at Aggregate Levels of Analysis: Implications for Organization Culture Research and the GLOBE Project. *The Leadership Quarterly, 17*(5), 506–521. doi:10.1016/j.leaqua.2006.07.001

Peterson, M. F., & Soendergaard, M. (in press). Traditions and transitions in quantitative societal culture research in organization studies. [in press]. *Organization Studies.*

Peterson, M. F., & Smith, P. B. (2000). Sources of meaning, organizations, and culture: Making sense of organizational events. In Ashkanasy, N., Wilderom, C. P. M., & Peterson, M. F. (Eds.), *Handbook of organizational culture and climate* (pp. 101–116). Thousand Oaks, CA: Sage.

Peterson, M. F., & Smith, P. B. (2008). Social structures and processes in cross cultural management. In Smith, P. B., Peterson, M. F., & Thomas, D. C. (Eds.), *Handbook of cross-cultural management research* (pp. 35–58). Thousand Oaks, CA: Sage Press.

Peterson, M. F., & Wood, R. (2008). Cognitive structures and processes in cross cultural management. In Smith, P. B., Peterson, M. F., & Thomas, D. C. (Eds.), *Handbook of cross-cultural management research*. Thousand Oaks, CA: Sage.

Petter, S., Straub, S., & Rai, A. (2007). Specification and Validation of Formative Constructs in IS Research. *Management Information Systems Quarterly, 31*(4), 623–656.

Pfeffer, J. (1982). *Organizations and Organization Theory*. Marshfield, MA: Pitman.

Pfeffer, J., & Leblebici, H. (1977). Information technology and organizational structure. *Pacific Sociological Review, 20*(2), 241–261.

Pfeffer, J., & Salancick, G. R. (1978). *The external control of organization: A resource dependency perspective*, New York, NY: Harper and Row Press.

Picot, A., & Wernick, C. (2007). The role of government in broadband access. *Telecommunications Policy, 31*(10-11), 660–674. doi:10.1016/j.telpol.2007.08.002

Pillai, R., Scandura, T., & Williams, E. (1999). Leadership and Organizational Justice: Similarities and Differences. *Journal of International Business Studies, 30*(4), 763–779. doi:10.1057/palgrave.jibs.8490838

Pipe, G. R. (2003). *Vietnam promotes ICT to accelerate development. I-WAYS, The Journal of E-Government Policy and Regulation, 26(1)*. Amsterdam, The Netherlands: ISO press.

Pitkow, J. E., & Kehoe, C. M. (1996). Emerging trends in the WWW user population. *Communications of the ACM, 39*(6), 106–108. doi:10.1145/228503.228525

Plouffe, C. R., Hulland, J. S., & Vandenbosch, M. (2001). Richness versus Parsimony in Modeling Technology Adoption Decisions--Understanding Merchant Adoption of a Smart Card-Based Payment System. *Information Systems Research, 12*(2), 208–222. doi:10.1287/isre.12.2.208.9697

Plouffe, C. R., Hulland, J. S., & Vand, M. (2001). Research report: Richness versus parsimony in modeling technology adoption decisions: Understanding merchant adoption of a smart card-based payment system. *Information Systems Research, 12*(2), 208–222. doi:10.1287/isre.12.2.208.9697

Png, I. P. L., Tan, B. C. Y., & Wee, K. L. (2001). Dimensions of National Culture and Corporate Adoption of IT Infrastructure. *IEEE Transactions on Engineering Management, 48*(1), 36–45. doi:10.1109/17.913164

Podsakoff, P. M., MacKenzie, S. B., Lee, J. Y., & Podsakoff, N. P. (2003). Common method biases in behavioral research: A critical review of the literature and recommended remedies. *The Journal of Applied Psychology, 88*(5), 879–903. doi:10.1037/0021-9010.88.5.879

Podsakoff, P., & Organ, M. (1986). Self-reports in organizational research: Problems and prospects. *Journal of Management, 12*(4), 531–544. doi:10.1177/014920638601200408

Podsakoff, P. M., & Organ, D. W. (1986). Self-reports in organizational research: Problems and prospects. *Journal of Management, 12*(4), 531–544. doi:10.1177/014920638601200408

Porter, M. E., & Millar, V. E. (1985). How Information Gives You Competitive Advantage. *Harvard Business Review, 63*(4), 149–160.

Poutsma, E. F., Van Uxem, F. W., & Walravens, A. H. C. M. (1987). *Process innovation and automation in small and medium sized business*. Delft, The Netherlands: Delft University Press.

Prahalad, C., & Doz, Y. (1987). *The Multinational Mission: Balancing Local Demands and Global Vision*. New York: The Free Press.

Prahalad, C. (1975). *The Strategic Process in a Multinational Corporation*. Unpublished doctoral dissertation, Harvard Business School, Boston.

Prais, S. J., & Winsten, C. B. (1954). Trend estimators and serial correlation. *Cowles Commission Discussion Paper*, 383.

Preece, J., Rogers, Y., Sharp, H., Benyon, D., Holland, S., & Carey, T. (1994). *Human-computer interaction.* Boston, MA: Addison-Wesley.

Premkumar, G., & King, W. (1994). Organizational Characteristics and Information Systems Planning: An Empirical Study. *Information Systems Research*, 5(2), 75–109. doi:10.1287/isre.5.2.75

Premkumar, G., & King, W. R. (1992). An empirical assessment of information systems planning and the role of information systems in organizations. *Journal of Management Information Systems*, 9(2), 99–125.

Premkumar, G., & Ramamurthy, K. (1995). The role of interorgainzational and organizational factors on the decision mode for adoption of interorganizational systems. *Decision Sciences*, 26(3), 303–336. doi:10.1111/j.1540-5915.1995.tb01431.x

Premkumar, G., Ramamurthy, K., & Crum, M. (1997). Determinants of EDI adoption in the transportation industry. *European Journal of Information Systems*, 6(2), 107–121. doi:10.1057/palgrave.ejis.3000260

Premkumar, G., & Roberts, M. (1999). Adoption of new information technologies in rural small business. *Omega*, 27(4), 467–484. doi:10.1016/S0305-0483(98)00071-1

Purvis, R. L., Sambamurthy, V., & Zmud, R. W. (2001). The assimilation of knowledge platforms in organizations: An empirical investigation. *Organization Science*, 12(2), 117–135. doi:10.1287/orsc.12.2.117.10115

Putnam, R. D. (1993). *Making democracy work: Civic traditions in modern Italy.* Princeton, NJ: Princeton University Press.

Qi, Y., & Zhang, Q. (2008). *Research on information sharing risk in supply chain management.* Paper presented at the 2008 International Conference on Wireless Communications, Networking and Mobile Computing (WiCOM 2008).

Qiu, J. L., & Hachigian, N. (2005). *A New Long March: E-Government in China.* Paris: OECD.

Quaddus, M., & Hofmeyer, G. (2007). An investigation into the factors influencing the adoption of B2B trading exchanges in small businesses. *European Journal of Information Systems*, 16(3), 202–215. doi:10.1057/palgrave.ejis.3000671

Quibria, M. G., Ahmed, S. N., Tschang, T., & Reyes-Macasaquit, M. L. (2003). Digital divide: Determinants and policies with special reference to Asia. *Journal of Asian Economics*, 13, 811–825. doi:10.1016/S1049-0078(02)00186-0

Quynh, N. (2006). *Digital review of Asia-Pacific 2005/2006: Vietnam.* Retrieved from http://www.digital-review.org/

Raghunathan, B., & Raghunathan, T. (1991). The Relationship between Information Systems Planning and Planning Effectiveness: An Empirical Analysis. *OMEGA: The International Journal of Management Science*, 19(1), 125–135. doi:10.1016/0305-0483(91)90022-L

Raghunathan, B., & Raghunathan, T. (1994). Adaptation of a Planning System Success Model to Information Systems Planning. *Information Systems Research*, 5(3), 326–340. doi:10.1287/isre.5.3.326

Rahim, M. A., & Magner, N. R. (1995). Confirmatory factor analysis of the styles of handling interpersonal conflict: First-order factor model and its invariance across groups. *The Journal of Applied Psychology*, 80(1), 122–132. doi:10.1037/0021-9010.80.1.122

Rainey, H. G. (1983). Public agencies and private firms: Incentive structures, goals and individual roles. *Administration & Society*, 15, 207–242. doi:10.1177/009539978301500203

Rajagopal, P. (2002). An innovation-diffusion view of implementation of enterprise resource planning (ERP) systems and development of a research model. *Information & Management*, 40(2), 87–114. doi:10.1016/S0378-7206(01)00135-5

Raman, K. S., & Yap, C. S. (1996). From a resource rich country to an information rich society: An evaluation on Information Technology policies in Malaysia. *Information Technology for Development*, 7(3), 109. doi:10.1080/02681102.1996.9525277

Ramingwong, S., & Sajeev, A. S. M. (2007). Offshore Outsourcing: The Risk of Keeping Mum. *Communications of the ACM, 50*(8), 101–103. doi:10.1145/1278201.1278230

Ranganathan, C., Dhaliwal, J. S., & Teo, T. S. H. (2004). Assimilation and diffusion of web technologies in supply-chain management: An examination of key drivers and performance impacts. *International Journal of Electronic Commerce, 9*(1), 127–161.

Rangaswamy, N. (2007). ICT for development and commerce: A case study of internet cafés in India. *Proceedings of the 9th International Conference on Social Implications of Computers in Developing Countries*, São Paulo, Brazil.

Ravichandran, T., & Rai, A. (2000). Quality Management in Systems Development: An Organizational System Perspective. *Management Information Systems Quarterly, 24*(3), 381–415. doi:10.2307/3250967

Raymond, L., Bergeron, F., & Blili, S. (2005). The assimilation of e-business in manufacturing SMEs: Determinants and effects on growth and internationalization. *Electronic Markets, 15*(2), 106–118. doi:10.1080/10196780500083761

Raymond, L. (1985). Organizational characteristics and MIS success in the context of small business. *Management Information Systems Quarterly, 9*(1), 37–60. doi:10.2307/249272

Raymond, E. S. (1999). *The cathedral and the bazaar: Musings on Linux and open source by an accidental revolutionary*. Sebastapol, CA: O'Reilly & Associates.

Reinig, B. A., Briggs, R. O., & de Vreede, G.-J. (2009). Satisfaction as a Function of Perceived Change in Likelihood of Goal Attainment: A Cross-Cultural Study. *International Journal of e-Collaboration, 5*(2), 61–74.

Reporters Without Borders. (2006). *Press freedom index 2006*. Retrieved from http://www.rsf.org/

Rice, R. E., & Katz, J. E. (2003). Comparing internet and mobile phone usage: Digital divides of usage, adoption, and dropouts. *Telecommunications Policy, 27*(8/9), 597–623. doi:10.1016/S0308-5961(03)00068-5

Riemenschneider, C. K., Harrison, D. A., & Mykytyn, P. P. (2003). Understanding IT adoption decisions in small business: Integrating current theories. *Information & Management, 40*(4), 269–285. doi:10.1016/S0378-7206(02)00010-1

Ringle, C. M., Wende, S., & Will, A. (2005). *SmartPLS 2.0 (beta)*. Retrieved from http://www.smartpls.de

Robertson, C. J., Gilley, K. M., Crittenden, V., & Crittenden, W. F. (2008). An analysis of the predictors of software piracy within Latin America. *Journal of Business Research, 61*(6), 651–656. doi:10.1016/j.jbusres.2007.06.042

Robertson, A., Soopramanien, D., & Fildes, R. (2007). Segmental new-product diffusion of residential broadband services. *Telecommunications Policy, 31*(5), 265–275. doi:10.1016/j.telpol.2007.03.006

Robichaux, B. P., & Cooper, R. B. (1998). GSS Participation: A Cultural Examination. *Information & Management, 33*(6), 287–290. doi:10.1016/S0378-7206(98)00033-0

Robison, K. K., & Crenshaw, E. M. (2002). Post-industrial transformations and cyber-space: A cross-national analysis of Internet development. *Social Science Research, 31*, 334–363. doi:10.1016/S0049-089X(02)00004-2

Roche, E. (1996). Closing the technology gap: Technological change in India computer industry: Global Information Technology and socio-economic development. *Journal of Global Information Management, 4*(3), 29.

Rocheleau, B. (2006). *Public management Information Systems*. Hershey, PA: IGI Global.

Rodriguez, F., & Wilson, E. J. (2000). *Are poor countries losing the information revolution?* (InfoDev Working Paper). Global Information and Communication Technologies Department.

Rodrik, D. (2003). Institutions, integration, and geography: In search of the deep determinants of economic growth. In Rodrik, D. (Ed.), *Search of prosperity: analytic country studies on growth*. Princeton, NJ: Princeton University Press.

Rogers, E. M. (1995). *Diffusion of innovations* (4th ed.). New York: The Free Press.

Rogers, E. M. (1995). *Diffusion of Innovations*. New York: The Free Press.

Rogers, W. H. (1993). Regression standard errors in clustered samples. *Stata Technical Bulletin, 53*, 32–35.

Rokeach, M. (1968). *Beliefs, attitudes, and values*. San Francisco, CA: Jossey-Bass.

Ronkainen, I. A., & Guerrero-Cusumano, J. L. (2001). Correlates of intellectual property violation. *Multinational Business Review, 9*(1), 59–65.

Rose, G., & Straub, D. W. (1998). Predicting General IT Use: Applying TAM to the Arabic World. *Journal of Global Information Management, 6*(3), 39–46.

Roth, K., & Morrison, A. J. (1990). An Empirical Analysis of the Integration-Responsiveness Framework in Global Industries. *Journal of International Business Studies, 21*(4), 541–605. doi:10.1057/palgrave.jibs.8490341

Ryan, R. M., & Deci, E. L. (2000). Intrinsic and extrinsic motivations: Classic definitions and new directions. *Contemporary Educational Psychology, 25*(1), 54–67. doi:10.1006/ceps.1999.1020

Salmenkaita, J., & Ahti Salo, A. (2002). Rationales for government intervention in the commercialization of new technologies. *Technology Analysis and Strategic Management, 14*(2), 183–200. doi:10.1080/09537320220133857

Sample, J. A. (1984). Nominal group technique: An alternative to brainstorming. *Journal of Extension, 22*(2).

Sanford, C., & Bhattacherjee, A. (2007). IT implementation in a developing country municipality: A sociocognitive analysis. *Journal of Global Information Management, 15*(3), 20. doi:10.4018/jgim.2007070102

Santos-Paulino, A., & Thirlwall, A. P. (2004). The impact of trade liberalisation on exports, imports and the balance of payments of developing economies. *The Economic Journal, 114*(493), 50–72. doi:10.1111/j.0013-0133.2004.00187.x

Sarker, S., & Lee, A. (2003). Using a case study to test the role of three key social enablers in ERP implementation. *Information & Management, 40*(8), 813–829. doi:10.1016/S0378-7206(02)00103-9

Schepers, J., & Wetzels, M. (2007). A meta-analysis of the technology acceptance model: Investigating subjective norm and moderation effects. *Information & Management, 44*(1), 90–103. doi:10.1016/j.im.2006.10.007

Scholl, H. J. (2005). *Organizational transformation through egovernment: Myth or reality?* Berlin, Germany: Springer Verlag.

Schware, R. (2007). Scaling up rural electronic governance initiatives. *Proceedings of the 1st International Conference on Theory and Practice of Electronic Governance*, Macao, Hong Kong.

Schwartz, S. H. (1999). A theory of cultural values and some implications for work. *Applied Psychology: An International Review, 48*(1), 12–47.

Schwartz, S. H. (2007). Universalism values and the inclusiveness of our moral universe. *Journal of Cross-Cultural Psychology, 38*(6), 711–728. doi:10.1177/0022022107308992

Schwartz, S. H., & Boehnke, K. (2004). Evaluating the structure of human values with confirmatory factor analysis. *Journal of Research in Personality, 38*(3), 230–255. doi:10.1016/S0092-6566(03)00069-2

Schwartz, S. H., & Sagiv, L. (1995). Identifying culture-specifics in the content and structure of values. *Journal of Cross-Cultural Psychology, 26*(1), 92–116. doi:10.1177/0022022195261007

Schwartz, S. H., Verkasalo, M., Antonovsky, A., & Sagiv, L. (1997). Value priorities and social desirability: Much substance, some style. *The British Journal of Social Psychology, 36*(1), 3–18. doi:10.1111/j.2044-8309.1997.tb01115.x

Schwartz, S. H. (1992). Universals in the content and structure of values: Theoretical advances and empirical tests in 20 countries. In M. Zanna (Ed.), *Advances in Experimental Social Psychology, 25*, 1-65. Orlando, FL: Academic Press.

Sciadas, G. (2006). *Digital review of Asia-Pacific 2005/2006: Vietnam*. Retrieved from http://www.digital-review.org/

Scott, J. E., & Kaindl, L. (2000). Enhancing functionality in an enterprise software package. *Information & Management*, *37*(3), 111–122. doi:10.1016/S0378-7206(99)00040-3

Scott, W. R. (2001). *Institutions and organizations.* Thousand Oaks, CA: Sage.

Segars, A., & Grover, V. (1998). Strategic Information Systems Planning Success: An Investigation of the Construct and Its Measurement. *Management Information Systems Quarterly*, *22*(2), 139–163. doi:10.2307/249393

Segars, A., & Grover, V. (1999). Profiles of Strategic Information Systems Planning. *Information Systems Research*, *10*(3), 199–232. doi:10.1287/isre.10.3.199

Segars, A. H., & Grover, V. (1993). Re-examining perceived ease of use and usefulness: A confirmatory factor analysis. *Management Information Systems Quarterly*, *17*(4), 517–525. doi:10.2307/249590

Sein, M., Ahmad, I., & Harindranath, G. (2008). *Sustaining ICT for development: The case of Grameenphone CIC*. Telektronikk.

Shan, W. (2001). The IT work force in China. *Communications of the ACM*, *44*(7), 76. doi:10.1145/379300.379319

Shang, R. A., Chen, Y. C., & Shen, L. (2005). Extrinsic versus intrinsic motivations for consumers to shop on-line. *Information & Management*, *42*(3), 401–413. doi:10.1016/j.im.2004.01.009

Sharifi, H., & Zarei, B. (2004). An adaptive approach for implementing e–government in IR Iran. *Journal of Government Information*, *30*(5-6), 600–619. doi:10.1016/j.jgi.2004.10.005

Shenkar, O., & von Glinow, M. A. (1994). Paradoxes of organizational theory and research: Using the case of China to illustrate national contingency. *Management Science*, *40*(1), 56–71. doi:10.1287/mnsc.40.1.56

Sherer, S. A. (2007). Comparative study of IT investment management processes in U.S. and Portugal. *Journal of Global Information Management*, *15*(3), 43–68.

Shi, J. L. (2004). Contract of Competitors. *Chinese Information System Engineering*, 74-78.

Shih, E., Kraemer, K. L., & Dedrick, J. (2008). IT diffusion in developing countries. *Communications of the ACM*, *51*(2), 43–48. doi:10.1145/1314215.1340913

Shih, E., Kraemer, K. L., & Dedrick, J. (2008). IT diffusion in developing economies: Policy issues and recommendations. *Communications of the ACM*, *51*(2), 43–48. doi:10.1145/1314215.1340913

Shin, S. K., Gopal, R. D., Sanders, G. L., & Whinston, A. B. (2004). Global software piracy revisited. *Communications of the ACM*, *47*(1), 103–107. doi:10.1145/962081.962088

Shore, B., & Venkatachalam, A. R. (1995). The role of national culture in systems analysis and design. *Journal of Global Information Management*, *3*(3), 5–14.

Siau, K., Nah, F. F.-H., & Ling, M. (2007). National Culture and Its Effects on Knowledge Communication in Online Virtual Communities. *International Journal of Electronic Business*, *5*(5), 518–532. doi:10.1504/IJEB.2007.015450

Siau, K., & Long, Y. (2005). Synthesizing e–government stage models – A meta-synthesis based on meta-ethnography approach. *Industrial Management & Data Systems*, *105*(4), 443–458. doi:10.1108/02635570510592352

Siau, K., & Long, L. (2006). Using social development lenses to understand e-government development. *Journal of Global Information Management*, *14*(1), 47–62.

Siddiqui, K. (1996). *Towards good governance in Bangladesh: Fifty unpleasant essays*. Dhaka, Bangladesh: University Press Limited.

Silva, L., & Figueroa, E. B. (2002). Institutional intervention and the expansion of ICTs in Latin America: The case of Chile. *Information Technology & People*, *15*(1), 8–25. doi:10.1108/09593840210421499

Simon, S. J. (2001). The impact of culture and gender on Web sites: An empirical study. *ACM SIGMIS Database*, *32*(1), 18–37. doi:10.1145/506740.506744

Singh, J., & Sirdeshmukh, D. (2000). Agency and trust mechanisms in consumer satisfaction and loyalty judgments. *Journal of the Academy of Marketing Science*, *28*(1), 150–167. doi:10.1177/0092070300281014

Singh, H., Das, A., & Joseph, D. (2007). Country-level determinants of e-government maturity. *Communications of the Association for Information Systems*, *20*, 632–648.

Sinkovics, R. R., Yamin, M., & Hossinger, M. (2007). Cultural Adaptation in Cross Border E-Commerce: A study of German Companies. *Journal of Electronic Commerce Research, 8*(4), 221–236.

Sinkula, J. M. (1994). Market information processing and organizational learning. *Journal of Marketing, 58*(1), 35–45. doi:10.2307/1252249

Sinkula, J. M., Baker, W. E., & Noordewier, T. (1997). A framework for market-based organizational learning: Linking values, knowledge, and behavior. *Journal of the Academy of Marketing Science, 25*(4), 305–318. doi:10.1177/0092070397254003

Sipior, J. C., Ward, B. T., Volonino, L., & Marzec, J. Z. (2003). A community initiative that diminished the digital divide. *Communications of the Association for Information Systems, 13*(1), 29–56.

Sir, J. C., Cho, C. H., Yang, H. J., An, I. H., & Kim, J. R. (2004). *2004 Survey on wireless internet use.* Seoul, Korea: Korea Network Information Center.

Sivo, S. A., Saunders, C., Chang, Q., & Jiang, J. J. (2006). How Low Should You Go? Low Response Rates and the Validity of Inference in IS Questionnaire Research. *Journal of the AIS, 7*(6), 361–414.

Smith, H. J. (2001). Information privacy and marketing: What the U.S. should (and shouldn't) learn from Europe. *California Management Review, 43*(2), 8–33.

Smith, P. B., Peterson, M. F., & Schwartz, S. H. (2002). Cultural values, sources of guidance, and their relevance to managerial behavior: A 47-nation study. *Journal of Cross-Cultural Psychology, 33*(2), 188–208. doi:10.1177/0022022102033002005

Soliman, K. S., & Janz, B. D. (2004). An exploratory study to identify the critical factors affecting the decision to establish internet-based interorganizational information systems. *Information & Management, 41*(6), 697–706. doi:10.1016/j.im.2003.06.001

Soper, D. S., Demirkan, H., Goul, M., & St. Louis, R. (2006). The impact of ICT expenditures on institutionalized democracy and foreign direct investment in developing economies. *Proceedings of the 39th Annual Hawaii International Conference on System Sciences* (HICSS'06).

Soyjaudah, K. M. S., Oolun, M. K., Jahmeerbacus, I., & Govinda, S. (2002). ICT development in Mauritius. *Proceedings of the 6th IEEE Africon Conference in Africa,* George, South Africa.

Spennemann, D. H. R., & Atkinson, J. S. (2003). A longitudinal study of the uptake of and confidence in using e-mail among parks management students. *Campus-Wide Information Systems, 23*(2), 55–67. doi:10.1108/10650740310467754

Spiller, J., Vlasic, A., & Yetton, P. (2007). Post-adoption behavior of users of internet service providers. *Information & Management, 44*(6), 513–523. doi:10.1016/j.im.2007.01.003

Srite, M., Thatcher, J. B., & Galy, E. (2008). Does Within-Culture Variation Matter? An Empirical Study of Computer Usage. *Journal of Global Information Management, 16*(1), 1–25.

Srite, M., & Karahanna, E. (2006). The Role of Espoused National Cultural Values in Technology Acceptance. *MIS Quarterly, 30*(3), 2006, 679-704.

Srivastava, S. C., & Teo, T. (2008). The relationship between e-government and national competitiveness: The moderating influence of environmental factors. *Communications of the Association for Information Systems, 23*(59).

Stanforth, C. (2006). Using actor-network theory to analyze e-government implementation in developing countries. *Information Technologies and International Development, 3*(3), 35–60. doi:10.1162/itid.2007.3.3.35

Staples, D., Sandy, P., & Seddon, P. (2004). Testing the Technology-to-Performance Chain Model. *Journal of Organizational and End User Computing, 16*(4), 17–36.

Stiglitz, J. (2008). Fostering an independent media with a diversity of views. In Islam, R. (Ed.), *Information and public choice: From media markets to policy making* (pp. 139–152). Washington, DC: World Bank.

Stimson, J. (1985). Regression in space and time: A statistical essay. *American Journal of Political Science, 29*(4), 914–947. doi:10.2307/2111187

Stinchcombe, A. L. (1997). On the virtues of the old institutionalism. *Annual Review of Sociology, 23*(1), 1–18. doi:10.1146/annurev.soc.23.1.1

Strang, D., & Meyer, J. W. (1993). Institutional conditions for diffusion. *Theory and Society*, *22*(4), 487–511. doi:10.1007/BF00993595

Stratman, J. K., & Roth, A. V. (2002). Enterprise resource planning (ERP) competence constructs: Two-stage multi-item scale development and validation. *Decision Sciences*, *33*(4), 601–628. doi:10.1111/j.1540-5915.2002.tb01658.x

Straub, D. (1994). The effect of culture on IT diffusion: E-mail and FAX in Japan and the U.S. *Information Systems Research*, *5*(1), 23–47. doi:10.1287/isre.5.1.23

Straub, D., Keil, M., & Brenner, W. (1997). Testing the Technology Acceptance Model Across Cultures: A Three Country Study. *Information & Management*, *33*(1), 1–11. doi:10.1016/S0378-7206(97)00026-8

Straub, D., Loch, K., Evaristo, R., Karahanna, E., & Srite, M. (2002). Toward a Theory-Based Measurement of Culture. *Journal of Global Information Management*, *10*(1), 13–23.

Straub, D. (1994). The effect of culture on IT diffusion: E-mail and FAX in Japan and the US. *Information Systems Research*, *5*(1), 23–47. doi:10.1287/isre.5.1.23

Straub, D., Loch, K., Evaristo, J., Karahanna, E., & Srite, M. (2002). Toward a Theory Based Measurement of Culture. *Journal of Global Information Management*, *10*(1), 13–23.

Straub, D. W., Loch, K. D., & Hill, C. E. (2001). Transfer of Information Technology to the Arab World: A Test of Cultural Influence Modeling. *Journal of Global Information Management*, *9*(4), 6–28.

Straub, D. W. (1994). The effect of culture on IT diffusion: E-mail and fax in Japan and the United States. *Information Systems Research*, *5*(1), 23–47. doi:10.1287/isre.5.1.23

Straub, D. W., Loch, K., Evaristo, R., Karahanna, E., & Srite, M. (2002). Toward a theory-based measurement of culture. *Journal of Global Information Management*, *10*(1), 13–23.

Straub, D. W., & Loch, K. D. (2006). Creating and developing a program of global research. *Journal of Global Information Management*, *14*(2), 1–28.

Straub, D., Loch, K. D., & Hill, C. E. (2001). Transfer of information technology to the Arab world: A test of cultural influence modeling. *Journal of Global Information Management*, *9*(4), 6–48. doi:10.4018/jgim.2001100101

Strauss, A., & Corbin, J. (1990). *Basics of qualitative research*. Newbury Park, CA: Sage.

Sturrock, F., & Kirwan, B. (1996). Interface Display Designs Based on Operator Knowledge Requirements. In Hanson, J. (Ed.), *Contemporary Ergonomics* (pp. 280–284). London: Taylor & Francis.

Sundararajan, A. (2004). Managing digital piracy: Pricing and protection. *Information Systems Research*, *15*(3), 287–308. doi:10.1287/isre.1040.0030

Suomi, R., & Kastu–Häikiö, M. (1998). Cost – and service effective solutions for local administration – The Finnish case. *Total Quality Management*, *9*, 335–346. doi:10.1080/0954412989180

Tajfel, H. (Ed.). (1978). *Differentiation between Social Groups: Studies in the Social Psychology of Intergroup Relations*. London: Academic Press.

Tan, B. C. Y., Smith, H. J., Keil, M., & Montealegre, R. (2003). Reporting Bad News about Software Projects: Impact of Organizational Climate and Information Asymmetry in an Individualistic and a Collectivistic Culture. *IEEE Transactions on Engineering Management*, *50*(1), 64–77. doi:10.1109/TEM.2002.808292

Tan, B. C. Y., Wei, K. K., Watson, R., & Walczuch, R. (1998). Reducing Status Effects with Computer-Mediated Communication: Evidence from Two Distinct National Cultures. *Journal of Management Information Systems*, *15*(1), 119–141.

Tan, J. (2002). Impact of ownership type on environment, strategy, and performance: Evidence from China. *Journal of Management Studies*, *39*(3), 333–354. doi:10.1111/1467-6486.00295

Tan, J., & Litschert, R. J. (1994). Environment-strategy relationship and its performance implications: An empirical study of Chinese electronics industry. *Strategic Management Journal*, *15*(1), 1–20. doi:10.1002/smj.4250150102

Tan, B. C. Y., Wei, K. K., Watson, R. T., Clapper, D. L., & McLean, E. (1998). Computer-mediated communication and majority influence: Assessing the impact in an individualistic and a collectivist culture. *Management Science*, *44*, 1263–1278. doi:10.1287/mnsc.44.9.1263

Tan, C.-W., Pan, S. L., & Lim, E. T. K. (2005). Managing stakeholder interests in e-government implementation: Lessons learned from a Singapore e-government project. *Journal of Global Information Management*, *13*(1), 31–53.

Tan, F., & Leewongcharoen, K. (2005). Factors contributing to IT industry success in developing economies: The case of Thailand. *Information Technology for Development*, *11*(2), 161–194. doi:10.1002/itdj.20009

Tan, A., & Ouyang, W. (2002). *Global and national factors affecting e-commerce diffusion in China.* University of California, Irvine. Retrieved Nov. 11, 2007, from www.crito.uci.edu/publications/pdf/GEC2_China.pdf

Tan, Z. (2004). Evolution of China's telecommunications manufacturing industry: Competition, strategy and government policy. *Communication & Strategies, 53.*

Tannenbaum, A. S., Kavcic, B., Rosner, M., Vianello, M., & Wieser, G. (1974). *Hierarchy in organizations.* San Francisco, CA: Jossey-Bass.

Taylor, S., & Todd, P. A. (1995). Understanding Information Technology Usage: A Test of Competing Models. *Information Systems Research*, *6*(2), 144–176. doi:10.1287/isre.6.2.144

Taylor, S., & Todd, P. A. (1995). Understanding Information Technology Usage: A Test of Competing Models. *Information Systems Research*, *6*(2), 144–176. doi:10.1287/isre.6.2.144

Taylor, S., & Todd, P. (1995). Understanding information technology use: a test of competing models. *Information Systems Research*, *6*(2), 144–176. doi:10.1287/isre.6.2.144

Taylor, S., & Todd, P. A. (1995). Understanding Information Technology usage: A test of competing models. *Information Systems Research*, *6*(2), 144–176. doi:10.1287/isre.6.2.144

Teboul, J., Chen, L., & Fritz, L. (1994). Intercultural Organizational Communication Research in Multinational Organizations. In Wiseman, R. L., & Shuter, R. (Eds.), *Communicating in Multinational Organizations* (pp. 12–29). Thousand Oaks, CA: Sage.

Teo, H. H., Wei, K. K., & Benbasat, I. (2003). Predicting Intention to Adopt Interorganizational Linkages: An Institutional Perspective. *Management Information Systems Quarterly*, *27*(1), 19–49.

Teo, T. S. H., Ang, J. S. K., & Pavri, F. N. (1997). The State of Strategic IS Planning Practices in Singapore. *Information & Management*, *33*(1), 13–23. doi:10.1016/S0378-7206(97)00033-5

Teo, T. S. H. (2001). Demographic and motivation variables associated with Internet usage activities. *Internet Research: Electronic Networking Applications and Policy*, *11*(2), 125–137. doi:10.1108/10662240110695089

Teo, T. S. H., Lim, V. K. G., & Lai, R. Y. C. (1999). Intrinsic and extrinsic motivation in Internet usage. *International Journal of Management Science (Omega)*, *27*(1), 25–37. doi:10.1016/S0305-0483(98)00028-0

Teo, T. S. H., & Pok, S. H. (2003). Adoption of WAP-enabled mobile phones among internet users. *International Journal of Management Science (Omega)*, *31*(6), 483–498. doi:10.1016/j.omega.2003.08.005

Thanacoody, R., Bartram, T., Barker, M., & Jacobs, K. (2006). Career progression among female academics: A comparative study of Australia and Mauritius. *Women in Management Review*, *21*(7), 536–553. doi:10.1108/09649420610692499

Thang, J. Y. L., Hong, S. J., & Tam, K. Y. (2006). The effect of post-adoption beliefs on the expectation-confirmation model for information technology continuance. *International Journal of Human-Computer Studies*, *64*(9), 799–810. doi:10.1016/j.ijhcs.2006.05.001

The World Bank. (2007). *World Bank development indicators (WDI).* CD-ROM.

The World Bank. (2006a). *Information and communications for development 2006: Global trends and policies.*

The World Bank. (2006b). *2006 information and communications for development: Global trends and policies.*

Thode, H. C. Jr. (2002). *Testing for normality.* New York, NY: Marcel Dekker. doi:10.1201/9780203910894

Thompson, J. (1967). *Organizations in Action: Social Science Bases of Administrative Theory.* New York: McGraw-Hill.

Thong, J. Y. L. (1999). An integrated model of information systems adoption in small business. *Journal of Management Information Systems, 15*(4), 187–214.

Thong, J. Y. L. (2001). Resource constrains and information systems implementation in Singaporean small business. *Omega, 29*(2), 143–156. doi:10.1016/S0305-0483(00)00035-9

Thong, J. Y. L., & Yap, C. S. (1995). CEO characteristics, organizational characteristics and information technology adoption in small businesses. *Omega, 23*(4), 429–442. doi:10.1016/0305-0483(95)00017-I

Thong, J. Y. L., Yap, C. S., & Raman, K. S. (1996). Top management support, external expertise and information systems implementation in small business. *Information Systems Research, 7*(2), 248–267. doi:10.1287/isre.7.2.248

Tian, Q. J., Ma, J., & Liu, O. (2002). A hybrid knowledge and model system for R&D project selection. *Expert Systems with Applications, 23*(2), 265–271. doi:10.1016/S0957-4174(02)00046-5

Tibi, B. (1990). *Islam and the Cultural Accommodation of Social Change* (Krojzl, C., Trans.). Boulder, CO: Westview Press.

Tipton, F. (2002). Bridging the digital divide in Southeast Asia. *ASEAN Economic Bulletin, 19*(1), 83–99. doi:10.1355/AE19-1F

Tiwana, A., & Bush, A. A. (2007). A comparison of transaction cost, agency, and knowledge-based theory predictors of IT outsourcing decisions: A U.S.-Japan cross-cultural field study. *Journal of Management Information Systems, 24*(1), 259–300. doi:10.2753/MIS0742-1222240108

Tolbert, C. J., Mossberger, K., & McNeal, R. (2008). Institutions, policy innovation, and e-government in the American states. *Public Administration Review, 68*(3), 549–563. doi:10.1111/j.1540-6210.2008.00890.x

Tornatzky, L. G., & Klein, R. J. (1982). Innovation characteristics and innovation adoption-implementation: A meta-analysis of findings. *IEEE Transactions on Engineering Management, 29*, 28–45.

Tornatzky, L. G., & Klein, K. J. (1982). Innovation characteristics and innovation adoption-implementation: a meta analysis of findings. *IEEE Transactions on Engineering Management, 29*(11), 28–45.

Tornatzky, L. G., & Fleischer, M. (1990). *The processes of technological innovation.* Lexington, MA: Lexington Books.

Treier, S., & Jackman, S. (2008). Democracy as a latent variable. *American Journal of Political Science, 52*(1), 201–217.

Triandis, H. C. (1977). *Interpersonal behavior.* Monterey, CA: Brooks/Cole Publishing Company.

Trkman, P., Jerman-Blazic, B., & Turk, T. (2008). Factors of broadband development and the design of a strategic policy framework. *Telecommunications Policy, 32*(2), 101–115. doi:10.1016/j.telpol.2007.11.001

Trompenaars, F., & Hampden-Turner, C. (1998). *Riding the Waves of Culture: Understanding Diversity in Global Business* (2nd ed.). New York (NY): McGraw-Hill.

Tsai, W. H., Chein, S. W., Fan, Y. W., & Cheng, J. M. (2005). Critical management issues in implementing ERP: Empirical evidence from Taiwanese firms. *International Journal of Services and Standard, 1*(3), 299–318. doi:10.1504/IJSS.2005.005802

Tsikriktsis, N. (2002). Does cultural influence website quality expectation? An empirical study. *Journal of Service Research, 5*(2), 101–112. doi:10.1177/109467002237490

Tsui, A. S. (2006). Contextualization in Chinese Management Research. *Management and Organization Review, 2*(1), 1–13. doi:10.1111/j.1740-8784.2006.00033.x

Tuttle, B., Harrell, A., & Harrison, P. (1997). Moral hazard, ethical considerations, and the decision to implement an information systems. *Journal of Management Information Systems, 13*(4), 27–44.

Umbach, J. M. (2004). Libraries: Bridges across the digital divide. *Feliciter, 50*(2), 44.

Umble, E. J., Haft, P. R., & Umble, M. M. (2003). Enterprise resource planning: Implementation procedures and critical success factors. *European Journal of Operational Research, 146*(2), 241–257. doi:10.1016/S0377-2217(02)00547-7

UN. (2005). *Global egovernment readiness report 2005- From egovernment to e-inclusion.* Department of Economic and Social Affairs Division for Public Administration and Development Management United Nations.

UN. (2008). *The United Nations e-Government Survey 2008: From e-Government to Connected Governance. Department of Economic and Social Affairs Division for Public Administration and Development Management.* United Nations.

UN ICT Task Force. (2005). *Information and communication technology for peace - The role of ICT in preventing, responding to and recovering from conflict.* UN ICT Task Force Series 11.

UN OHRLLS. (2011). *List of LDCs; United Nations Office of the High Representative for the Least Developed Countries.* Retrieved September 5, 2011, from http://www.unohrlls.org/en/ldc/related/59/

UN. (2010). *United nations global e-government survey: Leveraging e-government at a time of financial and economic crisis.* NY. Department of Economic and Social Affairs Division for Public Administration and Development Management, United Nations

UNCTAD. (2004). *World Investment Report 2004: The Shift towards Services.* Geneva, Switzerland: United Nations.

UNCTAD. (2007). *World investment prospects survey 2007–2009.* New York, NY: UN Publication.

UNCTAD. (2008). *World investment report: Transnational corporations and the infrastructure challenges.* New York, NY: UN Publication.

UNCTAD. (2010). *World investment report 2010.*

UNCTAD. (2011). *Global investment trend monitor, No. 6. April 27.*

UNCTAD. (2006a). *World investment report 2006: FDI from developing and transition economies: Implications for development.* UNCTAD Handbook of Statistics 2006-07. Retrieved from http://stats.unctad.org/fdi/ReportFolders/reportFolders.aspx?sCS_referer=&sCS_ChosenLang=en

UNCTAD. (2006b). *Information economy report 2006: The development perspective.*

UNCTAD. (2006c). *The digital divide report: ICT diffusion index 2005.* Retrieved from http://www.unctad.org/en/docs/iteipc20065_en.pdf

UNDP. (2001). *Making new technologies work for human development.* New York, NY: UNDP.

UNDP. (2009). *Overcoming barriers: Human mobility and development.*

UNDP. (2006). *Human development reports: Statistics.* Retrieved from http://hdr.undp.org/en/statistics/

UNFPA. (2009). *The state of world population 2009: Facing a changing world: Women, population and climate.* United Nations Population Fund. Retrieved December 9, 2010, from http://www.unfpa.org/swp/2009/en/pdf/EN_SOWP09.pdf

United Nations ICT Task Force. (2005). *Measuring ICT: The global status of ICT indicators: Partnership on measuring ICT for development.* New York, NY: United Nations ICT Task Force.

United Nations. (2005a). *The digital divide report: ICT diffusion index.* Paper presented at the United Nations Conference on Trade and Development. New York: UN.

United Nations. (2005b). *UN global e-government readiness report: From e-government to e-inclusion.* New York: UN.

United Nations. (2007). *Mauritius national ICT strategic plan final analysis report.* Retrieved from http://un.intnet.mu/undp/downloads/info/ social_development/nictsp/report/ Final%20Analysis%20Report%20Draft%20One1.zip

United States Central Intelligence Agency. (2011). *The world factbook of Mauritius.* Retrieved from https://www.cia.gov/library/publications/the-world-factbook/geos/mp.html

United States Department of Commerce. (1995). *Falling through the net: A survey of the "have nots" in rural and urban America*. Retrieved from http://www.ntia.doc.gov/ntiahome/ fallingthru.html

US Commercial Service. (2005). *Doing business in China: A country commercial guide for U.S. companies*. Retrieved Nov. 11, 2007, from www.buyusainfo.net/docs/x_7439525.pdf

US Department of State. (2007). *Background note: China*. Retrieved Nov. 11, 2007, from www.state.gov/r/pa/ei/bgn/18902.htm

Uzoka, F. E., Shemi, A. P., & Seleka, G. G. (2007). Behavioral influences on e-commerce adoption in a developing country context. *The Electronic Journal of Information Systems in Developing Countries*, *31*(4), 1–15.

Van De Ven, A., & Delbecq, A. (1971). Nominal versus interacting group processes for committee decision-making effectiveness. *Academy of Management Journal*, *14*(2), 203–212. doi:10.2307/255307

Van der Heijden, H. (2004). User acceptance of hedonic information systems. *Management Information Systems Quarterly*, *28*(4), 695–704.

Van Everdingen, Y., & Wierenga, B. (2002). Intra-firm adoption decision: Role of inter-firm and intra-firm variables. *European Management Journal*, *20*(6), 649–663. doi:10.1016/S0263-2373(02)00116-0

Van Slyke, C., Lou, H., Belanger, F., & Sridhar, V. (2004). *The influence of culture on consumer-oriented electronic commerce adoption*. Paper presented at the Proceedings of the 2004 Southern Association for Information Systems Conference.

Vavilov, S. (2006). *Investment development path in petroleum exporters*. University of Paris.

Veiga, O. F., Floyd, S., & Dechant, K. (2001). Towards modeling the effects of national culture on IT implementation and acceptance. *Journal of Information Technology*, *16*(3), 145–158. doi:10.1080/02683960110063654

Venkatesh, V., & Davis, F. D. (2000). A Theoretical Extension of the Technology Acceptance Model: Four Longitudinal Field Studies. *Management Science*, *46*(2), 186–204. doi:10.1287/mnsc.46.2.186.11926

Venkatesh, V., Davis, F. D., & Morris, M. G. (2007). Dead or Alive? The Development, Trajectory and Future of Technology Adoption Research. *Journal of AIS*, *8*(4), 267–286.

Venkatesh, V., & Morris, M. G. (2000). Why Don't Men Ever Stop To Ask For Directions? Gender, Social Influence, and Their Role in Technology Acceptance and Usage Behavior. *Management Information Systems Quarterly*, *24*(1), 115–239. doi:10.2307/3250981

Venkatesh, V., Morris, M. G., Davis, G. B., & Davis, F. D. (2003). User Acceptance of Information Technology: Toward a Unified View. *Management Information Systems Quarterly*, *27*(3), 425–478.

Venkatesh, V., & Davis, F. D. (2000). A Theoretical Extension of the Technology Acceptance Model: Four Longitudinal Field Studies. *Management Science*, *46*(2), 186–204. doi:10.1287/mnsc.46.2.186.11926

Venkatesh, V., Morris, M. G., Davis, G. B., & Davis, F. D. (2003). User Acceptance of Information Technology: Toward a Unified View. *Management Information Systems Quarterly*, *27*(3), 425–478.

Venkatesh, V., & Brown, S. A. (2001). A longitudinal investigation of personal computers in homes: adoption determinants and emerging challenges. *Management Information Systems Quarterly*, *25*(1), 71–102. doi:10.2307/3250959

Venkatesh, V., Ramesh, V., & Massey, A. P. (2003). Understanding usability in mobile commerce. *Communications of the ACM*, *46*(12), 53–56. doi:10.1145/953460.953488

Venkatesh, V. (2000). Determinants of perceived ease of use: Integrating control, intrinsic motivation, and emotion into the technology acceptance model. *Information Systems Research*, *11*(4), 342–365. doi:10.1287/isre.11.4.342.11872

Venkatesh, V., Morris, M., Davis, G., & Davis, F. (2003). User acceptance of information technology: Toward a unified view. *Management Information Systems Quarterly*, *27*(3), 425–478.

Venkatesh, V., & Davis, F. D. (2000). A theoretical extension of the technology acceptance model: Four longitudinal field studies. *Management Science*, *46*(2), 186–205. doi:10.1287/mnsc.46.2.186.11926

Venkatesh, V., & Morris, M. G. (2000). Why don't men ever stop to ask for directions? Gender, social influence, and their role in technology acceptance and usage behavior. *Management Information Systems Quarterly*, *24*(1), 115–139. doi:10.2307/3250981

Venkatesh, V., & Zhang, X. (2010). Unified theory of acceptance and use of technology: U.S. Vs. China. *Journal of Global Information Management*, *13*(1), 5–28.

Vixie, P. (1999). Software engineering. In Dibona, C., Ockman, S., & Stone, M. (Eds.), *Open sources: Voices from the open source revolution* (pp. 91–100). Sebastopol, CA: O'Reilly & Associates.

Vlahos, G. E., & Ferratt, T. W. (1995). Information technology use by managers in Greece to support decision making: Amount, perceived value, and satisfaction. *Information & Management*, *29*(6), 305–315. doi:10.1016/0378-7206(95)00037-1

Vlahos, G. E., Ferratt, T. W., & Knoepfle, G. (2000). *Use and perceived value of computer-based information systems in supporting the decision making of German managers.* Paper presented at the SIGCPR Evanston Illinois

Waarts, E., Everdingen, Y. M., & Hillegersberg, J. (2002). The dynamics of factors affecting the adoption of innovations. *Journal of Product Innovation Management*, *19*(6), 412–423. doi:10.1016/S0737-6782(02)00175-3

Walczuch, R. M., Singh, S. K., & Palmer, T. S. (1995). An Analysis of the Cultural Motivations for Transborder Data Flow Legislation. *Information Technology & People*, *8*(2), 37–57. doi:10.1108/09593849510087994

Walls, W. D., & Harvey, P. J. (2006). Digital pirates in practice: Analysis of market transactions in Hong Kong's pirate software arcades. *International Journal of Management*, *23*(2), 207–215.

Wallsten, S. (2005). Regulation and internet use in developing economies. *Economic Development and Cultural Change*, *53*, 501–523. doi:10.1086/425376

Walsham, G., Robey, D., & Sahay, S. (2007). Foreword: Special issue on information systems in developing countries. *MIS Quarterly*, *31*(2), 317–326.

Walsham, G., Robey, D., & Sahay, S. (2007). Foreword: Special issue on information systems in developing countries. *Management Information Systems Quarterly*, *31*(2), 317–326.

Walsham, G., & Sahay, S. (2006). Research on information systems in developing countries: Current landscape and future prospects. *Information Technology for Development*, *12*(1), 7–24. doi:10.1002/itdj.20020

Walsham, G. (2002). Cross-cultural software production and use: A structurational analysis. *Management Information Systems Quarterly*, *26*(2), 359–380. doi:10.2307/4132313

Walsham, G., Robey, D., & Sahay, S. (2007). Foreword: Special issue on information systems in developing countries. *Management Information Systems Quarterly*, *31*(2), 317–326.

Walsham, G., & Sahay, S. (2006). Research on information systems in developing countries: Current landscape and future prospects. *Information Technology for Development*, *12*(1), 7–24. doi:10.1002/itdj.20020

Wang, S., & Cheung, W. (2004). E-business adoption by travel agencies: Prime candidates for mobile e-business. *International Journal of Electronic Commerce*, *8*(3), 43–63.

Wang, T. G. E., Klein, G., & Jiang, J. J. (2006). ERP misfit: Country of origin and organizational factors. *Journal of Management Information Systems*, *23*(1), 263–292. doi:10.2753/MIS0742-1222230109

Wang, E. T. G., Barron, T., & Seidmann, A. (1997). Contracting structures for custom software development: The impacts of informational rents and uncertainty on internal development and outsourcing. *Management Science*, *43*(12), 1726–1744. doi:10.1287/mnsc.43.12.1726

Wang, C. H. (2004). Discussing on the manifestations, damages and characteristics of the phenomena of rent-seeking in scientific research field. *Science Research Management*, *25*(3), 131–136.

Wang, Q. (2006). Misconduct: China needs university ethics courses. *Nature*, *442*(7099), 132–132. doi:10.1038/442132b

Warschauer, M. (2003). *Technology and social exclusion: Rethinking the digital divide*. London, UK: MIT Press.

Warshaw, P. R. (1980). A New Model for Predicting Behavioral Intentions: An Alternative to Fishbein. *JMR, Journal of Marketing Research, 17*(2), 153–172. doi:10.2307/3150927

Watson, R., Ho, T., & Raman, K. (1994). Culture: A Fourth Dimension of Group Support Systems. *Communications of the ACM, 37*(10), 44–55. doi:10.1145/194313.194320

Watson, R. T. (1989). Key Issues in Information System Management: an Australian Perspective. *Australian Computer Journal, 21*(3), 118–129.

Weber, M. (1947). *The theory of social and economic organization*. New York, NY: Free Press.

Weisinger, J. Y., & Trauth, E. M. (2002). Situating culture in the global information sector. *Information Technology & People, 15*(4), 306–320. doi:10.1108/09593840210453106

Welsh, J. A., & White, J. F. (1981). A small business is not a little big business. *Harvard Business Review, 59*(4), 18–32.

West, D. M. (2005). *Digital government: Technology and public sector performance*. Princeton, NJ: Princeton University Press.

West, D. M. (2007). Global perspectives on e-government. In Mayer-Schonberger, V., & Lazer, D. (Eds.), *Governance and information technology: From electronic government to information government* (pp. 17–32). Cambridge, Massachusetts: MIT Press.

Westrup, C., Liu, W., El Sayed, H., & Al Jaghoub, S. (Eds.). (2003). *Taking culture seriously: ICTs, cultures and development*. Hampshire, UK: Ashgate.

Wheeler, B. (2004). The open source parade. *EDUCAUSE Review, 39*(5), 68–69.

Wickramasinghe, N. (2000). IS/IT as a tool to achieve goal alignment in the health care industry. *International Journal of Healthcare Technology and Management, 2*(1/4), 163–180. doi:10.1504/IJHTM.2000.001089

Wickramasinghe, N., & Silvers, J. B. (2003). IS/IT the prescription to enable medical group practices attain their goals. *Health Care Management Science, 6*(2), 75–86. doi:10.1023/A:1023376801767

Willcocks, L. (1994). Managing information systems in UK public administration: Issues and prospects. *Public Administration, 72*(Spring).

Williamson, O. E. (1999). Public and private bureaucracies: A transaction cost economics perspective. *Journal of Law Economics and Organization, 15*(1), 306–342. doi:10.1093/jleo/15.1.306

Wilson, E. J. III. (2004). *The information revolution and developing countries*. Cambridge, MA: MIT Press.

Wilson, J. Q. (1989). *Bureaucracy: What government agencies do and why they do it*. New York: Basic Books.

Wilson, E. J., III. (2003). Forms and dynamics of leadership for a knowledge society: The quad. *Center for International Development and Conflict Management, University of Maryland, College Park*. Retrieved from http://www.cidcm.umd.edu/leadership/quad2.pdf

Wiseman, C. (1985). *Strategy and Computers: Information Systems as Competitive Weapons*. New York: Jones-Irwin.

Wong, P.-K. (2002). ICT production and diffusion in Asia: Digital dividends or digital divide? *Information Economics and Policy, 14*, 167–187. doi:10.1016/S0167-6245(01)00065-8

Wooley, D. J., & Eining, M. M. (2006). Software piracy among accounting students: A longitudinal comparison of changes and sensitivity. *Journal of Information Systems, 20*(1), 49–63. doi:10.2308/jis.2006.20.1.49

World Bank. (2002). *The egovernment handbook for developing countries*. Center for Democracy and Technology.

World Bank. (2005). *Lessons learned from seventeen*. InfoDev Projects Gamos Limited, Big World.

World Bank Group. (2009). *Saudi Arabia Data Profile*. World Development Indicators Database.

World Bank. (2002). *Building institutions for markets: World Bank Development Report*. Washington, DC.

World Bank. (2004). *World Bank corporate corruption and ethics indices*. Retrieved from http://www.worldbank.org/wbi/governance/pdf/ETHICS.xls

World Bank. (2006). *World development indicators database*. Retrieved October 1, 2008, from http://web.worldbank.org

World Bank. (2008). *World Development Indicators* (available online at http://go.worldbank.org/RVW6YTLQH0).

World Economic Forum. (2006) *The global information technology report 2006-2007*. Retrieved from http://www.weforum.org/pdf/gitr/contents2007.pdf

World Trade Organization (WTO). (2007). *Risks lie ahead following stronger trade in 2006*. Retrieved Nov. 12, 2007, from http://www.wto.org/english/news_e/pres07_e/pr472_e.htm

Wresch, W. (2003). Initial e–commerce efforts in nine least developed countries: A review of national infrastructure, business approaches, and product selection. *Journal of Global Information Management, 11*(1), 67–78. doi:10.4018/jgim.2003040105

Wright, B. K. (2005). Researching Internet-based populations: Advantages and disadvantages of online survey research, online questionnaire authoring software packages, and web survey services. *Journal of Computer-Mediated Communication, 10*(3).

Wu, S.-Y., & Chen, P.-Y. (2008). Versioning and piracy control for digital information goods. *Operations Research, 56*(1), 157–174. doi:10.1287/opre.1070.0414

Wu, X. (2003a). Credit, science credit and its institutional structure. *Studies in Science of Science, 21*, 65–70.

Wu, D. J., Ding, M., & Hitt, L. M. (2004). Learning in ERP contracting: A principal-agent analysis. In *Proceedings of the Hawaii International Conference on System Sciences*.

Wu, X. (2003b). *The Institutional Analysis on Scientific-credit Issues of China*. Unpublished master's thesis, Zhejiang University, Hangzhou, China.

Xi, Y., & Wang, D. (2006, 0507-10-01). *Harmony configuration and Hexie management: An empirical examination of Chinese*. Paper presented at the 13th International Conference on Management Science and Engineering.

Xiaowen, F., Chan, S., Brzezinski, J., & Xu, S. (2005). Moderating effects of task type on wireless technology acceptance. *Journal of Management Information Systems, 22*(3), 123–157.

Xie, Y. Q., Huang, Y., Li, D. N., Jiang, J., & Li, D. (2008). Analysis of National Natural Science Foundation of China management in Hainan Medical College. *Science Foundation in China, 22*(1), 57–58.

Xin, H. (2006). Scientific misconduct: Scandals shake Chinese science. *Science, 312*(5779), 1464–1466. doi:10.1126/science.312.5779.1464

Xu, S., Zhu, K., & Gibbs, J. (2004). Global technology, local adoption: A cross-country investigation of internet adoption by companies in the Unites States and China. *Electronic Markets, 14*(1), 13–24. doi:10.1080/1019678042000175261

Xu, Q., & Ma, Q. (2008). Determinants of ERP implementation knowledge transfer. *Information & Management, 45*(8), 528–539. doi:10.1016/j.im.2008.08.004

Yap, C. S. (1990). Distinguishing characteristics of organizations using computers. *Information & Management, 18*(2), 97–107. doi:10.1016/0378-7206(90)90056-N

Yap, C. S., Soh, C. P. P., & Raman, K. S. (1992). Information systems success factors in small business. *Omega, 20*(5-6), 597–609. doi:10.1016/0305-0483(92)90005-R

Yap, A., Das, J., Burbridge, J., & Cort, K. (2006). A composite-model for e-commerce diffusion: Integrating cultural and socio-economic dimensions to the dynamics of diffusion. *Journal of Global Information Management, 14*(3), 17–36. doi:10.4018/jgim.2006070102

Yap, A., Das, J., Burbridge, J., & Cort, K. (2006). A composite-model for e-commerce diffusion: Integrating cultural and socio-economic dimensions to the dynamics of diffusion. *Journal of Global Information Management, 14*(3), 17–38. doi:10.4018/jgim.2006070102

Yavas, U. (1997). Management Know-How Transfer to Saudi Arabia: A Survey of Saudi Managers. *Industrial Management & Data Systems, 7*, 280–286. doi:10.1108/02635579710192959

Yavas, U., Luqmani, M., & Quraeshi, Z. A. (1992). Facilitating the Adoption of Information Technology in a Developing Country. *Information & Management, 23*(2), 75–82. doi:10.1016/0378-7206(92)90010-D

Ye, J. (2009). China's Internet Population Hits 338 Million. *The Wall Street Journal*. Retrieved from http://blogs.wsj.com/chinarealtime/2009/07/17/chinas-internet-population-hits-338-million/

Yi, M. Y., & Hwang, Y. (2003). Predicting the use of web-based information systems: self-efficacy, enjoyment, learning goal orientation, and the technology acceptance model. *International Journal of Human-Computer Studies*, *59*(4), 431–449. doi:10.1016/S1071-5819(03)00114-9

Yin, R. K. (1994). *Case study research: Design and methods* (2nd ed.). Thousand Oaks, CA: Sage Publications.

Yin, X. (2005). China. In Kuwayama, M., Ueki, Y., & Tsuji, T. (Eds.), *Information Technology for development of small and medium-sized exporters in Latin America and East Asia*. Chile: United Nations Publication.

Yoo, Y., & Alavi, M. (2001). Media and Group Cohesion: Relative Influences on Social Presence, Task Participation, and Group Consensus. *Management Information Systems Quarterly*, *25*(3), 371–390. doi:10.2307/3250922

Yoon, J., & Chae, M. (2009). Varying criticality of key success factors of national e-Strategy along the status of economic development of nations. *Government Information Quarterly*, *26*(1), 25–34. doi:10.1016/j.giq.2008.08.006

Yoshino, M. Y., & Rangan, U. S. (1995). *Strategic alliances – An entrepreneurial approach to globalization*. Boston, MA: Harvard Business Press.

Young, S., & Tavares, A. (2004). Centralization and Autonomy: Back to the Future. *International Business Review*, *13*(2), 215–237. doi:10.1016/j.ibusrev.2003.06.002

Zhang, N., Guo, X., Chen, G., & Chau, P. Y. K. (2009). Impact of Perceived Fit on e-Government User Evaluation: A Study with a Chinese Cultural Context. *Journal of Global Information Management*, *17*(1), 49–69.

Zhang, J. (2003). Overview and economic analysis on academic corrupt research. *Social Science of Beijing*, *3*, 105–110.

Zhang, J. (2005). Good Governance Through E-Governance? Assessing China's E-Government Strategy. *Journal of E-Government*, *2*(4), 39–71. doi:10.1300/J399v02n04_04

Zhang, M. (2005). Review and forecast of the system construction of scientific credit. *Scientific and Technology Progress and Policy*, *11*, 20–22.

Zhang, M. (2006). Bad credit behaviors in scientific research projects and their countermeasures. *Scientific Management Research*, *3*(24), 66–69.

Zhang, N., Guo, X., & Chen, G. (2006). *Extended Initial Technology Acceptance Model and Its Empirical Test*. Paper presented at the Fifth Wuhan International Conference on E-Business (WHICEB2006), Wuhan, China.

Zhang, N., Guo, X., Chen, G., & Song, G. (2008). *The Cultural Perspective of Mobile Government Terminal Acceptance – An Exploratory Study in China*. Paper presented at the 12th Pacific Asia Conference on Information Systems (PACIS 2008), Suzhou, China.

Zhang, N., Guo, X., Chen, G., & Song, G. (2009). *A MCT Acceptance Model from the Cultural Perspective and Its Empirical Test in the Mobile Municipal Administrative System Application*. Paper presented at the Eighth International Conference on Mobile Business.

Zhang, Q. H. (2002). ERP Crisis, ZDNet China. *eWeek, 22*. Retrieved from http://www.zdnet.com.cn/news/software/story/0,3800004741,39038580,00.htm

Zhao, H., Kim, S., Suh, T., & Du, J. (2007). Social institutional explanations of global internet diffusion: A cross-country analysis. *Journal of Global Information Management*, *15*(2), 28–55. doi:10.4018/jgim.2007040102

Zhao, R. (2003). Transition in R&D management control system: Case study of a biotechnology research institute in China. *The Journal of High Technology Management Research*, *14*(2), 213–229. doi:10.1016/S1047-8310(03)00022-1

Zhu, K., & Kraemer, K. L. (2005). Post-adoption variations in usage and value of e-business by organizations: Cross-country evidence from the retail industry. *Information Systems Research*, *16*(1), 61–84. doi:10.1287/isre.1050.0045

Zhu, K., Kraemer, K. L., Gurbaxani, V., & Xu, S. (2006). Migration to open-standard interorganizational systems: Network effects, switching costs, and path dependency. *MIS Quarterly*, *30*, 515–539.

Zhu, K., Kraemer, K. L., & Xu, S. (2003). E-business adoption by European firms: A cross-country assessment of the facilitators and inhibitors. *European Journal of Information Systems*, *12*(4), 251–268. doi:10.1057/palgrave.ejis.3000475

Zhu, K., Kraemer, K. L., Xu, S., & Dedrick, J. (2004). Information technology payoff in e-business environments: An international perspective on value creation of e-business in the financial services industry. *Journal of Management Information Systems*, *21*(1), 17–54.

Zhu, Z., & Gong, X. (2008). Basic research: Its impact on China's future. *Technology in Society*, *30*(3-4), 293–298. doi:10.1016/j.techsoc.2008.05.001

Zinatelli, N., Cragg, P. B., & Cavaye, A. L. M. (1996). End user computing sophistication and success in small firms. *European Journal of Information Systems*, *5*(3), 172–181. doi:10.1057/ejis.1996.23

Zogona, S., Willis, J. E., & MacKinnon, W. J. (1966). Group effectiveness in creative problem–solving tasks: An examination of relevant variables. *The Journal of Psychology*, *62*, 111–137.

Zweig, D., Chen, C., & Rosen, S. (2004). Globalization and transnational human capital: Overseas and returnee scholars to China. *The China Quarterly*, *179*, 735–757. doi:10.1017/S0305741004000566

About the Contributors

Felix B Tan is Professor of Information Systems and Chair of Business Information Systems discipline at Auckland University of Technology (New Zealand). He serves as the Editor-in-Chief of the *Journal of Global Information Management*. He is on the Executive Council and is Fellow of the Information Resources Management Association. He was also on the Council of the Association for Information System between 2003-2005. Dr. Tan's current research interests are in electronic commerce, global information management, business-IT alignment and the management of IT. Dr. Tan has published in *MIS Quarterly, Information & Management, IEEE Transactions on Engineering Management, IEEE Transactions on Personal Communications, Information Systems Journal, Journal of Information Technology, International Journal of Human-Computer Interaction, International Journal of Electronic Commerce*, as well as other journals and refereed conference proceedings. Dr. Tan has over 25 years experience in information systems management and consulting with large multinationals, as well as university teaching and research in Singapore, Canada and New Zealand.

* * *

Said S. Al-Gahtani is currently Professor of Computer Information Systems (CIS)/Management Information Systems (MIS) in the Department of Administrative Sciences at King Khalid University, Saudi Arabia. His research interests include IT Innovation and Diffusion, Information Technology Acceptance Modeling, End-User Computing, and Organizational Cross-Cultural Research, and his current research focuses on IT Acceptance Modeling, electronic commerce, and Internet issues. His work appears in *Information Resources Management Journal, IT for Development, Journal of Global Information Technology Management, Information & Management, Information Technology and People*, and *Journal of Global Information Management*.

Norm Althouse is a Senior Instructor of Business and Environment at the University of Calgary. His research interests include: team building, gender issues in the workplace, and managerial decision-making. He is active in the Academy of Management (Management Education Division) and has presented professional workshops at their conferences. He is the lead author on one of Canada's premier introduction to business textbook, "The Future of Business", currently in its 2nd edition that has been adopted by 19 Universities and Colleges in Canada.

Dolores Añón Higón is a Lecturer at the University of Valencia and research fellow at ERI-CES, Valencia, Spain. She received her M.Sc. and Ph.D. degrees from the University of Warwick, U.K. Her research interests include the internationalization of R&D, and the relationship between innovation, trade, information technology and productivity. She has published in *Industrial and Corporate Change*, *IEEE Transactions in Engineering Management*, *Information & Management*, *Journal of Global Information Management*, *Research Policy*, *World Economy*, and other journals.

Nicholas Athanassiou is an Associate Professor and Coordinator of International Business and Strategy at Northeastern University. He received his Ph.D. in International Business and Strategy from the University of South Carolina. His research interests focus on the international business capabilities of top management teams, on global innovation processes, and on knowledge acquisition and transfer issues. He has published numerous articles in publications such as the *Journal of International Business Studies*, *Strategic Management Review*, and *Management International Review*.

Bijan Azad is assistant professor at the Olayan School of Business, American University of Beirut. His research interests are on the organizational implementation of IT in the public sector, hospitals and private sector. He earned his Ph.D. and masters degrees from MIT, Cambridge, Massachusetts, U.S.A. Bijan Azad has over 20 years of experience in the systems integration industry before re-joining academe. His research has been published in *European Journal of Information Systems*, *Government Information Quarterly*, *Information Systems Journal*, *Journal of Strategic Information Systems*, *Academy of Management Best Paper Proceedings*, America's Conference on Information Systems Proceedings as well as several practitioner publications. He was the Organizational Communication and Information Systems Division's recipient of the Best Conference Paper Award at the Academy of Management 2008 Conference in Anaheim, California, USA.

She-I Chang is currently an associate professor at the Department of Accounting and Information Technology, CCU. Focusing on ERP systems, with a particular emphasis on the issues, challenges and benefits realization associated with ERP life cycle-wide implementation, management and support are his research interests. He also has interest in the application of qualitative research methodology. Currently at CCU, Taiwan, his extended research interest around the arena of information technology governance, information security management and computer auditing. He has presented and published his researches papers and articles at several IS conferences and journals.

Hun Choi is an assistant professor at the School of Business Administration, Catholic University of Busan, Korea. His research interests include end-user behaviors with ubiquitous computing services and design and evaluation of information technology systems.

Gudrun Curri joined the Faculty of Management at Dalhousie University as associate professor in 2000, and teaches organizational theory and design, organizational change, and international and inter-cultural management. In 2006 she introduced the Central East European Business Program for MBA students which includes a two-week study tour to CEE countries. Dr. Curri has published in national and international journals on issues relating to restructuring postsecondary systems and organizational change in universities in Australia, Germany and Canada. She has been an invited speaker at international and national conferences addressing the politics and challenges of organizational change. Prior to joining the Faculty of Management, Dr. Curri was a member of the senior university administration at two major Canadian Universities. Dr. Curri has joined the Strategic Networks Group, Inc., an international ICT consulting company in 2009 but will continue her relationship with Dalhousie University as an adjunct professor.

Samer Faraj is associate professor and holds the Canada Research Chair in Technology, Management & Healthcare at the Desautels Faculty of Management at McGill University. Previously, he was associate professor at the University of Maryland. His current research focuses on how IT transforms work and the provision of health care as well as the participation dynamics of online knowledge communities. His papers have been published in outlets such as: *Management Science, Organization Science, MIS Quarterly, Information Systems Research and Journal of Applied Psychology*. He is currently senior editor at Organization Science and serves on the editorial board of the Journal of AIS and Information and Organization.

Tony Feghali lectures in the Business Information and Decision Systems Track at the Olayan School of Business of the American University of Beirut. His research interests include the impact of electronic factors on social development and ethics. He is currently the director of the Entrepreneurship Initiative at the Olayan School of Business. Tony Feghali earned his PhD in 1991 from Purdue University.

Roya Gholami is a Lecturer in Operations and Information Management Group, Aston Business School, Birmingham, UK. Her current research interests are IT business value, IT for Green, IT adoption and diffusion, IT and Development. She has published in *IEEE Transactions on Engineering Management, Information & Management, Journal of Global Information Management, Information Resource Management Journal, World Economy, Technovation* and *Electronic Journal of Information Systems in Developing Countries*.

Shirley Gregor is the foundation Professor of Information Systems at the Australian National University, Canberra, where she is a Director of the National Centre for Information Systems Research. Professor Gregor's current research interests include the adoption and strategic use of information and communications technologies, intelligent systems and human-computer interface issues, and the theoretical foundations of information systems. Dr Gregor has led several large applied research projects funded by the Meat Research Corporation, the Department of Communications, Information Technology and the Arts, the Australian Research Council and AusAID. Professor Gregor spent a number of years in the computing industry in Australia and the United Kingdom before beginning an academic career. Dr Gregor's publications include 4 edited books, 15 book chapters and over 100 papers in conferences and journals. Professor Gregor was made an Officer of the Order of Australia in the Queen's Birthday Honour's list in June 2005 for services as an educator and researcher in the field of information systems and in the development of applications for electronic commerce in the agribusiness sector.

Jibao Gu is an associate professor of School of Management and vice dean of Graduate School at the University of Science and Technology of China. He received his Ph.D. in management science from the University of Science and Technology of China in 2004. He has published 50 papers about strategic management and science & technology management in Chinese journals. He is currently engaged in the research about leadership management.

Xunhua Guo received his Ph.D. from Tsinghua University and is now an assistant professor of information systems at the School of Economics and Management, Tsinghua University. He has published in journals including *Communications of the ACM, Information Sciences, Information Systems Frontiers* and *Journal of Enterprise Information Management*. His research interests include information systems adoption, organizational change, and data management.

Geoffrey S. Hubona currently holds a faculty appointment as an Associate Professor of computer information systems in the J. Mack Robinson College of Business at Georgia State University in Atlanta, Georgia. His research interests include the user acceptance of information technologies and methodological issues relating to quantitative research in MIS. In addition to the *Journal of Global Information Management*, he has published journal articles in *MIS Quarterly, ACM Transactions on Computer-Human Interaction, IEEE Transactions on Systems, Man and Cybernetics, Part A: Systems and Humans, DATA BASE for Advances in Information Systems, Information & Management, Information Technology & People*, among others.

Shin-Yuan Hung is a professor and the Chair of Information Management at National Chung Cheng University (CCU) in Taiwan. He was a visiting scholar of the MIS Department at the University of Arizona during summer 2007-spring 2008. He has published a number of papers in *Decision Support Systems, Information & Management, Electronic Commerce Research and Applications, Information Technology & People, Communications of the AIS, Expert Systems with Applications, Government Information Quarterly, Computer Standard and Interfaces, Industrial Management and Data Systems, Journal of Chinese Information Management*, among others.

Ahmed Imran is an Information System (IS) researcher whose research interest largely emerged from his personal experience that includes ICT adoption in public sector, e-government and ICT in developing countries. Along with his special skill and interest in Information Technology he had a range of challenging experiences in IT field. His extensive experience as an instructor and IT manager is invaluable for research in understanding and providing rich insight of the context in least developed courtiers. His PhD research gained an in-depth understanding of impediments to e-government in an LDC environment. Part of Ahmed's research has been successfully applied in Bangladesh through AusAID's Pubic Sector Linkage Program (PSLP) as 'eGovernment capacity building through knowledge transfer and best practice development in Bangladesh', which also received Vice Chancellor's award for Community outreach in 2010. Ahmed is actively involved in number of research projects at ANU and has published and presented his research in distinguished forums around the world.

A. Kankanhalli is Associate Professor in the Department of Information Systems at the National University of Singapore (NUS). She obtained her B. Tech. from the Indian Institute of Technology Delhi, M.S. from the Rensselaer Polytechnic Institute and Ph.D. from NUS. She had visiting stints at the University of California Berkeley and Indian Institute of Science. Dr. Kankanhalli has considerable work experience in industrial R&D and has consulted for organizations including World Bank and Bosch SEA. She conducts research in the areas of knowledge management, IT enabled organizational forms, and IT in service sectors, supported by government and industry grants. Her work has appeared in the *MIS Quarterly, Journal of Management Information Systems, IEEE Transactions on Engineering Management, Journal of AIS, International Journal of Human Computer Studies, Communications of the ACM, Decision Support Systems*, and the *International Conference on Information Systems* among others. She serves on several IS conference committees and on the editorial boards of MIS Quarterly, IEEE Transactions on Engineering Management, and Information and Management, among others. Dr. Kankanhalli received the ACM SIGMIS 2003 Best Doctoral Dissertation award. She has been listed among the leading IS researchers globally and in the Asia Pacific region.

Jinwoo Kim is Professor of HCI at the School of Business, Yonsei University. He is also working as the director of Human Computer Interaction Lab and Human Centered Innovation Center at Yonsei University. His research has been focused on innovative design and evaluation of user experiences of digital products, services and contents. His current research topics include designing a tangible social media service for tween generation users, building intergeneration mobile communication system for silver generation users, and designing a microblog service for emerging generation users.

Hak-Jin Kim is Professor of Management Science at the School of Business, Yonsei University. He graduated from Tepper school of business, Carnegie Mellon University. His research interest is application of quantitative models and methods to various business areas. He was originally trained in Operations Research, and conducted research about logic-based methods in optimization, constraint programming, and combinatorial optimization. His current interest has been propagated to constructing a constraint language in the Semantic Web, risk management in grid computing using MAD method, analyzing social networks using the Semantic Web technology, and designing telecommunication networks.

Robert Konopaske is an Associate Professor of Management at Texas State University. He received his Ph.D. from the University of Houston. His research interests include emerging global assignments and other IHRM-related topics, global relocation and mobility, and travel stress. He has published numerous articles in such journals as *Journal of Applied Psychology, Journal of Management Education, Management International Review*, and *Human Resource Management*. Dr. Konopaske has co-authored several popular management textbooks and has held national leadership positions with the Academy of Management's Human Resources Division.

Fujun Lai is an Associate Professor of Management at the University of Southern Mississippi. He received a Ph.D. from the City University of Hong Kong. His research interests include e-business, enterprise information systems, supply chain and logistics management, and quality management. He has published articles in *International Journal of Production Research, Communications of the ACM, Journal of Business Research*, and among others.

Albert L. Lederer is a Professor in management information systems at the Gatton College of the University of Kentucky. He holds BA in Psychology from the University of Cincinnati, an MS in Computer and Information Sciences from the Ohio State University, and a PhD in Industrial and Systems Engineering from Ohio State. He has over 10 years of industry experience in information systems. His research focuses on the management of information systems. His articles have appeared in *MIS Quarterly*, *Information Resources Management Journal*, *Journal of End User Computing*, *Journal of MIS*, *Information Systems Research*, *Decision Sciences*, and a variety of other academic and practitioner journals.

Pei-Ju Lee received her Master degree from the Department of Information Management at National Chung Cheng University, Taiwan.

Tomasz Lenartowicz is an Associate Professor of International Business at Florida Atlantic University. His research focuses on cross-cultural management, values, and practices. Since receiving his Ph.D. from the University of South Carolina, he has published numerous articles in journals such as the *Journal of International Business Studies*, *Management International Review*, and the *Journal of Marketing*.

Dong Li is a professor and the Chair of Management Science and Information Systems Department, Guanghua School of Management, Peking University. His Current research focuses on IT leadership, IT strategy planning, and IT business performance. His work has appeared in academic journals and International Conferences, include *Communications of Association Information Systems*, *Computers in Human Behavior*, *Frontiers of Business Research in China* etc. Professor Dong Li is serving as vice Chair of China Association of Information Systems, and member of Editorial Board of China Journal of Information Systems.

Dahui Li is an Associate Professor of MIS at the University of Minnesota Duluth. He received his Ph.D.in MIS from Texas Tech University. His current research focuses on business-to-consumer relationships, online community, and diffusion of technology innovation. He has had papers published in *Journal of the Association for Information Systems*, *Communications of the ACM*, *Decision Sciences*, *Decision Support Systems*, *International Journal of Electronic Commerce,* and among others.

Manlu Liu is an Assistant Professor in the E. Philip Saunders College of Business of Rochester Institute of Technology (RIT). She holds Ph.D. degree from Eller College of Management, University of Arizona. She holds MBA degree from the Hong Kong University of Science & Technology. She served as an Associate Professor in the School of Management at Zhejiang University in China. Her research interests include IT investment analysis, accounting information systems, service-oriented architecture and web services, open source/community source and virtual team. She has published research articles in SSCI-listed journals and major international conferences such as the *Journal of Global Information Management* (JGIM), *theInformation Conference on Information Systems* (ICIS), *the Americas Conference on Information Systems* (AMCIS), and several others. She is on the editorial review boards of *Electronic Government: An International Journal* (EG). She also served as a mini track co-chair for AMCIS in 2009, 2010 and 2011.

Jerry Luftman is the Executive Director of Graduate Information Systems Programs, and Distinguished Professor at Stevens Institute of Technology, Hoboken New Jersey. His career includes strategic positions in management (Information Technology, including being a CIO, and consultant), management consulting, Information Systems, and executive education. After a notable twenty-two year career with IBM, and over fifteen years at Stevens, Dr. Luftman's experience combines the strengths of practitioner, consultant, and academic. His framework for assessing IT-business alignment maturity is considered key in helping companies around the world understand, define, and scope an appropriate strategic planning direction that leverages Information Technology. Dr. Luftman is the founder and leader of Stevens graduate IS Programs; one of the largest in the world. He created and teaches a popular end-of Masters course on managing the IT resource, which explores how to be a successful IT executive. Dr. Luftman's newest book (one of thirteen) "Managing Information Technology Resources", has surpassed his initial best seller "Competing in the Information Age" which is still selling strong around the globe. His active membership in SIM includes being the VP of Academic Affairs for the SIM Executive Board. His advice is frequently requested as an executive mentor.

Jian Ma is a Professor in the Department of Information Systems, City University of Hong Kong. Prior to that, he was a Lecturer in the School of Computer Science and Engineering, the University of New South Wales (UNSW). He is an author/co-author of three books, 50 conference papaers and over 50 refereed articles in international journals including *IEEE Trans. on Systems*, *Men and Cybernetics*, *IEEE Trans. on Engineering Management*, *Journal of Management Information Systems*, *Information and Management*, and *Decision Support Systems*, etc.

Mark Meckler is an Associate Professor of Management at the University of Portland. He conducts research in three main areas of management. His strategy and innovation work is focused upon organizational knowledge and the cognitive issues impacting organizational knowledge flows, strategy implementation and innovation; his organization theory research is on the relationships between events, socially constructed theories, truth and knowledge and the impact these relationships have upon the scientific status of organization theory. Finally, his work on behavioral aspects of strategy implementation focuses on the physiological contract and motivation at work. Since receiving his Ph.D. from Florida Atlantic University in 2001, he has published in journals such as *Organization Studies*, the *Journal of Business Ethics* and the *Journal of Management Inquiry*.

Jie Mein Goh is a Ph.D. candidate at the Robert H. Smith School of Business, University of Maryland, College Park. Her current research interests center on health IT and strategic value of IT. She has published in Journal of Strategic Information Systems, and has presented at a variety of national and international conferences such as Academy of Management and INFORMS. Prior to her Ph.D. program, she earned her Masters and Bachelor's degree with First Class Honors in Computer and Information Sciences from the National University of Singapore.

Mark E. Mendenhall is the J. Burton Friesen Chair of Excellence in Business Leadership at the University of Tennessee-Chattanooga. His research interests include selecting and developing global leaders and expatriates in multinational corporations, managing human resources and sociocultural processes in mergers and acquisitions, and the implications of nonlinear systems theory for organizational change effectiveness. He has published a number of influential book chapters, books, and articles in journals such as the *Academy of Management Review*, the *Journal of International Business Studies*, and *Human Resource Management*.

Dinesh A. Mirchandani is an Associate Professor of Information Systems in the College of Business Administration at the University of Missouri St. Louis. He received a Ph.D. in Business Administration (MIS) from the University of Kentucky, an M.S in Electrical Engineering from Purdue University, and a Bachelor's Degree in Electronics Engineering (with honors) from the University of Bombay. His research interests include information systems planning and electronic business. His papers have been published in such academic journals as *Communications of the ACM, Omega, Information & Management, Journal of Organizational Computing and Electronic Commerce, International Journal of Electronic Commerce,* and *Information Systems Frontiers* among others.

Andrew A. Mogaji is a professor in the Department of Psychology at Benue State University, Nigeria. Topics in his publications include comparative management, education, and other areas of applied psychology.

Trevor T. Moores is an Associate Professor of Management Information Systems at the University of Nevada, Las Vegas. He received a B.A. in Combined Studies (Arts), majoring in Philosophy and Psychology, from Sunderland Polytechnic (UK), an M.Sc. in Intelligent Knowledge Based Systems from the University of Essex (UK), and a Ph.D. from the University of Aston (UK). His research interests include software measurement, software piracy, IS/IT ethics, self-efficacy, and issues of online trust. His work has appeared in *MIS Quarterly, Communications of the ACM, DATA BASE, European Journal of Information Systems, IEEE Transactions on Knowledge and Data Engineering,* and *Information & Management*.

L. G. Pee is an Assistant Professor in the Graduate School of Decision Science and Technology at the Tokyo Institute of Technology. She received her Bachelor in Computing and Ph.D. in Information Systems (IS) from the National University of Singapore. Her research interests are in knowledge management, adoption of emerging technology, and perceived security and risks. Her work has been published in journals such as *Information & Management, Journal of the Association for Information Systems,* and *IEEE Transactions on Engineering Management* and in proceedings of conferences such as the International Conference on Information Systems (ICIS) and Pacific Asia Conference on Information Systems (PACIS). Dr. Pee received the Best Paper Award of PACIS 2010.

Mark F. Peterson is the Internet Coast Adams Professor of Management and International Business at Florida Atlantic University. His principal interests are in questions of how culture and international relations affect the way organizations should be managed. He has published over 80 articles and chapters, a similar number of conference papers, and several books. Specific topics in his writings include the role different parties play in decision making in organizations throughout the world, the effects that culture has on the role stresses that managers experience, the way immigrant entrepreneur communities operate, and the way that intercultural relationships in multicultural teams and across hierarchical levels should function. Since receiving his Ph.D. from the University of Michigan in 1979, Prof. Peterson has held faculty positions at Wayne State University, the University of Miami, Texas Tech University and Florida Atlantic University.

Julie I. A. Rowney is a Professor in the Human Resources and Organizational Dynamics area at the University of Calgary, where she received her Ph.D. in psychology. She has received many awards and grants including a Killam Research Fellowship (1999), the Macleod Dixon International Achievement Award, the President's Faculty International Achievement Award (2004), the Alberta Centennial Award for Human Rights and Multiculturalism (2005), and she was named one of the top 40 scholars by the University of Calgary (2006). She has published research in numerous journals including the *Journal of International Management* and the *Women in Management Review.*

Wei Sha is an assistant professor for Department of Computer Science-Information Systems at Pittsburg State University. He received his doctorate degree from the University of Arkansas majoring in Information Systems. His dissertation is about the examination of the influence of web site design features on people's behavioral intentions in business-to-consumer ecommerce. Dr. Sha has published articles at Electronic Markets, Journal of Business and Leadership, Issues in Information Systems and conference proceedings. Dr. Sha is a member of the Association for Information Systems, the Decision Science Institute and International Association for Computer Information Systems.

Farid Shirazi is a Senior Researcher at Institute for Innovation and Technology Management (IITM) and Assistant Professor at Ted Rogers School of Information Technology Management, Ryerson University, Toronto, Canada. His research focuses mainly on the impact of ICTs on social and economic development within the context of developing countries. His research interests include also the impact of ICTs on environmental sustainability as well as the ethical and security perspectives associated with the introduction and use of ICTs in particular the e-government strategies. He has published in several journals and conference proceedings including LNCS book chapters; Associations *for Information Systems*; *Electronic Journal of Information Systems in Developing Countries*; *Journal of Telematics & Informatics*; *Journal of Information, Communication & Ethics in Society*; Journal *of Global Information Management (JGIM)* and *Journal of Information & Management (I&M).*

Stephanie J. Thomason is Associate Dean and Assistant Professor of Management at the University of Tampa. Since receiving her doctoral degree in 2007, Stephanie Thomason has published articles in journals such as the *International Journal of Selection and Assessment*, the *Journal of Cross-Cultural Psychology*, and the *International Journal of Innovation Management*. Her research interests include human resource selection and appraisal, cross-cultural differences in social structures, values and practices, and factors that relate to successful competition in small businesses.

Wen Tian is currently a Ph.D. candidate in Business and Management Team, USTC-CityU Joint Advanced Research Center. She received her Bachelor's degree in Department of Management Science, School of Management, University of Science and Technology of China in 2006. Her research interests include Knowledge Management, Social Network, Virtual Community and Business Intelligent applications.

Douglas R. Vogel is Chair Professor of Information Systems at the City University of Hong Kong and an Association for Information Systems (AIS) Fellow. He received his Ph.D. in Information Systems from the University of Minnesota in 1986. Professor Vogel has published widely and directed extensive research on group support systems, knowledge management and technology support for education. He has recently been recognized as the most cited IS author in Asia-Pacific. He is currently engaged in introducing mobile device and virtual world support for collaborative applications in educational systems. Additional detail can be found at http://www.is.cityu.edu.hk/staff/isdoug/cv/.

Jian Wang is a Professor of International Business at the University of International Business and Economics, China. His research interests include e-business, international business, and international trade law and practice. He is the Director of the International Business Bridge (IBB). He has extensively published in the areas of international business, electronic business, and others.

Wayne Wei Huang is Professor of Information Systems at College of Business, Ohio University, and Fellow of Harvard University, USA. He has worked full-time in research universities worldwide, including in Australia (University of New South Wales), Hong Kong, Singapore, and USA (Harvard University). Huang has had more than 20 years of full-time teaching experience in universities as well as a few years of IT industrial working experience. His research combines both quantitative and qualitative research methodologies. He has published more than 120 refereed research papers in international journals, books, and conference proceedings, including top-tier international MIS/IS journals such as *IEEE Transactions on Systems, Man, and Cybernetics*; *Journal of Management Information Systems (JMIS)*; *MIS Quarterly (MISQ)*; *Communications of ACM (CACM)*; *IEEE Transactions on Professional Communication*; *Information & Management (I&M)*; *Decision Support Systems (DSS)*; *ACM Transaction on Information Technology (ACM TOIT)*; and *European Journal of Information Systems (EJIS)*.

Elizabeth White Baker is currently an Assistant Professor in the Department of Economics and Business, Virginia Military Institute, Lexington, VA. Her current research interests include examining the influence of individual beliefs and characteristics on the use of information technology. She also studies strategic decision systems and the technology impact on mergers and acquisitions. Her work appears in *Information Technology and People, Information Systems Journal*, the *Journal of Decision Systems*, and *IEEE-Transactions on Engineering Management*.

Xiaobo Wu is professor of innovation management and strategy in School of Management at Zhejiang University. He is also the executive dean of School of Management and the director of National Institute for Innovation Management. Dr. Wu received his PhD in management science from Zhejiang University in 1992. He has been a visiting scholar at University of Cambridge, UK and the Fulbright Scholar in Sloan School of Management of MIT, USA. His research interests include technological innovation & management, global manufacturing & strategy and ICT & management change. Dr. Wu's research has appeared in journals such as *IEEE TEM, International Journal of Technology Management, Strategic Outsourcing: an International Journal and International Journal of Innovation and Technology Management*. Dr. Wu's research projects are mainly funded by National Natural Science Foundation of China, National Planning Office of Philosophy and Social Science of China, the Royal Society of UK, International Development Research Centre (IDRC-CRDI) of Canada and the EU Commission.

David C. Yen is currently a Raymond E. Glos Professor in Business and a Professor of MIS of the Department of Decision Sciences and Management Information Systems at Miami University. Professor Yen is active in research and has published books and articles which have appeared in *Communications of the ACM, Decision Support Systems, Information & Management, Information Sciences, Computer Standards and Interfaces, Government Information Quarterly, Information Society, Omega, International Journal of Organizational Computing and Electronic Commerce*, and *Communications of AIS* among others. Professor Yen's research interests include data communications, electronic/mobile commerce, database, and systems analysis and design.

Nan Zhang received his PhD from Tsinghua University and is currently a postdoctoral researcher in the School of Public Policy & Management, Tsinghua University. His research interests include e-government and mobile government, ICT adoption and diffusion, Cultural issues of Information management. His work has been published in journals like *Journal of Global Information Management, Information Systems Frontiers*, and *Tsinghua Science & Technology*.

J. Leon Zhao is Head and Chair Professor in Information Systems, City University of Hong Kong. He was Eller Professor in the Department of Management Information Systems, University of Arizona before January 2009. He holds Ph.D. and M.S. degrees from the Haas School of Business, UC Berkeley. His research is on information technology and management, with a particular focus on workflow technology and applications in knowledge distribution, e-learning, supply chain management, organizational performance management, and services computing. Leon's research has been supported by NSF, SAP, and other sponsors. Leon has served or has been serving as associate editor of *Information Systems Research, ACM Transactions on MIS, IEEE Transactions on Services Computing, Decision Support Systems, Electronic Commerce Research and Applications*, among other journals. He has co-edited several special issues in various IS journals including *Decision Support Systems and Information Systems Frontiers*. He received an IBM Faculty Award in 2005 for his work in business process management and services computing and was awarded a 2009 Chang Jiang Scholar Chair Professorship at Tsinghua University by the Ministry of Education of China.

Index